# Cultural Heritage Science

The preservation and interpretation of our cultural heritage is one of the major challenges of today's society. Cultural Heritage Science is a highly interdisciplinary book series covering all aspects of conservation, analysis and interpretation of artworks, objects and materials from our collective cultural heritage. The series focuses on science and conservation in three main fields

- Art technology
- Active conservation and restoration
- Preventive conservation and risk management

The series addresses conservators and conservation scientists at museums, institutes, universities and heritage organizations. It also provides valuable information for curators and decision makers at museums and heritage organizations. All titles in the book series will be peer reviewed. Titles will be published as printed books and as eBooks, opening up the opportunity to include electronic supplementary material (videos, high-resolution figures, special data formats, and access to databases).

Maarten R. van Bommel • Matthijs de Keijzer

# Bright Colours from the Past

## The History, Chemistry, Characterisation and Application of Synthetic Dyes between 1856 and 1914

Springer

Maarten R. van Bommel
Conservation and Restoration of Cultural Heritage
Department of Arts and Culture
Faculty of Humanities, and Analytical Sciences
Van't Hoff Institute for Molecular Sciences
Faculty of Science
University of Amsterdam
Amsterdam, The Netherlands

Matthijs de Keijzer
Cultural Heritage Laboratory
Cultural Heritage Agency of the Netherlands (retired)
Amsterdam, The Netherlands

ISSN 2366-6226 ISSN 2366-6234 (electronic)
Cultural Heritage Science
ISBN 978-3-031-88663-8 ISBN 978-3-031-88664-5 (eBook)
https://doi.org/10.1007/978-3-031-88664-5

Front cover: Émile Bernard (1868 – 1941), Hérodias, 1903, embroidery, 49 cm x 75.5 cm, v0110N1996, Van Gogh Museum, Amsterdam, the Netherlands.



This Springer imprint is published by the registered company Springer Nature Switzerland AG
The registered company address is: Gewerbestrasse 11, 6330 Cham, Switzerland

# Acknowledgements

Since we started the project in 2003, focusing on synthetic dyes developed in the second half of the nineteenth century up to the First World War, we have been supported by many institutions and persons. Firstly, we highly appreciate the support from the Instituut Collectie Nederland (ICN), later Rijksdienst voor het Cultureel Erfgoed (RCE, Cultural Heritage Agency of the Netherlands) for giving us time and means to investigate these dyes. Although originally developed as a 4-year project, it soon turned out that so much was still to be investigated that subsequent projects were developed and accepted. In addition to research time, equipment and consumables were allocated to the project to carry out scientific research, and budget was made available to visit libraries and archives and for participation in numerous conferences to distribute our knowledge.

Other important stakeholders are museums and other institutes allowing us to examine case studies, many of them are represented in this book. We developed a special relation with the University of Applied Arts Vienna in Austria, since we collaborated in many student projects. The authors would like to thank *Regina Hofmann-de Keijzer* and *Regina Knaller*, who formerly worked at this university for their cooperation in the project on early synthetic dyes.

We very much appreciate the collaboration during the project with institutions, museums, and colleagues who have contributed to the enrichment of this book through their cooperation in research projects and the provision of photographs.

Austria, in Bregenz: vorarlberg museum: *Michael Kasper, Tobias G. Natter, Theresia Anwander, Natalie* Ellwanger, and *Leonie Schwaerzler.* In Vienna: Institute of Conservation at the University of Applied Arts Vienna: *Martina Griesser-Stermscheg, Gabriela Krist, Edith Oberhumer, Leonie Schwaerzler* (née *Tscherner*), *Nina Zangerl, Stefan Oláh,* and *Georg Oberlechner*; Billardmuseum: *Heinrich Weingartner Sr.* and *Heinrich Weingartner Jr.;* Backhausen Archive (2006): *Peter Backhausen*; Backhausen Archive at the Leopold Museum: *Louise Kiesling* (†), *Ursula Oswald-Graf, Dominik Papst,* and *Klaus Pokorny*; Kunsthistorisches Museum Wien, Court Wardrobe ('Monturdepot'): *Monica Kurzel-Runtscheiner, Marianne Novotny-Kargl,* and *Michaela Morelli*; Liechtenstein: The Princely Collections, Vaduz–Vienna: *Stephan Koja, Johann Kräftner, Helga Musner, Michael*

*Schweller*, and *Robert Wald*; Schloß Schönbrunn Kultur-und Betriebsges.m.b.H./ Sammlung Bundesmobilienverwaltung: *Wolfgang Kippes*, *Elfriede Iby*, *Michael Wohlfahrt*, *Anja Hasenlechner*, *Anna Mader-Kratky*, and *Martin Mutschlechner*; Vienna Museum of Science and Technology: *Anne-Katrin Ebert, Luba Dovgan Nurse, Valentina Ljubic-Tobisch, Thomas Winkler, Chris Clouter,* and *Peter Sedlaczek*; Volkskundemuseum Wien: *Elisabeth Egger, Kathrin Pallestrang, Birgit Kainz, Peter Kainz, Christa Knott* and *Monika Maislinger;* and the artists *Monika Kus-Picco* and *Tanya Niedermüller* (†).

We also worked together with museums and institutions in the Netherlands, focusing on interesting case studies. We highly appreciated the collaboration with the Amsterdam Museum (Huis Willet-Holthuysen): *Jaap Boonstra*; the Rijksmuseum Amsterdam: *Suzan Meijer* and *Reinier Baarsen*; the Van Gogh Museum: *René Boitelle* and *Ella Hendriks*; and Bureau voor kleuronderzoek and restauratie: *Ruth Jongsma*, all located in Amsterdam. Furthermore, we studied objects from the Zuiderzeemuseum, Enkhuizen: *Marja Berkhout*; Kathedraalmuseum Sint. Bavo, Haarlem: *André Numan* and *Ingrid Henkemans*; Museum Paulina Bisdom van Vliet, Haastrecht: *Marco Bakker*; Parochie van Sint Jans Geboorte, Hoogwoud: *Bep de Haan-Appel*; and from the department collections of the RCE: *Ron Kievits*.

Many objects were studied by students from the University of Amsterdam, who contributed to background information of the objects studied and questioned us on all aspects. In no specific order: *Marijke de Bruijne*, *Federica van Adrichem*, *Jasmijn Groeneveld*, *Jettie Timmer,* and *Joni Steinman.*

A special attention deserves our former colleagues at the RCE, being involved in the research and sharing their know-how: *Han Neevel*, *Ineke Joosten*, *Sanne Berbers*, *Art Ness Proaño Gaibor*, *Muriel Geldof*, *Luc Megens, Suzan de Groot,* and *Klaas-Jan van den Berg*. We owe thanks to *Julian van den Berg*, the librarian of the Cultural Heritage Agency of the Netherlands, for his tireless help in finding digital publications.

We also like to acknowledge *Rinske de Vries* who re-analysed samples from reference materials of synthetic colourants which we show throughout this book, using the most recent developed chromatographic method.

Last, but certainly not least, we highly appreciate the feedback on our writing from *Jo Kirby*, formerly of the National Gallery of London. She reviewed all chapters several times, providing suggestion to improve the English so that it became much more readable and understandable. In many occasions, she asked for clarifications and came up with questions which we were forced to answer. In addition, she provided us access to historical sources which we could not access in advance. She has had a major influence on the improvement of this publication, encouraging us on several occasions to go deeper, and as such has been key to the development of this publication.

Maarten R. van Bommel
Matthijs de Keijzer

# Contents

# About the Authors

**Maarten R. van Bommel** (1968) is a full professor of conservation science at the University of Amsterdam (UvA), Faculty of Humanities, Amsterdam School of Heritage, Memory and Material Culture (AHM), Conservation and Restoration of Cultural Heritage (C&R), and at the Faculty of Science, Van 't Hoff Institute for Molecular Sciences (HIMS), Analytical Chemistry. From this unique position at two faculties, he bridges science and conservation. Before this appointment, he worked as (senior) scientist and program manager at The Netherlands Institute for Cultural Heritage.

Van Bommel's main research focus is on the material change of a wide range of heritage objects (paintings, furniture, plastics, paper, textiles, metal, glass, and ceramics), with a special attention to natural and synthetic dyes. He investigates the chemical composition and material change, aiming to understand the original appearance of heritage objects and to develop mitigation and preservation strategies for cultural heritage display, management, and conservation. This involves the use of advanced analytical methods, studying historical sources as a basis for model systems and historically informed reconstructions, and the development of visualisation techniques to show the presumed original appearance of objects of art. A common ground in all his projects is the development and application of a wide variety of analytical techniques to characterise material changes and understand degradation mechanisms. He has been awarded several national and international funded projects, has published over 125 peer-reviewed papers, chapters, books, and conference papers in both science and cultural heritage journals, and has contributed to museum catalogues and popular science publications.

**Matthijs de Keijzer** (1952) worked as a senior conservation scientist at the Cultural Heritage Laboratory of the Cultural Heritage Agency of the Netherlands in Amsterdam. Following a successful career in analytical chemistry in various industries, he turned his attention to the study of works of art. Working in interdisciplinary teams, his research interest lies in the microscopic and microchemical analysis of pigments on various cultural heritage objects to solve conservation and restoration problems. In the last decade of the twentieth century, his research focussed on the history of the modern pigments of that era. Since 2003, he has been involved in the project 'Early Synthetic Organic Dyes'. This project is based on a selection of well-known synthetic organic dyes, covering all dye classes and is focused on the period 1856–1914. His contribution in this project was to collect information by studying the original historical literature sources, including the international patent literature. He is an expert on early and modern synthetic pigments and dyes and their use in various techniques and has published and lectured widely in these fields. He has lectured at the University of Applied Arts Vienna, the University of Amsterdam, the George Enescu National University of Arts in Iasi, and the University of Gothenburg. He is also a co-author of the book *Pigmenten en Bindmiddelen* (*Pigments and Binding Media*).

# Chapter 1
# Introduction and Reading Information

In this book, we provide an overview of the most important synthetic dyes and dye classes developed and used in the second half of the nineteenth century up to the First World War. Some of these dyes had already been developed much earlier and many of them were also used after the First World War, not only as textile dyes, but also occasionally as organic pigments in paints, to stain furniture and to colour paper: for example, crêpe paper [1]. However, in this book we focus mainly on the use of synthetic colourants as textile dyes, although some other applications will be mentioned as well.

The publication is aimed at those involved in the field of research concerned with the conservation, preservation and presentation of cultural heritage objects: art historians, curators, conservators, conservation scientists and students from all disciplines, but it could also be useful to anyone interested in the history of synthetic colourants. The development of the synthetic dyes in the nineteenth century was rapid and paralleled the development of the understanding of organic chemistry. In addition to the academic interest in the synthetic dyes during the nineteenth century, the huge commercial market for dyes and colourants at that time acted as an important driving force.

The book starts with four general chapters, providing a brief overview of the history of the synthetic colourants; some information on the relevance of patents and international competition for the synthetic dye industry; several typical dye recipes, and an overview of analytical techniques used for their identification. However, the core of the book focuses on individual dyes or classes of dyes. In order to support readability, we have tried to create a common structure for the individual dye chapters. Each chapter starts with general information about the common name of the dye, the Colour Index name and number, the IUPAC name, the discoverer(s) and date(s) of discovery, contemporary and obsolete terminology when available, and the chemical formula and structure. We will use the English common name of the dye in the text. Next, chapters start with a short introduction and subsequently, the history of the dye or dye class is presented, including patent numbers, followed by



M. R. van Bommel, M. de Keijzer, *Bright Colours from the Past*, Cultural Heritage Science, https://doi.org/10.1007/978-3-031-88664-5_1

an overview of manufacturers and trade names; how the colourants can be applied; some properties and recipes; and we show an analytical result using liquid chromatography coupled to photo diode array detection. We conclude these chapters with a number of case studies, when available.

It is important to note that some dyes are so important that they could lead to a separate publication; here we will be more concise for those dyes. The same applies when we discuss a certain dye class: sometimes such a class of dyes can include dozens of different dyes, whereas we usually present a more general history of the class, followed by a few dyes which were important in our opinion, that is, often applied. Obviously, the importance of the dyes we select can lead to discussion, since as more research is done, we will learn more about which dyes were most often used.

We present the dyes in a more or less chronological order, bearing in mind that this is not that strict as the history of certain classes of dyes spans several decades. Although it is obviously possible to read the book from cover to cover, we have chosen to write the chapters in such a way that they can be read individually. In addition to the structured way in which we present the chapter, this also means that we had to make some choices about how to present the information. Firstly, in each chapter, when a person is mentioned in the introduction, we use the full name. Next, in the section about the history of a dye or dye class, when a person is mentioned for the first time, we provide also information on the nationality and profession, the full name and the dates of birth and death, when available. After this, we will only use the surname of the person when they are mentioned again. Secondly, we obtained information from the Colour Index, the third edition published by the Society of Dyers and Colourists in 1971 [2]. We will not provide the full references each time this is mentioned, but will simply refer to this as 'Colour Index', unless we refer to a different edition. Information obtained from the Colour Index includes fastness properties: when available, International Organisation for Standardisation (ISO) or American Association of Textile Chemists and Colorists (AATCC) values are given for light fastness or wash fastness. These values are presented as numbers: the lower the number, the less fast the dye. In some cases, different test methods have been used; when this is the case, it is noted in the text. For information on these fastness tests, the Colour Index can be consulted. Furthermore, we obtained a great deal of information from the publications by Dr. Adolf Lehne. He compiled an overview of the dyes available on the German market at the end of the nineteenth century. His first publication *Tabellarische Uebersicht über die künstlichen organischen Farbstoffe und ihre Anwendung in Färberei und Zeugdruck* dates from 1893 and contains information on trade names, dye recipes and fastness properties [3].

In an introductory chapter, Lehne described several properties such as wash fastness, rub fastness, resistance to acid and so forth. Here we report especially about his observations on light fastness. He uses the term *Wetterechtheit*, weathering, but between brackets mentions that this is focused on light fastness. He exposed pieces of dyed textiles to natural light, behind a window. A part of the textile was covered with cardboard and the colour changes were recorded from time to time, by comparing the exposed and covered parts of the textile with the naked eye. When possible, he compared the changes to dyed textiles using dyes for which the light fastness was already known. He noticed that, especially in spring, fading took place rapidly, and he recorded the weather conditions—sunny, rainy and so on—as well. He also added a comment that for dyes with a good light fastness, such as synthetic Indigo, lighter colours should be used, that is, textiles dyed using a low or medium concentration of dye. Finally, he reflected on his own research by stating that his tests could be of limited use for dyers and colourists, since light fastness does not mean the resistance to sunlight alone, but that it is a combination of light, air, heat, relative humidity and moisture. Obviously, standardised testing facilities did not yet exist.

Lehne also provided information on how to identify a dye by using different solvents, information we will not use as nowadays more advanced techniques are available. What is particularly interesting is that the books contain dyed samples, often of wool or silk, but sometimes also of cotton and linen. The importance of these sample books lies in the fact that he claimed that he had collected all the synthetic dyes available on the German market, which was of great importance as Germany was the main producer of these colourants at the time. The 1893 publication contains 324 samples; only 6 years later, in 1899, he produced a new volume with a further 186 samples, while in 1906 he published a third volume with another 135 samples. In our book, unless we use one of the other volumes, we refer to the 1893 edition when discussing information from Lehne.

Most of the recipes we discuss are taken from Lehne and described in Chap. 4. In this chapter, we present the full recipe, while in the individual dye chapters we give only a reference to that recipe, but add specific information when this has been provided by Lehne. Occasionally we have also provided other recipes.

Many dye manufacturers were in operation during the nineteenth century, but over time some of them merged or disappeared. To avoid repeating the full name continuously, we have decided to use abbreviations or acronyms instead. An overview of these manufacturers, the date of establishment, location(s) and their abbreviations or acronyms can be found in Appendix I. In addition, we provide several patents using different sources. The German patents are indicated by G.P., those from Britain by B.P., those from France by F.P. and the ones from the United States of America by U.S.P.

Finally, several case studies and analytical results using liquid chromatography coupled to photo diode array detection will be presented throughout the book. Most case studies were analysed when the authors both worked at the Cultural Heritage Agency of the Netherlands (Rijksdienst voor het Cultureel Erfgoed (RCE), formerly Instituut Collectie Nederland (ICN)) in Amsterdam, where we investigated synthetic colourants as part of several research projects and programmes. In several case studies, many dyes were identified and as such these could be described in different chapters. To avoid repetition, each case is only described in one chapter, but we refer to these case studies in other chapters where applicable. Since the case studies were investigated over a period of about two decades, the exact method of analysis has been changed. The case studies presented here are analysed using the methods presented in a 2007 publication, using different methods for basic, neutral and acid dyes [4]. At the time, there was no single technique which could be used for all types of dye. The analysis of reference materials has been done using a recently introduced method, offering a technique for the simultaneous analysis of a wide range of basic, neutral and acid dyes [5]. We therefore decided to show only the results obtained with the latter technique, as this also acts as a reference for conservation scientists. Although the chromatographic profile might differ when using a different chromatographic system, in our experience the spectra obtained using photo diode array detection are generally unaffected and can be used for preliminary identification. However, this should then be confirmed by the analysis of reference materials. The references materials we used are either obtained from the RCE reference collection (hence labelled RCE no. with a specific number so that they can be traced backed) or from modern manufacturers such as Sigma-Aldrich. Note, that the RCE reference are in general more representative as these are often obtained from dye manufacturers and contains side products and impurities, while those from modern manufacturers are in general pure.

Over the past 20 years, we have presented the results of our research on many different occasions, both to national and international audiences, and to both professionals and the general public. In particular, the presentations at the annual meetings of Dyes in History and Archaeology (DHA) are noteworthy, as they have often led to publications in the *Dyes in History and Archaeology* journals,[1] or, more recently, the journal *Heritage*. We have used this information in the current publication, but have updated it and added new information. In addition, we have described many new colourants that have not yet been published.

We wish the reader much joy reading this book.

## References


1. Ash, N., Homolka, S., Price, B., Sutherland, K.: Insights into the Materials and Techniques of James Castle. Book and Paper Group Annual. 31, 5–15 (2012)
2. Society of Dyers and Colourists and American Association of Textile Chemists and Colorists: The Colour Index. Society of Dyers and Colourists, Bradford (1971)
3. Lehne, A.: Tabellarische Uebersicht über die künstlichen organischen Farbstoffe und ihre Anwendung in Färberei und Zeugdruck. Verlag von Julius Springer, Berlin (1893)
4. Van Bommel, M.R., Vanden Berghe, I., Wallert, A.M., Boitelle, R., Wouters, J.: High-performance liquid chromatography and non-destructive three-dimensional fluorescence analysis of early synthetic dyes. J Chromatogr A. 1157, 260–272 (2007). https://doi.org/10.1016/j.chroma.2007.05.017
5. Astefanei, A., Adamson, B., Gaibor, A.N.P., van Bommel, M.R.: Simultaneous detection of a wide range of synthetic and natural dyes in artworks using UHPLC-PDA-HRMS. J. Chromatogr. A. 1740, 465562 (2025). https://doi.org/10.1016/j.chroma.2024.465562


[1] Dyes in History and Archaeology webpage (https://www.dyesinhistoryandarchaeology.com/).

# Chapter 2
# A Short History of Synthetic Colourants

## 2.1 Introduction[1]

The development of the synthetic dye industry aligns closely with the English Industrial Revolution. In the early 1790s, the Scottish chemist, inventor and mechanical engineer William Murdoch (1754–1839) illuminated his house in Redruth, Cornwall, using coal gas, extracted from coal, paving the way for gas street lighting in European cities after the post-Napoleonic wars. This resulted in the production of large amounts of coal tar, a by-product of the coal gas extraction, which initially lacked practical applications so was burned or dumped. Between 1826 and 1841, chemists explored coal tar and, interestingly, the natural dye Indigo, isolating various substances. Coal tar, rich in hydrocarbons like benzene, phenol and naphthalene, became the cornerstone for the synthetic dyes. Several chemists extracted a foul-smelling poisonous substance from coal tar and initially they did not realise that they all worked with the same product, which could be also isolated from natural Indigo.

Already in 1826, the German chemist and merchant Otto Unverdorben (1806–1873) obtained a novel substance by heating natural Indigo, which he named Crystallin. Next, in 1832, the German chemist Friedlieb Ferdinand Runge (1795–1867) began his experiments with coal tar and 2 years later, in 1834, he isolated a compound which he named Kyanol or Blauöl. Later on, in 1841, the German chemist Carl Julius Fritzsche (1808–1871) showed that by treating natural Indigo with potash, an oily substance was obtained which he named aniline. This was derived from the name of one of the Indigo-yielding plants, Anil indigo, a name used by the Portuguese, but originating from the Arabic word *an-nīl* (blue substance). In the same year, in 1841, the Russian organic chemist Nikolay Nikolaevich

[1] Much more information on the various synthetic organic colourants is presented in the following chapters, in which references to literature can be found.



M. R. van Bommel, M. de Keijzer, *Bright Colours from the Past*, Cultural Heritage Science, https://doi.org/10.1007/978-3-031-88664-5_2

Zinin (1812–1880) found that on reducing nitrobenzene using ammonium sulfide, a base was formed which he named Benzidam. Next, the German chemist August Wilhelm Hofmann (1818–1892) started his first research investigations in the laboratory of the German scientist Justus Liebig (1803–1873) at Giessen on organic bases obtained from coal tar. In 1843, in his first publication, he showed that these variously prepared substances were actually all the same, nowadays known as aniline [1–3].

Nitrobenzene was discovered in 1834 by the German chemist Eilhardt Mitscherlich (1794–1863), who treated benzene (from coal tar) with fuming nitric acid. In 1854, the French chemist Antoine Béchamp (1816–1908) greatly improved the production process of aniline by reducing nitrobenzene with iron filings in the presence of diluted hydrochloric acid. The Béchamp reaction was used to produce large amounts of aniline [4].

During the preparation of coke from coal gas, ammonia and coal tar were extracted from coal in addition to coke. 2,000 kg of coal contain about 100 kg of coal tar. The study of coal tar led to great difficulties, but gradual development enlarged the knowledge of its constituents. Coal tar consists of a mixture of numerous compounds; about 60 have been isolated.

Naphthalene, a white solid with a strange, pungent smell, was independently isolated by the distillation of coal tar by the British chemists William Thomas Brande (1788–1866), Alexander Garden (1757–1829) and John Kidd (1775–1851) in the early 1820s. The name naphthalene was given by Kidd and its chemical formula was determined by the English scientist Michael Faraday (1791–1867) in 1826 [5]. Anthracene was first isolated from coal tar in 1832 by the French chemists Augustin (Auguste) Laurent (1807–1853) and Jean-Baptiste Dumas (1800–1884). Laurent also made anthraquinone and phthalic acid for the first time in 1836 by oxidising anthracene.

In 1834, Runge discovered phenol in an impure form during the distillation of coal tar and called the substance Karbolsäure (coal-oil-acid, carbolic acid). Seven years later, in 1841, Laurent rediscovered it in pure form. In 1843, the French organic chemist Charles Fréderic Gerhardt (1816–1856) called it phenol.

Toluene was discovered in 1837 by the Polish chemist Filip N. Walter (1810–1847) and the French chemist Pierre Joseph Pelletier (1788–1842). Walter found the compound in the gas emitted by the bark of the maritime pine *Pinus pinaster* Ait. (syn. *Pinus maritima* var. *alterna* Dum.Cours.), native to the western and central Mediterranean region, and named it rétinnaphté from a combination of the French words resin and naphtha. The compound was rediscovered several times in later years and given a variety of names. In 1841, the French chemist Henri Étienne Sainte-Claire Deville (1818–1881) isolated it from Tolu balsam, an aromatic extract from the tropical Colombian tree *Myroxylon balsamum* (L.) Harms., Tolú being the town in Columbia (now Santiago de Tolú) where the trees were first found. Deville recognised that it was similar to Walter's rétinnaphté and benzene, hence he called it benzoène. In 1843, the Swedish chemist Jöns Jacob Berzelius (1779–1848) proposed the name toluin and by 1850 the French chemist Auguste André Thomas Cahours (1813–1891) came up with the name toluène.

The main constituent of coal tar is benzene. In 1825, Faraday discovered and identified benzene, giving it the name 'bicarburet of hydrogen'. Mitscherlich obtained it in 1833 by heating benzoic acid with an excess of lime (calcium hydroxide solution), named the compound obtained benzin and found a match with Faraday's bicarburet of hydrogen. This name was changed into benzol (in English benzene) by Liebig. In 1845, its presence in coal tar was recognised by the English chemist Charles Blachford Mansfield (1819–1855) while working under the direction of Hofmann. On the 11th of November 1847, Mansfield filed for and received, in May 1848, a patent for the fractional distillation of coal tar (B.P. 11.960). In 1849, Mansfield started the first industrial-scale production of benzene and in the late 1840s, he also produced nitrobenzene from benzene.

Only a trace of aniline is present in coal tar and the quantity was too small to make it worth extracting. When large quantities of aniline were required, benzene, a fairly abundant constituent of coal tar, had to be treated with concentrated nitric acid, then reducing the nitrobenzene formed into aniline.

Of less importance are other components present in coal tar: xylenes, cresols, benzonitrile and pyridine. 100 kg of coal tar include 2.0 kg of benzene, 0.5 kg of phenol, 0.5 kg of toluene, 5–6 kg of naphthalene and 0.6 kg of anthracene. Various chemical processes, such as nitration, sulfonation, halogenation, reduction, fusion, amidation, recrystallisation, oxidation, acylation, arylation, condensation, salting out and regrouping, allow the production of intermediates from these building blocks. The most important intermediates used in dye production are aniline, toluidine, xylidine, naphthylamines and anthraquinone [6].

## 2.2 The Theory of Chemical Structure by Friedrich August Kekulé

Chemistry in the early nineteenth century was largely empirical and there were only vague ideas about how the atoms in a molecule are bonded. This changed when the German chemist Friedrich August Kekulé (1829–1896) developed his carbon theory and put forward his benzene ring theory [7]. In the period 1857–1858, he recognised the ability of carbon atoms to link to each other, the tetravalence of carbon and the determination of the bonding order of all of the atoms in a molecule. It was not until 1864 that he formulated his famous benzene theory, suggesting that the structure of benzene contained a symmetrical ring of six carbon atoms with alternating single and double bonds. After Kekulé's benzene ring theory was published in 1865 and 1866, chemists had a useful model for determining the structure of dye molecules. His discoveries contributed significantly to the development of organic chemistry and, in particular, dye chemistry.

During a lecture in Berlin on the 11th of March 1890, organised by the German Chemical Society, celebrating the 25th anniversary of his first benzene paper, he reported that he had developed the ring shape of the benzene molecule after a reverie or waking dream of a snake seizing its own tail, an ancient symbol known as the Ouroboros or Endless knot, during his stay in Ghent, Belgium:

> 'I sat there and wrote in my textbook, but I couldn't do it properly; my mind was on other things. I turned my chair towards the fireplace and sank into a half-sleep. Again the atoms flickered before my eyes. This time, small groups remained modestly in the background. My mind's eye, sharpened by repeated visions of a similar kind, now distinguished larger structures of manifold shapes. Long rows, many times closer together; everything in motion, twisting and turning like a snake. And behold, what was that? One of the snakes caught hold of its own tail and the structure whirled mockingly before my eyes. I woke up as if by a flash of lightning, and this time too I spent the rest of the night working out the consequences of the hypothesis... Learn to dream, gentlemen' [8].[2]

At the onset of the synthetic dye development, chemistry relied heavily on trial and error, lacking a clear understanding of molecular structure: Kekulé's breakthroughs changed this. Published in 1865, his theory provided chemists with a valuable framework for unravelling the structures of dye molecules, marking a pivotal moment in synthetic dye chemistry. The history of the synthetic colourants reflects a dynamic interplay between scientific discovery, industrial innovation and cultural preferences, shaping the evolution of textile dyeing practices over time.

## 2.3 The First Steps: Semi-Synthetic Colourants

Before synthetic dyes were developed, various colourants marked the first attempts to develop new dyes for colouring textiles. Indigo carmine, a blue acid dye, was discovered in 1743 by the German lawyer Johann Christian Barth (c. 1700–1759). It simplified textile dyeing by transforming insoluble indigotin into water-soluble Indigo carmine. As such, a new dyeing methodology was developed, resulting in a much more rapid dyeing compared to the laborious procedures used for the natural colourants. Picric acid, a nitro dye, was already mentioned in alchemical writings around 1650, but was first synthesised by the Irish mineralogist and chemist (alchemist) Peter Woulfe (1727–1803) in 1771. It was obtained by treating natural Indigo with nitric acid, and it resulted in a new dye class. It gained prominence in the dyeing industry by the mid-nineteenth century, especially for silk. Prussian blue, an inorganic pigment and, as such, an exception, was discovered in 1706 by the German

---

[2] *'Da sasz ich und schrieb an meinem Lehrbuch; aber es ging nicht recht; mein Geist war bei anderen Dingen. Ich drehte den Stuhl nach dem Kamin und versank in Halbschlaf. Wieder gaukelten die Atome vor meinen Augen. Kleine Gruppen hielten sich diesmal bescheiden im Hintergrund. Mein geistiges Auge, durch wiederholte Gesichte ähnlicher Art geschärft, unterschied jetzt gröszere Gebilde von mannigfacher Gestaltung. Lange Reihen, vielfach dichter zusammengefügt; alles in Bewegung, schlangeartig sich windend und drehend. Und siehe, was war das? Eine der Schlangen erfaszte den eigenen Schwanz und höhnisch wirbelte das Gebilde vor meinen Augen. Wie durch einen Blitzstrahl erwachte ich, auch diesmal verbrachte ich den Rest der Nacht, um die Konsequenzen der Hypothese auszuarbeiten… Lernt zu träumen, meine Herren.'*

pigment and dye producer Johann Jacob Diesbach (c. 1670–1748) and used for dyeing silk and wool, particularly during the Napoleonic era due to a shortage of natural Indigo. It gained popularity until the 1860s when the first synthetic organic blue dyes were discovered. Pittacal, a blue acid dye, was accidentally discovered in 1832 by the German chemist Karl Ludwig Reichenbach (1788–1869), but never gained a widespread use despite being commercially available for over 20 years. Murexide, a red-purple mordant dye, was formed from uric acid and gained industrial significance in the mid-nineteenth century with the availability of guano as raw material. However, its popularity declined rapidly with the introduction of Perkin's Mauveine and the rise of the aniline dyes.

These semi-synthetic dyes revolutionised textile dyeing, offering new possibilities and overcoming the limitations of natural dyes by providing both new dyeing technology and a much more consistent supply. Their development was driven by a combination of scientific inquiry, industrial processes and market demands. However, their eventual decline in use was often due to the emergence of newer, more versatile synthetic dyes, changing fashion trends and improvements in dyeing technology.

## 2.4 The Exploration of the First Synthetic Dyes

Between 1856 and 1865, significant advancements led to the discovery of numerous synthetic dyes. The accidental creation of Mauveine in 1856 by the British student William Henry Perkin (1838–1907), while attempting to synthesise quinine, marked a pivotal moment. Perkin's oxidation of aniline resulted in a purple substance, which he successfully used to dye silk. Patenting his invention, Perkin established the dye firm Perkin & Sons, with his father and older brother, near London. By 1857, Mauveine became a sensation in Parisian fashion and quickly gained popularity in London the following year. Despite the fact that this dye was only produced for about a decade, it is considered as the first 'real' synthetic dye and as such marked the start of the synthetic dye era.

During the first 10 years, a flurry of advances in synthetic dye production revolutionised the textile industry. As examples, the red dye Safranine was discovered by the British chemist Charles Hanson Greville Williams (1829–1910) in 1859 and in 1867, the black dye Nigrosine was developed by the French chemist Jean Théodore Coupier (1820–1908). These dyes were of limited use. A more important development occurred in the same year as the creation of Mauveine, in 1856, when the Polish chemist Jakub Natanson (1832–1884) discovered Fuchsine, a red-violet dye, by reacting aniline with ethyl chloride, followed by the French chemist François-Emmanuel Verguin (1806–1864) in 1859, who used tin chloride instead of ethyl chloride. These marked significant milestones. The production of Fuchsine by the dye firm Renard Frères further propelled synthetic dye research. However, the monopolistic practices of the firm Société La Fuchsine and environmental issues related to arsenic derivatives led to its decline by 1868. Nonetheless, Fuchsine served as a key intermediate for other tri-arylmethane dyes, including Imperial violet (1859), Regina purple (1860), Hofmann's violet (1863), and Aniline blue (1860), leading to a surge in dye innovation. The discovery of Aniline blue and, in 1862, its transformation into soluble acid dyes by the British chemist Edward Chambers Nicholson (1827–1890), by treating it with fuming sulfuric acid, led to the development of the water-soluble mono-, di-, and tri-sulfonic acid dyes, such as Alkali blue, Methyl blue and Navy blue. BASF patented this sulfonation process in 1877, which further advanced acid dye production, especially for Fuchsine. The synthesis of Methyl violet in 1861 by the French chemist Charles Lauth (1836–1913), followed by its successful industrial production in 1866, replaced the Hofmann's violets.

Subsequent innovations, using Fuchsine and Methyl violet as starting materials, led to the discovery of the green dyes Aldehyde green, Iodine green, and Methyl green in the 1860s and early 1870s. The late 1870s witnessed the discovery of the basic green dyes of the Malachite series, including Malachite green (1878) and Brilliant green (1878).

Collaborations between companies like Bindschedler & Busch and BASF in 1883 resulted in the invention of Crystal violet and other tri-arylmethane dyes, along with the sulfonated Acid violets and Alkali violets. Such collaborations also led to the development of di-arylmethane dyes, exemplified by the yellow Auramine in 1884. These innovations not only expanded the colour palette available for textile dyeing, but also improved the production processes, leading to an increased efficiency and profitability in the dye industry.

## 2.5 Azo Reaction and the Development of Substantive Dyes

In 1858, the German chemist Johann Peter Griess (1829–1888) discovered the diazo coupling reaction, in which an aromatic amine treated with nitrous acid forms an unstable diazonium salt. This salt reacts with another aromatic amine to produce coloured azo products. Azo dyes, characterised by their azo group (–N=N–), became the largest synthetic dye class due to their versatility. Initially, the azo dyes faced scepticism, but the dye class expanded rapidly with the creation of Aniline yellow, Bismarck brown, and Chrysoidine in the early 1860s. The introduction of sulfanilic acid intermediates allowed the production of Orangés, Ponceaus, Fast reds, Congo red, and Benzopurpurine 4B, extending the colour range to red. Griess's discovery revolutionised dye chemistry, enabling the synthesis of a wide range of vibrant and fast coloured azo dyes.

In 1883, Griess's continued research on the diazo reaction led to the development of the direct azo dyes, eliminating the need for mordants to fix dyes on cotton and linen. Griess patented these dyes in 1883 and 1884. He shared his findings with the German industrialist Henry Theodore Böttinger (1848–1920) of the dye firm Bayer, who indirectly, via the German chemist Paul Seidler, passed it to the German industrial chemist Paul Böttiger. Böttiger refined Griess's invention, creating the first synthetic direct red dye by coupling 1-naphthionic acid with doubly diazotised benzidine. Bayer initially rejected the dye, but Böttiger patented it in 1884. The dye company AGFA later acquired the patent and marketed the dye as Congo red in 1885. Meanwhile, the German chemist Carl Duisberg (1861–1935) at Bayer achieved success with Benzopurpurine 4B, a red dye with better acid resistance, coupling 1-naphthionic acid with doubly diazotised *o*-tolidine.

## 2.6 The Introduction of Xanthenes

In 1871, the German chemist Adolf Baeyer (1835–1917) synthesised Fluorescein, a yellow acid dye, from phthalic anhydride and resorcinol with zinc chloride. Although not significant as a dye, it became a crucial building block in the development of similar dyes. Eosin, a red acid dye, was derived from Fluorescein through the complete bromination of the molecule by the German chemist and BASF research director Heinrich Caro (1834–1910), in 1873. After overcoming production problems, BASF marketed Eosin in 1874 without patent protection, enabling other companies in Germany, Switzerland and France to produce it. Some later, the dye class was extended by the French chemist Emilio Noelting (1851–1922) and the Swiss chemist Robert Gnehm (1852–1926) who discovered the Phloxines, the Erythrosines and their derivatives the Rose Bengals. In 1887, the Swiss chemist Maurice Cérésole (1860–1936) synthesised Rhodamine B, a basic dye, while the German chemist Heinrich August Bernthsen (1855–1931), working for BASF, discovered Rhodamine 6G in 1892. Rhodamine derivatives like Rhodamine 3B, S, G, and 3G followed, later serving as basic dyes or transformed into acid dyes called Violamines by Farbwerke ML&B.

## 2.7 Sulfur Dyes

Sulfur dyes emerged as the last significant innovation before the First World War, particularly applied for dyeing black. Before the sulfur dyes were discovered, various synthetic black dyes were used, such as Aniline black, which was introduced in 1859. Cachou de Laval, a predecessor of the sulfur dyes, was developed by the French chemists Louis Marie François Bretonnière (1842–1902) and his business partner Edouard Croissant in 1873 by heating sawdust with sodium sulfide. In 1893, the French chemist Henri Raymond Vidal (1862–1938) introduced the first black sulfur dye, Vidal black, produced by fusing *p*-aminophenol with sulfur and sodium sulfide. The German chemist Georg Kalischer (1873–1938) patented Immedial black FF in 1897, an improved black sulfur dye. Sulphur black T, a significant advancement made by the German chemists Bernhard Priebs and Oskar Kaltwasser in 1899, was marketed by AGFA. Further research led to the development of several coloured sulfur dyes between 1898 and 1904, notably Immedial pure blue, discovered in 1900 by the German chemists Arthur Weinberg (1860–1943) and Richard Herz (1867–1936), and marketed by the dye firm Cassella.

## 2.8 From Natural to Synthetic Origin, the Development of Alizarin and Indigotin

In the late 1860s, despite the availability of numerous synthetic dyes, interest in the natural dyes like Alizarin and Indigo persisted. The revelation in 1867 by the German chemist Carl Graebe (1841–1927) that Alizarin was a quinone, followed by its synthesis from anthracene by Graebe and the German chemist Carl Theodore Liebermann (1842–1914) in 1868, marked a breakthrough. Their synthesis disclosure prompted Perkin to devise a more economical production process. In 1869, both BASF and Perkin & Sons filed patents for the Alizarin production, but competition arose as Farbwerke ML&B and Bayer also entered the market. Despite an attempted market division, the competition for the Alizarin production intensified, leading to complex patent disputes among the key players. Alizarin was the first natural compound for which the chemical structure was elaborated and reproduced via synthesis.

In 1865, Baeyer initiated research on the structure and the synthesis of Indigo, revealing the formula to Caro in 1883. Despite discovering three synthesis processes between 1878 and 1882, safety concerns and costly starting materials hindered their commercial viability against the use of natural Indigo itself. The research of the German chemist Karl Heumann (1850–1894) led to a breakthrough in 1890, resulting in two synthesis methods. The second Heumann synthesis, based on *N*-phenylglycine-*o*-carboxylic acid (made from anthranilic acid) fused with caustic potash, proved economically feasible. BASF scaled up the production using this method in 1897. However, the process developed in 1901 by the German chemist Johannes Pfleger (1867–1957), who worked for Degussa in Frankfurt am Main, significantly reduced the production costs of synthetic Indigo. This Indigo production process was acquired by Farbwerke ML&B. Despite this, competition from Bayer, Kalle, and Cassella's inexpensive sulfur dyes, especially Cassella's Immedial pure blue, challenged synthetic Indigo. BASF's own competition emerged in 1901 when the French-German-Swiss chemist René Bohn (1862–1922) discovered Indanthrene blue RS, superior to synthetic Indigo for cotton and linen. Although BASF considered halting the production in 1960, demand surged in the late 1960s, particularly for dyeing blue jeans, maintaining indigo's relevance in the textile industry today.

## 2.9 To Conclude

The history of the synthetic colourants in a nutshell shows important developments during the nineteenth century. Prior to this development, scholars experimented with the treatment of natural colourants creating new dyes and dye mechanisms. Although the foundation of synthetic chemistry was already laid down in the first half of the nineteenth century, the main developments took place in the second half of the nineteenth century. This resulted in a rainbow of new dyes, a much more steady supply of colourants and new dyeing mechanisms, eliminating the need for the laborious dyeing procedures used for natural colourants. The dye industry at that time was of great commercial importance, which was stimulating for the discoverers of synthetic dyes since they could make great profits. As a result, the natural dye production industry in Europe collapsed within a few decades.

## References

1. Knecht, E.: The Chemistry of the Coal-tar Colours. George Bell and Sons, York Street, Convent Garden, London (1889)
2. Caro, H.: Referate, "Ueber die Entwicklung der Theerfarben-Industrie." Berichte der deutschen chemischen Gesellschaft zu Berlin. Commissionsverlag von R. Friedländer & Sohn. 25, 983–894 (1892)
3. Perkin, W.H.: Hofmann Memorial Lecture—'The Origin of the Coal-Tar Colour Industry, and the Contributions of Hofmann and his Pupils'. Journal of the Chemical Society, Transactions. 69, 597 (1896)
4. Béchamp, A.: De l'action des protosels de fer sur la nitronaphtaline et la nitrobenzine. Nouvelle méthode de formation des bases organiques artificielles de Zinin. Annales de Chimie et de Physique. Imprimerie de Mallet-Bachelier, Paris. 4, 186–196 (1854)
5. Kidd, J.: Observations on Naphthalene, a peculiar substance resembling a concrete essential oil, which is produced during the decomposition of coal-tar, by exposure to a red heat. Philosophical Transactions of the Royal Society of London. 111, 209–221 (1821)
6. Bucherer, H.Th.: Die Teerfarbstoffe mit besonderer Berücksichtigung der synthetischen Methoden. Vereinigung wissenschaftlicher Verleger, Berlin / Leipzig. 8–27 (1920)
7. Partington, J.R.: Kekulé. In: A History of Chemistry. pp. 533–565. Red Globe Press, London (1964)
8. Kekulé, A.: Benzolfest: Rede. Berichte der deutschen chemischen Gesellschaft zu Berlin. Commissionsverlag von R. Friedländer & Sohn, Berlin. 23, 1301–1311 (1890)

# Chapter 3
# Patents and International Competition

## 3.1 Introduction

At the rise of the era of the synthetic colourants, the attention of chemists and merchants in different countries was focused on the invention, production and sale of the dyes. In 1862, the German chemist August Wilhelm Hofmann (1818–1892) expressed firm confidence that Britain would be the leading producer of synthetic dyes for many years to come because of its coal reserves, its huge production of coal tar and its enormous market for textiles.[1] He was wrong [1]. Although Britain and France were the leading producers for the first decade or so of the developing synthetic dye industry, in the long run they became small players; it was Germany that came to dominate the global market before 1914. Initially German manufacturers copied British and French synthetic dyes, but when their patents expired, the leading German and Swiss firms entered the British and French markets successfully and at the outbreak of the First World War, the German dye industry possessed a world market share of about 85%. In 1913 the British and French global market shares had been reduced to 3.1% and 1.2% respectively. Switzerland had become the second-largest producer in the world with 6.2% and the United States of America produced 1.9% of the world total (see Fig. 3.1) [2].

There are several reasons why Germany became so dominant in the synthetic dye industry. It is mainly due to the differences in patent regulations, education and training, but also the policy to stimulate the synthetic dye industry, and the ways in which universities and industry collaborated with one another. This chapter gives a brief overview of the history of patent regulations, followed by an explanation of how, in different countries, research and development were stimulated.

[1] 'At no distant date…. Britain will be the greatest colour producing country in the world: Nay, by the strangest revolutions, she may, ere long, send her coal-derived blues to India growing India, her tar-distilled crimson to cochineal-producing Mexico and her fossil substitutes for quercitron and safflower to China and Japan, and the other countries whence these articles are now derived'.



M. R. van Bommel, M. de Keijzer, *Bright Colours from the Past*, Cultural Heritage Science, https://doi.org/10.1007/978-3-031-88664-5_3

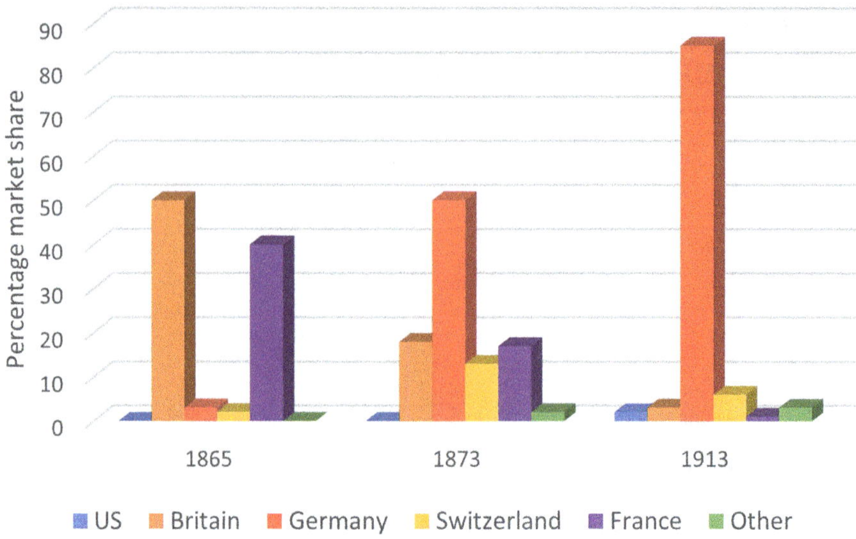


**Fig. 3.1** Percentage of the market division among the most important dye producing countries in the nineteenth century

## 3.2 A Brief History of Patents

There are several definitions of a patent. One of them reads: 'A patent is the granting of a property right by a sovereign authority to an inventor. This grant provides the inventor exclusive rights to the patented process, design, or invention for a designated period in exchange for a comprehensive disclosure of the invention' [3]. A patent is usually awarded for 10–20 years and is meant to stimulate invention. It allows the inventors to take commercial advantage of their discovery, for example by bringing a new dye into production, or by selling or licencing the patent to somebody else.

Some form of patent rights was already known in Ancient Greece. In the Greek colony of Sybaris in southern Italy—nowadays Calabria—annual culinary competitions were held from 500 BCE onwards. The Greek rhetorician and grammarian Athenaeus of Naucratis wrote in the third century CE, in his *Deipnosphistae* (Banquet of the Learned), 'if any confectioner or cook invented an exclusive dish, no one else was allowed to prepare it for one year', which can be considered a patent [4].

In medieval Europe, monarchs granted special rights to the people, so-called 'privileges', which could be used for monopolies over trade. In fact, weavers from the Low Countries were encouraged to relocate to England if they were fleeing from the French. As an example, in 1331 King Edward III of England (1312–1377) granted letters patent to John Kempe, a Flemish weaver of woollen cloths, to enter England with his company, which allowed him to practice and teach his trade. This patent was intended to encourage the growth of the English wool industry [5, 6]. However, this was not a patent for protection or an invention; it was a 'Letter of

Protection' to allow a foreign worker to ply his trade without the threat from rival local workers. The Letters Patent of Protection at this time were like passports which allowed them to come to England and practice their profession [7].

A remarkable grant was given by King Henry Vll of England (1467–1509) to the Italian navigator and explorer John Cabot (Giovanni Cabato) (c. 1450–1498/1499) in 1496. The King permits him, 'to sail, to conquer, to own heathen land, and to exclude others from so doing' [8]. Although, strictly speaking, this was not a patent, it awarded some rights to Cabot over others. In 1497 Cabot sailed with his ship the *Matthew* and landed on the coast of Newfoundland [9].

Other examples of early patents include an Italian patent awarded in 1416 in the medieval Venetian Republic to Francisco Petri for the fulling of wool, for a period of 50 years. An early patent for an industrial invention was granted in 1421 to the Florentine engineer and architect Filippo Brunelleschi (1377–1446), the builder of the cupola of the Duomo in Florence. He received a three-year patent for a barge with a hoisting device to transport slabs of marble from the quarries of Carrara, up the river Arno [10, 11].

The world's oldest Patent Law was developed in the Venetian Republic in 1474 [12, 13]. This statute had all the basic elements of a modern patent system, such as a requirement of novelty, proof of usefulness, and a requirement that the patentee describe and explain the invention. This is considered the start of the modern approach to patents.

## 3.3 Patent History in Dye Producing Countries

In Britain the first patent law, the Statute of Monopolies, was established in 1624, during the reign of King James l of England (1566–1625) [14–16]. This limited the power of the monarch to use the royal prerogative to make monopoly grants and, by establishing a basis on which patents for a new manufacture might be granted to its actual inventor, provided a legal basis for the patent system, as well as promoting industrial and economic interests. Early patents already date from 1617, however, including three examples dealing with pigments—smalt, white and red lead—and madder dye [17–19]. In principle monopolies were forbidden, unless they proposed real innovations, which could be monopolised for 14 years [7]. In 1624, the king's gardener William Shipman received a royal grant for the cultivation and preparation of madder by: 'sowing, setting, or planting of a herbage called madder' and of using 'the misterie, arte and science of breaking, dyeing, dressing and preparing the same for dyeing'. Shipman's madder was successfully sold to London dyers [20]. In the following two centuries, the patent system was developed further, but mainly by actions in court and by the involvement of lawyers, rather than by government regulations. As a result, the patents needed to be described in a very detailed manner. The patent system became more and more complicated and a new patent law was required. This was established in 1852 by the Patent Law Amendment Act. This not only simplified the patent application procedure, but also it became possible to make a single patent application covering the whole of the British Isles, rather than having to make a separate application in each of the countries of Britain. The patent office asked for a description of the invention and, once awarded, the patents were published and as such the information became available. In 1883, the Patents, Design, and Trade Marks Act was introduced, which resulted in an office that controlled an accurate description of the patent, but did not investigate whether a patent request dealt with a real innovation [21]. This improvement was introduced in 1902, when a new Patent Act was established, which ensured that the novelty of an invention was examined thoroughly before the patent was awarded [22].

The patent system in France was inspired by the developments in England [23]. There was also a strong belief in France that inventors should receive protection for their discoveries to ensure that they could profit from them, thus stimulating new inventions and, as a result, the economy. Prior to the French Revolution, the King awarded so-called 'privileges' to inventors, a system which, on the one hand, protected and stimulated inventions, but, on the other, was also criticised, as such a privilege should only be awarded to a genuine innovation. In the 1780s, the French administration supported the application of these privileges, which were a kind of patent, by simplifying the procedure, but also requested that the inventors should make their discoveries publicly available so that new inventions could be made. In January 1791, soon after the French Revolution, the first patent law was established in France, ending the recognition of the privileges. Initially, and this was comparable with the situation in Britain, no investigation was made to determine whether the patent was really innovative. This law was revised only 4 months later, in May 1791, as many complaints were being made that the privilege system still existed. The patents were called *´brevets d'invention´* to distinguish them clearly from the previously awarded privileges. Up to 1803, only a very limited number of patents were filled, about 15 a year, of which 19% came from the textile industry and 7.5% from the chemical industry. This could have been due to the high costs for the patent application to be made (300 francs for five years, 800 for 10 and 1,500 francs for a 15-year patent) and in the next decades several debates were held and attempts made to adjust the patent law. However, it took until 1844 before the French patent law was revised. The new law allowed patents for the invention of new industrial products. In France both the products and the processes could be patented, which was not the case in Britain. Another important development was that, although the cost of a five-year patent was increased to 500 francs, 1,000 francs for a 10-year patent, the cost for patent application was effectively reduced to 100 francs a year as the total cost could be spread over the lifetime of the patent, still too expensive for most people, but certainly more affordable for entrepreneurs [23]. The new law resulted in many more patent applications: up to about 200 per year between 1859 and 1861. One considerable problem was the short life time of many patents: about half of the patents failed in the first two years and only about 10% existed for longer than a decade [12]. The law also resulted in many court cases, where patents were disputed. To keep an overview and registration of all patents applied for, the Office Nationale de la Propriété Industrielle (ONPI) was established in 1902 [24, 25].

The situation in the United States of America developed in a similar manner. In the colonial period, no specific patent law existed, although inventors could ask the government for a commercial right for their products. At the end of the eighteenth century, after the United States of America had become independent, several states developed their own patent laws, but soon, in 1790, a federal law was introduced [26]. The law awarded a 14-year patent, without any right of extension. The first patent, signed by the first President of the United States, George Washington (1732–1799), was granted on July 31, 1790 to the American inventor Samuel Hopkins (1743–1818) for a method of producing potash, an ingredient used as fertilizer and as an ingredient in textile dyeing. Up to 1793, only 57 patents were awarded and the 1790 law was replaced with a new one, making it more explicit that the topic of a patent should be 'any new and useful art, machine, manufacture or composition of matter and any new and useful improvement on any art, machine, manufacture or composition of matter' [27]. The patent process became simpler and stated that the innovation should be 'sufficiently useful and important', which is obviously a vague description. Interestingly, this resulted in 10,000 patents being awarded up to the introduction of a new patent law in 1836. This number was so high that the patent officers could not evaluate the applied patents accurately; as a result, the quality of the patents decreased. The law introduced in 1836 resulted in an official patent office, which became better organised. In addition, patents became available in libraries so that everybody could access them. It also made it possible to extend a patent for another 7 years, after the first term of 14 years [28]. Although several adjustments were made to this law, it remained active for the rest of the nineteenth century [29].

Whereas in Britain, France and the United States of America patent laws were already developed in the eighteenth century, this was not the case in Germany and Switzerland. Before the unification of Germany, each state had its own regulations and licences were in general awarded by the ruler of the respective states, although some larger German states had patent laws. On the 18th of January 1871, the German Empire was established and several years later, in 1877, an all-German patent law was introduced with the establishment of the Imperial Patent Office. The first German patent was granted one day later to the German manufacturer Johann Zeltner (1805–1882) of the firm Nürnberger Ultramarinfabrik, for patenting the manufacture of a red ultramarine pigment. In a way similar to the situation in other countries, the patents were to be made publicly accessible by publishing them in the Imperial Law Gazette (*Reichsgesetzblatt*). Clearly there was a great need for the possibility of obtaining a patent, as shown by the number of applications: already in 1877 3,212 applications were filed, of which only 190 were granted, and in 1,890, 4,680 patents were granted [30–32]. Interestingly, compared to Britain and the United States of America, in Germany the processes were patented, but not the products, while in France, both processes and products could be patented. This means that a newly invented product could be made by competitors as long as they used a different route for its manufacture, which must have stimulated invention, especially when a competitor was able to develop a more advantageous synthesis process for synthetic dyes. This lasted till 1891 when the German patent law was revised and, as a consequence, products could also be patented.

The situation in Switzerland was different: there was a strong opposition towards a patent law [33]. Switzerland was formed in 1291 as a union of three confederations, each of which operated relatively independently. This lasted till the occupation by the French in 1797–1798 when the first central government of the Helvetic Republic was installed; this functioned as a French client state. The confederation was restored under the Act of Mediation by Napoleon in 1803, but after the fall of the Napoleonic empire, the central government collapsed. It was not until 1848 that a new central government was established, in which 22 cantons were united. Although each canton gave up some of its sovereignty to allow the introduction of national laws, it was not until 1888 that the first national Swiss patent law was enacted to protect products. Prohibition on patenting processes, rather than just the products, continued until 1907, when the Swiss government finally bowed to German pressure. This pressure was effective because Switzerland was poor in the appropriate natural resources and was therefore dependent on German supplies of coal-tar distillates and chemicals needed to produce synthetic dyes [33, 34]. The most famous patent employee is the German-born theoretical physicist Albert Einstein (1879–1955), who worked at the Swiss Patent Office between 1902 and 1909.

Two examples of how patents could help to profit from inventions are the syntheses of Alizarin and Eosin. Alizarin is the main constituent of the dye extracted from madder roots (*Rubia tinctorum* L.) and had already been used for centuries. The search to synthesise this dye led to two patents. The German chemists, Carl Graebe (1841–1927), Carl Theodore Liebermann (1842–1914) and Heinrich Caro (1834–1910), working at BASF, applied for a patent for their Alizarin production process in Britain on the 25th of June in 1869 (B.P. 1.936). Only 1 day later, the British chemist William Henry Perkin (1838–1907) filed a patent application for a similar method (B.P. 1.948). In the end both patents were granted. By March 1870 they had come to an agreement over cross-licensing of the June 1869 patents: Perkin would monopolise the British supply and BASF would control the mainland Europe and the North American sectors. The production at BASF had been held up due to the Franco-Prussian war of 1870–1871, and Perkin found that he had the world market to himself for almost one year [35].

Another example was the discovery of Eosin by Caro in 1873. He was repeatedly disappointed by earlier attempts in 1869 to obtain German patents for the production of benzene and synthetic Alizarin. Because of these bad experiences, Caro decided to keep the production process of Eosin a secret and it was not patented, even in Britain and other countries. But it turned out that this was not an adequate protection. The German chemist Carl Alexander Martius (1838–1920), a co-founder of AGFA, purchased some BASF Eosin and gave it to Hofmann. Unfortunately for Caro, Hofmann determined the chemical constitution of Eosin in 1875. As a result, other firms produced this dye as well [36, 37].

## 3.4 Patent Affairs and Conventions

From the first day onwards, patent struggles between dye factories about synthetic dyes were an issue. Perkin & Sons, Simpson, Maule & Nicholson, Renard Frères and Société La Fuchsine attempted to control the British and French markets with monopolies based on their patents. In 1856 Perkin patented Mauveine in Britain, but he also wanted a patent in France. His application was declared invalid because it was filed more than 6 months after the British one. However, the demand for Mauveine was growing fast and French dye firms could not supply enough, allowing Perkin & Sons to receive large orders. Perkin & Sons did not want competitors on the British market for Mauveine and in 1858 Perkin refused to grant Simpson, Maule & Nicholson a licence to make Mauveine because he feared competition.

But the market for red dyes was much larger than that for purple. To infringe upon the patent for Fuchsine would be more lucrative. Simpson, Maule & Nicholson obtained the monopoly in Britain and made it known publicly that they would take to court any firm that made Fuchsine. From 1861 to 1865 they became involved in a series of very expensive patents trials; Simpson, Maule & Nicholson alone spent £30,000 which nowadays would be equivalent to £1.6 million. Trials were so expensive because they dragged on for a long time and both sides called in numerous expert witnesses. The most important trial was against Read Holliday. Simpson, Maule & Nicholson won the early trials, but by 1865 their patent was declared invalid, and the firm lost enormous profits.

To secure a good long-term performance, the best a firm could do was to license its patents even when in the short run it could make higher profits by keeping the synthesis route and product secret. In the early 1860s, Renard Frères in France were closely connected with Simpson, Maule & Nicholson, who in 1862 were granted a licence to produce Aniline blue in Britain. Sometime later, the firm also acquired the licence to manufacture Hofmann's violet, which reduced the demand for Mauveine. In this way Simpson, Maule & Nicholson provided a portfolio of the most important synthetics at the time. By this strategy it became the largest dye manufacturer in the early 1860s [38, 39].

## 3.5 Methylene Blue Patents and Patent Suits

### *Methylene Blue*

Colour Index name: Basic Blue 9; Colour Index number: 52015.
IUPAC name: 3,7-bis(Dimethylamino)-phenothiazin-5-ium chloride.
Discoverer: Heinrich Caro, 1876
Contemporary terminology: Methylene blue (En), Bleu methylène (Fr), Methylenblau (Ge), Methyleenblauw (NL).
Chemical formula: $C_{16}H_{18}N_3SCl$ (Fig. 3.2).

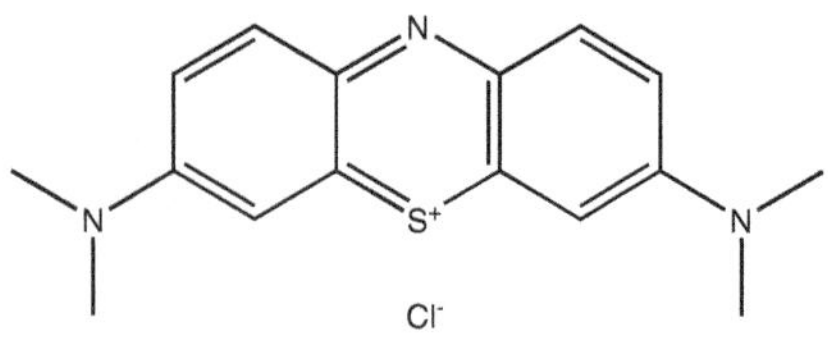


**Fig. 3.2** Structural formula of Methylene blue

An example of a patent battle is the development of Methylene blue (C.I. name: Basic Blue 9; C.I. number: 52015), the most important dye of the thiazine class. The development of this blue dye can be studied by patents and by a series of patent suits. In 1876, Heinrich Caro prepared Methylene blue for which BASF obtained its first German patent (G.P. 1.886) on December 15, 1877. Methylene blue was prepared using *p*-nitroso-*N, N*-dimethyl-*p*-phenylenediamine in the presence of hydrogen sulfide.

At first, the structure and intermediates of Methylene blue were not known, and therefore the reaction mechanism was unclear. This limited understanding complicated the defense of Caro's patent against circumvention. The situation became problematic when German dye competitors patented new processes for Methylene blue. In 1881 an AGFA patent was declared invalid by expert witnesses, a decision accepted by both parties without litigation. BASF received support from the German chemists Adolf Baeyer and Friedrich August Kekulé.

Some later, new synthesis routes developed by the German chemist Wilhelm Majert were a real danger for BASF. The Berlin firm Ewer & Pick used Majert's processes and patented methods for producing BASF's Methylene blue and Auramine. This led to BASF's purchase of the Ewer & Pick patents in 1886 for a considerable amount of money.

The German dye company Oehler, where the German chemist Eduard Hepp (1851–1917) worked, had a valid patent for another Methylene blue process (U.S.P. 270.311, B.P. 4.048 and G.P. 24.125, all filed in 1882). Oehler marketed this dye as Ethylene blue. In Germany, BASF could not stop its production because the German patent law protected only processes, and not the product. In the United States of America, the situation was different since products could be patented. To prevent the sale of Ethylene blue, BASF needed to demonstrate that Oehler's product was identical to its own. To achieve this, BASF sent Magnus Boesler, a former employee of Caro's laboratory, to the United States of America and supported him with analytical work that was carried out in Ludwigshafen am Rhein in February 1884 (Table 3.1).

In 1883, the patent landscape shifted due to research performed by the German chemist Heinrich August Bernthsen (1855–1931), who worked as an external employee of BASF. Bernthsen demonstrated in 1884 that Methylene blue was a phenothiazine derivative, but a new technical synthesis attempt failed due to a low yield. Using this knowledge, Carl Roth from Berlin developed a new process and patented it (G.P. 38.573, filed in 1885). In 1887, Methylene blue was also patented in the United States of America by the German chemist Eduard Ullrich (1839–1917),

**Table 3.1** Discoverers and patents of Methylene blue

| Name<br>C.I. name<br>C.I. number | Discoverers | Year of discovery | Patents |
|---|---|---|---|
| Methylene Blue<br>Basic Blue 9<br>52015 | H. Caro (BASF) | 1876 | G.P. 1.886/1876<br>B.P. 3.751/1877<br>B.P. 10.314/1888<br>U.S.P. 204.796/1878<br>F.P. 122.720 |
| | A. Bernthsen (BASF) | 1887 | G.P. 45.839/1887<br>G.P. 47.374/1888<br>B.P. 8.221/1893 |
| | E. Hepp (Oehler) | 1882 | U.S.P. 270.311/1882<br>G.P. 24.125/1882<br>B.P. 4048/1882 |
| | C. Roth (Farbwerke ML&B) | 1885 | G.P. 38.573/1885<br>F.P. 173.137/1885<br>B.P. 43/1886<br>G.P. 39.757/1886<br>G.P. 46.805/1888 |
| | E. Ullrich (Farbwerke ML&B) | 1887 | U.S.P. 362.592/1887<br>U.S.P. 366.639/1887<br>U.S.P. 366.640/1887<br>U.S.P. 384.480/1888<br>F.P. 181.827 |

assignor to Carl Roth in Berlin and Farbwerke ML&B (U.S.P. 362.592, U.S.P. 366.639 and U.S.P. 366.640, all filed in 1887, and U.S.P. 384.480, filed in 1888).

Some later Bernthsen discovered a new synthesis for Methylene blue, based on the structural elucidation of Methylene red, a by-product of the Methylene blue production process (G.P. 45.839, filed in 1887). Farbwerke ML&B challenged this patent, leading to litigation that resulted in an agreement in 1888. Farbwerke ML&B received a market share of 27%, the chemical firm Georg Carl Zimmer in Mannheim received a market share of 18%, allowing BASF to secure 55% of the market share. During the litigation Kékulé served again as the expert witness for BASF, while the German chemists Johannes Wislicenus (1835–1902) and Hermann Emil Fischer (1852–1919) represented Farbwerke ML&B. Despite Farbwerke ML&B's earlier patent (G.P. 38.573, filed in 1885), BASF managed to capture a larger portion of the market, showing the role of scientific expertise and of a strong patent department in defending patents in court [40].

The number of agreements and conventions was modest during the 1870s, but firms had learned from the past. The uncertainty about the outcome of lengthy processes fostered the willingness to compromise, to license and to develop cartels of products, markets and prices; the number of conventions increased in the 1880s as a result. An important example, as discussed earlier, was the agreement on synthetic Alizarin by which Perkin & Sons and BASF shared the market.

Another important convention was that for Benzopurpurine in the mid-1880s. The first synthetic direct cotton dye was Congo red, introduced by AGFA, followed by Bayer's Benzopurpurine 4B. When Bayer filed its patent, AGFA filed its opposition against it because the chemical structure of the two dyes differed only slightly. The German court ruled that Benzopurpurine was not patentable. Bayer's management advised strongly against appealing against the judgement in favour of AGFA. As a result, both firms agreed to form a cartel to commercialise both dyes [41, 42].

## *Ivan Levinstein*

A very important person with respect to international patents was the German-born British chemist Ivan Levinstein (1845–1916). After his chemistry studies at the Berlin University, Levinstein moved to Britain where he founded a dye factory in Blackley near Manchester in 1865. Between 1866 and 1875 he did not file patents; in the period 1876–1883 the firm received eight British patents and in the last period, up to 1914, he pursued the strategy of patenting as much as possible, especially azo dyes, just as the German firms did. This strategy gave the firm 67 British patents between 1888 and 1914.

The Germans considered that many of the azo dye patents were infringements upon their British patents, which led to lengthy and expensive litigation. In the 1880s, BASF fought the biggest patent battle with Levinstein over the azo dye Fast red A. The case dragged on for 6 years and the BASF was declared victorious after the affair had been waged all the way from the High Court to the Court of Appeal, and finally to the House of Lords.

Interestingly, in 1890 Levinstein Limited entered into a relationship with Bayer and AGFA. Because of the difficulty of raising capital in Britain for building a new factory, Levinstein sold two-thirds of his company, one-third to Bayer and the other third to AGFA. By this act, the firm essentially became an outpost of the German synthetic dye industry. Besides capital, Levinstein also received assistance. Bayer and AGFA gave access to all their patents for the lucrative direct cotton dyes to Levinstein Limited. In return, they obtained a fixed share of profit from their sales and maintained a monopoly on the direct azo dyes. Because of this relationship, Levinstein also fought patent suits between 1890 and 1895 against Read Holliday and other British firms on behalf of his German partners to protect the monopoly. When their partnership agreement was up for renewal in 1895, Bayer and AGFA sold back their two-thirds stake in the Levinstein dye firm.

From the early 1880s there were moves to reform the British patent law. In the late 1890s Levinstein stepped up his campaign of patent reform and protectionism for the British dye-makers. In 1907 the new British patent act became law and the German companies were forced to build their own production in Britain to preserve their patents. As a result, the Union of Three (BASF, Bayer and AGFA) and the Association of Three (Farbwerke ML&B, Cassella and Kalle) both built a plant in the Liverpool area. This was not what Levinstein had hoped for. After the German firms refused to buy dye intermediates from Levinstein, starting in 1908, because they were upset by his strategic actions in connection with the British patent law, the firm faced a financial crisis, particularly in 1910–1911. Still, Levinstein Limited was the largest producer of dyes in Britain from the 1880s until the First World War and possessed a global market share of 1% [40].

## 3.6 International Competition

In the first decade of the development of the synthetic dye industry in Europe and the United States of America, Britain and France were the leading producers. This next part will concentrate on the international competition between the five important synthetic dye producing countries and proof will be given explaining why Germany came to dominate this industry in comparison to the other countries before the First World War.

### *France*

After the invention of Mauveine in 1856, purple became the leading fashion colour in Paris and London. Perkin did not license his patented process to other firms in Britain or abroad, so new progress was made elsewhere. In France, dyers in Lyon specialised in dyeing silks were soon experimenting with aniline, resulting in the discovery of Fuchsine which was produced by the factory Renard Frères in 1859. The commercial success of this dye attracted other producers in France. Renard Frères attempted to control the French markets with monopolies based on their patents, in which they not only claimed the manufacturing process of Fuchsine, but also the application of the dye to textile dyeing and printing. Renard Frères did not want to supply silk dyers in Lyon and Saint-Etienne with Fuchsine because it would harm their own economic interests. These cases outraged Renard's rivals, and in 1861, 117 Lyon firms petitioned the French Minister of Agriculture, Commerce and Public Works to establish an impartial scientific commission to resolve the questions raised on the Fuchsine disputes. Renard Frères rejected this proposal and argued that it should be judged by normal courts. At the end of 1863, Renard Frères & Franc entered into an alliance with the Crédit Lyonnais bank to set up the controlling and anticompetitive firm Société La Fuchsine. This firm convinced the courts to grant it a master patent for Fuchsine, which was not only a product, but also a key intermediate for other dyes, such as Aniline blue and Hofmann's violet. As a result, the firm was able to demand that the French police shut down many rival producers. As a consequence, many chemists and firms moved to Switzerland [38, 43].

An innovator of circumventing the monopoly was the French chemist Jean Théodore Coupier (1820–1908). In 1866, he patented a method to produce Fuchsine by heating pure toluidine alone, without using aniline. The commercial exploitation was obstructed by Société La Fuchsine, who argued for priority by holding up the Franc patent of 1859 for the production of Fuchsine from toluidine and cumidine. Coupier instituted legal action to establish the nullity of this patent. Experts were commissioned and their conclusions were favourable to Coupier. Société La Fuchsine appealed and the Court at Lyon reversed the earlier decision. Now Coupier in his turn appealed, but without success [38, 44]. In 1869 the owners of the firm Meister, Lucius & Brüning (later Farbwerke ML&B) visited him and wanted to buy his process. The negotiations had no result. However, in January 1873 the German chemist Johann Adolf Brüning (1837–1884) writes that in October 1872 Farbwerke ML&B was successful in producing Fuchsine by the nitrobenzene method [45–48]. As soon as 1868, only 5 years after its foundation, Société La Fuchsine collapsed. Its competitive activity ended, but the damage was already done. Due to all the patent disputes, innovation was slowed down, which was one of the reasons why France played a minor role in the synthetic dye industry at that time. There were two other main factors why the French industry lost their leading position. Firstly, France had been the leading country in the development of chemistry until Germany surpassed it in the 1830s. By the time the dye industry started there were two important centres of organic chemistry, one in Paris and one in Lyon. The most influential teacher in the field of industrial organic chemistry was the French chemist Théophile-Jules Pelouze (1807–1867). His private laboratory in Paris had been a centre for the formation of dozens of young industrial chemists, among them the Swede Alfred Nobel (1833–1896). When Pelouze passed away in 1867 there was no one to continue his laboratory. As a result, the number of potential entrepreneurs in France decreased dramatically. The centralised French academic system was much more vulnerable to a temporary dramatic loss in quality and output than the decentralised German system. The other factor was that German companies had become very strong by the late 1860s and exported products to France. German firms started to patent new dye innovations in France, bought French factories and produced the patented dye in France in small quantities to circumvent the patent laws, while exporting large quantities from their German firms [49].

## *Switzerland*

The Swiss dye industry was, to some extent, a child of the French patent system. The Société La Fuchsine monopoly forced many French chemists, and even firms, to relocate. In effect, the French patent system was responsible for a major review of the state of affairs. At least seven firms closed down their production in France and moved to Switzerland, which had no patent law at that time. They settled in the textile and dye city Basel. The French chemist Alexandre Clavel (1805–1873) started the company that was to become CIBA in 1884. The factory Müller-Pack in Basel, which originally sold dye wares on behalf of the Geigy family, was taken over by the Swiss dye manufacturer Johann Rudolf Geigy-Merian (1830–1917) in

1862 and the Swiss chemists Alfred Kern (1850–1893) and Edouard Sandoz (1853–1928) established the dye firm Kern & Sandoz in 1886. The Swiss firms began by manufacturing bulk dye products such as Fuchsine and synthetic Alizarin, but soon they noticed that they were unable to compete with Germany because they relied on intermediates from German firms. In response, these firms shifted their production to high quality dyes and pharmaceuticals, mostly for export. Another reason was the presence of two research centres in organic and industrial chemistry in Zurich and Geneva. These strategies were so successful that, by the 1890s, Switzerland was already the world's second largest dye producer [43].

### *The United States of America*

While Europe was experiencing the first impact of the aniline colours, the United States of America was troubled by the Civil War between 1861 and 1865. Moreover, the market for synthetic dyes in the United States of America in the early years was much smaller than that in Europe. Initially, synthetic dyes were very expensive and only applied to luxury textiles such as silk. In Europe, a network of leading silk dyeing firms was developed in the 1860s, but the United States of America was not included in this network through which knowledge of the early aniline dyes was transmitted. Secondly, the American universities offered almost no advanced training in synthetic organic chemistry before 1914. This made it difficult for American firms to build up the necessary expertise to compete in the international dye industry. To compensate for this, Americans went to Germany to receive graduate training in organic chemistry. The American system of higher education made great advances in the period from 1890 until 1914: the number of US bachelor's degrees and doctorates in chemistry increased by 400%, but none of the PhDs was in synthetic organic chemistry. As a result, the United States of America did not possess a large number of people who had the skills to start the production of organic intermediate chemicals or synthetic dyes. Before 1914 the United States of America relied heavily on imported German intermediates. With the outbreak of the First World War, there were shortages of intermediates and the price of dyes increased considerably. In 1916 home-based production was encouraged by the Revenue Act, which provided a protective barrier for foreign products for a period of 5 years. From that time onwards, large dye firms were founded [2].

### *Germany*

The reasons for Germany's success in the 1870s can be attributed to several factors. Before the creation of an all-German Patent Law, the 39 sovereign states were not willing to give up their own patent rules. This meant that anyone could enter the synthetic dye business and could imitate the products of other producers, both nationally and internationally, until the mid-1860s. After the formation of the German Empire in 1871, the right to pass patent laws now rested with the national government and in 1877 an all-German patent system was created, as described above. The number of dye factories and patent filings between 1874 and 1913 shows that Germany was superior to the other dyeing producing countries.

In addition to the patent system, the collaboration between industry and academia and the political situation in each country were also different, which was to the advantage of Germany. In Germany the interaction between scientific education and the dye industry was far more developed than in the other countries. Firstly, the German system of higher technical and scientific education was superior to the British system due to the higher level of financial support by the German governments. At the turn of the twentieth century Prussia, the largest state in Germany, spent almost 18 times more on universities than the whole of Britain. Unlike the United States of America, where many private universities existed, German universities and research institutes were public entities, funded by public money. This meant that increased contributions to universities had to pass through parliament. Government officials and parliaments had to be convinced that it was advantageous to spend more money on education. But German universities were often educating more scientists than society needed in the 1840s and 1850s and many of them went to Britain because jobs for chemists in Germany were scarcer than qualified applicants. Some of them were involved in the developments of synthetic dyes in Britain and they then returned to their homeland in the 1860s to contribute to the rise of new firms, such as Bayer, Farbwerke ML&B, BASF and AGFA. In Germany an academic-industrial network developed and the laboratories of Hofmann, Baeyer and Emil Fischer became the leading centres in organic chemistry where new knowledge was created at a faster rate than in Britain, France and the United States of America. The industrial-academic collaboration led to successes, such as the discovery of synthetic Alizarin and Eosin [2].

The German dye industry employed three strategies: collective organisations for the mobilisation of support; lobbying parliament directly, and the creation of private-public academic partnerships. A good example of how dye firms were engaged in the mobilisation of support is Carl Duisberg's draft of a petition, written in the name of the Club of German Chemists (Verein Deutscher Chemiker). The aim of the petition was to convince the Prussian authorities to appoint more professors in chemistry. When Henry Theodore Böttinger (1848–1920), one of Bayer's directors, won a seat in the Prussian parliament, he used this platform to advance the educational goals of the dye industry directly. His friend Friedrich Theodor Althoff (1839–1908), head of the university department in the Prussian Ministry of Education, was responsible for the appointments at Prussian universities and technical colleges. In this way Böttinger coordinated his parliamentary proposals with the key players inside the Prussian government to increase educational spending. Böttinger and Althoff worked together on numerous projects to upgrade German science and engineering education [2]. At the turn of the century, they created a new organisational structure in the German research and teaching system, a public-private partnership to sponsor industrially relevant applied research. An example is that the dye industry representatives formed the largest contingent of industrialists on the board of the Kaiser Wilhelm Institute for Chemistry and made substantial financial contributions. Every large dye firm was involved via German representatives: Adolf Brüning (Farbwerke ML&B), Franz Oppenheim (AGFA), Henry Theodore Böttinger (Bayer), Heinrich Brunck (BASF), Carl Duisberg (Bayer) and Herman Edmund ter Meer (ter Meer) joined leading academics like Emil Fischer and Carl Theodore Liebermann as well as government officials on the board [49].

### *Britain*

Britain certainly had a significant number of scientists with an international reputation in the nineteenth century; however, eminent British figures such as Michael Faraday, Humphry Davy, John Dalton and William Wollaston did not create research schools as the German chemists Justus von Liebig, Friedrich Wöhler, August Friedrich Kekulé, Adolph Wilhelm Hermann Kolbe, Robert Bunsen, August Wilhelm Hofmann, Emil Erlenmeyer, Adolf Baeyer, Hermann Emil Louis Fischer, Otto Nikolaus Witt and others did in Germany. Britain excelled in producing many world-class 'private' scientists, who were often associated with learned societies and academies, but did not develop the industrial framework for producing the large number of qualified students necessary who could be employed by industry. The overproduction of chemists in Germany was initially worth a fortune to Britain, as they could make up for the lack of local training. When the expansion of the German dye and academic chemistry created excellent opportunities for German chemists at home, however, it became difficult for British firms to hire German talent. In Britain as well as in Germany, powerful people in government, industry, and the academic world resisted the expansion of science and technology teaching. It was unfortunate for Britain that these forces were much stronger there than in Germany. Although the British student population increased, German universities had a much larger output of scientists. The German system of higher technical and scientific education was superior to the British one and behind this lies the fact that German governments provided a higher level of financial support [2].

## 3.7 To Conclude

From the history of the regulations for patents in the respective countries, some conclusions can be drawn. Firstly, Britain and France had had a system of protection for inventions since the Middle Ages and developed their first patent laws in the 17th and 18th centuries, which was also the case for the United States of America. Germany and Switzerland did not have a unified patent law until the end of the nineteenth century. Although on the one hand, this was a disadvantage for inventors since their discoveries were less well protected, on the other hand it was also an advantage, since products could be copied and the one who could produce the merchandise most cost-effectively could profit the most. In addition, in Germany initially only processes could be patented, not products, which was not the case in all the other countries. In Britain, the United States of America and Switzerland only the products could be patented, while in France, both products and processes could be patented. That meant that in Germany, anyone who could develop another route for the synthesis of a synthetic colourant was allowed to bring the product onto the market; this was not the case in the other countries. The effect of this might have been to slow down innovation in these countries, as developing a new, maybe more cost-effective, procedure for synthesis was not useful since the resulting product could not be sold.

Another aspect was the way in which education was organised. Britain did not foster strong collaboration between academia and industry at the same rate as Germany and therefore lost more and more market share to the German firms. The same applies to France, where it seems that inventors were more interested in protecting their patents than working on new discoveries. The centralised education systems were not helpful either. The United States of America was initially very much dependent on German chemists.

On the eve of World War 1, the total worldwide production of synthetic dyes amounted to 160,000 tons. The United States of America produced 3,000 tons. The three leading British dye firms, Ivan Levinstein, Read Holliday and the British Alizarine Company, all together produced 4,000 tons of dyes, whereas Germany produced 140,000 tons [50]. Thus, the British global market share was between 1.8% and 3.1% in 1913.

The British chemist and physicist William Crookes (1832–1919) complained that Great Britain relied on Germany for all synthetic dyes by saying:

> 'Like William Perkin I personally aspire to metamorphose lower into higher. His transforming coal-tar into brilliant dyes has come for me of late to symbolise chemistry at its most profound and true creating radiance out of basest residue. It is infinitely satisfying to see what was black dreck serve the art of dyeing.
>
> William Perkin's ingenious transformation could have benefited the British nation. But unfortunately, Great Britain still relies on Germany for all synthetic dyes.
>
> The coal-tars of the Ruhr and Rhine metamorphosed in industrial bulk into aniline. If more of us in Britain had done work in the processes pioneered by William Perkin we would not now as a nation still have to rely on Germany for all synthetic dyes.
>
> Like many physical or chemical invention pioneered by the British I could mention Perkin's valuable synthetic dyes which will always, for yours truly, symbolise the magic of chemistry, Germans monopolise' [50, 51].

## References

1. Hofmann, A.W.: Colouring Derivatives of Organic Matter, Recent and Fossilized. Report On The Chemical And Pharmaceutical Products And Processes Exhibited In Class II. Section A. Of The International Exhibition Held At London In 1862. (1862)
2. Murmann, J.P.: Knowledge and competitive advantage. Cambridge University Press, Cambridge (2003)

3. What Is a Patent in Simple Terms? With Examples, https://www.investopedia.com/terms/p/patent.asp
4. Yonge C.D. The Deipnosophists, or, Banquet of the Learned of Athenæus. p. 835, Henry G. Bohn, London (1854)
5. Calendar of Patent Rolls, Edward III: Volume 2, 1330-1334, http://www.british-history.ac.uk/series/calendar-of-patent-rolls-edward-iii/vol2/pp.145–169, (1893)
6. Lambert, B., Pajic, M.: Immigration and the Common Profit: Native Cloth Workers, Flemish Exiles, and Royal Policy in Fourteenth-Century London. Journal of British Studies. 55, 633–657 (2016). https://doi.org/10.1017/jbr.2016.75
7. Wyndham Hulme, E.: The History of the Patent System under the Prerogative and at Common Law. Law Quarterly Review. 46, 141–154 (1896)
8. Biggar, H.P. ed: The Precursors of Jacques Cartier 1497-1534: A Collection of Documents relating to the Early History of the Dominion of Canada. Government Printing Bureau, Ottawa (1911)
9. Wilson, I.: John Cabot and the Matthew. Breakwater Books Ltd., Saint John's (2001)
10. Brown, R.: The Patent: from Classical Antiquity to Modern Industry, https://nbindustrial.org/blog/patent-history
11. Kurz, P.: Brunelleschi and Galilei: super-early patents in Florence and Venice. Il Nuovo Saggiatore. 40, (2024)
12. Schippel, H.: Die Anfänge des Erfindungsschutzes in Venedig. In: Lindgren, U. (ed.) Europäische Technik im Mittelalter. 800 bis 1400. Tradition und Innovation. pp. 1748–1749. Gebr. Mann Verlag, Berlin (2001)
13. Mola, L.: Inventors, patents and the market for innovations in Renaissance Italy. In: Guagnini, A. and Mola, L. (eds.) Italian Technology from the Renaissance to the Twentieth Century. pp. 7–34. Bloomsbury Academic, London (2014)
14. An Act concerning Monopolies and Dispensations with Penal Laws and the forfeitures thereof, 1624, 21 Jac.I, c.3. (1624)
15. Fox, H.G.: Monopolies and Patents: A Study on the History and Future of the Patent Monopoly. University of Toronto Press, Toronto (1947)
16. MacLeod, C.: Inventing the Industrial Revolution: the English patent system, 1660-1800. Cambridge University Press, Cambridge (1988)
17. Woodcroft, B.: Titles of Patents of Invention, Chronologically Arranged: From March 2, 1617 (14 James I.) to October 1, 1852 (16 Victoriae). G.E. Eyre & W. Spottiswoode, London (1854)
18. Costaras, N.: Early modern blues: the smalt patent in context. In: Kirby, J., Nash, S., and Cannon, J. (eds.) Trade in Artists' Materials: Markets and Commerce in Europe to 1700. pp. 401–414. Archetype Publications Ltd., London (2010)
19. Bruce, J. ed: Calendar of State Papers Domestic: Charles I, 1635–1636. London (1866)
20. Armstrong, E.F.: Pigments Of Other Days—II. J R Soc Arts. 87, 444–445 (1939)
21. Munro, J.E.C.: The Patents, Designs, and Trade Marks Act, 1883 (46 & 47 Vict. c. 57) with rules and instructions, together with pleadings, orders, and precedents. Stevens and Sons Publishers and Booksellers, London (1884)
22. Patents Act, 1902. (1902)
23. Galvez-Behar, G.: The Patent System during the French Industrial Revolution: Institutional Change and Economic Effects. Jahrbuch für Wirtschaftsgeschichte / Economic History Yearbook. 60, 31–56 (2019). https://doi.org/10.1515/jbwg-2019-0003
24. Galvez-Behar, G.: Des médiateurs au coeur du système d'innovation : les agents de brevets en France (1870–1914). In: Corcy, M.-S., Douyère-Demeulenaere, C., and Hilaire-Pérez, L. (eds.) Les archives de l'invention. Ecrits, objets et images de l'activité inventive. pp. 437–447. CNRS; Université Toulouse-Le Mirail (2006)
25. Godelier, É.: La République des inventeurs: Propriété et organisation de l'innovation en France (1791–1922) [The Republic of Inventors: Intellectual Property and the Organization of Innovation in France (1791–1922)]. By Gabriel Galvez-Behar. Presses Universitaires de Rennes. Bus Hist Rev. 85, 396–398 (2011). https://doi.org/10.1017/S0007680511000511

26. The First United States Patent Statute. Patent Act of 1790, Ch. 7, 1 Stat. 109 (April 10, 1790) CHAP. VII.—An ACT to promote the Progress of useful Arts. US (1790)
27. Patent Act of 1793, Ch. 11, 1 Stat. 318–323 (February 21, 1793). US (1793)
28. Patent Act of 1836, Ch. 357, 5 Stat. 117 (July 4, 1836). US (1836)
29. Lamoreaux, N.R., Sokoloff, K.L., Sutthiphisal, D.: Patent Alchemy: The Market for Technology in US History. Bus Hist Rev. 87, 3–38 (2013). https://doi.org/10.1017/S0007680513000123
30. Burhop, C., Wolf, N.: The German Market for Patents during the "Second Industrialization," 1884-1913: A Gravity Approach. Bus Hist Rev. 87, 69–93 (2013)
31. Streb, J., Baten, J., Yin, S.: Technological and geographical knowledge spillover in the German empire 1877–1918. Econ Hist Rev. 59, 347–373 (2006). https://doi.org/10.1111/j.1468-0289.2006.00347.x
32. Meyer-Thurow, G.: The Industrialization of Invention: A Case Study from the German Chemical Industry. Isis. 73, 363–381 (1982)
33. Ritter, D.S.: Switzerland's Patent Law History. Fordham Intellectual Property, Media & Entertainment Law Journal. 14, 463 (2004)
34. Dutfield, G., Suthersanen, U.: Harmonisation or Differentiation in Intellectual Property Protection? The Lessons of History. Quaker United Nations Office. 9–10 (2004)
35. Morris, P., Travis, A.: A History of the International Dyestuff Industry. American Dyestuff Reporter. 81, (1992)
36. Hewitt, T.C.: Synthetic Colouring Matters, Dyestuffs derived from Pyridine, Quinoline, Acridine and Xanthene. Longmans, Green & Co., New York/Toronto/Bombay/Calcutta/Madras (1922)
37. Hofmann, A.W.: Ueber das Eosin. Berichte der deutschen chemischen Gesellschaft zu Berlin. Ferd. Dümmler's Verlagsbuchhandlung, Berlin. 8, 62–66 (1875). https://doi.org/10.1002/cber.18750080120
38. van den Belt, H.: Why Monopoly Failed: The Rise and Fall of Société La Fuchsine. The British Journal for the History of Science. 25, 45–63 (1992)
39. Travis, A.S.: The Rainbow Makers: The Origins of the Synthetic Dyestuffs Industry in Western Europe. Lehigh University Press, Bethlehem, PA (1993)
40. Reinhardt, C., Travis, A.S.: Heinrich Caro and the Creation of Modern Chemical Industry. Kluwer Academic Publishers, Dordrecht / Boston / London (2000)
41. Werkgeschichte, 1863-1938. In: I.G. Farbenindustrie Aktiengesellschaft, Werke: Farbenfabriken vorm. Friedr. Bayer & Co. pp. 76–78. F. Bruckmann KG., München (1938)
42. Sonnemann, R., Etzold, H.: Patent und Monopol: Eine Studie zur Herausbildung von Monopolen unter dem Einfluß des Patentwesens. Jahrbuch für Wirtschaftsgeschichte / Economic History Yearbook. 6, 121–159 (1965). https://doi.org/10.1524/jbwg.1965.6.1.121
43. Murmann, J., Homburg, E.: Comparing Evolutionary Dynamics Across Different National Settings: The Case of The Synthetic Dye Industry, 1857–1914. J Evol Econ. 11, 177–205 (2001). https://doi.org/10.1007/PL00003863
44. Wagner, J.R.: Vl Gruppe, Technologie der Gespinnstfasern, Theerfarbstoffe. Jahres-Bericht über die Leistungen der Chemischen Technologie für 1869, Verlag von Otto Wigand, Leipzig. 568–570 (1870)
45. Farbwerke Hoechst Aktiengesellschaft vormals Meister Lucius & Brüning Dokumente aus Hoechster Archiven: Neunzig Jahre Fuchsin in Hoechst. Beiträge zur Geschichte der Chemie. 3, 34 (1965)
46. Brüning, A.: Darstellung des Fuchsins. Berichte der deutschen chemischen Gesellschaft zu Berlin. 6, 25–26 (1873). https://doi.org/10.1002/cber.187300601111
47. Coupier: Ueber die neue Methode zur Darstellung des Fuchsins, nach Brüning. Berichte der deutschen chemischen Gesellschaft zu Berlin. Ferd. Dümmler's Verlagsbuchhandlung, Berlin. 6, 423 (1873). https://doi.org/10.1002/cber.187300601140

48. Brüning, A.: Antwort auf Coupiers Bemerkungen über die Darstellung des Fuchsin ohne Arsensäure. Berichte der deutschen chemischen Gesellschaft zu Berlin. Ferd. Dümmler's Verlagsbuchhandlung, Berlin. 6, 1072 (1873). https://doi.org/10.1002/cber.18730060259
49. Wissenschaftszentrum Berlin für Sozialforschung, Wettbewerb und Institutionen, WZB-Mitteilungen. WZB-Mitteilungen. 98, 41–43 (2002)
50. Garfield, S.: Chapter 7: The terrible glare. In: Mauve: How one Man invented a Colour that changed the World. pp. 78, 83–84. Faber and Faber Ltd., London (2000)
51. Harrison, T.: Square Rounds. Faber and Faber Ltd, London (1992)

# Chapter 4
# Dye Recipes for Synthetic Colourants

## 4.1 Introduction

The introduction of synthetic colourants did not only offer a new range of colours, but also new methods to dye textiles. In this chapter, various dye procedures and some recipes are described. Dyeing with natural colourants was a time-consuming process. In the first step, the dyes needed to be extracted from plant materials, or from insects or molluscs, a process which could easily take hours and, when fermentation was required, even several days. Using natural dyes, three main dyeing processes were used: direct dyeing, mordant dyeing and vat dyeing [1, 2]. Whereas direct dyeing can be a rather quick process, the most sustainable dyeing processes, mordant dyeing and vat dyeing, were quite laborious. This changed dramatically with the introduction of synthetic dyes, especially since acid and basic dyes were easy to dissolve and did not require any extraction from a source material, as is the case with natural dyes. Furthermore, dyeing could be done immediately and in a very short time. Nevertheless, dyeing procedures originating from natural colourants remained important for some synthetic colourants and are still applicable; hence, these will be described here as well.

Dyeing is normally done at elevated temperatures, between 40 °C and the boiling point (100 °C), which makes the dye easier to dissolve and ensures that the textile fibre swells and become more receptive to the dye. It must be noted that, especially at high temperatures, the textile can be affected in a negative way. To avoid a temperature shock, the textile is often brought into the dye bath at a lower temperature, between room temperature and 40 °C, and slowly brought to the desired temperature. Many recipes for both natural and synthetic dyes are performed at a temperature not higher than 60–80 °C, although it is also feasible to dye at boiling temperature. The addition of salt to a dye bath is presented in several recipes where synthetic colourants are used. This addition basically ‘pushes’ ionic dyes out of the dye bath into the textile since it enhances the affinity of the dye for the textile fibre.



M. R. van Bommel, M. de Keijzer, *Bright Colours from the Past*, Cultural Heritage Science, https://doi.org/10.1007/978-3-031-88664-5_4

It can even stimulate an exhaustion of the dye bath by reducing the solubility of the dye, ensuring that no dye is lost [3]. Although nowadays, this is especially applied for cotton dyeing, the recipes presented below show that during the nineteenth century salts were also added for other textiles.

The recipes presented are extracted from the 1893 book by Adolf Lehne [4]. In the first section of the book, he provided general recipes for dyeing wool, silk, cotton and linen and also for printing, using numbers for each recipe. These numbers were used when the application of the dye was described: that is, for each dye for which a fabric sample is present, information is provided about the recipe used for dyeing by means of these numbers. For example, for Picric acid, the recipe for wool was V3 and for silk V8, as can be observed in Fig. 4.1.

However, for each dye present in Lehne's publication, additional information is often given. Sometimes specific recipes are reported and for all textile samples present, the dye concentration is given. Note that these concentrations are used for the samples present in the book; the dye concentration can be adjusted according to the desired shade. Concentrations of many ingredients are sometimes presented in °Bé, which stands for degrees Baumé. This was a scale developed in the eighteenth century by the French pharmacist Antoine Baumé (1728–1804) and used to measure the density of solutions. In the *CRC Handbook for Chemistry and Physics* conversion tables show the relationship between degrees Baumé and specific gravity and, in some cases the w/w percentage of several solutions [5]. However, the number of solutions for which this information is available is limited due to the fact that the degrees Baumé scale is obsolete nowadays. Another useful source, the *Handbook of Chemistry* written by the chemistry professor Norbert Adolph Lange (1892–1970), contains multiple tables in which degrees Baumé are converted into specific gravity and w/w percentage [6]. Unfortunately, not all compounds are listed and in these

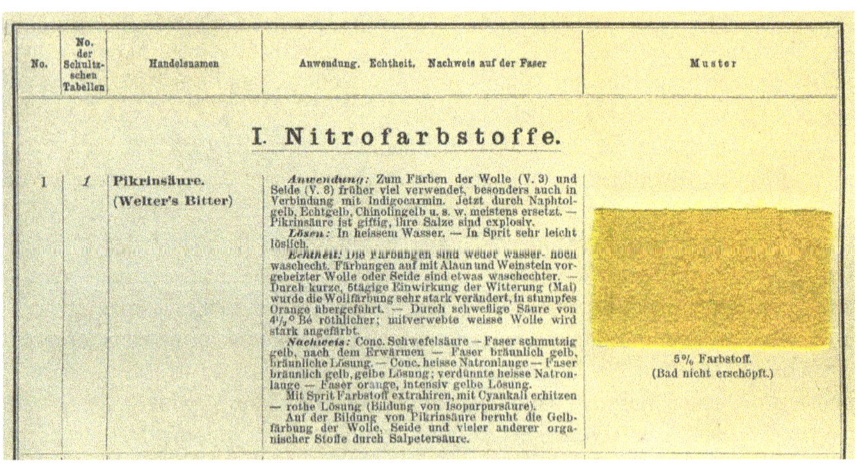

| No. | No. der Schultz-schen Tabellen | Handelsnamen | Anwendung. Echtheit. Nachweis auf der Faser | Muster |
|---|---|---|---|---|
| | | | **I. Nitrofarbstoffe.** | |
| 1 | *1* | **Pikrinsäure.** (Welter's Bitter) | *Anwendung:* Zum Färben der Wolle (V. 3) und Seide (V. 8) früher viel verwendet, besonders auch in Verbindung mit Indigocarmin. Jetzt durch Naphtolgelb, Echtgelb, Chinolingelb u. s. w. meistens ersetzt. — Pikrinsäure ist giftig, ihre Salze sind explosiv. *Lösen:* In heissem Wasser. — In Sprit sehr leicht löslich. *Echtheit:* Die Färbungen sind weder wasser- noch waschecht. Färbungen auf mit Alaun und Weinstein vorgebeizter Wolle oder Seide sind etwas waschechter. — Durch kurze, 5tägige Einwirkung der Witterung (Mai) wurde die Wollfärbung sehr stark verändert, in stumpfes Orange übergeführt. — Durch schweflige Säure von $4^1/_2$° Bé röthlicher; mitverwebte weisse Wolle wird stark angefärbt. *Nachweis:* Conc. Schwefelsäure — Faser schmutzig gelb, nach dem Erwärmen — Faser bräunlich gelb, bräunliche Lösung. — Conc. heisse Natronlauge — Faser bräunlich gelb, gelbe Lösung; verdünnte heisse Natronlauge — Faser orange, intensiv gelbe Lösung. Mit Sprit Farbstoff extrahiren, mit Cyankali erhitzen — rothe Lösung (Bildung von Isopurpursäure). Auf der Bildung von Pikrinsäure beruht die Gelbfärbung der Wolle, Seide und vieler anderer organischer Stoffe durch Salpetersäure. | 5% Farbstoff. (Bad nicht erschöpft.) |

**Fig. 4.1** Information provided by Adolf Lehne regarding Picric acid. Photograph: Maarten R. van Bommel

cases only the concentration in degrees Baumé is given in this chapter: the actual concentration must be determined experimentally. It must be noted that in the publication by Lehne, degrees Baumé values are given without much precision, so this can lead to small deviations. Furthermore, Lange also reported the issue that in different countries the scales used were slightly different, and probably the method used to determine specific gravity differed as well, which might also lead to small deviations. However, in the recipes presented it is not expected that these small deviations have a significant effect on the dyeing.

Most fabric samples present in Lehne's book are dyed wool while recipes for other textiles, such as silk, cotton, linen and jute, are given as well. For wool, typically a dye concentration between 0.5–2% is used (sometimes even higher). For other fibre materials, the concentration of the dyes used is not always given. As a rule of thumb, for silk a dye concentration twice that used for wool is applied. For cotton and linen, for which the same recipe is used, this concentration can go up to 5%.

The general recipes are listed below, and we refer to them in the respective dye chapters. In some cases, the recipe applied deviates from the standard recipe and in those cases the deviations are mentioned in the subsequent chapters. For some dyes, recipes from other sources are presented as well, but many recipes are known, and each manufacturer had their own procedures, which were also adapted over time. As such, this chapter is by no means comprehensive as that is beyond the scope of this publication, but the recipes given are typical of those used in the late 19th and early 20th centuries.

## 4.2 Direct Dyeing

The easiest way to dye textiles is by direct dyeing, a procedure which applies to both natural and synthetic colourants. First, the dye is dissolved, either after extraction from natural sources or directly in the case of synthetic colourants. When the dye is brought into solution, the textile can be added and dyeing takes place by hydrophobic interaction and hydrogen bridges, which are formed between the dye and the textile fibre. When applicable, ionic interaction can also play a role, which enhances the bonding strength. Direct dyeing does not require other ingredients or other subsequent steps, although salts can be added to enhance the affinity between the dye and fibre. Often the dye bath was heated to temperatures varying from lukewarm to the boiling point. Although this method of dyeing is quite straightforward and relatively fast, the wash fastness and light fastness of fabrics dyed by direct dyeing are relatively poor. Direct dyeing is typically done for cellulosic textiles, such as cotton or linen, although protein fibres (wool and silk) can be dyed directly as well. It must be noted that several direct dyes can be also applied as mordant dyes; mordanting usually improves the fastness properties.

There are just a few recipes for direct dyes given in the publication by Lehne. For dyeing wool in a neutral bath a short recipe is described (recipe V2)[1]: '[For] dyes that dye in a neutral bath such as Congo red, Sulfonazurine, Diamine fast red, etc. Add 15–20% sodium sulfate and dye for one hour [at] boiling [temperature]; dye as condensed as possible, i.e. use little water.'

For silk, no recipes are presented for direct dyes, but for cotton and linen, a rather extensive dye recipe is reported (recipe V12)[2]: '[For] direct dyeing, substantive dyes such as Congo red, Toluene orange, Mikado brown, Chrysamine, Chrysophenine, Azo blue, Erika, Diamine black etc. Add the textile to the [dye] bath at about 60 °C and boil for ¾ of an hour after the addition of 5% soft soap (3% curd soap), 20% sodium chloride or sodium sulfate. Wring out (spin) [the textile] and, without rinsing, dry at a moderate heat. The addition of water is limited as much as possible so that not too much dye is kept in the solution. Larger amounts of sodium chloride or sodium sulfate often negatively affect the beauty of the dyeing; they are only useful in cases where the baths cannot be kept and used for further dyeing. Hard water is purified by first adding soap, boiling it and skimming off the lime soap that has been separated on top.

Subsequent treatment of the dyeing with an alkaline solution of Turkey red oil makes it more vivid and, in some cases, more lightfast. Otto N. Witt, on the other hand, observed a reduction in the light fastness by a Turkey red oil treatment. (Private communication). Subsequent boiling in a diluted copper vitriol [i.e. copper sulfate] solution improves the fastness of some dyeings, e.g. those with Benzoazurine.

---

[1] In neutralem Bad färbende Farbstoffe wie Congoroth, Sulfonazurin, Diaminechtroth u.s.w.

Unter Zusatz von 15–20% Glaubersalz kochend in einer Stunde ausfärben; möglichst kurzes Bad d. h. wenig Wasser.

[2] Direct färbende, substantive Farbstoffe wie Congoroth, Toluylenorange, Mikadobraun, Chrysamin, Chrysophenin, Azoblau, Erika, Diaminschwarz u.s.w.

Bei etwa 60 °C eingehen und in ¾ Stunden kochend ausfärben nach Zusatz von 5% Schmierseife (3% Kernseife), 20% Kochsalz oder Glaubersalz. Abwinden (schleudern) und, ohne zu spülen, bei mässiger Wärme trocknen. Der Wasserzusatz wird möglichst beschränkt, damit nicht zu viel Farbstoff in Lösung gehalten wird. Grössere Mengen Kochsalz oder Glaubersalz beeinträchtigen häufig die Schönheit der Färbung, sie sind nur in solchen Fällen zweckdienlich, wenn die Bäder nicht aufbewahrt und für weitere Färbungen verwendet werden können. - Hartes Wasser wird gereinigt, indem zunächst die Seife zugesetzt, gekocht und die oben abgeschiedene Kalkseife abgeschöpft wird. Nachträgliche Behandlung der Färbungen mit einer alkalischen Türkischrothöllösung macht dieselben etwas lebhafter und in einigen Fällen auch lichtechter. Otto N. Witt hat dagegen eine Erniedrigung der Lichtechtheit durch Türkischrothöl beobachtet. (Privatmittheilung.) Nachträgliches Kochen in einer verdünnten siedenden Kupfervitriollösung ertheilt manchen Färbungen z.B. solchen mit Benzoazurin grössere Echtheit. Die Seife im Färbebad wird häufig weggelassen, die Waare nachträglich geseift und ohne zu spülen getrocknet. Auch Zusätze von Soda, Pottasche, phosphorsaurem Natron, Wasserglas u.s.w. zu dem Färbebad sind üblich.

The soap in the dye bath is often omitted, and the goods are subsequently soaped and dried without rinsing. Additions of soda, potash, phosphate of soda, water glass, etc., to the dye bath, are common.'

The codes V2 and V12 are used in the following chapters when these general recipes for direct dyes are applicable. If there is additional information on specific dyes, this is also noted.

## 4.3 Mordant Dyeing

A more complicated dyeing method is the so-called mordant dye process. In this case, the dye does not bind directly to the textile fibre, but via a coordination metal. These coordination metals, such as aluminium, iron, tin, copper and chromium, form a bridge between the textile fibre and the dye molecules [2]. Since this is a very strong bond, the wash fastness is significantly improved and, in addition, the light fastness is in general better as well. Mordanting can be done prior to or after dyeing, called pre-mordanting and post-mordanting respectively, and a combination of both pre- and post-mordanting is possible as well. Pre-mordanting is done by dissolving the mordant first; a widely used mordant is potash alum, i.e. potassium aluminium sulfate ($KAl(SO_4)_2{\cdot}12H_2O$). Other important mordants are iron sulfate ($FeSO_4{\cdot}7H_2O$), tin chloride (either $SnCl_2$ or $SnCl_4$), copper sulfate ($CuSO_4{\cdot}5H_2O$) and sodium or potassium dichromate ($Na_2Cr_2O_7$ or $K_2Cr_2O_7$). The textile is then added. Mordanting is often done at temperatures between 60 and 80 °C; usually an hour is sufficient. After mordanting, the bath is allowed to cool down to room temperature. The homogeneity of the mordanting process is improved by leaving the textile in the mordant bath overnight at room temperature. The textile is subsequently rinsed with water, after which dyeing can be performed.

It must be noted that most natural dyes belong to the class of mordant dyes, while many of the synthetic colourants are acid or basic dyes. However, some of the synthetic dyes can be applied as mordant dyes. The most important example is synthetic Alizarin, which is described in Chap. 20. One of the natural sources for Alizarin is madder root (*Rubia tinctorum* L.) in which it is the main component. Obviously, the mordant process used for natural dyes such as Madder could be also used for synthetic Alizarin. However, in Lehne's book, the general recipes show the use of chromium as a mordant, while for Madder usually aluminium was used. Interestingly, in the case of synthetic Alizarin, both chromium and aluminium are mentioned by Lehne. For the last named, no recipe is given in the general recipe chapter, but only in the text next to the textile sample. This suggests that not many dyes other than Alizarin were dyed on an alum mordant.

Using chromium as a mordant, Lehne gave the following recipe for wool (recipe V4)[3]: '[For] mordant dyes. Chromium mordant using potassium dichromate (potassium chromium).

1. Boil for 1½ h with 3% potassium dichromate (or sodium dichromate) and 2½% cream of tartar [potassium hydrogen tartrate, $KC_4H_5O_6$]; rinse. Cream of tartar is often replaced by 1–1½% sulfuric acid or a cream of tartar preparation (sodium disulfate) or 1½–2% oxalic acid; using potassium dichromate alone is not advisable because in this case the reduction of chromate takes place, which is harmful for the wool. For light colour dyeings, a weaker mordant, e.g. with 2% chromium potash and 1½% cream of tartar, is sufficient. Very hard water is also unfavourable for mordanting; it is improved with acetic acid (cf. V. 1).[4]
2. Dyeing. After adding 1–2 L of 45% acetic acid to 1000 L of water, dissolve the dye, i.e. mix it with water, pour through a fine sieve into the bath, add the textile at 40 °C, bring to the boil slowly and boil for 1¼ h. (Alizarine blue, Coerulein and Anthracene brown, if saturated, require 2½ h of boiling for full development).'

Some other recipes for wool are given in which chromium is used, but no recipes for the use of an aluminium salt as a mordant are presented. For silk, both aluminium and chromium are used. A BASF recipe using chromium as a mordant is as follows (recipe V9A)[5]: '[For] mordant dyes. Chromium mordant. Add the textile to a chromium chloride solution of 20 °Bé for 12 h at about 40 °C. Move [the textile] 4–5 times; soak it; rinse well and wring out or spin. Move the textile about 5 times in a water-glass solution [$Na_2Si_8O_2$] of 1% and rinse. Dye in a bast soap[6] bath treated with acetic acid. Add

[3] Beizenfarbstoffe. Chrombeize mit Kaliumbichromat (Chromkali).

1. Ansieden 1½ Stunden mit 3% Kaliumbichromat (oder Natriumbichromat) und 2½% Weinstein, spülen. Weinstein wird vielfach durch 1-1½% Schwefelsäure oder Weinsteinpräparat (Natriumbisulfat) oder 1½–2% Oxalsäure ersetzt; mit Kaliumbichromat allein anzusieden ist nicht empfehlenswerth, weil in diesem Falle die Reduction des Chromats auf Kosten der Wolle erfolgt. Für helle Färbungen genügt eine schwächere Beize, z.B. mit 2% Chromkali und 1½% Weinstein. - Sehr hartes Wasser wirkt auch bei dem Beizen ungünstig, es wird mit Essigsäure (vgl. V. 1) verbessert.
2. Färben. Nach Zusatz von 1–2 Liter Essigsäure von 8 °Bé zu je 1000 Liter den Farbstoff, gelöst bezw. mit Wasser verrührt, durch ein feines Sieb in das Bad giessen, bei 40 °C. eingehen, langsam zum Kochen bringen, 1¼ Stunden kochen [Alizarinblau, Coerulein und Anthracenbraun erfordern, falls es sich um satte Färbungen handelt, zu deren vollständigen Entwicklung 2½ stündiges Kochen].

[4] See recipe for basic dyeing.

[5] Beizenfarbstoffe. Chrombeize. (Bad. Anilin- und Sodafabrik).

In Chromchlorid von 20 °Bé 12 Stunden bei etwa 40 °C. 4–5 Mal umziehen, einlegen, gut spülen, abwinden oder schleudern. In Wasserglaslösung von 1 °Bé etwa 5 Mal umziehen, spülen. Ausfärben in mit Essigsäure gebrochenem Bastseifenbad. Bei 40 °C eingehen, langsam in ca. 40 Minuten zum Kochen treiben, 1–1½ Stunden auf dem kochenden Bade lassen. Spülen, mit einer Lösung von 5 g guter Kernseife im Liter kochend seifen, spülen und mit 20–30 g Essigsäure im Liter lauwarmen Wassers aviviren.

[6] Bast soap is the soap used to remove the sericin from the silk (peeling, degumming).

the textile at 40 °C, bring it slowly to the boil in about 40 min and leave it in the boiling bath for 1-1½ h. Rinse; boil it in a solution of 5 g good curd soap per litre; rinse and finish it with 20–30 g acetic acid per litre of lukewarm water.' Although the conversion of 20 °Bé chromium chloride could not be found in literature, based on similar solutions a concentration of about 18% could have been used. Knecht mentions that basic chromic chloride, $CrCI(OH)_2$, has been introduced by BASF as a mordant for cotton and silk. It is prepared by saturating a solution of chromic chloride with chromic hydroxide. The mordant can be also used on cotton [7].

A mordant recipe for dyeing cotton and linen is presented as well, namely recipe V13A[7]: '[For] mordant dyes: chrome mordant. Soak the textile in the chrome mordant G.A.I. (chromic acid chrome oxide) for 12 hours. Wring out (spin) and treat for ½ hour in a solution of 30 g calcined soda per litre at 60 °C; rinse well. For light colours, it is sufficient to use the chrome mordant G.A.I. of 3 °Bé and a treatment with a solution of 7½ g calcined soda per litre. Others use double sulfuric acid chromium (chromium bisulfite) as a mordant: Padding, drying in the 'Hot flue', fixing in the small 'Mather-Platt', stripping in a diluted chalk bath and washing.' Although the conversion of 3 °Bé chrome mordant cannot be made, since it is unclear which chrome solution was used, it is likely that a concentration of 3% was used.

In this recipe, the terms 'Klotzen', 'Hot flue' and 'Mather-Platt' occur. 'Klotzen' (padding) means that textiles are moved through a dye bath either by human muscle power or by padding machines (Foulard machines). Two other machines were used in textile dyeing at that time: the 'Hot-flue' for drying and the 'Mather-Platt-quick steamer' for fixing the dyes [8].

The codes V4, V9A and V13A are used in the following chapters when these general recipes for mordant dyes are applicable. If there is additional information on specific dyes, this is also noted.

## 4.4 Vat Dyeing

Only two vat dyes, Indophenol white and Indigo salt, are mentioned in the first two editions of Lehne's book. Indophenol white is a dye which was used in combination with natural Indigo and Indigo salt is Indigo carmine, a water-soluble acid dye. An important vat dye is synthetic Indigo, but this was not introduced onto the market before 1897. In the 1893 and 1899 editions of Lehne, therefore, no recipes for synthetic Indigo are present. We will discuss recipe no. 125 in Lehne's third edition of 1906 in the chapter on synthetic Indigo.

---

[7] Beizenfarbstoffe, Chrombeize.

In Chrombeize G.A.I. (Chromsaures Chromoxyd vgl. 9B. [D.R.P. 45998 Farbwerke in Höchst a/M.]) 12 Stunden einlegen. Abwinden (schleudern) und ½ Stunde in einer Lösung von 30 g calcinirte Soda im Liter bei 60 °C. behandeln; gut spülen. - Für helle Farben genügen Chrombeize G.A.I. von 3 Bé und Passage durch eine Lösung von 7½ g calc. Soda im Liter. Andere wenden als Mordant doppelt schwefligsaures Chrom (Chrombisulfit) an: Klotzen, Trocknen in der Hot flue, Fixiren im kleinen Mather-Platt, Abziehen im leichten Kreidebad und Waschen.

## 4.5 Acid Dyeing

Many of the synthetic colourants belong to the group of acid dyes, so called because they dye in a slightly acidic dye bath, although many acid dyes also colour textiles in a neutral bath. Since the synthetic colourants do not require extraction, the first step is just dissolving the dye. The dye bath is acidified using, for example, diluted sulfuric acid after which the textile can be added. Dyeing is usually performed at elevated temperatures. A typical dye recipe for wool is given by Lehne and reads as follows (recipe V3A)[8]: '[For] dyes that dye in an acid bath, such as Acid fuchsine, Ponceau, Fast red etc. Add 15–20% sodium sulfate and the dye solution [to the water]. Add the textile at about 70 °C with gradual addition of 5% concentrated sulfuric acid [i.e. 96%], which has previously been heavily diluted; dye at boiling for about ½ hour. Sodium sulfate and sulfuric acid are often replaced by 12–15% cream of tartar (sodium disulfate).'

Another typical recipe for acid dyeing of wool is reported by Lehne that reads (recipe V3B)[9]: 'Add 90–100% sodium sulfate [e.g. 30 kg of sodium sulfate per 1,000 litres of water]. Add the textile at 90 °C, with the addition of 5% concentrated sulfuric acid [i.e. 96%]; dye in a boiling bath for about ½ hour. This process enables uniform colouring. In practice, this is even possible with dyes which otherwise produce uneven colouring. The baths are, of course, kept and reinforced with sodium sulfate from time to time.'

For silk, the following recipe is given by Lehne (recipe V8)[10]: '[For] dyes that dye in an acid bath, such as Fast red, Ponceau, Naphthol yellow S, Quinoline yellow etc. To 5 kilos of silk, approximately 120 litres of water and 30 litres bast soap is added. Add so much sulfuric acid until the reaction is clearly acidic. Add a third of the dye, then add the textile at 50 °C and step-by-step the rest of the dye; dye for about ½ hour in a boiling bath; and top up with sulfuric acid if necessary. Wash, treat with about 10 cc of 96% sulfuric acid in approximately 120 litres of water and dry.'

---

[8] In saurem Bad färbende Farbstoffe wie Säurefuchsin, Ponceau, Echtroth u.s.w.

15–20% Glaubersalz und die Farblösung zugeben. Bei etwa 70 ° C. eingehen, unter allmäligem Zusatz von 5% Schwefelsäure von 66 °Bé, welche vorher stark verdünnt wird, in ca. ½ Stunde kochend ausfärben. Glaubersalz und Schwefelsäure werden häufig durch 12–15% Weinsteinpräparat (Natriumbisulfat) ersetzt.

[9] 90–100% Glaubersalz [auf 1000 Liter Wasser beispielsweise 30 kg Glaubersalz] zusetzen. Bei 90 °C. eingehen, unter Zusatz von 5% Schwefelsäure von 66 °Bé kochend in etwa ½ Stunde ausfärben. Nach diesem Verfahren gelingt es, in der Praxis auch mit Farbstoffen, welche sonst leicht unegal anfallen, gleichmässige Färbungen zu erzielen. Die Bäder werden selbstverständlich aufbewahrt und von Zeit zu Zeit mit Glaubersalz verstärkt.

[10] In saurem Bad färbende Farbstoffe wie Echtroth, Ponceau, Naphtolgelb S, Chinolingelb u.s.w.

5 Kilo Seide, ca. 120 Liter Wasser, 30 Liter Bastseife. Schwefelsäure bis zur deutlich sauren Reaction beifügen. Ein Drittel des Farbstoffs zugeben, bei 50 °C. eingeben, Rest des Farbstoffs allmälig beifügen, kochend in etwa ½ Stunde ausfärben; nach Bedarf Schwefelsäure nachsetzen. Waschen, aviviren mit ca. 10 cc Schwefelsäure von 66 °Bé in ca. 120 Liter Wasser, trocknen.

Lehne also describes a short recipe for jute using acid dyes on an alum mordant (V18)[11]: '[For] Ponceau, Fast red, Water blue. Dye for 1 hour at 90 °C after addition of 5% alum and 1% [concentrated, 96%] sulfuric acid. (The bath is usually not exhausted; acid dyes are less recommendable for dyeing jute).'

The codes V3A, V3B, V8 and V18 are used in the following chapters when these general recipes for acid dyes are applicable. If there is additional information on specific dyes, this is also noted.

## 4.6 Basic Dyeing

Another important dye class is that of the basic dyes, which can dye under alkaline conditions, although neutral and even slightly acidic conditions can also be used for these dyes, which in fact is often the case. Except for the pH of the dye bath, the procedure is similar to the acid dyes.

A typical dye recipe for wool in neutral conditions is given by Lehne (recipe V1)[12]: '[For] dyes that dye in a neutral bath, such as Fuchsine, Methyl violet, etc. Add the dye solution, add the textile at 50 °C, heating it up to about 95 °C; dye for about ¾ hours. Hard water can be improved before dyeing preferably by adding 1–2 litres of concentrated acetic acid (that is, diluted to 45%) per 1,000 litres of water.'

For silk, the following recipe is given (recipe V7)[13]: '[For] dyes which dye in a neutral or slightly acetic acid bath, such as Fuchsine, Methyl violet etc. For 5 kilos of silk, use approximately 120 litres of water and 20 litres of bast soap. Add acetic acid until the reaction is slightly acidic. Dye at about 85 °C with gradual addition of the dye solution. Rinse well and finish within about 10 minutes with 20 cc acetic acid in 120 litres of water; dry without rinsing.'

---

[11] Ponceau, Echtroth, Wasserblau u.s.w.

Ausfärben in 1 Stunde bei 90 °C. nach Zusatz von 5% Alaun und 1% Schwefelsäure von 66 °Bé. (Das Bad wird gewöhnlich nicht erschöpft; Säurefarbstoffe sind zum Färben der Jute wenig empfehlenswerth.)

[12] In neutralem Bad färbende Farbstoffe wie Fuchsin, Methylviolett u.s.w.

Farblösung zugeben, bei 50 °C. eingehen, unter Erwärmen auf etwa 95 °C. in ungefähr ¾ Stunden ausfärben. Hartes Wasser wird vor dem Färben, am besten durch Zusatz von 1–2 Liter Essigsäure von 8 °Bé (etwa 45%) auf je 1000 Liter Wasser, verbessert.

[13] In neutralem oder schwach essigsaurem Bad färbende Farbstoffe, wie Fuchsin, Methyl violett u.s.w.

5 Kilo Seide, ca. 120 Liter Wasser, 20 Liter Bastseife, Essigsäure bis zur schwach sauren Reaction hinzufügen. Bei ca. 85 °C. unter allmäligem Zusatz der Farbstofflösung ausfärben. Gut spülen und in etwa 10 Minuten aviviren mit etwa 20 cc Essigsäure in 120 Liter Wasser; ohne zu spülen, trocknen.

Basic dyes can also be used to dye cellulosic textiles, such as cotton and linen, after tannin mordanting. The recipe given by Lehne explains the tannin mordanting process. Regarding the dyeing, only a sentence is added in this recipe. It states that dyeing is done by gradually adding the dye solution at 40 °C for 30 min (recipe V11).[14] The lack of details in this recipe suggests that dyeing cellulosic textiles is a simple process or widely known.

'Tannin mordant: Dissolve 5% tannin in enough hot water to allow the cotton to be handled, add the textile at 50–60 °C, move it several times; leave it in the mordant bath for about 6 hours, moving it 4-5 times; spin or wring out. Soak in a 30 °C bath containing 2% tartar emetic [antimony potassium tartrate, $K_2Sb_2(C_4H_2O_6)_2$] for 15–20 minutes, then rinse well. Dye in about ½ hour with gradual addition of the dye solution at about 40 °C.'

As with the acid dyes, Lehne gives a short recipe for the dyeing of jute (V15)[15]: '[For dyeing] Fuchsine, Methyl violet, Diamond green, Auramine, etc., dyeing [is performed] with gradual addition of the dye solution in about ¾ of an hour at 50 °C.'

The codes V1, V7, V11 and V15 are used in the following chapters when these general recipes for basic dyes are applicable. If there is additional information on specific dyes, this is also noted.

## 4.7 Fabric Printing Recipes

In addition to the dye recipes for wool, silk, cotton, linen and jute, Lehne gives several recipes to print wool, silk and cotton as well. In general, they require a pretreatment of the textile and the dye is dissolved and mixed with a binding medium such as gum Arabic, and then applied onto the textile.

For printing silk and wool, the same recipes are given. However, wool requires a bleaching treatment which goes as follows[16]: 'The well-cleaned wool is chlorinated before printing and is also often prepared with metal salts such as sodium stannate

---

[14] Fuchsin, Methylenblau, Bismarckbraun u.s.w.

Beize: 5% Tannin in soviel heissem Wasser, wie zum Hantiren der Baumwolle erforderlich ist, lösen, bei 50–60 °C. eingehen, einige Mal umziehen, etwa 6 Stunden in der Beize liegen lassen, 4–5 Mal umziehen, schleudern oder abwinden. In einem 30 °C. warmen Bade, welches 2% Brechweinstein enthält, 15–20 Minuten umziehen, sodann gut spulen. Ausfarben in etwa ½ Stunde unter allmäligem Zusatz der Farblösung bei ca. 40 °C.

[15] Fuchsin, Methylviolett, Diamantgrün, Auramin u.s.w.

Ausfärben unter allmäligem Zusatz der Farblösung in etwa ¾ Stunden bei 50 °C.

[16] Die gut gereinigte Wolle wird vor dem Bedrucken gechlort und ausserdem auch häufig noch mit Metallsalzen wie mit zinnsaurem Natron und essigsaurer Thonerde vorbereitet. Auf unvorbereiteter Wolle werden die Farbstoffe fast ohne Ausnahme nur oberflächlich fixirt und die Farben fallen meistens ungleichmässig und unansehnlich aus. Das Chloren kann in folgender Weise ausgeführt werden. Die Stücke werden geklotzt in einer kalten verdünnten Clorkalklösung, hergestellt aus 10 Liter Chlorkalklösung von 5 °Bé und 100 Liter Wasser, hierauf passirt man durch Salzsäure von 1 °Bé, wäscht sorgfältig und trocknet. Die Vorschrift, die Salzsäure dem Chlorkalkbad selbst allmälig während der Durchnahme der Stücke zuzufügen, ist nicht zweckmässig; es werden auf diese Weise grössere Mengen Chlorkalk als nöthig zersetzt und dadurch - abgesehen von der Vergeudung des Materials - die Arbeiter stark belästigt.

and aluminium acetate. On unprepared wool, the dyes are almost without exception only superficially fixed and the colours usually turn out uneven and unsightly. Chlorination can be carried out as follows. The pieces are added to a cold, diluted chlorinated lime solution [calcium hypochloride] prepared from 10 litres of chlorinated lime solution at 5 °Bé and 100 litres of water, then passed through a 1.4% hydrochloric acid bath, rinsed thoroughly and dried. It is not advisable to add the hydrochloric acid to the chlorinated lime bath during the treatment of the textiles; in this way, larger quantities of chlorinated lime than necessary are decomposed which, apart from wasting the material, is a major [health] risk for the workers.'

Next some recipes are given for printing both wool and silk. A recipe using basic dyes (V19) reads as follows[17]: '[For] Fuchsine, Victoria blue, Methyl violet, Malachite green etc. Dissolve the dye in hot water, thicken it with gum thickener (1 part gum Arabic in 1 part water); add 40 cc acetic acid at 52% to 1 litre of printing ink; print; wrap in moistened, but not too wet cloths and steam for 1 hour without or with a very weak overpressure of ¼ atmospheric pressure, and wash.'

For the xanthene dyes the following recipe is given (V20)[18]: '[For] Eosin, Rose Bengal, Rhodamine, etc. Add 10 g tartaric acid per litre instead of 40 cc acetic acid to the printing ink, otherwise proceed as in V19.'

The recipe to print with acid dyes only provides information about the ingredients, thereby suggesting that the printing is similar as for the other dyes (V21)[19]: '[For] Acid fuchsine, Ponceau, Tartrazine, Naphthol black, etc. [Ingredients] for one litre of printing ink: 12 g tartrate preparation (sodium bisulfate) or 8 g of concentrated sulfuric acid [i.e. 96%] or 10 g oxalic acid.'

The recipe for cotton is somewhat more complicated, using tannins. The recipe V23A reads as follows[20]: '[For] Safranine, New fuchsine, Methylene blue, Thioflavin T, Induline, etc. Dissolve approximately 10 g of dye in approximately 120 cc of hot water. Add 800 g thickener and 30 g tannin dissolved in 30 cc of 52% acetic acid. Recipe for the thickener: boil 160 g wheat starch with 700 g water, add 200 g of mucilage of tragacanth [60 g of tragacanth gum in one litre of water] and 150 g of

[17] Fuchsin, Victoriablau, Methylviolett, Malachitgrün u.s.w.

Farbstoff in heissem Wasser lösen, mit Gummiverdickung (1 Theil Gummi arabicum in 1 Theil Wasser) verdicken, auf 1 Liter Druckfarbe 40 cc Essigsäure von 8 °Bé zusetzen, drucken, in angefeuchtete nicht zu nasse Tücher einschlagen und 1 Stunde ohne oder mit ganz schwachem Überdruck von ¼ Atmosphäre dämpfen, waschen.

[18] Eosin, Rose bengale, Rhodamin u.s.w.

Der Druckfarbe werden pro Liter 10 g Weinsäure anstatt 40 cc Essigsäure beigefügt, sonst verfahren wie bei 19.

[19] Säurefuchsin, Ponceau, Tartrazin, Naphtolschwarz u.s.w.

Im Liter Druckfarbe: 12 g Weinsteinpräparat (Natriumbisulfat) oder 8 g Schwefelsäure von 66 °Bé oder 10 g Oxalsäure.

[20] Safranin, Neu-Fuchsin, Methylenblau, Thioflavin T, Indulin u.s.w.

Druckfarbe ca. 10 g Farbstoff in etwa 120 cc Wasser heiss lösen. 800 g Verdickung und 30 g Tannin gelöst in.30 cc Essigsäure von 8 °Bé hinzufügen.

Verdickung: 160 g Weizenstärke mit 700 g Wasser kochen, 200 g Traganthschleim [60 g Traganth - Gummi im Liter Wasser], 150 g Essigsäure von 8 °Bé hinzufügen. Nach dem Drucken 1 Stunde ohne Druck dämpfen, durch eine 50 °C. warme Brechweinsteinlösung (10 g im Liter) ziehen, seifen und spülen.

52% acetic acid. After printing, steam for 1 hour without pressure, pass through a 50 °C warm tartar emetic solution (10 g per litre), soap and rinse.'

The second recipe for printing cotton contains less information (V23B)[21]: 'Dissolve the dye in 120 g of 52% acetic acid, 780 g gum thickener (1 part gum in 1 part water) and 50 g tannin dissolved in 100 g water. Steaming and further treatment as in V23A.'

The codes V19, V20, V21, V23A and V23A are used in the following chapters when these general recipes for printing are applicable. If there is additional information on specific dyes, this is also noted.

## References

1. Hofenk de Graaff, J.H.: The Colourful Past Origins, Chemistry and Identification of Natural Dyestuffs. Abegg-Stiftung / Archetype Publications Ltd, Riggisberg / London (2004)
2. Kirby, J., van Bommel, M.R., Verhecken, A.: Natural Colorants for Dyeing and Lake Pigments : practical recipes and their historical sources. Archetype Publications Ltd., London (2014)
3. Wolela, A.D.: Effect and Role of Salt in Cellulosic Fabric Dyeing. Advance Research in Textile Engineering. 6, 1061–1066 (2021)
4. Lehne, A.: Tabellarische Uebersicht über die künstlichen organischen Farbstoffe und ihre Anwendung in Färberei und Zeugdruck. Verlag von Julius Springer, Berlin (1893)
5. Hayes, W.M. ed: CRC Handbook of Chemistry and Physics. CRC Press Boca Raton / London / New York (2016)
6. Lange, N.A.: Handbook of Chemistry. New York / Toronto / London (1923)
7. Knecht, E., Rawson, C., Loewenthal, R.: A Manual of Dyeing: for the Use of practical Dyers, Manufacturers, Students, and all interested in the Art of Dyeing, Vol. I. Charles Griffin and Company Ltd., London (1893)
8. Herzog, R.O. ed: Technologie der Textilfasern, Vol. IV, Part 3. Verlag vor Julius Springer, Berlin (1928)

[21] Farbstoff in 120 g Essigsäure von 8 °Bé heiss lösen, 780 g Gummiverdickung (1 Theil Gummi in 1 Theil Wasser) und 50 g Tannin in 100 g Wasser gelöst hinzufügen. Dämpfen und weitere Behandlung wie in 23A.

# Chapter 5
# Chemical Analysis of Synthetic Colourants

## 5.1 Introduction

Many scientific articles concerning the analysis of organic colourants in cultural heritage objects state that this is a challenging task, and this is correct for several reasons. In particular, there is a wide range of organic colourants that often occur in mixtures. Dyes are mixed to obtain a certain shade or hue, but even when a single dye is applied, side products can be present and characterisation of these is sometimes useful to identify the origin of the colourant. In addition, many organic colourants are not stable and degrade, especially under the influence of (UV) light. The amount of dye required to colour a textile is low, often between 1 and 2%, which might affect the possibilities for detection and identification of the material present.

Yet, there can be multiple reasons to investigate the composition of dyes in samples from historical objects. Often, there is a general interest in the manufacturing process. The identification of dyes and other ingredients can provide information about dyeing processes and sometimes the origin and date of the object. Synthetic dyes in particular offer ample opportunities for dating since most of them are patented and the date of discovery is known. When a synthetic dye is identified, a terminus post quem date can be established, as the textile should have been dyed after the discovery of the dye. It should be noted, however, that it is difficult to determine the terminus ante quem date from this, as some dyes were used decades after their invention. Nevertheless, for some dyes it is known that they were replaced after a certain time; in such a case an indication of the dating can be given.

Another reason for the identification of dyes could be to understand more about the current condition of the artefact: the dye can be faded and by identifying the remaining dye, a clue about the original colour can be obtained. Saying this, the chemistry of degradation is still poorly understood and as such it remains rather difficult to determine how an object originally appeared. In principle, identification of the dye can also support conservation. For most synthetic dyes, information is



M. R. van Bommel, M. de Keijzer, *Bright Colours from the Past*, Cultural Heritage Science, https://doi.org/10.1007/978-3-031-88664-5_5

available on their light fastness and it is possible to predict how the colour will develop when an object is exhibited over a long period of time. In addition, the identification of dyes can also be used to understand the wash fastness of a colourant. It should be noted that spot tests can sometimes provide more information on wash fastness than dye identification, as not only the type of dye, but also the dyeing process, can affect the interaction with the textile fibre.

To avoid disturbing the integrity of the object, sampling should be avoided, and non-invasive analysis should be the aim. Although techniques are available which can be applied directly to an area of the object, the signal obtained can be masked by the substrate, i.e. the textile fibre. When mixtures are present, a combined response is obtained from which it may be difficult to determine the identity of the individual components, especially when one colourant shows a (much) stronger response than another, due either to the chemical nature of this component, or because it is present in abundance. As a result, the compounds present in lower concentrations can easily be missed. A second issue with non-invasive analysis is that compounds with the same chromophore show a similar response, making it complicated to differentiate between these compounds.

A second strategy is to sample an object, but to perform an analysis which is non-destructive for the sample. After analysis, the sample can either be analysed by another technique, or stored for future research. Sampling has the advantage that techniques can be used which cannot be applied directly on the object, due to the analytical conditions and, for example, the geometry of the object. In addition, it is not always feasible to bring the analytical technique to the object, or vice versa, while a sample can be easily transported. Non-destructive chemical analysis on a sample is often performed as the first step in the laboratory, using complementary techniques to obtain the maximum amount of information from a single sample.

Finally, the sample can be subjected to analysis which destroys it; the dyes are extracted from the textile fibre and techniques are used which separate all colourants present in the extract, allowing identification of the main and minor components.

In the field of cultural heritage, sampling is preferably avoided, not only because sampling affects the object, but also since it can be considered as a spot analysis. This means that the results obtained provide information about the colourants present in the sample, but this cannot always be translated to the entire object. Besides, many textile objects are composed of multiple colours, so multiple samples may be required. If time and the availability of techniques and staff allow, a multi-analytical approach should be chosen. In that case, it is possible to start with non-invasive analysis of multiple areas on the object, which can also be helpful to determine where to take a sample and to avoid taking multiple samples from the same colour, containing the same colourants. Next, when samples are taken, non-destructive analysis should be performed to achieve the maximum information before the samples are subjected to destructive analysis. Saying this, this is the ideal situation, assuming that there are no time constraints and unlimited resources. Often, very specific questions are asked which require just one or two techniques.

Although it may seem that the chemical analysis of dyed textiles should focus on the organic colourants, it is of utmost importance to use techniques which can provide information on inorganic elements present in the object. Several dyes require a so-called mordant, a coordination metal which forms a bridge between the dye and the textile fibre, improving the wash fastness. In some cases, these metals can have an influence on the overall colour. In addition, inorganic analysis may also be an indication of other treatments a textile has gone through.

This chapter will describe the techniques which can be used to analyse dyed textiles. The concept of each technique is briefly explained, and its possibilities and limitations are outlined. This chapter will not be a comprehensive state-of-the-art; review papers are regularly published and as such are much more up-to-date [1, 2]. Rather, this chapter is intended to provide an overview to conservators and art historians so that they can think about questions to ask and discuss with the scientists who will execute most of the analysis.

## 5.2 Visual Examination and Microscopy

Chemical analysis of a cultural heritage object should always be preceded by a careful visual examination of the object. During the examination, the various areas of an object can be studied and the number of locations to be analysed can be determined according to the research question(s). It is of importance to determine the condition of the (various sections of the) object, for example to observe the discolouration, bleeding, degradation or contamination. In addition, information about the object's manufacturing process can help to select areas for investigation. For example, the weft and warp are often dyed differently, and sometimes yarns are composed of multicoloured fibres. Although this examination can be done with the naked eye, a magnification glass (a loop) or a microscope provides more detailed information. A stereo microscope is very helpful, but is not always available. A digital microscope connected to a laptop via a USB connection is also a good option. Not only is this relatively cheap and portable, so it can be used on site, but it also allows making photomicrographs that can be stored. Visual examination is sometimes supported by UV illumination, to differentiate further between differently coloured yarns. Visual examination and microscopy can be done without taking a sample. However, if an analysis is planned for which a sample is required, the sample can (or even should) be examined under a microscope, as this often provides information that may be missed by examining the object. It must be noted that dyes cannot be identified by visual examination or microscopy.

## 5.3 Fibre Optic Reflectance Spectroscopy (FORS) and Spectral Imaging

A subsequent step for studying dyed textiles is Fibre Optic Reflectance Spectroscopy (FORS), which can examine the object directly in a non-invasive manner. The technique uses a light source to project light through an optical fibre onto the surface and a second optical fibre to measure the reflectance, i.e. the radiation that is not absorbed by the surface. Although it might be thought that only the visible part of the spectrum is useful for the identification of the colourants, spectra are often recorded from the UV region up to the near infrared, as this part of the electromagnetic spectrum can also provide information. FORS is a fast and comparatively inexpensive technique, so it is relatively easy to analyse several parts of the object.

Textiles often have coarse surfaces which might affect the spectrum collected, hence doing some repetitive analysis to ensure a correct spectrum is obtained might be required. The spot size, which is between 1 and 5 mm, has to be taken into account. This is not an issue if the surface to be analysed is homogeneous, but can be problematic if small details need to be investigated, or if differently coloured yarns are present: for example, a twisted yarn, or if the weft and warp are dyed differently. The main drawback of FORS is the lack of spectral resolution. As a consequence, the molecular information obtained is rather limited: at best, a distinction can be made between dye classes, but the identification of the final dye is difficult. When mixtures are present, or the colourant is (partly) degraded, a combined spectrum is obtained making identification even more complicated. However, as it is non-invasive as far as the object is concerned, it is an interesting tool for quickly testing multiple areas for comparison. If different areas show the same reflectance spectrum, it may be decided to take only one sample, on the assumption that this sample is representative of these different areas. Reflectance can also be measured in imaging mode, either by multispectral imaging, collecting the reflectance using different bands of the electromagnetic spectrum, or by spectral imaging, using cameras which can record a spectrum from each pixel. This provides an overview of the dye distribution, although the identification of the dyes present is still very complicated.

## 5.4 Fluorescence Spectroscopy

Another technique that can be applied directly to the object, without sampling, is fluorescence spectroscopy, which is in some cases more specific than FORS. Fluorescence is a process which consists of several steps. An irradiation source is applied, illuminating the area of interest. The dyes present absorb a part of the spectrum, causing the molecules to become excited as they absorb photons. This absorption can be measured indirectly by reflectance spectroscopy (which measures the non-absorbed part) as mentioned above, but some dyes, depending on their molecular structure, show fluorescence. This is due to the fact that when excited, some of the energy absorbed is lost through vibration and heat release, followed by the loss of the remaining energy through the emission of a photon. Since the energy of the absorbed photon differs from the energy of the emitted photon, the wavelength of the absorption (also called excitation) and emission differs and can be recorded. Not all molecules show this phenomenon, but some dyes are known to fluoresce. However, fluorescence can be greatly affected by other materials present and when the analysed area contains a mixture of dyes, a combined spectrum is obtained. Often one dye will fluoresce very strongly while the other(s) will not, or will only weakly fluoresce and will therefore not be detected. Due to these drawbacks, this technique is not widely used for the identification of dyes.

## 5.5 Optical 'Fingerprint' Techniques

While the techniques described above are mainly applied directly to the object, there are also spectroscopic techniques that require samples, but leave the sample intact during analysis. A technique sometimes mentioned, but not often applied to characterise dyes, is Fourier Transform Infrared Spectroscopy (FT-IR). Using this technique, infrared light is projected onto the analysed surface. The part of the spectrum which is absorbed depends on the chemical bonds present in a molecule. In particular, the technique can determine which functional groups are present, providing a 'fingerprint' of the molecule of interest. Identification can be performed by comparison with known reference materials. However, when FT-IR is carried out on the dyed textile fibre, the fibre itself also contributes to the signal and often dominates the spectrum. This has a major impact on the applicability of the technique for dye characterisation on textiles. The dye can be extracted from the textile fibre and measured separately, but that requires an additional step.

As an alternative, Raman spectroscopy provides information which is partly overlapping with FT-IR and partly complementary. Raman uses lasers to irradiate the sample and, instead of absorption, the scattering of this light is measured. In a similar way to FT-IR, it delivers information about the functional groups of a molecule in particular and identification is done using databases containing spectra from reference materials. As a consequence, dye classes can be differentiated and often the specific dye can be identified. When mixtures are present, a combined response is obtained, which may complicate the identification, especially since some molecules are much better Raman scatterers than others and the resulting spectrum is thus predominantly from such a molecule, with the consequence that some colourants can be missed without realising it.

Although the functional groups of the textile fibre can also influence the spectrum, the response of the textiles is reduced by selecting the correct lasers, especially those emitting visible or near infrared light. The laser can be rather intense and can damage the sample; however, scientists operating these systems are well aware of this and, based on their experience, will select the correct energy of the laser beam. One of the main problems is that the Raman signal is quite weak, and is especially hampered by fluorescence, which is a much stronger effect and can obscure the Raman spectrum. Fluorescence is less if a laser is chosen that emits visible or near-infrared light, but it can still be a problem. Currently, new tools are being brought on the market circumventing the fluorescence problem, for example Sequentially Shifted Excitation (SSE™), which mitigates the fluorescence and seems promising for dye analysis [3].

Raman spectroscopy can be performed with bench-top equipment, but nowadays more and more portable systems are available which can be brought to the object in storage or the conservation studio. Although not all dyes can be identified using this technique, especially as it is hampered by the fluorescence problem and by the fact that it might be difficult to identify different dyes in a mixture, it is being applied more and more since it is non-destructive, relatively quick and, when a portable system is available, can be done on-site. To overcome the issues with fluorescence, Surface Enhanced Raman Scattering (SERS) can be performed. For this technique, a sample is required: the colourant is extracted from the textile fibre and brought into contact with silver or gold colloids, i.e. in solution. Although the process is not fully understood, it has the advantage that the fluorescence is suppressed and especially the Raman response is significantly enhanced. This enhancement will be even stronger if resonance occurs when selecting a laser with a wavelength near the absorption maximum of the component of interest, a technique called Surface Enhanced Resonance Raman Scattering (SERRS). As a result, the detection limit is significantly improved and minute samples can be analysed. It must be noticed that when mixtures are present, a combined spectrum is obtained which complicates the identification, especially when one component dominates in the spectrum.

## 5.6 X-Ray Fluorescence Spectroscopy (XRF)

Although, the focus of this chapter is on organic colourants, it is important to study the presence of inorganic elements as well. Firstly, they can be part of the dyeing process when mordants are used as fixation agents. Secondly, the presence of inorganic elements can provide information about potential treatments that may have been carried out in the past. X-Ray Fluorescence Spectroscopy (XRF) is an excellent technique for this purpose, which can be applied directly on the object and is non-invasive. The fluorescence process is comparable to that described earlier, but now the excitation of an electron in a chemical element is done by means of X-rays. When the electron returns to the ground state, it emits a photon which can be recorded. Each element shows a different profile and hence can be identified. Lighter elements, aluminium and below in the Periodic Table of Chemical Elements, cannot be detected by XRF, although some systems are equipped with a so-called helium flush, which allows the detection of aluminium. In some cases, the signal from an element can overlap with another, so care should be taken during identification, but in general the interpretation is rather straightforward. Handheld XRF systems are available which can be brought to the object in storage or in the conservation studio, and since it is a very quick technique, multiple sample spots can be measured in a short time. These handheld systems have a relatively large spot size of between 5 and 8 mm, which complicates the analysis of small details. More advanced XRF systems are available with a spot size as small as 0.1 mm, but these are usually operated in the laboratory, which means that the object must be transported. The detection limit of XRF is about 1%, which means that elements present at a low concentration might be missed. Besides, since the technique only provide information about the elements present, organic colourants cannot be detected.

## 5.7 Scanning Electron Microscopy: Energy Dispersive X-ray Spectroscopy (SEM-EDX)

Another important technique is Scanning Electron Microscopy coupled to Energy Dispersive X-ray Spectroscopy (SEM-EDX, sometimes also abbreviated as SEM-EDS). The analysis is carried out in a vacuum chamber, so samples have to be taken unless the object is small and can be placed in the vacuum chamber (10–20 cm), for example when dealing with textile fragments. The sample is bombarded with electrons, which interact with the surface providing different signals. One of these signals is due to secondary electrons; the incoming electrons hit the electrons in the outer shells of the elements at the surface of the object, from which a very detailed topography of the surface can be obtained. Magnifications of up to three million times can be obtained, although often in the field of cultural heritage magnifications of a factor of 1,000–10,000 is sufficient. By measuring secondary electrons, a very high resolution and depth of field can be obtained. This provides information about the condition of the textile fibre: for example, to determine contamination and degradation by the presence of cracks and tears in the fibre. Textile fibres can be identified to a certain extent since wool, silk, cotton and linen can be differentiated, which represents the majority of the natural fibres.

A second technique is to measure backscattered electrons. These are incoming electrons that are reflected by the elements. Elements with high atomic numbers, so the heavier elements, reflect more strongly. As a result, it is feasible to record the distribution of light and heavy elements, although it does not lead to their identification.

Identification can be done since, in a way similar to XRF, the incoming electrons can excite the elements present. When they return to their ground state, they emit a photon which can be measured using an EDX detector. Since the analysis is performed in a vacuum, the lighter elements can also be detected, but it is not feasible to identify organic colourants. Apart from carrying out a measurement on a small spot of the sample, elemental mapping can also be performed, showing the distribution of different elements across the sample. Some SEM systems are not equipped with an EDX detector; in that case elemental analysis is not feasible, but at a high magnification the technique can be used to study the surface of the sample.

## 5.8 Full Characterisation of Organic Colourants

The techniques described so far do not separate the colourants from the textile fibre or from each other. To improve identification, separation techniques are available. In this case the colourants are extracted from the substrate using organic solvents, such as dimethyl sulfoxide (DMSO), or acids such as oxalic acid and hydrochloric acid. As such, these are destructive techniques which not only require a sample, but since the sample will be consumed, no subsequent analysis is possible. Therefore, it should be the last technique in line. Nevertheless, separation of the dyes offers the possibility of differentiating the components with very similar molecular structures and often compounds present at a low concentration can be detected. Although there are different separation techniques available, (ultra) high-performance liquid chromatography (HPLC or U(H)PLC) is the working horse for the analysis of organic colourants.

A typical HPLC system comprises an analytical column packed with stationary phase particles, usually 3–10 μm in diameter. The column dimensions range from an internal diameter of 2–4.6 mm and a length of 5–30 cm. Smaller columns (50 μm internal diameter) require specialised equipment, while larger columns (10 mm internal diameter) are typically employed for preparative purposes. In this system, a solvent (called the mobile phase), is pumped through the column under high pressure (150–400 bar). Smaller particles and longer columns enhance the resolution, but require a higher pressure. Especially when particles between 1 and 2 μm are used, the efficiency is increased, hence this is called U(H)PLC. The injection loop introduces the sample, often via an autosampler. Components in the sample interact with the stationary phase in the column, leading to separation based on their properties. Organic colourants, which are hydrophobic, strongly absorb to a hydrophobic stationary phase like $C_{18}$, resulting in retention; the more hydrophobic the compound, the later it elutes from the analytical column. The separation can be improved and speeded up by using gradients. In this case the mobile phase is a relatively hydrophilic solvent (e.g. 10% methanol in water) at the start of the analysis and gradually increases to a hydrophobic one (e.g. 100% methanol or acetonitrile), to facilitate optimal separation. For synthetic colourants, present as ions in the solution, counter-ions such as tetra-butyl ammonium hydroxide, tri-ethyl amine or ammonium formate are employed to improve the separation. After the separation, the components eluting from the analytical column are guided to a detector.

One of the most frequently used detectors is the Photo Diode Array Detector (PDA or DAD), which records the absorption of the components eluting from the analytical column across UV to visible wavelengths (200–800 nm). This range is particularly effective for organic colourants, as they absorb UV-Vis light. Although UV absorption does not influence the observed colour, it is recorded because some components with a similar visual absorption show distinct differences in the UV range. The UV-Vis spectrum can help with the deduction of the colour and the discrimination between dye classes, but identification is typically performed using known reference materials analysed under identical conditions. By recording and storing UV-Vis spectra and retention times in a searchable database, components in a sample can be compared and identified. The strength of this technique lies in its ability to distinguish between components with both similar and very different structures, since the combination of retention time and spectrum is used. In addition, it is a very sensitive technique able to detect both main and minor components in complex mixtures, which can be created during the synthesis (i.e. side products) or by deliberate mixing of dyes to obtain a certain hue, and degradation products.

Although HPLC-PDA can identify many organic colourants, this is not always feasible due to a lack of reference materials, unknown dyes, degraded materials, or complex dye compositions. When further identification is needed, Mass Spectrometry (MS) can be employed. Using a volatile mobile phase, HPLC can be coupled with MS via an Electrospray Ionisation Interface (ESI), resulting in HPLC-ESI-MS. In ESI, the analytical column's outlet is connected to the MS, and a high voltage is applied to ionise the components as the solvent evaporates. This allows each component separated by HPLC to enter the MS. Inside the MS, the components are accelerated and deflected based on their mass and charge, providing an accurate mass determination and a molecular composition. The combination of the retention time, the PDA spectrum and the molecular mass can often identify the component [4]. If necessary, further analysis using MS-MS can fragment the molecule, and identifying these fragments can lead to a final identification.

Another technique which can be used, is Gas Chromatography-Mass Spectrometry (GC-MS). In this case, the colourants need to be converted into volatile compounds by derivatisation. Since that is an additional step, GC-MS is still not often applied to identify dyes.

## 5.9 Sampling and Sample Preparation

Chromatography and mass spectrometry require a sample as mentioned before. This demands a careful consideration of sample size and location, which is particularly important in cultural heritage research. Minimising the sample size is essential to preserve valuable artefacts, but excessively small samples can lead to inconclusive results, wasting both time and the sample. Determining the right sample size depends on the amount of the organic colourant present, which is difficult to estimate prior to analysis and often relies on the judgement of the expert. Equally important is the choice of a suitable sampling location. Ideally, samples should be taken from inconspicuous areas that are less exposed to light, as these regions typically contain higher concentrations of organic colourants. For textiles, suitable spots for example include the back of tapestries or the inside of costumes where often a loose yarn can be found. For paintings, samples are best taken from under the frame or over the edge of the canvas or panel. If that is not feasible, a small sample can be taken by a scalpel from an area which is already (slightly) damaged, such as from a crack. Sampling an intact paint surface should be avoided. For furniture, the inside, bottom, or beneath fittings and marquetry are recommended areas. Generally, a small surface scraping is sufficient, bearing in mind that most organic colourants are in the top layer of the wood. Uncoloured samples should also be collected to account for natural wood colourants, allowing for a more accurate analysis by subtracting these from the results of coloured samples.

After sampling, the colourants should be brought into solution. This means that the colourants must be extracted using acids, such as hydrochloric acid or oxalic acid, and/or hydrophobic solvents like methanol, acetonitrile, dimethylformamide (DMF), or dimethyl sulfoxide (DMSO). Acids break bonds between the mordant dyes and the coordination metals, while hydrophobic solvents dissolve the colourants directly applied to wood or difficult-to-dissolve indigoids. Some colourants do not resist strong acids, such as hydrochloric acid and therefore milder acids such as formic acid and oxalic acid can be utilised. However, in our experience, synthetic colourants generally can resist strong acids. It should be noted that many synthetic dyes do not require a mordant and can often be extracted in the first step with an organic modifier. Any remaining dyes can be extracted with acids in the second step. The extraction process is usually conducted at elevated temperatures to increase extraction efficiency. Below a typical sample preparation procedure is given, used for the samples from the case studies presented in this book, but other procedures are available as well and can provide similar results.

## 5.10 A Typical Extraction Protocol for Organic Colourants [5]

Samples taken from the case studies in this book are subjected to the following procedure. All samples are investigated under a digital microscope (brand Dino-lite) prior to analysis with a maximum magnification of 50× to observe the colour and to identify whether the sample contain fibres or other material with a different colour to the overall impression. If so, attempts are made to separate this and to analyse only the part responsible for the overall impression, or to analyse differently coloured parts of the sample separately. This investigation is non-destructive and has therefore no influence on the dye analysis. Microphotographs are taken of all samples.

Next, the sample is first extracted by the addition of 50 μL dimethyl sulfoxide (DMSO) to the sample in a 1 mL vial. For very small samples, a 300 μL small insert vial can be used as an alternative, adding just 20 μL solvent. The vial is heated at 80 °C for 10 min in a water bath, *au-bain-marie*. Next, the DMSO fraction is removed and set apart. The remaining sample is then hydrolysed by the addition of 50 μL reagent[1] (water/methanol/hydrochloric acid) to the sample in the vial. The vial is heated for 10 min in a water bath, *au-bain-marie*, at 100 °C to extract and dissolve the remaining dye. After the hydrolysis, the sample is evaporated to dryness and dissolved in the first DMSO fraction. The two extractions are thus combined. The samples are centrifuged for 10 min at 6,000 rpm to remove precipitation. If necessary, i.e. when particles are present, the solvent is transferred to a new vial, followed

[1] For a milder extraction a solution of 1/1/2 or 0.5M oxalic acid/acetone/water/methanol mixture (1:30:40:30, V/V/V/V) is added instead, heating *au-bain-marie* at 80 °C for 10 min).

by another 10 min of centrifugation at 6,000 rpm. The sample is analysed by HPLC-PDA by injection of 1–5 μL sample. The procedure described is employed for samples from furniture, textiles and inks. For paint samples, the first DMSO extraction step is omitted since the DMSO will not affect the binding medium.

## 5.11 Possibilities and Limitations of Full Characterisation Techniques

(U)HPLC and MS are powerful tools for identifying minute amounts of organic colourants that are undetectable by the naked eye. Their key advantage lies in their ability to separate a wide range of components, allowing the detection of not only the primary colourants but also side products, which helps to identify the origin of the dyes. This ability to separate is crucial because samples often contain mixtures of dyes used to create different colours and shades. The spectroscopic techniques like Raman, UV-Vis, and FT-IR spectroscopy struggle with such mixtures as they do not separate the dyes from their substrates before identification, making the process challenging and often ineffective. This limitation is particularly evident in degraded or faded samples with low concentrations of the main colourant, which can still be detected by HPLC and MS.

Despite their advantages, the techniques have some drawbacks. Small samples are required for analysis, which can be a problem when dealing with valuable artefacts. The process, including sample preparation and interpretation, is time-consuming, even with fully automated systems. Additionally, the analytical systems are expensive and require skilled scientists to operate them. Accurate analysis also requires an appropriate reference collection.

Nonetheless, the advantages of (U)HPLC and MS in the separation and identification of dyes in micro-samples outweigh these disadvantages, leading to their increasing application in the identification of organic dyes. Their ability to detect and differentiate between complex mixtures of dyes make them indispensable in the study and the preservation of cultural heritage objects.

## 5.12 To Conclude

As shown, many analytical techniques are available which can determine the dye class or identify the individual colourants present in complex mixtures. When dye analysis is requested, it is essential that a discussion takes place between the conservator and/or art historian and the scientist to clarify the research question(s) and to discuss the best analytical strategy. It is important to recognise that sometimes questions are asked which cannot be answered using dye analysis alone: it often needs to be supplemented with information about the historical context, and dye reconstructions can be informative as well. It is clear that there are a number of techniques available that can be used in a non-invasive manner which is preferred, but they generally provide less information. It is therefore important not only to discuss the research question, but also to consider in advance what will be done with the answer(s) to that question(s), how it will help the conservation and the technical art history of cultural heritage objects, and whether it is worth the effort.

## References

1. Tamburini, D., Sabatini, F., Berbers, S., van Bommel, M.R., Degano, I.: An Introduction and Recent Advances in the Analytical Study of Early Synthetic Dyes and Organic Pigments in Cultural Heritage. Heritage. 7, 1969–2010 (2024). https://doi.org/10.3390/heritage7040094
2. Degano, I., Ribechini, E., Modugno, F., Colombini, M.P.: Analytical methods for the characterization of organic dyes in artworks and in historical textiles. Appl Spectrosc Rev. 44, 363–410 (2009). https://doi.org/10.1080/05704920902937876
3. Handheld Raman Spectrometer, https://www.bruker.com/en/products-and-solutions/infrared-and-raman/raman-spectrometers/bravo-handheld-raman-spectrometer.html
4. Astefanei, A., Adamson, B., Gaibor, A.N.P., van Bommel, M.R.: Simultaneous detection of a wide range of synthetic and natural dyes in artworks using UHPLC-PDA-HRMS. J. Chromatogr. A. 1740, 465562 (2025). https://doi.org/10.1016/j.chroma.2024.465562
5. Albano Serrano, A., van den Doel, A., van Bommel, M.R., Hallett, J., Joosten, I., Van den Berg, K.J.: Investigation of crimson-dyed fibres for a new approach on the characterization of cochineal and kermes dyes in historical textiles. Anal Chim Acta. 897, 116–127 (2015). https://doi.org/10.1016/j.aca.2015.09.046

# Chapter 6
# Pittacal: An Unusual Blue Dye

## 6.1 Introduction

Pittacal, one of the first semi-synthetic dyes, was discovered by accident. Karl Ludwig Reichenbach owned the metallurgical factories in Blansko in Moravia (then in the Austrian Empire, nowadays the Czech Republic) jointly with Count Hugo Franz zu Salm-Reifferscheidt of Vienna (1776–1836). His position secured him excellent opportunities for conducting large-scale experimental research. In the 1830s, Reichenbach was engaged in research on compounds isolated from beechwood tar produced in the charcoal ovens of the factories. In 1832, he accidentally discovered 'the first synthetic dye' to be produced commercially: Pittacal. Reichenbach is also recognised as being the discoverer of a number of important hydrocarbon compounds of beechwood tar, such as kerosene, phenol, paraffin (1830), picamar, eupion (1831) and creosote (1832), also extracted from coal tar [1, 2]. In 1832 and 1833 Reichenbach published his fundamental investigations into the alkali-soluble components of beechwood tar. In 1839, after the king of Württemberg had made him a baron (Freiherr), he disposed of most of his industrial interests and retired to his castle at Reisenberg near Vienna to devote himself entirely to science. Reichenbach lost much of his wealth in unwise investments, and his last years were spent as a recluse in his castle.

## 6.2 Pittacal

Colour Index name: Acid dye; Colour Index number: 43875
IUPAC name: 4,4′-((4-imino-3,5-dimethoxycyclohexa-2,5-dien-1-ylidene)methylene)bis(2,6-dimethoxyaniline)
Discoverer: Karl Ludwig Reichenbach, 1832
Contemporary terminology: Pittacal (En), Pittacall (Fr), Pittakall (Ge), Pittacal (NL)



M. R. van Bommel, M. de Keijzer, *Bright Colours from the Past*, Cultural Heritage Science, https://doi.org/10.1007/978-3-031-88664-5_6

Obsolete terminology: Eupittonic acid, Eupitton, Eupiton (En), Eupittonsäure, Eupittonschwarz, Hexaoxyaurin, Hexaoxydioxyfochson, Hexamethoxyparafuchsine and Hexamethoxypararosaniline (Ge)
Chemical formula: $C_{25}H_{29}N_3O_6$ (Fig. 6.1)

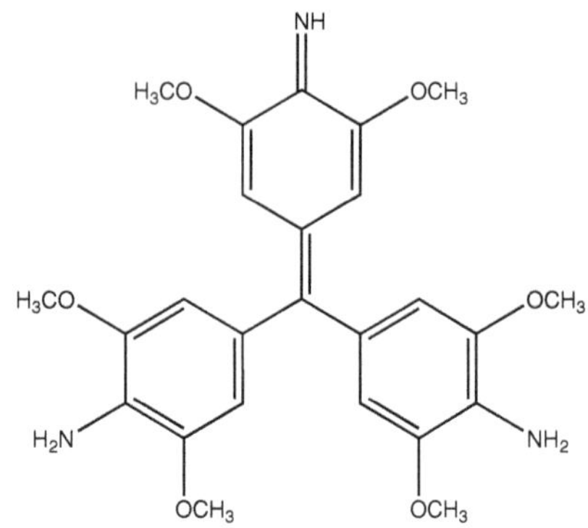


**Fig. 6.1** Structural formula of Pittacal

## 6.3 History

The discovery of the blue acid dye Pittacal is amusing and has an unusual origin. According to history, the German chemist Karl Ludwig Reichenbach (1788–1869) had a problem with dogs: they used the wooden fence around his house as a convenient urinating post. In 1832, Reichenbach applied his recently discovered creosote to the wooden fence around his property as a deterrent, hoping that the penetrating and unpleasant smell would hold back the dogs from urinating on the fence. This strategy failed, but he noted that the dogs' urine had reacted with the creosote, turning the fence an intense dark-blue colour. He named the new substance Pittacal (from the Greek words *pítta*, tar, and *kállos,* beautiful). In 1833 Reichenbach described beechwood tar, which is an oily substance that boils at a higher temperature than creosote. In 1835, Reichenbach was able to produce pure Pittacal by treating beechwood tar with barium oxide and, with alumina as a mordant, used it to dye fabrics [3–5]. Although sold commercially as a dye for over 20 years, Pittacal never became popular.

In the mid-1870s Pittacal was almost forgotten when it was, one can say, rediscovered by the factory owner Alfred Grätzel in Hanover, Germany, while he was working on the processing of beechwood tar oil to produce a dye he named Violacein [6]. Grätzel observed the formation of products similar to Pittacal which were

**Table 6.1** Discoverers and patents of Pittacal

| Discoverers | Patents | Patent title |
|---|---|---|
| Karl Ludwig Reichenbach | – | – |
| R. Gottheil | G.P. 9.328/1878 | Verfahren zur Darstellung von Eupittonsäure aus Holzteer |
| C. Timmermann and R. Grätzel | G.P. 13.787/1880 | Verfahren und Apparat zur Darstellung von Eupittonsäureaus Holzteer |

subsequently investigated by the German chemists Carl Theodore Liebermann (1842–1904) and August Wilhelm Hofmann (1818–1892) [7–9]. Liebermann investigated the dye he received from Grätzel, for whom it was not clear if it was the same as Reichenbach's Pittacal. He concluded that, according to the behaviour, Pittacal seems to be a higher oxidation product of Violacein and the latter is a higher oxidation product of eupitton [10]. In order to avoid errors, and in memory of Reichenbach's Pittacal, Liebermann proposed renaming the dye eupitton [4, 6, 9, 10]. Eupitton, also named eupittone or eupittonic acid, is an orange-yellow crystalline compound formed by the oxidation of beechwood tar oil, containing pyrogallol and methyl pyrogallol ethers; precipitated with heavy metals, like lead or tin, it gives a blue dye [4]. But this compound is also called Pittacal! In the second half of the 1870s, there seemed to be uncertainty about the different chemical names. Liebermann writes that he doubts the importance of Violacein for the time being [10]. However, Grätzel writes that as far as the technical importance of the dye is concerned, the judgement made by the future might perhaps be described as premature [9].

At the end of 1877, Grätzel wrote a letter to the German chemist Emil Erlenmeyer (1825–1909) about his new blue dye. He asked Erlenmeyer to examine the dye to see if it had any practical use [11]. In 1878, the chemist R. Gottheil from Berlin patented a process for the preparation of eupittonic acid from wood tar described in G.P. 9.328. In the year 1880, the first supplementary patent (G.P. 13.787) to G.P. 9.328 was filed by Christian Timmermann of Hanover and Richard Grätzel[1] of Braunschweig (see Table 6.1) for the preparation and equipment to produce eupittonic acid from wood tar. Nearly the same text for the preparation of eupittonic acid from wood tar is written in Gottheil's patent.

The chemical composition of Pittacal, whose method of production was kept secret by Reichenbach, was clarified by the investigations of Hofmann [12–14]. Hofmann received samples of eupitton from Liebermann and Grätzel and showed that this component is related to the blue dye Grätzel prepared from beechwood tar [14]. In 1878, Hofmann showed that the bitter-tasting oil which Reichenbach called picamar ('pitch', 'bitter'), a colourless hydrocarbon oil extracted from creosote of beechwood tar, consists of pyrogallol dimethyl ether (i.e. 2,6 dimethoxyphenol) and its homologues [1, 4, 15]. The following year, in 1879, he proved that the reaction of 2,6-dimethoxyphenol with 4-methyl-2,6-dimethoxybenzene (see Fig. 6.2)

[1] It is currently unclear if Alfred and Richard Grätzel are family or that they coincidentally share the same surname.

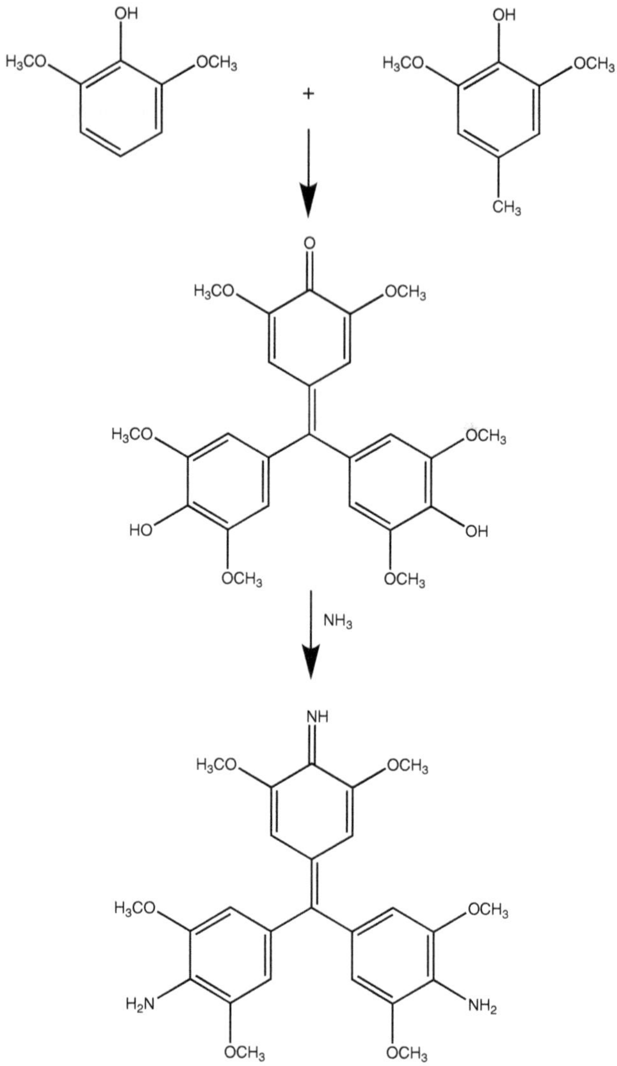


**Fig. 6.2** The reaction of 2,6-dimethoxyphenol with 4-methyl-2,6-dimethoxybenzene forming Pittacal (after [1])

forms the yellow crystalline substance eupittonic acid (eupitton). Subsequent reaction of eupittonic acid with ammonia yields the blue dye hexamethoxypararosaniline, that is, Pittacal. Hofmann prepared eupittonic acid by heating pyrogallol dimethyl ether and methylpyrogallol dimethyl ether with sodium hydroxide at 200–220 °C [13]. Furthermore, he showed that the eupittonic acid contained in Reichenbach's Pittacal is formed in a manner analogous to the formation of pararosaniline from aniline and toluidine. He thus regarded the compound as hexamethoxyrosolic acid, but it is now considered to be hexamethylaurine [1, 13, 16]. The last time Pittacal and eupitton derivatives were discussed was in a lecture by Liebermann in 1901 [17].

## 6.4 Applications

Reichenbach said that Pittacal could become important for Europe, which does not produce natural Indigo: 'Preliminary dyeing experiments which I have made show that Pittacal can be fixed quite well on cotton and linen by means of acetate of alumina in the presence of a tin mordant, and gives these materials a permanent blue colouration, which is resistant to light, water, soap, ammonia, urine and wine' [16]. Pittacal was used to dye wool and silk in an acidic solution, producing an orange colour. In an alkaline solution, and especially in the presence of a tin mordant, a blue-violet colour is obtained. This colour change occurs with the slightest amount of acid or alkali; hence it was suggested by Reichenbach as an indicator reagent because it is even more sensitive than litmus.

## 6.5 Properties

The pure compound consists of orange-yellow needles which yield a dark blue solution with alkalis. The colour of Pittacal is similar to that of Indigo, so that it can easily be confused with it, but Pittacal differs significantly in its behaviour with reagents. Pittacal is insoluble in water, alcohol and ether, soluble in acids and precipitated as a blue dye in fabrics mordanted with alumina. The colour of Pittacal is quite lightfast, does not change in the air and is fast to washing [3, 5]. In the Colour Index, no information is presented regarding the fastness properties of Pittacal.

## 6.6 Recipes

As described above, Pittacal was, so to speak, rediscovered in the mid-1870s by the firm of Adolf Grätzel in Hanover and sold under his trade name Violacein. In an appendix on coal-tar colours, he mentions dyeing methods with Pittacal on silk, wool, linen and cotton [8].

'My dyeing method consists in dissolving the dye in alcohol and stirring the concentrated solution into cold water, and heating the water, stirring constantly, until the alcohol has evaporated, where the dye adheres to the fabric. I first dye wool and silk with this dye liquor, and then, after dyeing, I let the fabric run through a boiling mordant of lead acetate and finally through a boiling soap bath. A shorter time of dyeing is required when the dye-bath is more concentrated. The unused dye is not lost, but is still retained and during dyeing experiments I have often used the same liquor ten times or more.'

Adolf Grätzel also describes a dyeing method for linen and cotton: 'In the case of linen and also cotton, I first mordanted the fabrics with lead acetate, then pulled them through the dye liquor, to which I added a little borax, and finally washed them in boiling soapy water. The alcoholic solution is brown, the dyed fabrics come out of the liquor orange to brown, then turn blue on treatment with a mordant or an alkali.

Wool and silk can also be dyed well from an acetic, sulfuric, or hydrochloric acid solution, and the colour is fast to air and light if the fabric is subsequently drawn through ammonia water until the acid is neutralized; but the dye is too sparingly soluble in acids, and one does not get the rich colours as from an alcoholic solution. The dye is extraordinarily sensitive to acids, but the colour can be immediately changed with alkali; it is therefore not destroyed by acids and resists Javel water (eau de Javelle, sodium hypochlorite (NaClO), called bleach or liquid bleach) almost better than indigo.'

## References

1. Kaufmann, G.B.: Pittacal - the first synthetic dyestuff. J Chem Educ. 54, 753 (1977)
2. Reichenbach, K.L.F. von: Pittacal. Berzelius Jahresberichte über die Fortschritte der physischen Wissenschaften, Verlag Heinrich Laupp, Tübingen. 14, 358 (1835)
3. Berzelius, J.J.: Lehrbuch der Chemie. Arnoldsche Buchhandlung, Dresden, Leipzig (1838)
4. Nietzki, R.H.: Chemie der Organischen Farbstoffe. Verlag von Julius Springer, Berlin. 4, 159 (1901)
5. Leuchs, J.C.: Vollständige Farben- und Färbekunde oder Beschreibung und Anleitung zur Bereitung und zum Gebrauche aller färbenden und farbigen Körper. Verlag von C. Leuchs & Comp., Nürnberg. 1, 232 (1857)
6. Liebermann, C.: Vorläufige Mittheilung: Ueber Pittakal. Berichte der deutschen chemischen Gesellschaft zu Berlin. Ferd. Dümmler's Verlagsbuchhandlung, Berlin. 9, 334–337 (1876)
7. Kurze Bericht über die Fortschritte der chemischen Grossindustrie. Zeitschrift für das chemische Grossgewerbe. Verlag von Robert Opperheim, Berlin. 1, 204 (1877)
8. Grätzel, A.: 6. Anhang zu den Theerfarben. Jahres-Bericht über die Leistungen der Chemischen Technologie für das Jahr 1877, Verlag von Otto Wigand, Leipzig. 23, 940–942 (1878)
9. Grätzel, A.: Ueber Eupitton und Pittakall. Berichte der deutschen chemischen Gesellschaft zu Berlin. Ferd. Dümmler's Verlagsbuchhandlung, Berlin. 11, 2085–2087 (1878)
10. Liebermann, C.: Bemerkung über Violacein und Eupitton. Berichte der deutschen chemischen Gesellschaft zu Berlin. Ferd. Dümmler's Verlagsbuchhandlung, Berlin. 11, 1104–1106 (1878)
11. Letter from Adolf Grätzel to Emil Erlenmeyer, Hannover, 30.12.1877-05.01.1878. Inhaltsangabe: ´Darstellung eines neuen blauen Farbstoffes. Bitte an Erlenmeyer, den Farbstoff zu untersuchen, um festzustellen, ob er eine praktische Verwendung hat´.

12. Hofmann, A.W.: Nachträgliches über Pittakall und Eupittonsäure. Berichte der deutschen chemischen Gesellschaft zu Berlin. Ferd. Dümmler's Verlagsbuchhandlung, Berlin. 12, 2216–2222 (1879)
13. Hofmann, A.W.: Ueber die Methylpyrogallussäure und über die Bildung des Pittakalls. Berichte der deutschen chemischen Gesellschaft zu Berlin. Ferd. Dümmler's Verlagsbuchhandlung, Berlin. 12, 1371–1385 (1879). https://doi.org/10.1002/cber.18790120239
14. Hofmann, A.W.: Ueber Farbabkömmlinge der Pyrogallussäure-Aether. Berichte der deutschen chemischen Gesellschaft zu Berlin. Ferd. Dümmler's Verlagsbuchhandlung, Berlin. 11, 1455–1461 (1878)
15. Hofmann, A.W.: Ueber dreisäurige Phenole im Buchenholz-theeröl und über den Ursprung des Cedrirets. Berichte der deutschen chemischen Gesellschaft zu Berlin. Ferd. Dümmler's Verlagsbuchhandlung, Berlin. 11, 329–338 (1878)
16. Ferzak, F.: Karl Freiherr von Reichenbach. FFWASP EBOOKS (2017)
17. Liebermann, C., Wiedermann, F.: Ueber Eupittonderivate. Berichte der deutschen chemischen Gesellschaft zu Berlin. Ferd. Dümmler's Verlagsbuchhandlung, Berlin. 34, 1031–1040 (1901). https://doi.org/10.1002/cber.190103401169

# Chapter 7
# Murexide: A Purple Dye From Guano

## 7.1 Introduction

Murexide was the last-discovered semi-synthetic dye. Strange and fantastic stories circulated about the origin of Murexide: it was said that it was made from snake excrement. This is actually true; in 1818 William Prout obtained the dye from boa constrictor excrement. Twenty years later Justus Liebig and Friedrich Wöhler renamed the red-purple dye Murexide, an allusion to the murex mollusc which was the source of Tyrian purple in Antiquity. Tyrian purple originates from mollusc species such as *Bolinus brandaris* (Linnaeus, 1758), which contains a precursor in the glands which forms brominated indigotin. It was very expensive and only used for precious textiles. The name Murexide is connected to the scene from a myth in the *Onomasticon* of Julius Pollux, a second-century CE Graeco-Roman scholar and professor of rhetoric. In Pollux's story, Hercules, a divine hero in Greek mythology, and his dog were walking on the beach on their way to court a nymph named Tyro. The dog bit a sea snail and the snail's blood dyed the dog's mouth purple. Seeing this, the nymph demanded a gown of the same colour.

## 7.2 Murexide

Colour Index name: Mordant dye; Colour Index number: 56085
IUPAC name: Ammonium 2,6-dioxo-5-[(2,4,6-trioxo-5-hexahydropyrimidinylidene) amino]-3H-pyrimidin-4-olate
Discoverers: Carl Wilhelm Scheele, 1776; William Prout, 1818; Justus Liebig and Friedrich Wöhler, 1838



M. R. van Bommel, M. de Keijzer, *Bright Colours from the Past*, Cultural Heritage Science, https://doi.org/10.1007/978-3-031-88664-5_7

Contemporary terminology: Murexide (En), Murexide (Fr), Murexid (Ge), Murexide (NL)
Obsolete terminology: Purpuric acid ammonium salt, Purpuric acid, Ammonium purpurate, MX, Ammonium salt of purpuric acid (En), Rouge de Naples (Fr), Purpursaures Ammoniak (Ge), Purpurzure ammonia (NL)
Chemical formula: $C_8H_8N_6O_6$ (Fig. 7.1)

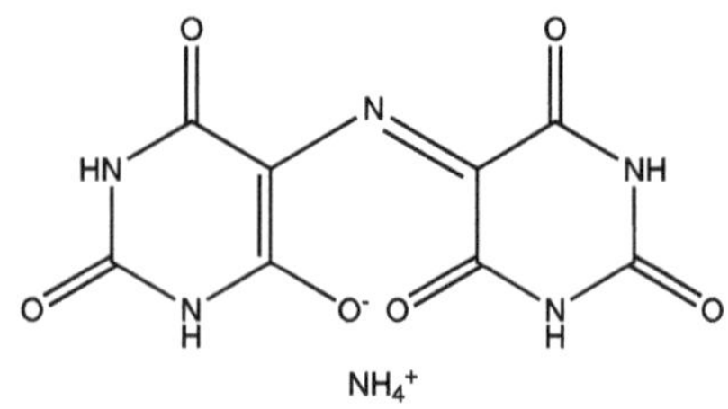


**Fig. 7.1** Structural formula of Murexide

## 7.3 History

In 1776, the German-Swedish chemist and pharmacist Carl Wilhelm Scheele (1742–1786) obtained an intense red-purple colour when he first treated uric acid, extracted from human *calculi* (human kidney stones, *calculus vesicae*), with nitric acid. After evaporation, this colour formed a precipitate with ammonia. Murexide was the first semi-synthetic dye which was used as a mordant dye.

Further research in 1818 by the English medical doctor William Prout (1785–1850) discovered a richer source of uric acid by using boa constrictor excrement, which can contain 90% of ammonium acid urate. The uric acid with nitric acid gave the colourless product alloxan, which he called erythric acid. The addition of ammonia gave a purple product which he named purpuric acid. He noted that it could be used as a purple dye for wool and some of the salts as artists' pigments [1–4]. The scarcity of materials hindered an industrial development; a more concentrated uric acid source was required to make the dye commercially interesting. This was found in guano.

Twenty years later, in 1838, the German chemists Justus Liebig (1803–1873) and Friedrich Wöhler (1800–1882) in Giessen, Germany, investigated Murexide obtained from snake excrement. They overcame the shortage of raw materials by using guano, the dried excrement of seabirds. Both chemists were the first to experiment with guano, which they obtained from William Kind, a pharmacist from Bremen, Germany. They received 100 pounds by weight (approximately 50 kg) of Valparaíso (in other words, guano) named after a seaport in Chile [5]. Guano came exclusively from the Chincha Islands, off the coast of Peru, and was imported as fertiliser. Its uric acid content varied from very little to as much as 20%, which was sufficient for the industrial production of Murexide as a dye. Unfortunately, the concentration of uric acid decreased as the years went by since less rich sources were mined. Duck or chicken excrement could also be used, but the uric acid content was not very high. The first shipment of Peruvian guano was consigned to a Liverpool merchant in 1835, but in 1841 it was still little known: only 1,700 tons (1,723.3 metric tons) were imported. Twelve years later, in 1841, the import amounted to 220,000 tons (223,531 metric tons). In the 1840s guano was used as a source of saltpetre for gunpowder and as an agricultural fertiliser. The price in England was about £10 per ton [6].

Liebig and Wöhler introduced the term Murexide, an allusion to the shellfish purple dye Tyrian purple, obtained from the hypobranchial glands of *Murex* species (family Muricidae). The colours were so similar that some believed that Murexide was identical to shellfish purple [7].

The uric acid from guano led to the industrial production of Murexide. Applying it to natural fibres started in 1853 at Mulhouse, France and within a year or two in Manchester, England, from where its use spread over to Europe. The Swiss chemist and colourist Frédérick Sacc (1819–1890) made some trials which he communicated to the French industrial chemist Jules-Albert Schlumberger (1804–1892), who then read a paper on the subject to the Société Industrielle, Mulhouse. The manufacture was attempted almost at once, and the dye was taken up both by Schlumberger's firm, Depouilly Frères, and by Dolfuss-Mieg et Compagnie. It did not take long before the dye was actually produced by the firm of Zuber & Cie, at Rixheim, on the eastern outskirts of Mulhouse, while the two Mulhouse firms concentrated on its application.

Murexide was not easy to use. The first application was only for wool, using a tin salt as mordant. In July 1855 Depouilly Frères patented the dyeing of silk with a mercuric chloride mordant (F.P. 24.089), and in August 1856 the French chemist Charles Lauth (1836–1913), working for the same firm, extended the application to cotton using a lead acetate mordant.

In Manchester, a major producer of Murexide was the British brewer Robert Rumney (1811–1872) at the Ardwick Chemical Works, Gorton. In 1845 he bought together with William Hadfield the Ardwick Chemical Works from 'a chemical vessel and firebrick manufacturer'. In the 1850s Hadfield left the partnership. The British calico-printer Edmund Potter (1802–1883), the founder of Edmund Potter & Co calico printing works, showed how to produce Murexide by treating uric acid with nitric acid and ammonia, and Rumney exploited Potter's discovery commercially. Later, in 1858 and 1859, Rumney filed two English 'application' patents (B.P. 1.421 in 1858 and B.P. 40 in 1859) (see Table 7.1). A little later Murexide was also produced by the German chemist Carl Theodor Wilhelm Goldschmidt (1817–1875) in Berlin and Jacob Braun in Prague [6]. Around 1857–1859 Rumney was processing 12 cwt of Murexide a week (a hundredweight in Britain is a unit of mass equal to 112 pounds), which he sold for 30 shillings (£1 10 s) per lb (1 lb = 1 pound $\approx$ 0.454 kg). The guano would have cost him £6,000 per year and sales of Murexide brought him about £100,000 for that period, which is not a bad profit.

Table 7.1 Discoverers and patents of Murexide

| Name<br>C.I. name<br>C.I. number | Year of discovery | Discoverers | Patents |
|---|---|---|---|
| Murexide | 1776 | Carl Wilhelm Scheele | – |
| Mordant dye | 1818 | William Prout | – |
| 56085 | 1838 | Justus Liebig and Friedrich Wöhler | – |
| | | Depouilly Frères | F.P. 24.089/1855 |
| | | Robert Rumney | B.P. 1.421/1858 |
| | | Robert Rumney | B.P. 40/1859 |

To produce Murexide on a large scale as well as in the laboratory is delicate and undoubtedly difficult to carry out. The conditions for success must be determined with care and precision, because of the extreme facility with which this substance changes as soon as one exceeds the suitable proportions of the reacting bodies and the desired degree of heat. One recipe describes the preparation of Murexide as a paste:

> 'Place a series of large bowls of 5 or 6 L capacity side by side and charge each with 1 L of nitric acid. Add uric acid in small portions one at time, using the precautions indicated for the preparation of alloxan. The flasks are very large to contain the large amount of foam that forms. Each bowl will use about 1 kg of uric acid. The operation is complete in 10 or 12 h and gives a brown liquid that contains urea nitrate, alloxan, alloxantin and parabanic acid, with a deposit of unreacted uric acid. The simultaneous presence of alloxan and alloxantin is very favourable to the formation of the red compound. These two compounds form a crystalline crust on the surface of the liquid. This acid solution is evaporated on a steam bath, in a boiler of enamelled cast iron, at a temperature that does not exceed 80 °C. The colour slowly becomes red and finally, as it thickens, a dark brown. The paste hardens after cooling. During the concentration, urea decomposes, and the ammonia derived from it reacts with the mixture of alloxan and alloxantin, giving murexide. It is advantageous to neutralise the free nitric acid, before evaporation, either with ammonia, or with sodium carbonate, in the second case it forms sodium purpurate' [4, 8].

The popularity of Murexide declined rapidly with the introduction of Perkin's Mauveine. It is often argued that the introduction of the aniline dyes caused the disappearance of the expensive Murexide. However, a decrease in the uric acid content of guano, which seems to have occurred during this time, was responsible for this decline and therefore prevented a lower murexide price. In addition, the stability of Murexide was poor. These factors, combined with the discovery and rapid exploitation of aniline dyes, led to the rapid demise of Murexide as a dye. Nevertheless, Murexide was a popular dye throughout Europe in the period 1853–1869.

## 7.4 Applications

Murexide was used to dye wool, silk and cotton, and in calico printing, but as a dye it was not easy to use. Tin, mercury and lead salts were used to produce purple colours and with alum a scarlet red colour was obtained. The French chemist Paul

Schützenberger (1829–1897) described industrial process recipes for preparing Murexide from uric acid. In the beginning, Murexide was sold as a paste, then as a powder, but currently it is delivered as crystals.

## 7.5 Properties

Murexide in its dry state has the appearance of a reddish-purple powder. It is slightly soluble in water and dissolves better in hot water. It is insoluble in alcohol and ether. Although Murexide is quite lightfast, it faded due to the acidity of sulfurous atmospheres and the colour bleached if washed in very hot water. In the Colour Index, no fastness properties are given.

## 7.6 Recipes

In the Dutch periodical series *Nederlandsche Maatschappij ter Bevordering van Nijverheid*, dyeing recipes with Murexide are mentioned in the volume for 1861, applying it to the natural fibres silk, wool and cotton. The translated recipes are as follows:

> 'One can dye silk purple-red, by dipping it in a mixture of murexide and sublimate (Kwik(I) chloride ($Hg_2Cl_2$) i.e. mercury chloride), both dissolved in water.'
>
> Wool cannot be dyed with murexide in the same simple manner, for to bring out the colour one must use an acid, e.g., oxalic acid. This can be done effectively in the following way: 'The well-washed and rinsed wool is hung in a strong solution of murexide for a long time, then wrung out and dried. The wool is now put into a bath that consists of a solution of 60 g of sublimate and 75 g of acetic soda in 10 L of water, heated to 40–50 °C.'
>
> To dye cotton with Murexide, according to Lauth, it is dipped first in a murexide solution, that in a bath of lead salts and finally in a solution of sublimate and acetic soda. For cotton printing, thickened nitric acid lead oxide mixed with Murexide can serve. After printing, the cotton is dried and put in a mixture of 100 L of water, 1 kg of sublimate and 1 kg of acetic soda. In this way one can obtain beautifully coloured samples on a white ground. (*Dingler's Polyt. Journ. Bd. 145, S. 137*)

> Oscar Meister gave a description of the French method of printing with Murexide, and then added some important remarks. The cotton fabrics that are to be dyed red are covered with salts of tin, lead and mercury and then coloured with Murexide. A yellow colour is obtained by first printing the fabrics with zinc salt. For printing, a mixture of Murexide and sodium lead oxide with gum is used to make a paste. The printed goods are exposed to ammonia vapours for half a minute to one minute at 70 °C. They are then placed in a sublimate bath and then in a second bath consisting of an aqueous solution of sublimate, acetic acid and acetic soda.
>
> Meister remarks, that all the colours, which one obtains with murexide, are definitely unreal, for they are neither resistant to the action of light, nor to soap, alkalis and acids (*Dingler's Polyt. Journ. Bd. 145, S. 15*) [9].

No recipes for Murexide were given by Lehne, suggesting that it was not in use anymore at the end of the nineteenth century.

## 7.7 Identification Using UHPLC-PDA

In Fig. 7.2, the analysis of Murexide (RCE no. 7966) is shown. Murexide elutes almost unretained.

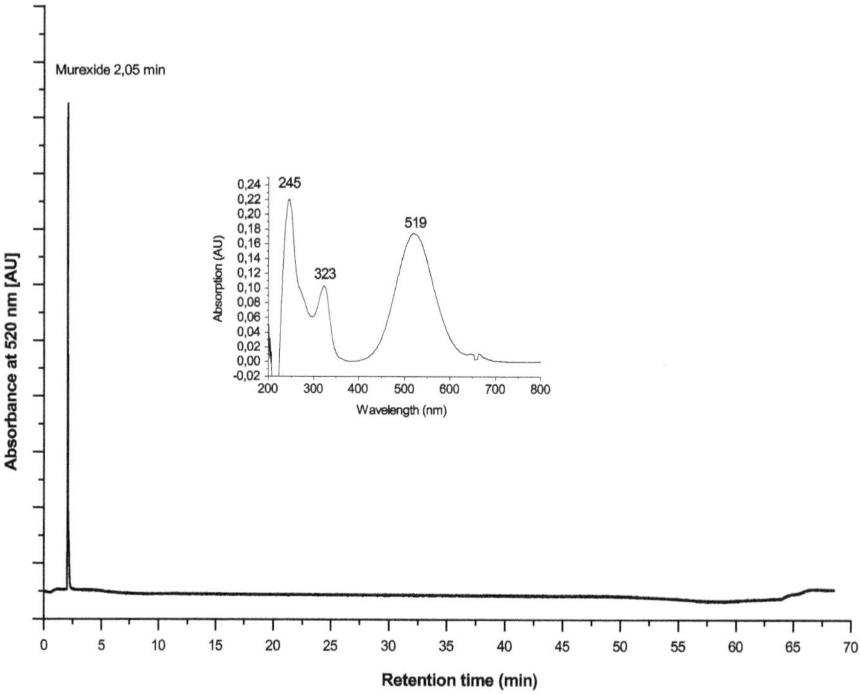


**Fig. 7.2** Chromatogram and spectrum obtained from Murexide (RCE no. 7966)

## References

1. Prout, W.: Some observations on the action of nitric acid on lithic or uric acid. Annals of Philosophy, or, Magazine of Chemistry, Mineralogy, Mechanics, Natural History, Agriculture, and the Arts. 14, 363–365 (1818)
2. Prout, W.: XXII. Description of an acid principle prepared from the lithic or uric acid. Philos Trans R Soc Lond. 108, 420–428 (1818). https://doi.org/10.1098/rstl.1818.0024
3. Nietzki, R.: Chemie der Organischen Farbstoffe. Verlag von Julius Springer. 331–332 (1901)
4. Cooksey, C., Dronsfield, A.: Pre-Perkin Synthetic Routes to Purple. Journal of Dyes in History and Archaeology. 19, 119–122 (2003)
5. Wöhler, F., Liebig, J.: Untersuchung über die Natur der Harnsäure. Annalen der Pharmacie. Universitäts-Buchhandlung von C.F. Winter, Heidelberg. 25, 319–327 (1838). https://doi.org/10.1002/jlac.18380260302
6. Farrar, W.V.: Synthetic dyes before 1860. Endeavour. 33, 149–155 (1974)

7. Napier, J.: Manufacturing Arts in Ancient Times, with Special Reference to Bible History. 257 (1874)
8. Schützenberger, P.: Traité des matières colorantes, comprenant leurs applications à la teinture et à l'impression et des notices sur les fibres textiles les épaississants et les mordants. Victor Masson et fils, Paris (1867)
9. Murexid (Purpurzure ammonia) als kleurstof voor zijde, wol en katoen. Mededeelingen, Tijdschrift Nijverheid, derde reeks - deel ll (van de geheele reeks deel XXlV). 18 (1861)

# Chapter 8
# Prussian Blue: An Inorganic Pigment Used as a Textile Dye

## 8.1 Introduction

Most dyes, of both natural and synthetic origin, are organic compounds and dissolve well in the medium they are applied in, normally water from the dye bath. One exception is Prussian blue, which is inorganic in nature. Although mainly used as an artists' pigment, it can also be used to dye textiles.

Prussian blue, discovered by Johann Jacob Diesbach in 1706, was the first modern synthetic inorganic pigment. In the nineteenth century, it became popular in France for dyeing silk and wool, especially after Napoleon's Continental Blockade (1806), which disrupted the access to natural Indigo. After Napoleon's fall, Prussian blue remained popular, despite the return of natural Indigo. By the 1860s, however, new synthetic organic blue dyes led to a decline in Prussian blue's use as a dye.

## 8.2 Prussian Blue

Colour Index name: Pigment Blue 27; Colour Index number: 77510
UIPAC name: Iron (III) hexacyanoferrate (II)
Discoverer: Johann Jacob Diesbach, 1706
Contemporary terminology: Prussian blue (En), Bleu de Prusse (Fr), Preußischblau (Ge), Pruisisch blauw (NL)
Obsolete terminology: Bronze blue, Chinese blue, Iron blue, Raymond's blue, Turnbull's blue, Williamson's blue, New blue, Persian blue, French blue, Paris blue (En), Marie-Louise bleu, Bleu Raymond, Bleu de Milori (Fr), Berliner Blau, Diesbach Blau, Gmelin Blau, Hamburg Blau, Zwickauer Blau, Mineral blau, Modeblau, Erlangen Blau (Ge), Antwerps blauw, Haarlems blauw (NL)
Chemical formula: $Fe_4[Fe(CN)_6]_3 \cdot nH_2O$ (Fig. 8.1)



M. R. van Bommel, M. de Keijzer, *Bright Colours from the Past*, Cultural Heritage Science, https://doi.org/10.1007/978-3-031-88664-5_8

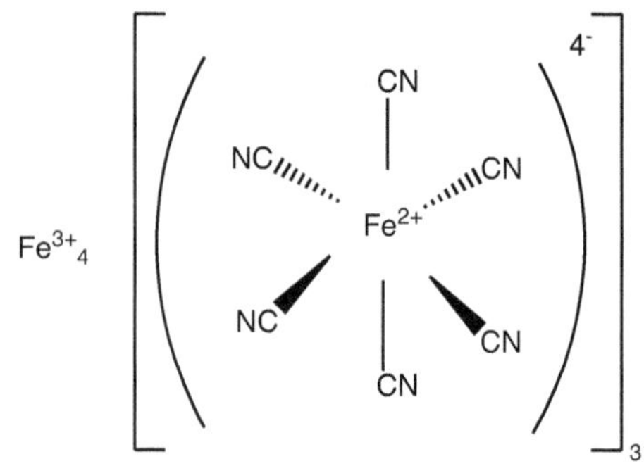


**Fig. 8.1** Structural formula of Prussian blue

## 8.3 History

Prussian blue is often referred to as the world's first modern synthetic inorganic pigment. In 1706, the Swiss pigment and dye producer Johann Jacob Diesbach in Berlin wanted to prepare a Florentine lake by the precipitation of an extract of crushed cochineal insects with alum, iron vitriol (English vitriol, ferrous sulfate, iron(II) sulfate) and potash, which happened to be tainted with blood. With this alkali, Diesbach got, instead of the expected red pigment, the blue colourant iron(III) ferrocyanide now known as Prusssian blue [1–3]. Prussian blue became popular in artists' paints and after 1730 was rapidly taken up as a substitute for the expensive pigment natural ultramarine blue and as an alternative for natural Indigo, which is not very stable when applied as a pigment [4].

Prussian blue was not only used as a pigment, but in the nineteenth century also for dyeing silk, wool, linen and cotton, especially in France. Already in the mid-eighteenth century, the French chemists Father François Ménon, Jean Hellot (1685–1766), Pierre Joseph Macquer (1718–1784) and Charles Antonin Le Pileur d'Apligny had proposed the use of Prussian blue in textile dyeing. In the second of two memoirs by Ménon, published on the 23th of March 1749, he described the dyeing of cloth with Prussian blue. His efforts failed with cotton, but worked well with wool, although the blues were not very uniform. The dyeing of silk with Prussian blue turned out to be much easier [5, 6]. In 1753, Macquer mordanted different fibres in alum and a sulfuric acid solution, the last also known as oil of vitriol. Afterwards they were dipped in a prussiate mixture, in other words, a solution of ferrocyanide. Positive results were obtained for linen, cotton and silk, but for wool, unfortunately, the blue colour was uneven and the cloth was 'rough to the touch'. [5, 7] The French dyers were not interested in his process because at this time huge amounts of natural Indigo were cheaply imported to France from the French colonies of Saint-Domingo and Guadeloupe (the French Antilles) [5]. This changed due to French warfare after the Revolution. In 1793, Great Britain imposed a naval blockade on French port cities, which was intended to cut France off from its overseas trade. After the defeat of his fleet by the English at Trafalgar in 1805, the French Emperor Napoleon Bonaparte (1769–1821) erected a Continental Blockade (the Continental System) in November 1806 in order to keep Great Britain's goods and ships away from the European continent. As a result, the import of cotton and natural Indigo became very difficult and was discontinued after 1806. A shortage arose that required a replacement product for natural Indigo. In June 1810, Napoleon Bonaparte instituted three prizes to stimulate the search for alternatives for Indigo: the French government awarded a prize of 25,000 francs to the person who made a successful process to dye wool and silk with Prussian blue for the uniforms of the Grande Armée [5, 8].

In 1811, the French chemist Jean Michel Raymond (also called Raymond-Latour, 1766–1837), professor at the Ecole Polytechnique in Paris, succeeded in dyeing silk with Prussian blue, but not wool; however, he was awarded 8000 francs of the prize as an incentive. His winning process involved mordanting the textile with iron prior to dipping it in an acid solution of yellow prussiate of potash (potassium hexacyanoferrate(II), $K_4[Fe(CN)_6] \cdot 3H_2O$, potassium ferrocyanide). This process proved problematic for dyeing wool, and especially wool cloth: the difficulty was to fix ferric iron on woollen fibres while preserving the quality of the wool. The solution was to follow the dyeing by a chlorine (bleach) bath, a procedure developed by his son P. Raymond in 1822, and later modified by the addition of tartaric acid in a bath containing a ferric iron salt [5]. He communicated his results to the *Bulletin de la Société d'Encouragement pour l'Industrie Nationale* in 1822 and showed his first specimens of dyed cloth at the 1823 Exhibition of Products of French Industry (Exposition des produits de l'industrie française) in Paris. For this process he received a silver medal. In 1828, a full memoir was published describing the application of the process to cloth, in addition to wool fleece [5, 9–12].

In France, Prussian blue was named *Bleu de France* (French blue) when applied to wool. The shades achieved using Prussian blue, which was also called Bleu Raymond, were also appreciated after the fall of Napoleon Bonaparte in 1814–1815, although natural Indigo, this time from the English North and South Carolinas, was imported into France again. From 1811 to 1860, Prussian blue was produced industrially for textile dyeing.

## 8.4 The Decline of Prussian Blue

At the beginning of the 1860s, the use of Prussian blue as a dye decreased significantly after the discovery of several blue synthetic organic dyes, starting with Azuline, one of the first synthetic blue dyes. Azuline is made from phenol, first oxidising it with oxalic acid and sulfuric acid to make a red substance called Aurin (C.I. name: Solvent dye; C.I. number: 43800). Rosolic acid had already been discovered in 1834 by the German chemist Friedlieb Ferdinand Runge (1794–1867), derived from phenol extracted from coal-tar, but as rather an impure product. He named it *Rosölsäure* or *Rosaölsäure* (red oil acid) [13].

Twenty-five years later, in 1859, the German-English chemist Hugo Müller (1833–1915), who lived in England, reinvestigated rosolic acid by mixing phenol with calcium carbonate and by exposing the mixture to air for a prolonged time

[14]. In the same year, the German chemists Adolph Wilhelm Hermann Kolbe (1818–1884) and Rudolf Schmitt (1830–1898) obtained the acid, which they referred to as a red dye, by heating oxalic acid and creosote (which contains phenol) in the presence of concentrated sulfuric acid. They did not patent the process, but details were published two years later [15].

Probably, Kolbe and Schmitt's process was taken over by the French chemist Jules Persoz (1837–1927), who lived in Paris. Persoz developed a commercially successful process for rosolic acid and by reacting it with ammonia, the red dye coralline or *péonine* was obtained. Subsequently, treatment of péonine with aniline gave the blue dye Azuline, also known as *blue d'acide phénique* [16, 17]. The French silk dyeing industrialist Jean-Aimé Marnas (1828–1908) noted that on a trip to Paris in July 1860 he had received a sample of the blue colourant from Persoz. The Lyonnaise company of Guinon, Marnas & Bonnet bought the process from Persoz, established a patent ('Preparation of a red colouring matter', F.P. 2.130, by François Bonnet, Nicolas-Philibert Guinon, Jean-Aimé Marnas and William Spence; and 'Preparation of a blue colouring matter', F.P. 2.132, by François Bonnet, Nicolas-Philibert Guinon, Jean-Aimé Marnas and William Spence, both issued on the 28th of July 1862), kept the preparation process a secret and started dyeing silk in 1860. The name Azuline was a combination of 'azure' and 'aniline'. Azuline was introduced into England in the second half of 1861 [18]. It became popular for colouring silk and in France was selling for 450 francs per kilogram in 1862 [19]. The blue dye hardly dissolves in water, but does dissolve in alcohol and is colourfast.

Azuline was soon to be replaced by Aniline blue (Lyon blue), which was developed in 1861, obtained by heating Fuchsine with an excess of aniline and discovered by the French chemists Adam Charles Girard (1837–1918) and Georges de Laire (1836–1908). This was followed by a series of water-soluble blue acid dyes (see Chap. 13). By 1865, as a result of all these developments, Raymond bleu dyeing at Guinon, Marnas & Bonnet was just a memory [5].

## 8.5 Applications

Prussian blue was and is mainly used in watercolour and oil painting, but also in the production of wallpaper, coloured paper, in letterpress (letterpress printing), lithography (printing from a stone or metal plate), in blue prints, cyanotypes and in ink manufacture. For textile purposes the blue pigment was used for silk and wool dyeing; for cotton dyeing Prussian blue was applied onto the fabrics. In Italy especially, it was also used for dyeing velvet. The colourant is formed on the textile fibre. This can be done in two ways: for silk and cotton, iron oxide is used, which acts as a mordant. After this, the textile is dyed with potassium hexacyanoferrate(II). For dyeing wool, potassium hexacyanoferrate(II) is precipitated onto the fibre in an acid dye-bath. The precipitate is not water-soluble so is relatively wax-like.

## 8.6 Properties

Prussian blue is fairly permanent to light and air. It is resistant to acid, but in alkaline solutions it breaks down, leaving a brown colour on the fibre. This is due to the conversion of iron into trivalent iron hydroxide. In the Colour Index, no information is given about the light fastness as a textile dye; as a pigment it is categorised as ´excellent´. However, it must be noted that when mixed with extenders or white pigments discolouration can occur, especially when the blue pigment is present at a low concentration [20, 21].

## 8.7 Recipes

Recipes for dyeing with Prussian blue were published in several manuscripts, brochures and dye books between 1811 and 1834. Jean Michel Raymond published his process in several brochures and in the *Bulletin de la Société d'Encouragement pour l'Industrie Nationale* in the years 1811 and 1814 [9, 10]. His French publications are translated into English in the book *The Repertory of Arts, Manufactures, and Agriculture* (London, 1815) [11, 12, 22]. The processes on wool and wool cloth were published by his son P. Raymond in 1828 [23–25]. In the 1834 edition of his book *A Practical Treatise on Dying Woollen, Cotton, and Silk* (first published in New York in 1823), the textile dyer William Partridge from Gloucestershire, who emigrated to the United States of America in 1807, describes briefly the Prussian blue dyeing process published by J.M. Raymond and gives recipes for dyeing Prussian blue on cotton and silk [26].

Referring to Jean Michel Raymond, William Partridge writes: *Prussian-blue, by Mons. Raymond.* Previous to the year 1811, the silks dyed blue were dull; but, in that year, M. Raymond invented a method of giving silk a deep and brilliant colour, which is now generally adopted, and is known by his name. Process as follows:

> 'When the silk has been cleansed, immerse it for a quarter of an hour in water containing about one-twentieth part of its weight of the sulphate of the peroxyde of iron, at the ordinary temperature, wash, and hold it for half an hour in a bath, nearly boiling, of soap and water; wash it again, and put it in a cold and very weak solution of prussiate of potash, soured by sulphuric acid, or by muriatic acid. As soon as it is immersed it becomes blue, and nothing more is wanting, than, in about a quarter of an hour, to wash and dry it.' [26]

*To dye Prussian blue on woollens:*

> 'Prepare the cloth by passing it through lime-water, at a gentle boiling heat. Wash well, cuttle up, and let lie till wanted. Prepare a fresh liquor, with one ounce of prussiate of potash to each pound of cloth. The same proportions may be used for yarns. Boil gently for ten minutes, wind up, and let drain on the reel. Prepare a fresh liquor, using one ounce (28.35 gram) of nitrate of iron to each pound (454 gram) of cloth, or yarn. This will give about a half-blue, and if required darker, use more nitrate of iron. Wind up and let drain. It has then to be run through a third liquor, in which has been dissolved one ounce and a half of what is called the mordant, to each pound of cloth, or yarn. This is to make the colour even, and

to give a bright clear lustre. There should be at least one gallon of water to every ounce of mordant. The temperature of the liquor should not exceed two hundred degrees (sic. Fahrenheit). Turn the cloth in this rapidly for ten minutes, wash well in clear water, and tenter as quick as possible.' [26]

*To dye cotton and silk a Prussian-blue:*

'Steep the yarn or cloth in a tub filled two-thirds with water, put therein twenty pounds of the nitrate of iron, or in proportion to the shade of blue that is wanted, or the pounds of stuff to be dyed; try two waters to one of the liquor; if too strong a shade is produced, add more water, or the reverse, steep well therein, wring out, and dry; then have a tub of water, say to hold four gallons, put therein two pounds of prussiate of potash, which dissolve; add thereto a pound or so of oil of vitriol. Before running the goods through the prussiate of potash liquor, run them through a solution of chloride of lime, prepared as follows: Stir two pounds of chloride of lime into four gallons of water, let stand to settle, and pour the clear liquor into one hogshead of water, through which run the goods; then run them through the prussiate liquor, when they will assume the blue shade, light or dark, as may be wanted; these shades however, must be regulated by using more or less of the nitrate of iron. The French writers recommend two thirds of nitrate of iron, and one third copperas, in preference to using all nitrate of iron. After the colour has been obtained, it may be raised two or three shades, by running it through a weak solution of water, slightly impregnated with ammonia.' [26]

This recipe states that it could be adapted according to the desired colour intensity, as described a few years earlier by P. Raymond. It also shows that people in different countries had different ideas about dyeing with Prussian blue.

In the period 1825–1826, the Austrian dyer Josef Zötl (1806–1871), who came from the state of Upper Austria, travelled during his journeyman years halfway across Europe and collected hundreds of dyeing recipes. One year later, in 1827, he compiled his handwritten recipe book for textile dyeing and printing, which included 385 recipes. Most recipes concern dyeing on silk, cotton, linen and wool, and printing on cotton and linen, describing 24 dyeing materials. It is interesting that he also came into contact with the more modern materials of his time, such as Indigo carmine for dyeing and Prussian blue for printing. He states that Prussian blue obtained as a commercial product does not have a good quality because it contains *Thonerde*, an aluminium-containing clay. Therefore, he made the blue pigment himself [27]. Thirty years later, in 1858–1859, his son Josef Zötl junior (1836–1925) made his journeyman years through Austria and Hungary. Before that time, in 1857, so during his apprenticeship, he wrote a recipe booklet giving instructions for dyeing and printing. In Zötl junior's booklet, Prussian blue was reported for dyeing on cotton, but not for printing [28].

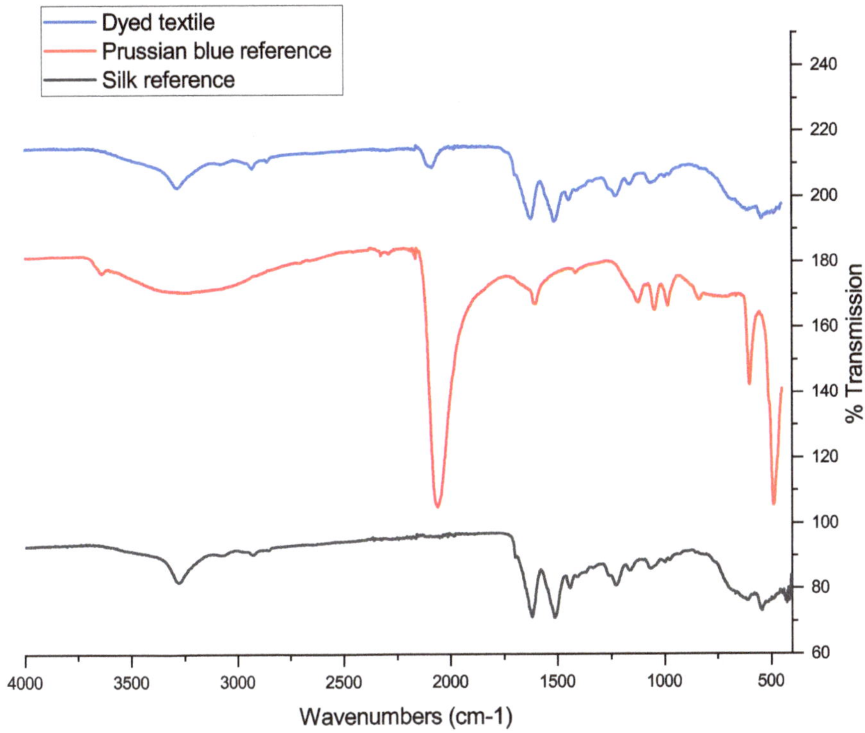


**Fig. 8.2** ATR-FTIR spectra of a blue-white striped dress (blue line), Prussian blue reference (red line) and undyed silk (black line). (Courtesy Suzan de Groot, Cultural Heritage Agency of the Netherlands)

## 8.8 Identification

Since Prussian blue is not soluble, it cannot be analysed using liquid chromatography. Prussian blue can be identified by microchemical analysis (a spot test). A sample of a blue thread treated with a drop of a dilute sodium hydroxide solution turns the blue colour brown and after acidification it should be blue again. By Fourier Transform Infrared Spectroscopy (FT-IR) Prussian blue can be unambiguously identified by its characteristic cyanide bond (CN bond) at 2080–2090 $cm^{-1}$ (see Fig. 8.2).

Another method to analyse Prussian blue is by Scanning Electron Microscopy coupled to Energy Dispersive X-ray Spectroscopy (SEM-EDX). In a blue thread, the chemical elements iron, potassium and aluminium should then be identified. It must be noted that determining the elemental composition does not lead to a conclusive identification of Prussian blue.

## 8.9 Case Studies

### *Folding-Screens from LIECHTENSTEIN. The Princely Collections, Vaduz–Vienna*

The Princely Collections, Vaduz–Vienna LIECHTENSTEIN own a set of folding screens with Chinese "Baizitu" (Pictures of a Hundred Boys) embroidery (see Fig. 8.3). Originally the embroidered panels, dated around 1780, were mounted as scroll panels in a Chinese interior. The embroidery was transported to Europe and became part of the collection of the Lednice Castle in the Czech Republic. In the silk embroidery threads, dye analysis showed that natural dyes were present, from the identification of carthamin (indicating Safflower), berberin (Amor cork tree, Chinese barberry species), rutin and quercetin (probably Chinese yellow berries) and indigotin, indirubin and isatin (indicating the use of an Indigo plant). These dyes all originate from Asia, suggesting local production. In the late 1840s, the folding screens on wooden frames were manufactured using the embroidered panels.

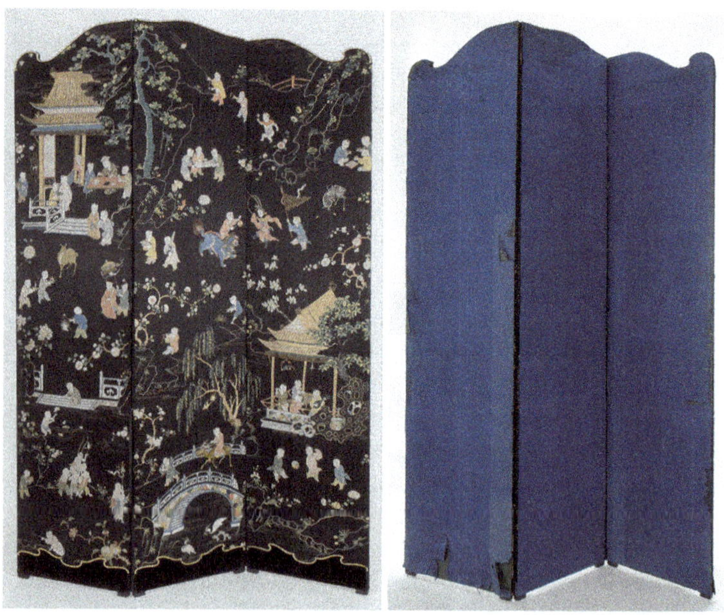

**Fig. 8.3** (Left) Folding screen with Chinese "Baizitu" (Pictures of a Hundred Boys) embroidery, textile dated c. 1780 (after conservation). (LIECHTENSTEIN: The Princely Collections, Vaduz–Vienna (inv. MO 1904/MO 1905). © Institute of Conservation, University of Applied Arts Vienna. Photograph: Edith Oberhumer and Stefan Oláh). (Right) Back of the folding screen with a green cotton lining and a blue silk lining, dated 1847 (before conservation). (LIECHTENSTEIN, The Princely Collections, Vaduz–Vienna (inv. MO 1904/MO 1905). ©Institute of Conservation, University of Applied Arts Vienna. Photograph: Edith Oberhumer and Stefan Oláh)

Due to their size of 270 cm height and 190 cm width, as well as the construction with wood, each screen weighs about 100 kg. The imperial upholsterer Friedrich Stöger brought them to completion on the 26th of July 1847. The back sides of the wooden frames were first covered with a green cotton cloth. Examination of the cotton showed an intense bleeding of a blue colour. This fabric was dyed with the natural yellow dye Weld (*Reseda luteola* L.) and Indigo carmine, as proven by dye analysis (see Fig. 8.3). Then a blue silk weave, dyed with Prussian blue, was mounted at the backs and afterwards the Chinese embroideries were applied on the front of the frames. As opposed to the Chinese embroideries, the green and blue fabrics were apparently produced in Europe [29].

### *Dresses from the Rijksmuseum Amsterdam: Project Majestic and Heavenly*

The Rijksmuseum Amsterdam has a substantial collection of nineteenth-century costumes (see Fig. 8.4). Dutch textile objects produced in the second half of the nineteenth century were the subject of a master's thesis at the University of Amsterdam in 2011 to gain insight into which synthetic organic dyes were used and when. The choice of the textile objects focused mainly on the colour blue, with a

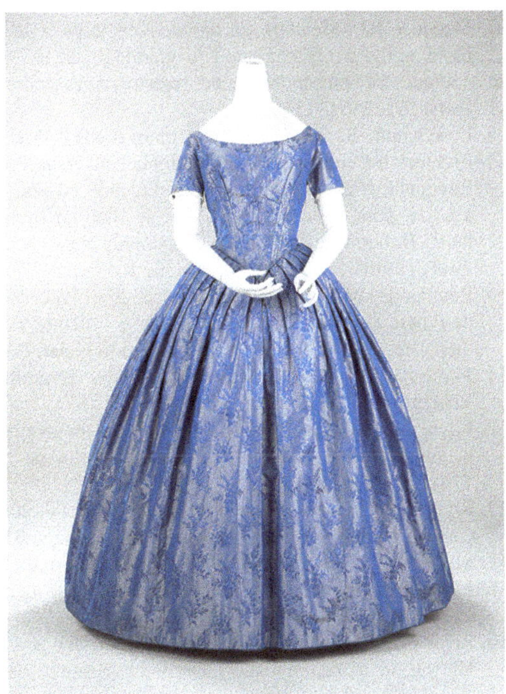

**Fig. 8.4** Blue changeant silk gown with woven-in motif consisting of flower branches, 1850, RK-15416. (Collection Rijksmuseum, Amsterdam the Netherlands. Photograph: Rik Klein Gotink)

delineation regarding dating, and fall within the time period 1850–1875. In four gowns, a checked gown (RBK-1975-132) from the period 1860–1865, a gown dated 1850 with woven flowers (RBK-15416), a striped gown (RBK-14670) from 1868 and a gown with a blue fringe (BK-TN 1199) without date, Prussian blue was identified. The last striped gown in particular shows that, even in the late 1860s, Prussian blue was used in the textile industry [30].

## References

1. Bartoll, J., Jackisch, B.: Prussian Blue: A Chronology of the Early Years. Zeitschrift für Kunsttechnologie und Konservierung/Journal for Art Technology and Conservation. 24, 88–102 (2010)
2. Harley, R.D.: Artists' Pigments c. 1600-1835: A Study in English Documentary Sources. Butterworth Scientific, London (1982)
3. Berrie, B.H.: Prussian Blue. In: Fitzhugh, E.W. (ed.) Artists' Pigments: A Handbook of their History and Characteristics. Volume 3. pp. 193–217. National Gallery of Art, Archetype Publications Ltd., Washington, London (1997)
4. van Eikema Hommes, M.: Changing Pictures: Discoloration in 15th-17th-century Oil Paintings. Archetype Publications Ltd., London (2004)
5. Delamare, F.: Blue Pigments: 5000 Years of Art and Industry. Archetype Publications Ltd., London (2013)
6. Ménon, F.: Second mémoire sur le bleu de Prusse. Mémoires de Mathématique et de Physique présentés à l'Académie Royale des Sciences par divers Sçavants, & lûs dans les Assemblées (Sçavants étrangers). 1, 573–592 (1750)
7. Macquer, P.J.: Mémoire sur une nouvelle espèce de teinture bleue, dans laquelle il n'entre ni pastel ni indigo. Histoire de l'Académie royale des sciences … avec les mémoires de mathématique & de physique tirez des registres de l'Académie Royale des Sciences de l'année 1749. Paris, 255–256 (1753)
8. Lowengard, S.: Techniques and Innovations - Prussian Blue. In: The Creation of Color in Eighteenth-Century Europe. Columbia University Press, New York (2006)
9. Raymond, J.M.: Description d'un Procédé invente par M. Raymond, professeur de chimie à Lyon, pour teindre la Soie avec le Bleu de Prusse, d'une manière égale, solide et brillante. Bulletin de la Société d'Encouragement pour l'Industrie Nationale (1st series). 13, 29–41 (1814)
10. Raymond, J.M.: Description raisonnée d'un Procédé sur et facile pour teindre la Soie en bleu de Prusse d'une manière égale, solide et brillante, dans les nuances les plus foncées, sans lui oter aucune de ses qualities. Imprimerie Imperial, Paris (1811)
11. The Repertory of Arts, Manufactures, and Agriculture, Volume CLX. printed for J. Wyatt, Repertory Office, Hatton Garden (1815)
12. Raymond, J.M.: Description of a Process for dyeing Silk of a Prussian Blue, so as to give a uniform, firm, and bright Colour. Bulletin de la Société D'Encouragement. 112–119, 179–190 (1815)
13. Runge, F.F.: Ueber einige Produkte der Steinkohlendestillation. Annalen der Physik und Chemie. Verlag Johann Ambrosius Barth, Leipzig. 31, 70 (1834)
14. Müller, H.: Note on rosolic acid. Quarterly Journal of the Chemical Society. 11, 1–5 (1859)
15. Kolbe, H., Schmitt, R.: Rother Farbstoff aus dem Kreosot. Annalen der Chemie. 119, 169–172 (1861)
16. Reinhardt, C., Travis, A.S.: Heinrich Caro and the Creation of Modern Chemical Industry. Kluwer Academic Publishers, Dordrecht/Boston/London (2000)

17. The New Blue Dye, “Azuline.” The Chemical News. III, 128–129 (1861)
18. Perkin, W.H.: XIX. On colouring matters derived from coal-tar. Quarterly Journal of the Chemical Society of London. 14, 230 (1862)
19. Morris, P.J.T., Travis, A.S.: A history of the international dyestuff industry. American Dyestuff Reporter. 81, 3–6 (1992)
20. Kirby, J.: Fading and colour change of Prussian blue: occurrences and early reports. National Gallery Technical Bulletin. 14, 62–71 (1993)
21. Kirby, J., Saunders, D.: Fading and colour change of Prussian blue: methods of manufacture and the influence of extenders. National Gallery Technical Bulletin. 25, 73–99 (2004)
22. Raymond, J.M.: Additions to the Description of the Process for dyeing Silk of a Prussian Blue´. Published by the Government. Archives des Decouvertes. 247–250 (1815)
23. Raymond, P.: De la teinture des Laines au moyen du bleu de Prusse. Annales de Chimie et de Physique. 39, 44–77 (1828)
24. Raymond, P.: On the dyeing of Wool with Prussian Blue. Quarterly Journal of Science, Literature and Art. 427–428 (1828)
25. Raymond, P.: Extrait d’un Mémoire sur la teinture des laines au moyen du bleu de Prusse par M. Raymond fils. Bulletin de la Société d’Encouragement pour l’Industrie Nationale (1st series). 27, 312–316 (1828)
26. Partridge, W.: A Practical Treatise of Dying Woollen, Cotton, and Silk. New York (1834)
27. Zötl, J.: Handwritten recipe book. Archive Färbermuseum Gutau, Freystadt (1827)
28. Zötl, J.: Handwritten recipe book. Archive Färbermuseum Gutau, Freystadt (1857)
29. Oberhumer, E., van Bommel, M.R., de Keijzer, M., de Groot, S., Erlach, R., Hofmann-de Keijzer, R., Knaller, R.: One hundred boys, one hundred challenges: The examination and conservation of two Viennese folding screens decorated with a Chinese silk embroidery. Studies in Conservation. 59, 119–123 (2014)
30. Steinmann, J.: Masterscriptie Majestueus & Hemels Negentiende-eeuwse blauwe kleurstoffen in de Rijksmuseum kostuumcollectie, (2011)

# Chapter 9
# Indigo Carmine: The First Semi-Synthetic Dye

## 9.1 Introduction

Indigo carmine is a dye that, following its introduction, was often used to dye textiles and to stain leather and furniture despite its poor properties. The development of this semi-synthetic dye cannot be underestimated since it caused a revolution in dyeing technology. Until then, Indigo, the only stable blue dye, was extracted from natural sources such as Woad (*Isatis tinctoria* L.) or Indigo (*Indigofera tinctoria* L.), so-called vat dyes. The dye extraction was very laborious and included complicated fermentation processes, a procedure which could easily take days or even weeks. Dyeing was also very time-consuming as the insoluble main component, indigotin, needs to be converted into the soluble leuco-indigotin form first [1]. Indigo carmine, however, is a water-soluble acid dye. As such, the dye can be easily dissolved and it binds directly to the textile fibre; the dyeing itself only takes minutes, a tremendous advantage compared to the laborious procedure required for Indigo. As a result, Indigo carmine was often used to dye blue or mixed with yellow dyes to obtain green.

## 9.2 Indigo Carmine

Colour Index name: Acid Blue 74; Colour Index number: 73015
Colour Index name: Natural Blue 2; Colour Index number: 75781
IUPAC name: Disodium [2(2′)E]-3,3′-dioxo-1,1′,3,3′-tetrahydro[2,2′-biindolylidene]-5,5′-disulfonate
Discoverer: Johann Christian Barth, 1743
Contemporary terminology: Indigo carmine (En), Bleu de Saxone (Fr), Indigokarmin (Ge), Indigokarmijn (NL)



M. R. van Bommel, M. de Keijzer, *Bright Colours from the Past*, Cultural Heritage Science, https://doi.org/10.1007/978-3-031-88664-5_9

Obsolete terminology: Saxon blue, Saxe blue, Intense blue, Chemick blue (En), Sächsisches Blau (Ge), Indigo extract, Saksisch blauw, Saksiesch blaauw, Chymick, Murabba (NL)
Chemical formula: $C_{16}H_8N_2Na_2O_8S_2$ (Fig. 9.1)

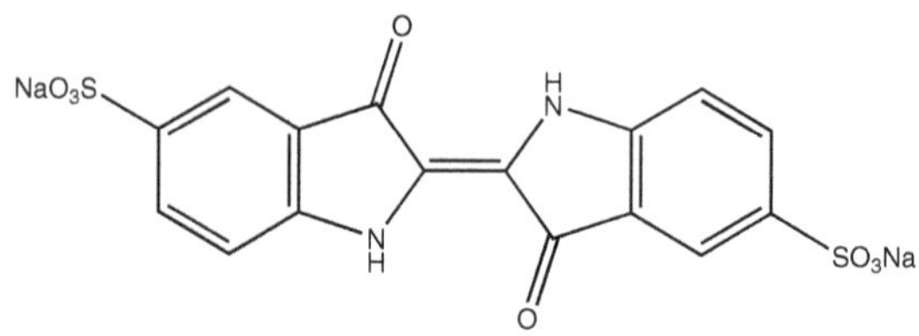


**Fig. 9.1** Structural formula of Indigo carmine

## 9.3 History

The German lawyer Johann Christian Barth (*c.* 1700–1759), the discoverer of Indigo carmine, was born in the town of Grossenhain, near the cities of Meissen and Dresden in Germany. In his youth he attended the protestant boarding school for boys (Fürstenschule) in Schulpforta, which was established in 1543 by Duke Moritz of Saxony, and afterwards he studied law at the University of Leipzig. After his studies he settled as a lawyer in Grossenhain until his death in 1759. He held an appointment as a tax collector, but throughout his entire lifetime he showed a greater interest in chemistry and physics. With the help of a dyer, Barth discovered Indigo carmine in 1743, for which he received a prize and the title Counsellor (Königlich Polnischer und Kurfürstlich Sächsischer Bergrat) in 1746 [2]. Indigo carmine was synthesised by treating natural Indigo with sulfuric acid, preferably the highly concentrated sort made in Saxony by distilling iron sulfate. Barth named the dye Sächsisches Blau (Saxon blue), after his native region Saxony [3].

Barth already possessed a dye factory in 1744, as became clear from archival research in the Sächsisches Hauptstaatsarchiv in Dresden, which revealed a document dated 1744–1745 [4]. The archival sources do not clarify which dyes were used in this dye factory, but his affinity with dyeing technology is clearly demonstrated. On the eighth of June 1744, the town of Grossenhain was almost completely devastated by a great fire and Barth lost his factory. To rebuild his factory, he wanted to buy a parcel of land near a ditch from the miller's wife, Christiane Erdmuthe of Grossenhain. The document indicates a dispute and the two neighbours did not reach an agreement. From that time on, it remains unclear how Barth continued dyeing. However, his son, Johann Carl Barth (1725–1801), started a cotton factory in his birth-place Grossenhain in 1763 and accomplished the Saxon blue dyeing process discovered by his father for dyeing wool and linen. He received the title Kurfürstlich Sächsischen Hofkommissar, i.e. Court Commissioner.

Already in 1749, several problems connected with Indigo carmine were reported by the German doctor and natural scientist Gottfried Michael Kortum. Kortum (1699–1749) born in Quedlinburg, Germany, lived in Bielitz, nowadays Bielsko-Biała, Poland. The main issue, he indicated, was the use of the destructive ingredients, which led to brittle woollen cloths. In addition, he mentioned that the use of rain water and cold water in general caused spots on the fabrics [5]. Furthermore, in his 1749 book he reported that only a few dyers living in Saxony knew the preparation of the Saxon colours in 1745, two years after the discovery of Saxon blue. They kept it a secret and even early in 1748 dyers were willing to pay a significant sum of money to obtain knowledge about the dyeing technology, not only in Germany but also in other countries. The breakthrough came during the 1748 Easter Trade Fair in Leipzig, where a foreign visitor and his friend revealed information about the dyeing method in exchange for money. Kortum wrote that a foreigner 'O.' was a close friend to a dyer 'R.', who was not successful in dyeing with the Saxon colours. For this reason, the foreigner 'O.' went to a Saxon dyer, who showed him the technology. After telling his close friend 'R.' the trick, both went to the trade fair. Here they talked about the dyeing method in exchange for money. The foreigner 'O.' left Leipzig in a hurry and on his way to Holland he sold the method to anyone interested in it. The identities of foreigner 'O.' and dyer 'R.' remain unclear.

In 1748 the process was bought by the wool-dyers of Norwich and patented in England. On the eighth of August 1748, B.P. 635, for 'Dyeing Green and Blue Saxon Colours´ was issued to the Southwark dyer George Spence, the warehouseman Charles Dolby and the merchant John Christopher Weguelin, both from London. In this first English patent for Indigo carmine it is written: '…they had their knowledge at a great expense and charge, by the means and assistance of a foreigner, who had sold all his interest in the art or mystery'. [6] The patent further explains the production to prepare green and blue Saxon colours and the dyeing processes. For Saxon green a natural yellow dye, either Old fustic (*Maclura tinctoria* (L.) D.Don ex G.Don) or Young fustic (*Cotinus coggygria* Scop.), is boiled with Indigo carmine.

Early in 1749, Jean Leonard Roederer, a dyer in Strasbourg, sent several lengths of blue and green woollen cloth to Paris, hoping that the cloths would be made into hunting costumes for the King, Louis XV, and his son, the Dauphin. The French chemist Jean Hellot (1685–1766), the government inspector of the dyeing industries at the Académie des sciences de l'institut de France (Academy of Sciences) in Paris, tested the qualities of the fabrics and noticed that by boiling them in soap the blue colour was lost. He recognised that an uncommon material or process had been used and this inspired him to further investigations. Under Hellot's guidance, the process to create Indigo carmine was established and a pamphlet describing it was printed in 1750. Publication meant that the technique to create the Saxon colours was no longer a secret kept from French dyers, but available to anyone who found the publication [7].

In 1754 the dye was sold under its modern name Indigo carmine. Even with its difficulties, the blue colourant offered new possibilities for dyeing and printing, such as those described in the 1770 B.P. 951. This patent, entitled 'Colours for

Dyeing Paper or Leather', was issued to the gentlemen Charles Fearne, of the Inner Temple, London, and James Gray, of Saint Margaret's, Westminster:

> 'The blue colour for leather is produced from a solution of indigo in oil of vitriol (fuming sulfuric acid); the green is an admixture of the blue with any yellow tincture at pleasure. The blue and green colours for papers are prepared in the same manner; the purples are produced by the admixture of logwood with the blue tincture, or from cochineal with sal martis (ferrous sulfate) or tincture of steel (an alcoholic solution of iron chloride).'

The Dutch dyer G. Boot reported several important facts about Indigo carmine in his book *De Verwer* (*The Dyer*). He described the preparation and discovery of Indigo carmine by Barth, the dyeing processes for wool and silk and wrote that the dyers were aware of the poor properties of Indigo carmine. Furthermore, he mentions that the preparation process was a secret in the beginning, but that it has been generally known for fifty years [8]. From this survey of literature it can be concluded that, from 1770, Indigo carmine was used for dyeing. Its early use has been confirmed by chemical analysis, as will be discussed below.

The use of Indigo carmine decreased in the second half of the nineteenth century because after the 1850s the first synthetic organic blue dyes, especially those of the tri-arylmethane class, were discovered. In 1860 the French chemists Adam Charles Girard (1837–1918) and Georges de Laire (1836–1908) obtained Aniline blue, the first synthetic blue dye, by heating Fuchsine with an excess of aniline. Two years later, in 1862, the British chemist Edward Chambers Nicholson (1827–1890) discovered that by treating the basic dye Aniline blue with fuming sulfuric acid, a series of water-soluble mono-, di- and tri-sulfonic acid dyes are formed. Alkali blue, Methyl blue, Water blue, Cotton blue, Navy blue and Soluble blue are examples of those synthesised and these are far better adapted for dyeing wool than Aniline blue, which was not soluble in water. Concentrated sulfuric acid converted the water-insoluble basic dye Aniline blue into a number of water-soluble acid dyes, which resembles the process developed by Barth converting the insoluble indigotin into water-soluble Indigo carmine.

However, Indigo carmine was still being produced. Indigo carmine was mentioned in the 1871 *Muster-Zeitung*, a German dyers' periodical, and also a piece of blue cloth was attached. The cloth was first dyed with the water-soluble Indigo carmine two tones lighter than the desired shade and afterwards with the alcohol-soluble Aniline blue. The colour produced has the advantage that in gaslight the colour stays blue, whereas the blue colour from pure Indigo carmine always shows a greenish tone [8]. The dye was also produced in the United States. The chemist William J. Matheson, the American sales agent for the Frankfurt company Cassella from 1876, founded a dye company in New York with manufacturing plants at 12th Street, Brooklyn, erected around 1885, and in Long Island. The Brooklyn and Long Island plants mainly produced chemicals for textiles, including both natural and synthetic dyes, among them Indigo carmine. In two English sample books of the 1890s, one produced by the company Wood & Bedford in Leeds (*c.* 1890) and the other the third volume of Knecht, Rawson and Loewenthal's *Manual of Dyeing for the Use of practical Dyers, Manufacturers, Students, and All interested in the Art of Dyeing* (London, 1893), Indigo carmine is included [9].

Indigo carmine can also be produced by treating synthetic Indigo with concentrated sulfuric acid. Soon after the indigo syntheses of the German chemist Karl

Heumann (1850–1894), German companies patented 'artificial' Indigo carmine by using the same starting materials or derivatives for the production of synthetic Indigo (see Table 9.1). In G.P. 63.218, filed on the 9th of December 1890 by Bayer in Elberfeld, a new chemical route is introduced: by treating *N*-phenylglycine with fuming sulfuric acid at room temperature Indigo carmine was obtained [10].

The German chemist Bernhard Heymann (1861–1933) wrote that this method for the preparation of Indigo carmine had been filed by Bayer in patents in all countries in December 1890. More patents describing the same procedure followed. As an example, G.P. 68.372, claimed by BASF on the 18th of January 1891, is based on the derivatives of *N*-phenylglycine (methyl-phenylglycine and ethyl-phenylglycine) and fuming sulfuric acid. The dyes prepared from both derivatives dye wool a greenish-blue colour. In B.P. 8.726, issued to BASF in 1890, the same components for obtaining Indigo carmine are described.

In May 1890, the German chemist Rudolf Knietsch (1854–1906), who worked for BASF, experimented with Indigo and fuming sulfuric acid during his work on the synthesis of indigotin. BASF deposited this process on the 24th of July 1890 through a solicitor. Encouraged by the results, Knietsch set up a small production for Indigo carmine in the factory building LU 36. However, in the long run this did not give the desired results: in practice, the Indigo carmine obtained in this way could not compete with natural Indigo and the production was stopped at the end of 1891. It remains unclear why Indigo carmine could not compete with Indigo: it could be related to its poor light fastness and wash fastness properties, the colour obtained, or for a commercial reason. U.S.P 524.256, filed on the seventh of August 1894 by Knietsch suggests that he must have followed a different path to success, but for this there is no evidence. The patent relates to the manufacture of the blue colouring matter, soluble in water, which dyes animal fibres from an acid bath. This patent also describes the fact that that Indigo carmine is obtained by treating *N*-phenylglycine with fuming sulfuric acid at room temperature [11]. The American patent refers to several European patents on Indigo carmine, dated 1891–1892, without specifying the title and the starting material (see Table 9.1).

In G.P. 143.141, filed on the 24th of July 1901 by Kalle, *N*-phenylglycine sulfo-carboxylic acid, a derivative of *N*-phenylglycine-*o*-carboxylic acid, is fused with caustic alkali, followed by oxidation at 300–340 °C to obtain Indigo carmine.

Finally, B.W. Gerland in Accrington, England, applied for a patent (G.P. 180.097) on the 30th of December 1903. Here, one part of pulverised natural or synthetic Indigo is mixed with one part of fuming sulfuric acid at room temperature. The mixture is then heated to 140 °C. After cooling, the melt sets, is easy to crumble and is put into tin cans. To protect the tin from the acidic material a bitumen layer is attached to the inside of the cans.

In the third volume of *Fortschritte der Theerfarbenfabrikation und verwandter Industriezweige* (1890–1894), the German chemist Paul Friedländer (1857–1923) writes that the German processes, filed in G.P. 63.218 and G.P. 68.372, achieved no practical interest because the yield of Indigo carmine was too low and as a result it played a secondary role compared to Indigo [12]. In the seventh volume of 1902–1904 he comments on G.P. 143.141, writing that this process does not come into consideration for large-scale production either [13]. Other manufacturers

**Table 9.1** Patents for Indigo carmine [14]

| Discoverer, company | Patent | Title, reference | Starting material | | | | | |
|---|---|---|---|---|---|---|---|---|
| | | | Natural indigo | Natural or synthetic indigo | *N*-phenylglycine | Derivatives of *N*-phenylglycine | Derivative of *N*-phenylglycine-*o*-carboxylic acid | Not specified |
| George Spence, Charles Dolby and John Christopher Weguelin | English patent no. 6358<br>August 1748 | Dyeing Green and Blue Saxon Colours | + | | | | | |
| Charles Fearne and James Gray | English patent no. 951<br>25 May 1770 | Colours of Dyeing Paper or Leather | + | | | | | |
| Farbenfabriken vorm. Friedrich Bayer & Co. | G.P. no. 63.218<br>9 December 1890 | Verfahren zur Darstellung von künstlichem Indigocarmin | | | + | | | |
| BASF | G.P. no. 68.372<br>18 January 1891 | Verfahren zur Darstellung von Farbstoffen der Indigoreihe | | | | + | | |
| BASF | B. P. no. 8.726<br>Complete Specification.<br>Example 4<br>2 March 1891 | Preparation of artificial indigo, in the form soluble in water, by the action of fuming sulfuric acid upon phenyl-glycol | | | + | + | | |
| BASF | Belgian patent no. 93.863<br>28 February 1891 | Reference in<br>U.S. patent no. 524.256 | | | | | | + |
| BASF | Italian patent XXV, 29.239<br>and LVIl, 257<br>28 March 1891 | Reference in<br>U.S. patent no. 524.256 | | | | | | + |

| | | | | | | | | |
|---|---|---|---|---|---|---|---|---|
| BASF | Spanish patent no. 11.774<br>25 April 1891 | Reference in<br>U.S. patent no. 524.256 | | | | | | + |
| BASF | F.P. no. 206.567<br>22 May 1891 | Reference in<br>U.S. patent no. 524.256 | | | | | | + |
| BASF | Austrian patent no. 41 and<br>No. 2.153<br>6 August 1891 | Reference in<br>U.S. patent no. 524.256 | | | | | | + |
| BASF | Russian patent no. 10.700<br>3 October 1892 | Reference in<br>U.S. patent no. 524.256 | | | | | | + |
| Rudolf Knietsch, BASF | U.S.P. no. 524.256<br>7 August 1894 | Blue dye | | | + | | | |
| Kalle & Co. | G.P. no. 143.141<br>24 July 1901 | Verfahren zur Darstellung von<br>Indigodisulfosäure und Indigo | | | | | + | |
| Dr. B.W. Gerland | G.P. no. 180.097<br>30 December 1903 | Verfahren zur Darstellung von löslichem<br>Indigo (Indigodisulfosäure) | + | + | | | | |

changed their products as well; for example, Farbwerke ML&B successfully replaced Indigo carmine by the sulfonated greenish-blue tri-arylmethane dyes: the Patent blues discovered in 1888. The name 'Indigo carmine' maybe lasted longer and was used for other products as well. In sample books dating from 1922, used by salesmen from the Berlin company AGFA, 'Indigo carmine blue A' was mentioned for dyeing silk, leather and wool. However, HPLC analyses of this sample identified mainly the blue-green sulfonated tri-arylmethane dye Patent blue V. Obviously in the twentieth century, the term 'Indigo carmine' was used to characterise the colour shade rather than the chemical product.

## 9.4 Properties

Indigo carmine is a blue powder with a purplish lustre which dyes wool and silk a splendid blue colour. Chemically speaking, Indigo carmine is indigotin disulfonic acid: two sulfonic acid groups are attached to the indigo molecule. The consequence and advantage is that the dye is water-soluble, which makes the dyeing easy, in contrast to the vat dyeing process with Indigo. The disadvantage is that the fastness to washing is low as it runs in water, giving a blue colour. Unfortunately, the fastness to light is also much reduced compared to Indigo and the colour changes from blue to green to yellowish during the fading process. Lehne describes the fastness of Indigo carmine[1]: 'After 5 days exposure to the elements, in June, the colour has not changed much, after 3 weeks much greener and duller'. According to the Colour Index, the light fastness of Indigo carmine is very low, ISO 1–2.

## 9.5 Recipes for Dyeing

The earliest recipes found for the preparation of and for dyeing with the Saxon colours are mentioned in a Dutch hand-written manuscript from 1775. Dye recipes are also given in the 1789 book *Anfangsgründe der ökonomischen und technischen Chymie* by the German physicist and chemist Georg Adolph Suckow (1751–1813), in the 1794 *Wiener Farbenkabinett*, in the 1798 *The Dyer's Companion* by Elijah Bemiss and in the 1804 book *Allgemeine Grundsätze der Bleichkunst* by the German pharmacist and chemist Sigismund Friedrich Hermbstädt (1760–1833). Most recipes are published in manuscripts and dye books in the last quarter of the 18th and throughout the 19th centuries, especially between 1790 and 1840. In the last two decades of the nineteenth century Indigo carmine was often mentioned in textile books, lexicons, encyclopaedias and technical and chemical dictionaries (see Table 9.2).

[1] Künstlicher Indigocarmin in Teig: Echtheit: ʻNach 5 tägiger Einwirkung der Witterung (Juni) wenig verändert, nach 3 Wochen viel grünlicher und matterˊ.

**Table 9.2** Indigo carmine mentioned in various sources [14]

| Year | Author, source | Title |
|---|---|---|
| 1749 | Gottfried Michael Kortum | *Neue Versuche der Färbekunst betreffend die, bisher unter dem Namen 'Sans pareille de Saxe' bekannten blauen und grünen Farben* |
| 1750 | Jean Hellot | 'Rapport sur la vert de Saxe du sr. Roederer', in the folder *Vert de Saxe—Sieur Roderer de Strasburg* |
| 1750 | Jean Hellot | 'Sur le beau vert de Saxe' in the folder *Vert de Saxe—Sieur Roderer de Strasburg* |
| 1750 | Jean Hellot | *Manière de teindre un drap blanc en verd, nommé vert de saxe*, Paris |
| 1751 | Johann Heinrich Gottlob von Justi | *Das entdeckte Geheimniss der neuen sächsischen Farben, benebst einigen Betrachtungen von dem Vorzug und der Theorie dieser neuen Färbekunst* |
| 1775 | Dutch hand written manuscript | – |
| 1780–1782 | – | *Acta die von Johann Carl Barthen zu Gr.Hayn erfundene blaue Farbe* |
| 1789 | Georg Adolph Suckow | *Anfangsgründe der ökonomischen und technischen Chymie* |
| 1794 | – | *Wiener Farbenkabinett: Oder vollständiges Musterbuch aller Natur- Grund- und Zusammensetzungsfarben* |
| 1798, 1806, 1815 | Elijah Bemiss | *The Dyer's companion* |
| 1804 | Sigismund Friedrich Hermbstädt | *Allgemeine Grundsätze der Bleichkunst* |
| 1813 | Edward Bancroft | *Experimental researches concerning the philosophy of permanent Colours* |
| 1820 | G. Boot | *De Verwer* |
| 1823 | William Partridge | *A practical treatise on dying of Woollen, cotton and skein silk* |
| 1827 | Josef Zötl | Handwritten recipe book |
| 1837 | J. Dumas | *Annalen der Chemie* |
| 1857 | Johann Carl Leuchs | *Vollständige Farben-und Färbekunde (Erster band)* |
| 1857 | Johann Carl Leuchs | *Vollständige Farben-und Färbekunde (Zweiter band)* |
| 1860 | Ludwig Brade | *Das illustrierte Buchbinderbuch* |
| 1863 | Capron de dole and Dussaunce | Blues & carmines of indigo |
| 1871 | W. Grüne and H. Grothe (Eds.) | *Muster-Zeitung* |
| c. 1890 | – | Sample book of the firm Wood & Bedford, Leeds |
| 1893 | Edmund Knecht, Christopher Rawson and Richard Loewenthal | *Manual of dyeing: For the use of practical dyers, manufacturers, students, and all interested in the art of dyeing* |

Studies of Indigo carmine recipes show that generally one pound[2] of natural Indigo is mixed with four pounds of oil of vitriol (concentrated sulfuric acid) and two ounces of fine salt, probably sodium chloride. An early example for its preparation is written in the 1798 book *The Dyer's Companion*, by the American dyer Elijah Bemiss in Waterford (Connecticut):

> 'Take one pound of good flotong indigo pulverised, four pounds of oil of vitriol, and two ounces of fine salt—put this in a stone pot (or some earthen vessel) that will contain six times the quantity of this compound, or it will be liable to rise and run over. First put in the vitriol, then the indigo, then the salt, stir this continually one hour, or till it gets pretty well settled and cool—for it will boil and foment in a terrible manner. Let it stand four days or a week, covered close, stirring it now and then, as is most convenient.' [15]

In recipes the quantity for oil of vitriol varies up to as much as eight parts compared to one part of natural Indigo. But dyers warn against too high an amount of oil of vitriol since the dyed cloths will be damaged and become brittle. In recipes sometimes the dye prepared by using natural Indigo and oil of vitriol is misleadingly termed Prussian blue.

A typical dye recipe was found in Bemiss:

> 'The dyeing process starts by filling a copper vat with fair water, heat it to nearly boiling hot, then add your blueing (as mentioned before) a little and stir it well the water, run your cloth, roll it out, air, and add your compound little by little, till your colour pleases. – You may make this dye, and shade you wish of this kind of blue, and very bright this dye.' [16]

The dyeing can be done either with or without the use of alum and cream of tartar (potassium bitartrate), but using a mordant seems to improve the fastness. However, the Upper-Austrian master-dyer Josef Zötl (1806–1871) in the town of Freistadt, Austria, mentions in his hand-written recipe book of 1827 that for the preparation of Indigo carmine to produce blue, green and black colours for dyeing silk, linen, cotton and printing, additives are completely useless [17].

Lehne writes in his 1893 book:

> 'The artificial indigo carmine in paste is identical to the indigo carmine made from indigo and is dyed like this on wool according to the dyeing recipe V3A.
>
> Add 15–20% sodium sulfate (Glauber's salt) to the colour solution. Increase the temperature at about 70 °C, with gradual addition of 5% sulfuric acid obtained from concentrated sulfuric acid, which is diluted beforehand, and boil for about half an hour. Sodium sulfate and sulfuric acid are often replaced by 12–15% of pure cream of tartar (sodium bisulfate). The nuances obtained are considerably purer than those made by ordinary indigo carmine.'[3]

[2] 1 American or British pound = 453.6 g; 1 American or British ounce = 28.35 g.

[3] Künstlicher Indigocarmin in Teig: Der künstliche Indigocarmin in Teig ist identisch mit dem aus Indigo hergestellten Indigocarmin und wird wie dieser an Wolle nach V.3A gefärbt. Die damit erhältlichen Nuancen sind erheblich reiner wie die mittels gewöhnlichem Indigocarmin.

Färbevorschrift: 15–20% Glaubersalz und die Farblösung zugeben. Bei etwa 70 °C. eingehen, unter allmäligem Zusatz von 5% Schwefelsäure von 60 °Bé, welche vorher stark verdünnt wird, in ca. ½ Stunde kochend ausfärben. Glaubersalz und Schwefelsäure werden häufig durch 12–15% Weinsteinpräparat (Natriumbisulfat) ersetzt.

## 9.6 Identification Using UHPLC-PDA

In general, Indigo carmine consists of a mixture of two components, mono-and di-sulfonated indigotin. Sometimes, traces of indigotin and/or equivalents are found as well. A chromatogram typical for Indigo carmine (Sigma-Aldrich CAS 860-22-0) is shown in Fig. 9.2, including the PDA spectra obtained from both components.

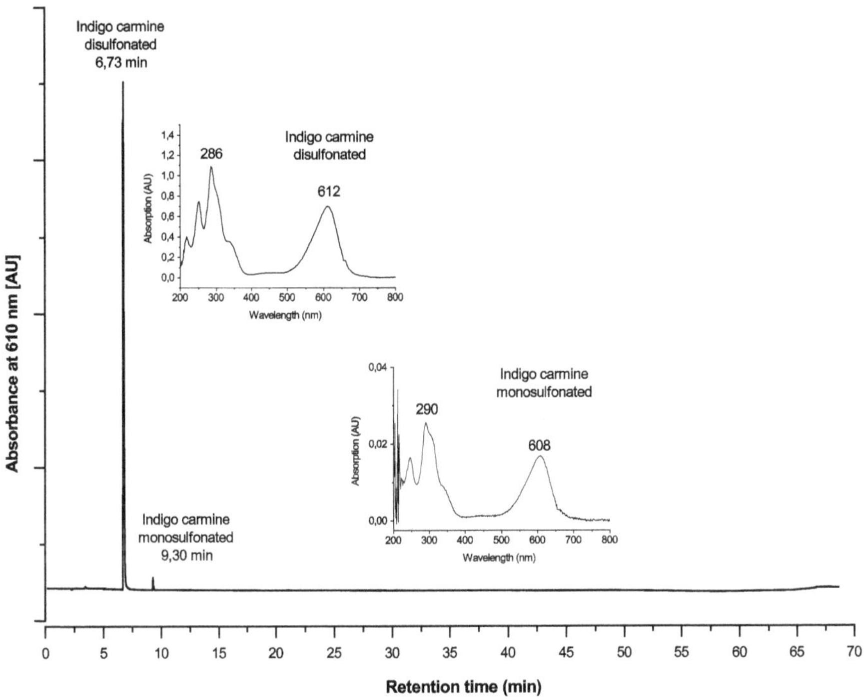


**Fig. 9.2** Chromatogram and spectra of Indigo carmine (Sigma-Aldrich CAS 860-22-0)

## 9.7 Case Studies

Indigo carmine was mainly used in smaller workshops, especially for dyeing silk, wool and leather, in printing and even for colouring soap and to create different colours of blue, green, violet and black. It was not recommended for cotton and linen because of their higher sensitivity to acids. Indigo carmine has been identified in textile objects dating from the last quarter of the eighteenth century till the beginning of the twentieth century (see Table 9.4). Indigo carmine was also found in three embroideries designed by Émile Bernard; these are discussed in Chap. 13.

## *Samplers from the Zuiderzeemuseum in Enkhuizen, the Netherlands*

Within our research we have investigated several samplers. The oldest samplers in which we detected Indigo carmine were dated 1814 and 1819. A strong indication of the presence of a synthetic dye was that these samplers showed a blue bleeding on white paper when treated with water. For textile conservators it is important to prove that late eighteenth- and nineteenth-century yarns, woven fabrics and textiles contain Indigo carmine as particular care needs to be taken in their conservation because of the sensitivity of Indigo carmine to washing.

Samplers from the Zuiderzeemuseum in Enkhuizen were also analysed. An example is shown in Fig. 9.3. These samplers, dated between 1842 and 1849, are made of white linen on which figures and lettering have been embroidered with coloured wool and silk. In every sampler, the colour of the blue threads had run into the background fabric, forming blue-coloured halos on the white linen (see Fig. 9.4). In view of the dating of the samplers it could be expected that natural Indigo (*Indigofera* species) or Logwood (*Haematoxylum campechianum* L.) had been used

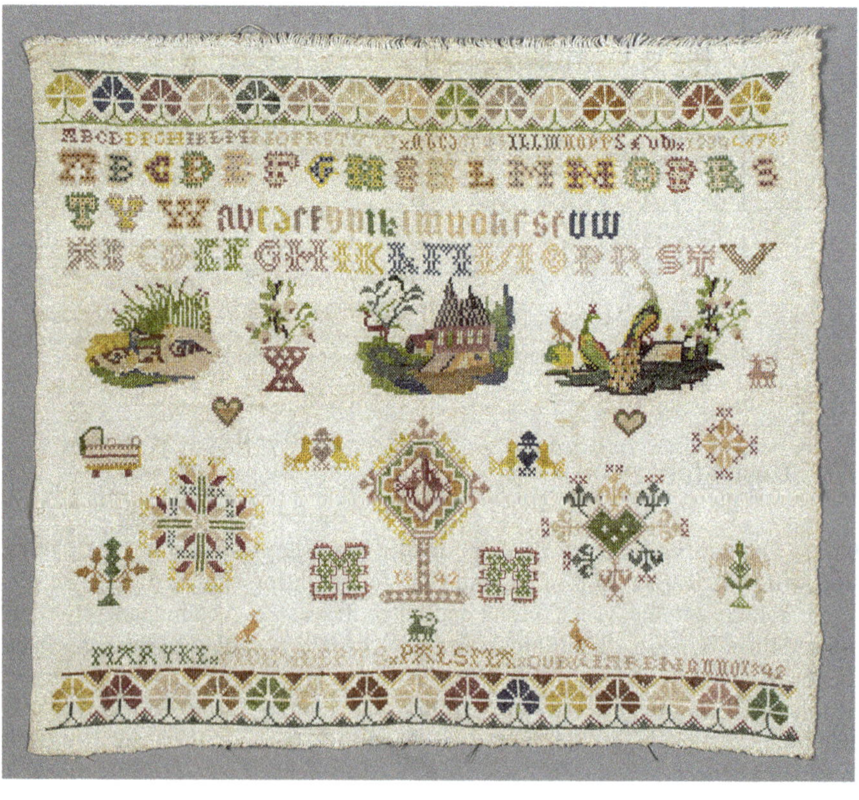

**Fig. 9.3** Sampler after treatment, 1842, Maryke Meinderts Palsma, Friesland, Silk on linen 38.0 cm H × 44.0 cm W, Collection Zuiderzeemuseum Enkhuizen. Photograph: Wim Zandbergen

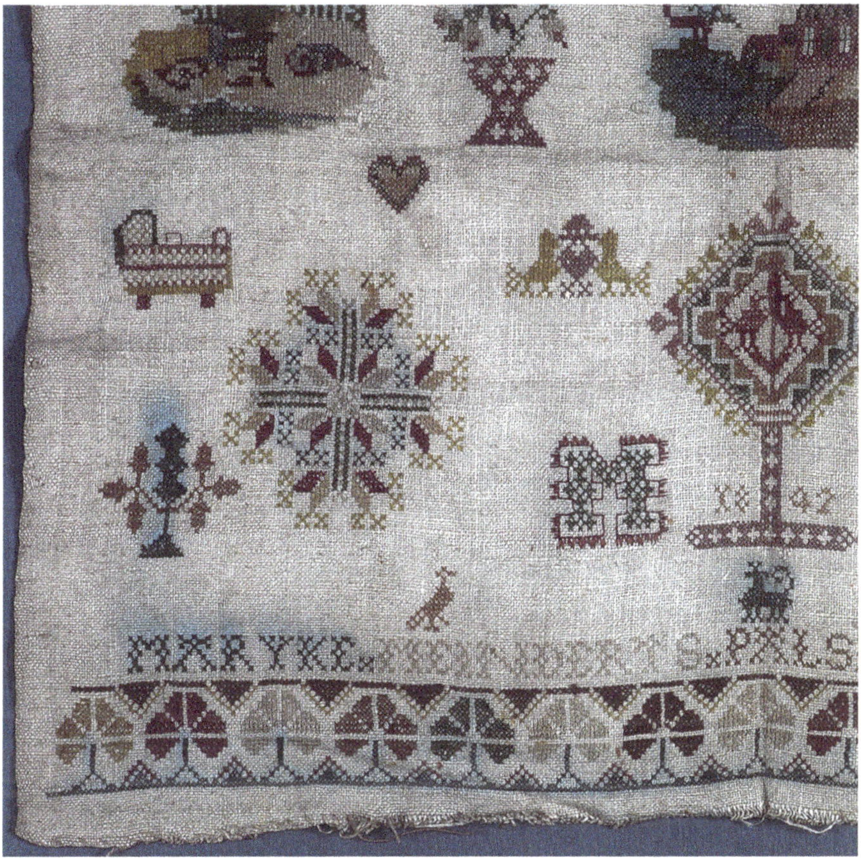

**Fig. 9.4** Detail of sampler prior to treatment, 1842, Maryke Meinderts Palsma, Friesland, Silk on linen 38.0 cm H × 44.0 cm W, Collection Zuiderzeemuseum Enkhuizen. Photograph: E. Klusman

to dye the blue threads. However, Logwood was used for dyeing black. Furthermore, natural Indigo is relatively fast to light and both natural Indigo and Logwood are fast to washing, thus the running of the blue did not fit into this pattern. Analysis showed the presence of Indigo carmine in every sampler [18]. The same phenomenon has been observed in the sampler by Barbara Greil, dated 1849, in the Krahuletz-Museum in Eggenburg, Austria, and in the sampler by Anneliza Keachie, dated August 1844, from a private collection.

### *Ball Gown of Princess Elisabeth in Bavaria—Kunsthistorisches Museum, Court Wardrobe ("Monturdepot"), Vienna*

In 1854, Princess Elisabeth of Bavaria (1837–1898), considered one of the most beautiful women in the world, married Emperor Franz Joseph (1830–1916) and thus became Empress of Austria. Traditionally, she is said to have worn this ball gown at

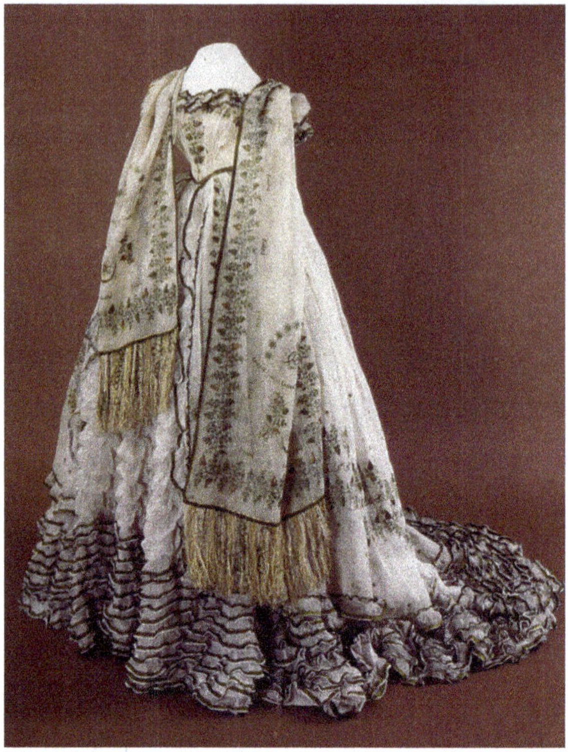

**Fig. 9.5** Spinster party dress of Empress Elisabeth of Austria (1854). © KHM Museumsverband/ Monturdepot, Austria

her spinster party ('Polterabend'); however, there is no evidence of such a celebration in written sources (see Fig. 9.5). The white cotton organza dress (Inv. nr. N 140) is decorated with embroideries made from green silk and gold threads and with green silk ribbons (see Fig. 9.6) [19]. Presumably due to the influence of moisture, a blue dye has leaked out of the ribbons (see Fig. 9.7). This was not the case in the embroideries. In both applications the natural yellow dye Weld (*Reseda luteola* L.) was detected. The blue dye of the embroideries was identified as indigotin. The silk ribbons, however, were dyed with Indigo carmine, which explains the bleeding of the blue dye.

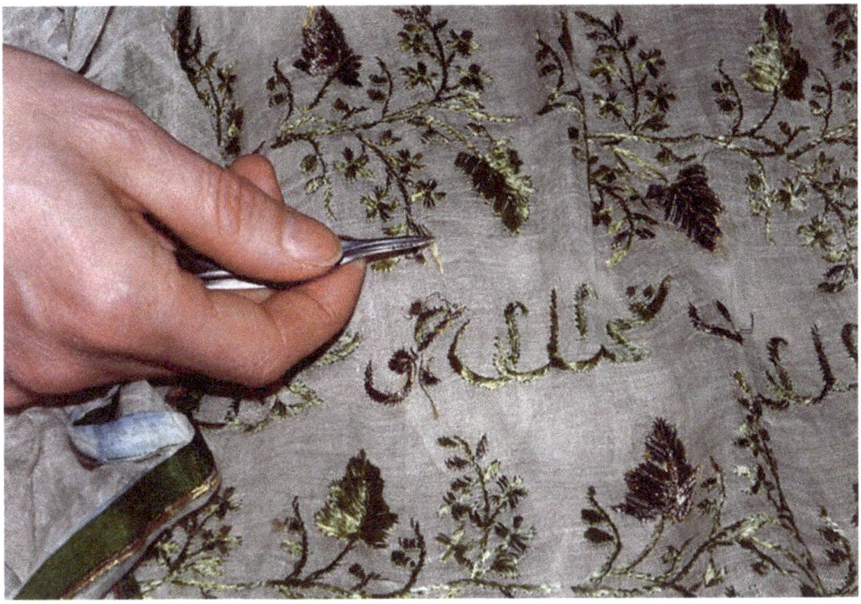

**Fig. 9.6** Detail of the cotton organza with green non-bleeding silk embroidery and a green silk ribbon with blue bleeding of Indigo carmine. © KHM Museumsverband/Monturdepot, Austria

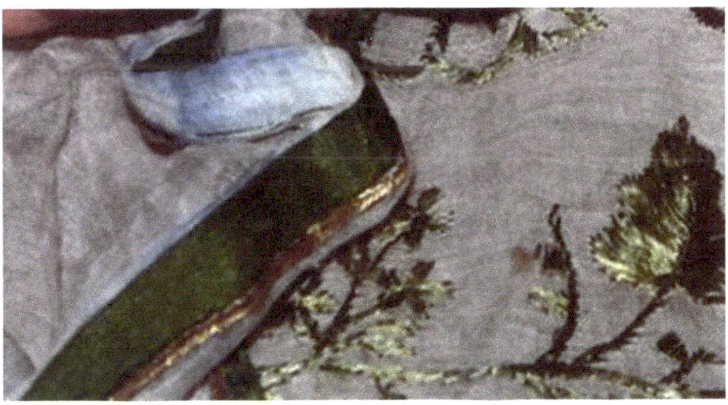

**Fig. 9.7** Blue bleeding of Indigo carmine. © KHM Museumsverband/Monturdepot, Austria

## *Habit à la française (Tail Coat) from LIECHTENSTEIN. The Princely Collections, Vaduz-Vienna*

An early object in which Indigo carmine was found is a tail coat or frock (*Habit à la française*, Inv. 201), dated between 1780 and 1790, from LIECHTENSTEIN. The Princely Collections, Vaduz–Vienna (see Figs. 9.8a and 9.8b). Originally the silk velvet had a violet-black colour with a turquoise weft insertion, but the colours in the left sleeve in particular have faded almost completely (see Fig. 9.9) [20]. This

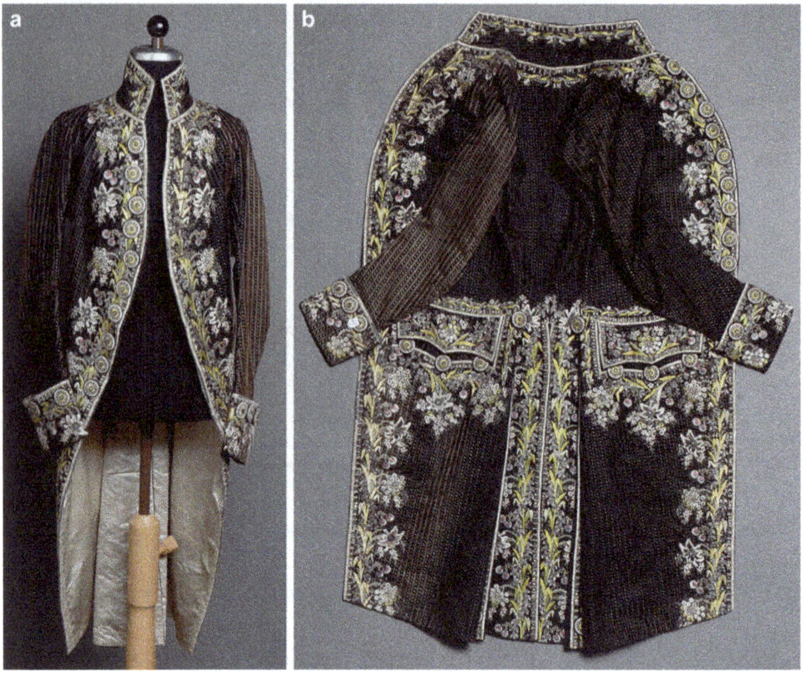


**Figs. 9.8a and b** Habit à la française (tail coat, frock, Inv. 201), dated 1780/90 (after conservation). LIECHTENSTEIN. The Princely Collections, Vaduz–Vienna. © Institute of Conservation, University of Applied Arts Vienna. Photographs: Nina Zangerl and Stefan Oláh

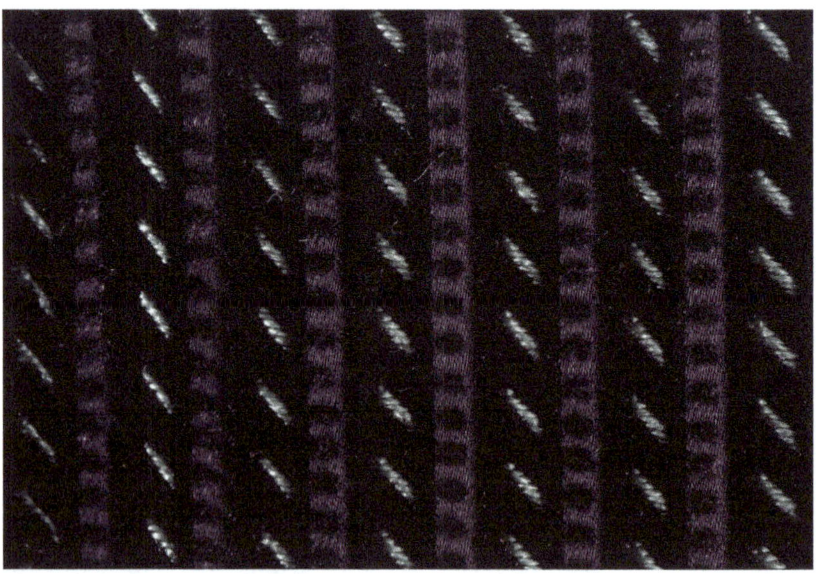

**Fig. 9.9** Pattern of the velvet, detail from an unfaded part. The turquoise weft is dyed with Indigo carmine. © Institute of Conservation, University of Applied Arts Vienna. Photograph: Nina Zangerl and Stefan Oláh

can be explained by the results of the dye analysis: the violet was dyed with the natural dyes Indigo and Logwood, the turquoise weft with the natural dye Barberry (*Berberis vulgaris* L.) and Indigo carmine. Logwood, Barberry and Indigo carmine especially show a poor light fastness.

## *Anatolian Kilim*

Indigo carmine was used for dyeing wool for oriental Persian and Turkish carpets during the second half of the nineteenth century. It was introduced by European firms that until the First World War influenced the carpet production in Persia and Turkey to a considerable extent. Dyers came from Europe to Turkey to support the dyeing and carpet production industry. A good example how two methods of dating complement each other is the research performed on a kilim. The 'dovetailed' tapestry weave with its unique design is very rare in Anatolia and a date between the sixteenth and eighteenth century for the kilim had been assumed. However, the radio-carbon dating result placed the kilim with the highest probability into the nineteenth century, which was disillusioning. In an attempt to confirm or contradict the result, attention was drawn to the unusual olive-green in the centre of the eight-pointed star. Dye analysis indicated that Indigo carmine was used, which means a date after 1743 and presumably later, similar to the radio-carbon dating [21].

## *Billardmuseum Weingartner in Vienna*

A special application of Indigo carmine was to dye the cloth of billiard tables. Heinrich Weingartner (1939–2025) was an Austrian carom player, 15-times Austrian national champion with international success, founder of a billiard café, a billiard school and until 2019 editor-in-chief of a professional magazine. For many years he collected historical billiard artefacts and in 1992 he opened the Billardmuseum Weingartner in Vienna. The museum has a collection of eleven antique billiard tables, which are now covered with modern green cloth (see Fig. 9.10). In the pockets and along the cushions, remnants of original bluish-green cloth were left. HPLC analyses indeed showed that some of the old billiard cloths were dyed with Indigo carmine and the natural yellow dye Old fustic (*Maclura tinctoria* (L.) D.Don ex G.Don) (see Fig. 9.11). Billiard cloths of later origin, dyed with Quinoline yellow, Flavazine L and Patent blue V, must have been produced after 1888 and 1892 respectively (see Table 9.3).

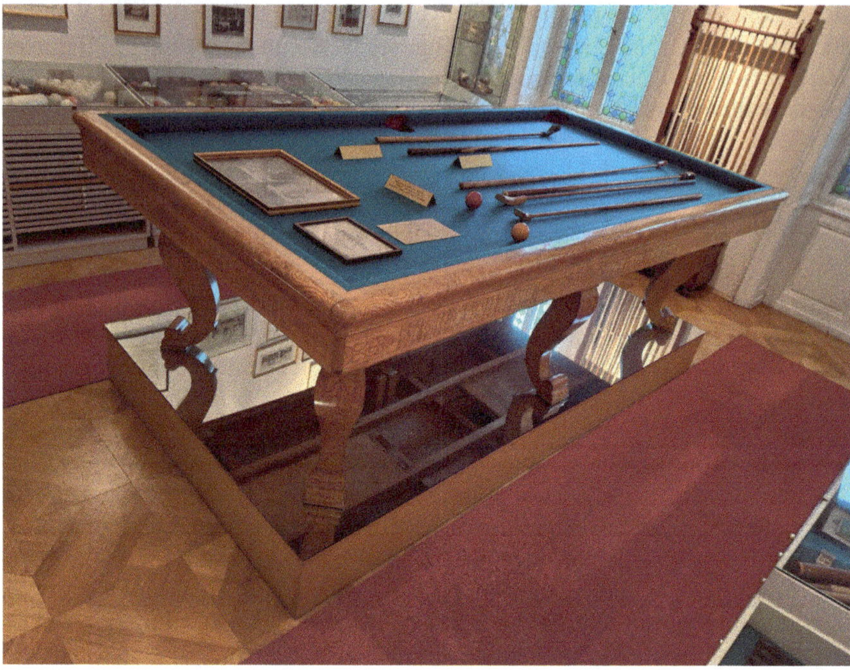

**Fig. 9.10** Antique billiard table, Billardmuseum Weingartner, Vienna, Austria. Photograph: Regina Hofmann-de Keijzer

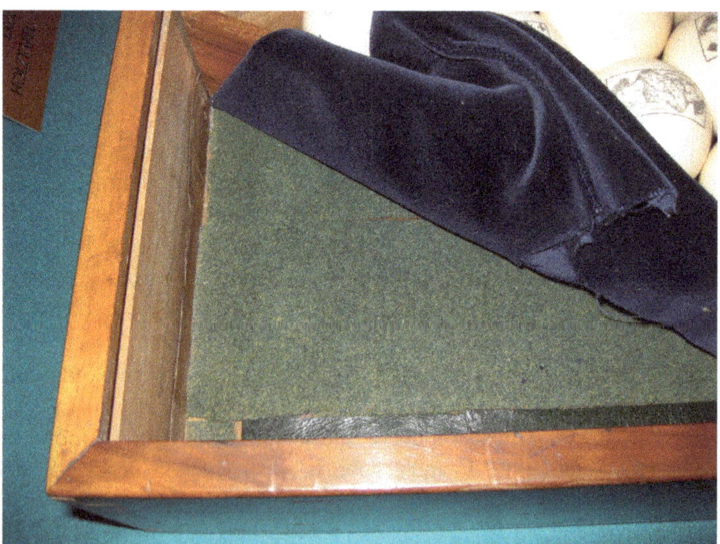

**Fig. 9.11** Remains of an old green billiard cloth (sample B8). Billardmuseum Weingartner, Vienna, Austria. Photograph: Regina Hofmann-de Keijzer

**Table 9.3** Dye analysis of old green billiard cloths

| Sample no. | Billiard cloth | HPLC results | Dating (museum) / Dating (based on dye analysis) |
|---|---|---|---|
| B 8 | Remains of an old green billiard cloth from a box with a set of ivory billiard balls with tarot figures | Indigo carmine (1743)<br>Old fustic (natural dye) | 1770–1810<br>After 1743 |
| B 7 | Remnants of a green billiard cloth from a cushion stuffed with horsehair | Indigo carmine (1743)<br>Old fustic (natural dye) | Around 1860<br>After 1743 |
| B 2 | Remnant of a green billiard cloth from a cushion | Indigo carmine (1743)<br>Old fustic (natural dye) | Around 1890<br>After 1743 |
| B 1 | Green sample from a roll of billiard cloth | Quinoline yellow (1882)<br>Patent blue equivalent | Around 1900<br>After 1882 |
| B 3 | Remnants of a green billiard cloth from Verviers in Belgium on a board from Oosterhout in the Netherlands | Patent blue V (1888)<br>Unknown yellow | Around 1900<br>After 1888 |
| B 4b | Loose fibres from remnants of a green billiard cloth from a short cushion | Quinoline yellow (1882)<br>Patent blue V (1888) | Around 1870<br>After 1888 |
| B 4a | Loose fibres from remnants of a green billiard cloth from a short cushion | Flavazine L equivalent<br>Unknown blue | Around 1870<br>Unknown |

**Table 9.4** Identification of Indigo carmine in dated textile objects. Some case studies in this table will be described elsewhere in this publication

| Textile object | Dating | Museum |
|---|---|---|
| Habit à la française (tail coat) | 1780–1790 | LIECHTENSTEIN. The Princely Collections, Vaduz–Vienna, Austria |
| Silk-satin curtains and hangings of the canal-house of Willem Philip kops in Haarlem | 1790–1794 | Rijksmuseum, Amsterdam, the Netherlands |
| Four samplers | 1842–1849 | Zuiderzeemuseum, Enkhuizen, the Netherlands |
| Sampler by Barbara Greil | 1849 | Krahuletz museum, Eggenburg, Austria |
| Folding screens | 1847 | LIECHTENSTEIN. The Princely Collections, Vaduz–Vienna, Austria |
| Floral silk upholstery on a set of six armchairs (*divan de milieu*) made by the cabinet-makers Horrix | 1852 | Rijksmuseum, Amsterdam, the Netherlands |
| Spinster party dress, ball gown of Princess Elisabeth in Bavaria | 1854 | Kunsthistorisches museum, court wardrobe ("Monturdepot"), Vienna, Austria |
| Carpet | 1865–1866 | Museum Willet-Holthuysen, Amsterdam, the Netherlands |
| Embroidery *two Breton women* by Émile Bernard | 1892 | Van Gogh museum, Amsterdam, the Netherlands |
| Embroidery *Herodias* by Émile Bernard | 1903 | Van Gogh museum, Amsterdam, the Netherlands |
| Embroidery *plants and flowers on Orange fond* by Émile Bernard | 1927 | Van Gogh museum, Amsterdam, the Netherlands |

## *The Kops Chamber in the Rijksmuseum Amsterdam, the Netherlands*

In 1790 the merchant and art-collector Willem Philip Kops (1755–1805) asked the architect Abraham van der Hart (1747–1820) to rebuild his canal-house in Haarlem and to redecorate the interior (see Fig. 9.12). The interior has been preserved: the wooden panelling, the pale green-blue silken satin curtains and hangings, the carpet and the furniture (see Fig. 9.13). It is one of the best preserved rooms between the Neo-Classicism and Empire periods in the Netherlands. The silk showed the characteristic light greenish-blue colour of Indigo carmine and HPLC analysis confirmed this assumption. In addition, ellagic acid was identified, which indicates the use of tannins, presumably as a weighting agent.

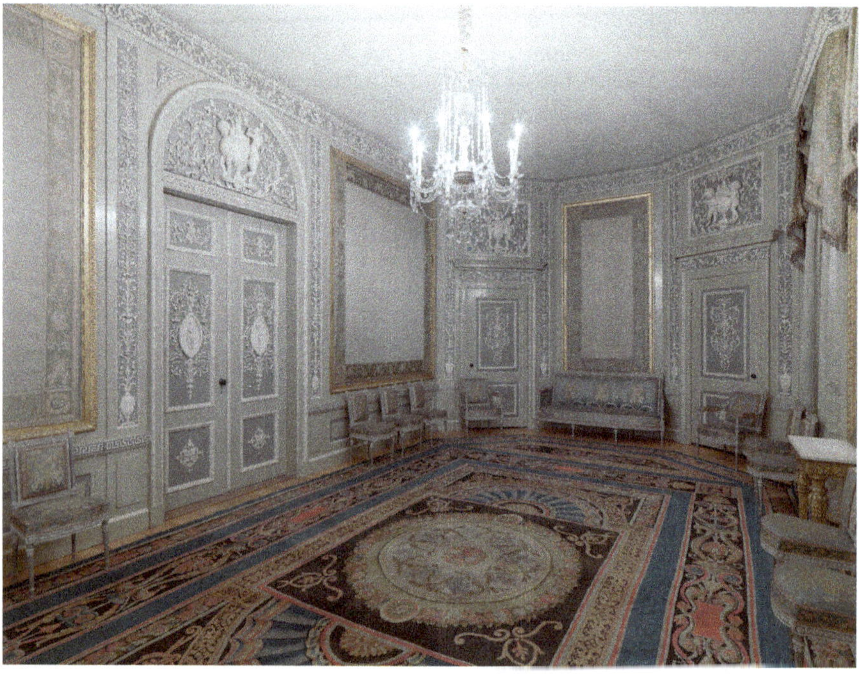

**Fig. 9.12** Interior Kops chamber, reinstalled 2003, around 1790, BK-15613, Collection Rijksmuseum, Amsterdam, the Netherlands. Photograph: Rik Klein Gotink

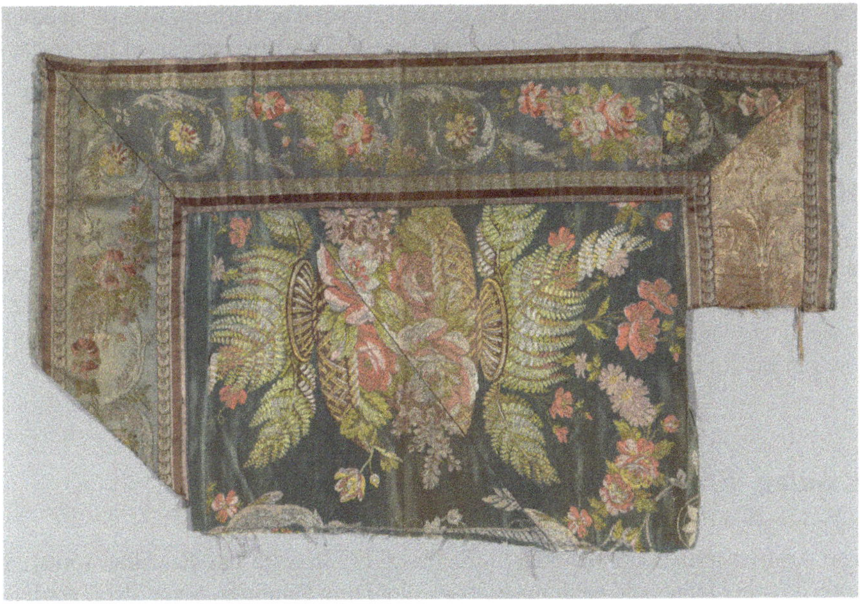

**Fig. 9.13** Fragment of one of the curtains used in the Kops chamber, 1984, BK-15613-18, Collection Rijksmuseum, Amsterdam, the Netherlands. Photograph: Rik Klein Gotink

## *A Set of Six Armchairs, divan de milieu, Rijksmuseum Amsterdam, the Netherlands*

A divan de milieu, a set of six armchairs, was made by the Dutch cabinet-makers Horrix in The Hague, around 1852. The upholstery is a silken structure with yellow flowers on a blue ground. The upholstery showed severe water damage in the form of dark rings of grime. To remove or to reduce these rings one had to wash them, but washing experiments with different detergents showed that the light- and dark-blue colours were immediately lost. Dye analysis showed that the upholstery was dyed with the natural yellow dye Weld (*Reseda luteola* L.) and Indigo carmine.

## *Morocco Leather Wing-Armchairs*

The princess wagon compartment of the Portuguese Royal Train (Museu Nacional Ferroviáta, Entroncamento, Portugal), which belonged to King Luis I of Portugal (1838–1889; reign 1861–1889) contains wing-armchairs and wall panels upholstered with a dark-green Morocco leather. Morocco leather is goatskin tanned with oak galls or sumac leaves, which has been widely used for upholstery, bookbinding, footwear and other decorative artefacts. Commonly, this type of leather presents the grain layer, the outer layer of the skin, dyed here in a dark-green colour. Descriptions from the nineteenth century for the production of dark-green leather include dyeing with a mixture of the natural dyes Fustic and Logwood or Barberry root (or another yellow dye) and Indigo carmine. The dark-green colour here consists of Indigo carmine and a yellow dye [22].

## *Modern Applications of Indigo Carmine*

Nowadays Indigo carmine is used as a colourant for food (E132), cosmetics and the aluminium salt is used for colouring pharmaceuticals. As an example, in the pharmaceutical Viagra the aluminium salt of Indigo carmine is used for the blue colour, one can say for a sort of a restoration.

## *Painting 'Erection Day' (2023) by Monika Kus-Picco*

The Austrian artist Monika Kus-Picco (born 1973), who has Brazilian roots, is known for her large- and medium-format Color Drug Paintings. Since 2018, she has been working with expired pharmaceutical products and liquids instead of industrially produced paints. The pigments and colours she produces from medicines, including Viagra (Sildenafil), are often associated with people and their stories, with death and suffering or healing. Working on the floor, she grinds and pulverises pills, opens capsules and mixes them with medical liquids or disinfectants and water before she applies them directly as a colour pigment onto a white primed canvas. During the drying process, colour changes occur between different substances [23]. The idea behind the painting 'Erection Day' (2023) is to draw attention to the American Election Day and the voting system there (see Fig. 9.14).

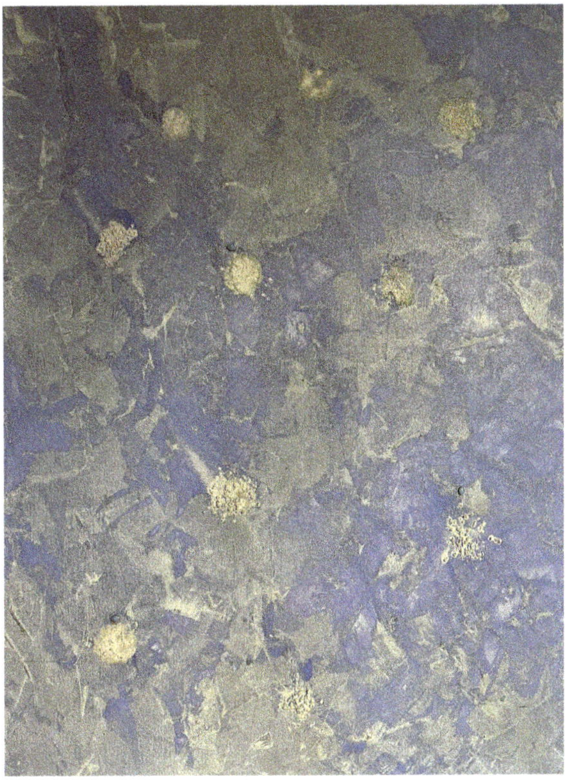

**Fig 9.14** 'Erection Day', 2023, Sildenafil on Canvas, 140 × 100 cm. © Monika Kus-Picco

## *Beschuit Met Muisjes*

In the Netherlands, it is a custom at the birth of a baby to eat Beschuit met Muisjes ('Mice on top of rusk'). Muisjes, meaning 'little mice' in Dutch, are made of anise seed—aniseed—sprinkles with a sugared and coloured outer layer. They are so named because the aniseed sprinkles are shaped like little mice, with the stem of the seed resembling a tail. The anise in the Muisjes was thought to stimulate lactation and they symbolised fertility. As early as the seventeenth century, the parents of a new-born gave Beschuit with a layer of butter and Muisjes to the baby's visitors at their home. This tradition continues today and Beschuit met Muisjes are brought by the older siblings to share at school. Every supermarket in the Netherlands sells boxes of Muisjes and the food manufacturer De Ruijter is currently the only company producing them: they have been making them since 1860. Before the twentieth century, visitors to a newborn baby got only white Muisjes; this then changed to a mixture of pink and white Muisjes for both boys and girls. With the introduction of blue Muisjes, pink Muisjes are now reserved for baby girls and blue for baby boys. The pink Muisjes are coloured with the natural dye carminic acid (E127) gained from the dye insect Cochineal (*Dactylopius coccus* Costa, 1829). The blue colour consists of Indigo carmine.

## References


1. Hartl, A., Proaño Gaibor, A.N., van Bommel, M.R., Hofmann-de Keijzer, R.: Searching for blue: Experiments with woad fermentation vats and an explanation of the colours through dye analysis. J Archaeol Sci Rep. 2, 9–39 (2015). https://doi.org/10.1016/j.jasrep.2014.12.001
2. Hanspach, D., Porada, H.T.: Grossenhainer Pflege: eine landeskundliche Bestandsaufnahme im Raum Grossenhain und Radeburg. (2008)
3. Farrar, W.V.: Synthetic dyes before 1860. Endeavour. 33, 149–155 (1974)
4. Lentherey Acta, Weigerung Christiane Erdmuthe Müllers aus Großenhain, ihren Garten an Christian Barthen zur Anlegung einer Schönfärberei abzutreten. Sächsisches Hauptstaatsarchiv, Dresden (1744)
5. Kortum, G.M.: Neue Versuche der Färbekunst: betreffend die, bisher unter dem Namen Sans pareille de Saxe bekannten blauen und grünen Farben, 10.25673/94787, (1749)
6. Spence, G., Dolby, C., Weguelin, J.C.: Dyeing green and blue Saxon colours, (1748)
7. Lowengard, S.: Techniques and Innovations - Saxon Blue and Saxon Green. In: The Creation of Color in Eighteenth-Century Europe. Columbia University Press, New York (2006)
8. Grüne, W., Grothe, H. eds: 2. Indigoblau übersetzt mit Anilinblau. Muster-Zeitung, Zeitschrift für Färberei, Druckerei, Bleicherei, Appretur von Gespinsten, Geweben, Papieren etc. und für die gesamte Farbenanwendung. 34 (1871)
9. Knecht, E., Rawson, C., Loewenthal, R.: A Manual of Dyeing: for the Use of practical Dyers, Manufacturers, Students, and all interested in the Art of Dyeing, Vol. I. Charles Griffin and Company Ltd., London (1893)
10. Heymann, B.: Ueber eine Synthese von Indigodisulfosäure (Indigocarmin). Berichte der deutschen chemischen Gesellschaft zu Berlin. Ferd. Dümmler´s Verlagsbuchhandlung, Berlin. 24, 1476–1478 (1891). https://doi.org/10.1002/cber.189102401254
11. Knietsch, R.: Zur Geschichte und Chemie der Synthese von Indigosulfosäuren (Indigocarmin). Berichte der deutschen chemischen Gesellschaft zu Berlin. Ferd. Dümmler´s Verlagsbuchhandlung, Berlin. 24, 2086–2089 (1891). https://doi.org/10.1002/cber.189102401348
12. Friedlander, P.: Fortschritte der Theerfarbenfabrikation und verwandter Industriezweige. Dritter Theil, 1890-1894. Verlag von Julius Springer, Berlin (1896)
13. Friedlander, P.: Fortschritte der Theerfarbenfabrikation und verwandter Industriezweige. Siebenter Theil, 1902-1904. Verlag von Julius Springer, Berlin (1905)
14. de Keijzer, M., van Bommel, M.R., Hofmann-de Keijzer, R., Knaller, R., Oberhumer, E.: Indigo carmine: Understanding a problematic blue dye. In: The International Institute for Conservation of Historic and Artistic Works, Contributions to the Vienna Congress. pp. 89–90. Vienna (2012)
15. Bemiss, E.: The Dyer's Companion. New York (1806)
16. Bemiss, E.: The Dyer's Companion. Dover publications, INC. New York (1973)
17. Zötl, J.: Handwritten recipe book. Archive Färbermuseum Gutau, Freystadt (1827)
18. Hofenk de Graaf, J.H.: The Colourful Past, Origins, Chemistry and Identification of Natural Dyestuffs. Abegg Stiftung / Archetype Publications Ltd., London (2004)
19. Kurzel-Runtscheiner, M.: Sisis Polterabendkleid. Ein prominentes Kulturdenkmal und seine fragwürdige Vermarktung. Restauro. Forum für Restauratoren, Konservatoren und Denkmalpfleger. 2, 130–134 (2005)
20. Zangerl, N.: Ein Herrenrock des 18. Jahrhunderts aus der Fürstlichen Sammlung Liechtenstein. Konservierung und Restaurierung, Student project. University of Applied Arts Vienna, Institute of Conservation, (2009)
21. Rageth, J.: Anatolian Kilims & Radiocarbon Dating: A New Approach to Dating Anatolian Kilims. Jürg Rageth und Freunde des Orientteppichs, Basel (1999)
22. Falcão, L., Eduarda, M., Araújo, M., Vaz, P.D.: Study of nineteenth Century Green Morocco Leather Upholstery. In: Dyes in History & Archaeology. pp. 60–61. Abstract book, Posters, Lisbon (2010)
23. Monika Kus-Pico, https://kus-picco.com/

# Chapter 10
# The Nitro Dyes

## 10.1 Introduction

The nitro dyes belong to the first known group of synthetic organic dyes and were of limited use. The shade range is restricted to brown, yellow, orange and green hues. They are characterised by the presence of nitro ($NO_2$) groups. Picric acid was originally obtained by treating natural Indigo with nitric acid, resulting in a new dye class. After the invention of this semi-synthetic dye, fully synthetic routes were discovered using products originating from the coal tar industry.

In general, nitro dyes are prepared by the action of nitric acid on phenol, as in the case of Picric acid, and naphthols, which are used to prepare Martius yellow and Naphthol yellow S. Despite the fact that the nitro dye class did not find a very wide application, Picric acid, Martius yellow and Naphthol yellow S were often used. The yellow nitro dyes absorb directly to animal fibres, such as silk and wool, and also to plant fibres although in that case the light and wash fastness properties are less. While Picric acid was of great importance as a dye, by the end of the nineteenth century its use largely shifted to the manufacture of explosives, with names such as Melinite, Lyddite and Schimose. The chemical structure of Picric acid shows similarity with trinitrotoluene, also known as TNT. The second dye of this class is Martius yellow, applied in textile dyeing and for colouring soap. The last important dye is Naphthol yellow S, which was extensively used for dyeing animal textile fibres, since this dye is faster than the other nitro dyes. The nitro dyes were at one time very largely used by dyers for the purpose of dyeing wool and silk. Their fugitive character led in recent times to their replacement by the yellow azo dyes.



M. R. van Bommel, M. de Keijzer, *Bright Colours from the Past*, Cultural Heritage Science, https://doi.org/10.1007/978-3-031-88664-5_10

## 10.2 Picric Acid

Colour Index name: Acid dye; Colour Index number: 10305
IUPAC name: 2,4,6-Trinitrophenol (TNP)
Discoverer: Peter Woulfe, 1771
Contemporary terminology: Picric acid (En), Acide picrique (Fr), Pikrinsäure (Ge), Pikrinezuur (NL)
Obsolete terminology: Welter's bitter, Indigo bitter (En), Trinitrophenol (Ge). Explosives containing Picric acid are called Melinite, Lyddite, Schimose, Ecrasite and Dunnite
Chemical formula: $C_6H_3N_3O_7$ (Fig. 10.1)

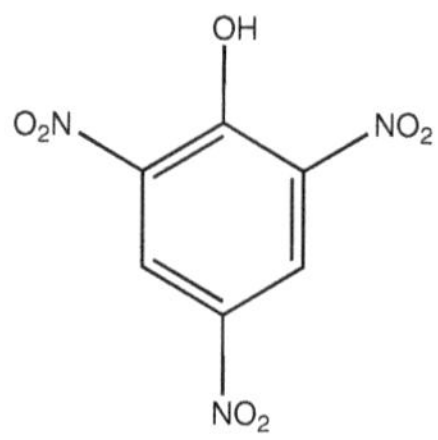


**Fig. 10.1** Structural formula of Picric acid

### *History*

Picric acid was actually the first real synthetic dye produced and was probably already mentioned in the alchemical writings of the German-Dutch alchemist and chemist Johann Rudolph Glauber (1604–1670). Around 1650, he applied concentrated nitric acid to sheep's wool. The result was a strongly yellow-coloured acidic solution and he added potash for neutralisation. As a result, potassium picrate was formed. An alcoholic potassium picrate solution ('*Tinctura nitri Glauberi*') was used as a pharmaceutical well into the eighteenth century. In 1771, Picric acid was obtained by the Irish mineralogist and chemist (alchemist) Peter Woulfe (1727–1803) from the reaction between natural Indigo (*Indigofera tinctoria* L.) and concentrated nitric acid which was published in the *Philosophical Transactions of the Royal Society*, hence he is known as the discoverer [1]. In 1799, the French chemist Jean-Joseph Welter (1763–1852), the owner of a chemical factory in Valenciennes, produced Picric acid by treating silk with nitric acid, followed by neutralisation with potash. Welter also discovered that the potassium picrate formed could explode. He named his product Welter's bitter because of its taste, which was the old-fashioned name for Picric acid [2]. The name Picric acid (from the Greek word *pikros*, 'bitter') was given by the French chemist Jean Baptiste André Dumas (1800–1884) in 1841, because of the extremely bitter taste of its yellow aqueous solution [3]. In the same year, 1841, the French chemist Auguste Laurent (1807–1853) had designed a method of isolating crude phenol from coal tar and obtained a more practical preparation of Picric acid [4, 5]. He dissolved phenol in sulfuric acid and the resultant mixture of phenol sulfonic acids was added to nitric acid. After cooling and adding water, Picric acid crystallised and precipitated [6]. In 1842, he clarified the correct chemical constitution of phenol.

## *Synthesis of Picric Acid*

Picric acid is prepared from phenol by reacting phenol with concentrated sulfuric acid and the resulting phenol-2,4-disulfonic acid is then treated with concentrated nitric acid (see Fig. 10.2). During this reaction nitro groups are introduced and the sulfonic acid groups are displaced, resulting in 2,4,6-trinitrophenol, also known as Picric acid. Another synthesis route is the direct nitration of 2,4-dinitrophenol with nitric acid.

Picric acid was made on a large scale from July 1845 onwards by the French chemist Nicolas-Philibert Guinon (1807–1885), of the firm of Guinon in Lyon, for dyeing silk. He made it by nitration of a crude phenol fraction from coal tar, using just nitric acid; it was several years before the more efficient method of sulfonation followed by nitration was employed. Certainly, the only samples of dyeing with 'Picrique acid' at the Great Exhibition of 1851 at the Crystal Palace in Hyde Park, London, were from Guinon's firm. It is clear from Guinon's account that his phenol was so impure that he had much trouble in isolating Picric acid from resinous by-products [7].

An alternative raw material was a cheap phenolic resin from the Australian plant *Xanthorrhoea hastilis* Pers. (nowadays *Xanthorrhoea resinosa* Pers.), originally suggested by the Scottish chemist John Stenhouse (1809–1880) but it is not clear to what extent this was used in practice. It seems that the British chemist John Dale (1815–1889), the co-founder of Roberts, Dale & Company in Manchester, had commenced the manufacture of Picric acid somewhat before 1860, first from this Australian plant and later more advantageously from phenol, which was commercially available.

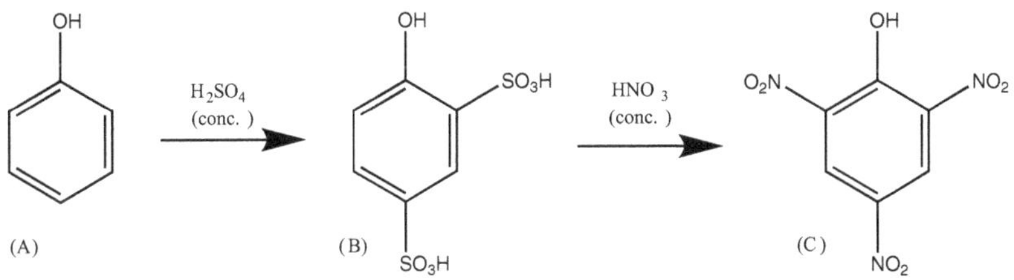


**Fig. 10.2** Synthesis of Picric acid with phenol as starting material. (**A**) = phenol, (**B**) = phenol-2,4-disulfonic acid, (**C**) = Picric acid

The first known tar distiller was the German chemist Ernst Sell (1808–1854) from Offenbach am Main, who founded his company the Sellsche Asphaltfabrik in 1842 and produced phenol in 1846. The phenol he produced was sufficiently pure to crystallise under cold conditions, that is, in cold weather. The technique of tar distillation was initially quite simple: tar was distilled several times and the individual fractions were collected using differences in boiling point. The quantities obtained were still very limited; Sell's largest distillation boiler had a volume of only about 400 L [8]. The distance from Lyon to Offenbach am Main, and the fact that Guinon started operations in 1845, makes it likely that there was an earlier French supplier of phenol.

Probably the first German manufacturer of synthetic dyes was the chemist Karl Gottlieb Reinhard Oehler's (1797–1874) firm in Offenbach am Main. Oehler had purchased the coal tar distillery from Sell in 1850, who sold it for health reasons, and 1 year later he produced Picric acid derived from coal tar products [9]. Picric acid was also manufactured by another firm, located in Lyon. The French chemist François-Emmanuel Verguin (1814–1864) entered the service of the firm of Louis Raffard et Cie about 1855 in order to establish a Picric acid plant at Givray near Péage de Roussillon.

Guinon had noticed that Picric acid would also dye wool, a fact of greater interest in the North of England and of potential economic importance. This was recognised by the English chemist Frederick Crace Calvert (1819–1873), who lived in France from around 1836 until 1846. When he returned to England, he settled in Manchester as a consulting chemist and was appointed honorary professor of chemistry at the Royal Manchester Institution. He made phenol from 1847 onwards, using fractions from the tar distillery of Samuel Clift, and Picric acid in 1849, probably in association with the firm Roberts, Dale & Company. However, it seems that Calvert's production was not on a large scale until he started his own firm in 1859. In 1859, the French silk dyeing industrialist Jean-Aimé Marnas (1828–1908) of the Lyonnaise firm of Guinon, Marnas & Bonnet visited Manchester and asked Crace Calvert to supply the firm with a purer variety of phenol than had hitherto been available. A decade later, in 1869, the German chemists Rudolf Schmitt (1830–1898) and L. Glutz obtained Picric acid by heating phenol sulfonic acid with concentrated nitric acid [10].

## *Trade Names*

In the literature many patents related to Picric acid are dated after the late 1880s (see Table 10.1). The yellow dye was put on the market as Pikrinsäure by C. vom Bauer in Elberfeld, Durand-Huguenin in Basel and de Haën in List near Hanover, and as Picric acid by Robert Graeser near Ruabon in North Wales, Read Holliday in Huddersfield and Charles Lowe in Manchester.

**Table 10.1** Discoverers and patents of Picric acid

| Name<br>C.I. name<br>C.I. number | Discoverers | Year of discovery | Patents |
|---|---|---|---|
| Picric acid<br>Acid dye<br>10305 | P. Woulfe<br>J. J. Welter<br>A. Laurent<br>A. Laurent<br>R. Schmitt and L. Glutz | 1771<br>1799<br>1841<br>1855<br>1869 | –<br>–<br>–<br>–<br>– |
| | | | G.P. 51.321/1889<br>G.P. 51.603/1889<br>G.P. 67.074/1891<br>B.P. 16.371/1900<br>USP. 666.627/1901<br>F.P. 303.683<br>G.P. 125.096/1900<br>B.P. 16.628/1900<br>F.P. 304.224<br>G.P. 126.197/1900<br>B.P. 17.541/1906<br>USP. 923.761/1906<br>F.P. 380.121<br>G.P. 194.883/1906<br>G.P. 214.045/1907<br>B.P. 16.607/1915<br>B.P. 126.062, B.P. 126.082<br>B.P. 126.675, B.P. 126.676<br>B.P. 128.303, B.P. 131.403 |

## *Applications*

Although Picric acid has poor light fastness and wash fastness properties, it was used often in north-Western Europe. Picric acid was used to dye wool, silk and leather. It is not used to dye cotton. Picric acid gives a unique, bright canary-yellow colour, especially to silk. Due to its greenish tone, Picric acid is mostly used for enhancing other colours or for producing mixed colours. It is combined with Methyl green for greenish-yellow, with Indigo carmine or Aniline blue for green, with Methyl violet for olive green and with Fuchsine for red. According to Knecht, Picric acid was still in use in the 1930s for dyeing silk in greenish-yellow shades [6]. The standard method to create a green colour on textiles at that time was the use of the natural dyes Weld (*Reseda luteola* L.) and natural Indigo (*Indigofera* sp.), which was a time-consuming activity; a mixture of Picric acid with natural Indigo or Indigo carmine was therefore a welcome alternative.

Picric acid was also used in conjunction with the pigment Prussian blue to give a much admired green tint (*Vert lumière*) on silk, which exactly imitated the expensive Chinese green Lo-kao, an unusual and rare, and thus highly desirable, natural green dye. Lo-kao was made from *Hong-pi-lo-chou* (*Rhamnus utiles* Decne), which is similar to common buckthorn (*Rhamnus cathartica* L.), by a process of alkaline oxidation leading to a pentahydroxy-methylanthraquinone. From the early 1850s, the very costly green dye Lo-kao (533 Fr/kg or £10/lb) was imported to Europe. The Lyonnaise firm Guinon started commercial silk dyeing using Lo-kao in 1855 and the colour was marketed under the names Vert-Venus, Vert-Azof and Vert lumière [7, 11]. Since the production of Lo-kao was complicated and initially expensive, the use of a cheap alternative from the firm Guinon, Marnas & Bonnet in Lyon to dye silk, first with Prussian blue and then with Picric acid was welcomed. These dyeings were particularly admired in gas or candlelight. While the greens produced with natural Indigo and Picric acid appeared blue, those produced with Prussian blue and Picric acid remained green. The scale of production was quite large. In 1867, Crace Calvert estimated a total production volume of 80–100,000 lb, his own firm making about 16,000 lb; each pound would dye 70–100 lb of silk or 40–50 lb of wool. From about £1/lb the price had dropped to 3 s/lb. Perkin's Mauveine marked the beginning of the end of the natural dyes, however the first natural dye to be replaced was not Indigo, Madder or Cochineal, but Lo-kao.

## Properties

Picric acid has a strong dyeing power; it dyes a quantity of silk equivalent to a thousand times its own weight an intense yellow. Lehne wrote on the light fastness of Picric acid:[1] 'After 6 days of light exposure in May, the colour has drastically changed and turned into a dull orange'. In an acidic dye bath, Picric acid produces a beautiful greenish yellow on wool and silk and despite its bad fastness properties—it even sublimed out of the fibre in summer temperatures—it was widely used, especially for silk dyeing. To produce better properties on wool, alum or potassium dichromate are sometimes used as mordants. Picric acid is rarely used for plant fibres and then only in conjunction with mordants, such as alum. The yellow dye is also not fast to washing: it bleeds when washed with water. When used as a mordant dye, for example with alum and cream of tartar on wool or silk, the wash fastness improves. However, since the wash fastness and light fastness were very poor, it lost its importance as a dye after the discovery of the nitro dye Naphthol yellow S. No properties on Picric acid are given in the Colour index.

[1] Pikrinsäure: Echtheit: 'Durch kurze, 6 tägige Einwirkung der Witterung (Mai) wurde die Wollfärbung sehr stark verändert, in stumpfes Orange überfuhrt'.

## Recipes

Typical dye recipes for wool suggest using 1–5% Picric acid solution containing 2–3% sulfuric acid or 6–7.5% alum: add the wool cold, heat to the boil in 15 min, dye for 15 min at the boil and dry without rinsing. For silk, less acid and a lower temperature of 50–60 °C were recommended [12]. Lehne proposed recipe V3 for dyeing wool and V8 for dyeing silk. The concentration of Picric acid required is 5% of the weight of wool to be dyed. He mentioned that it is often used in combination with Indigo carmine.

## Identification Using UHPLC-PDA

Analysis of Picric acid (RCE no. 1657) using UHPLC-PDA results in a clear, single peak, as presented in Fig. 10.3.

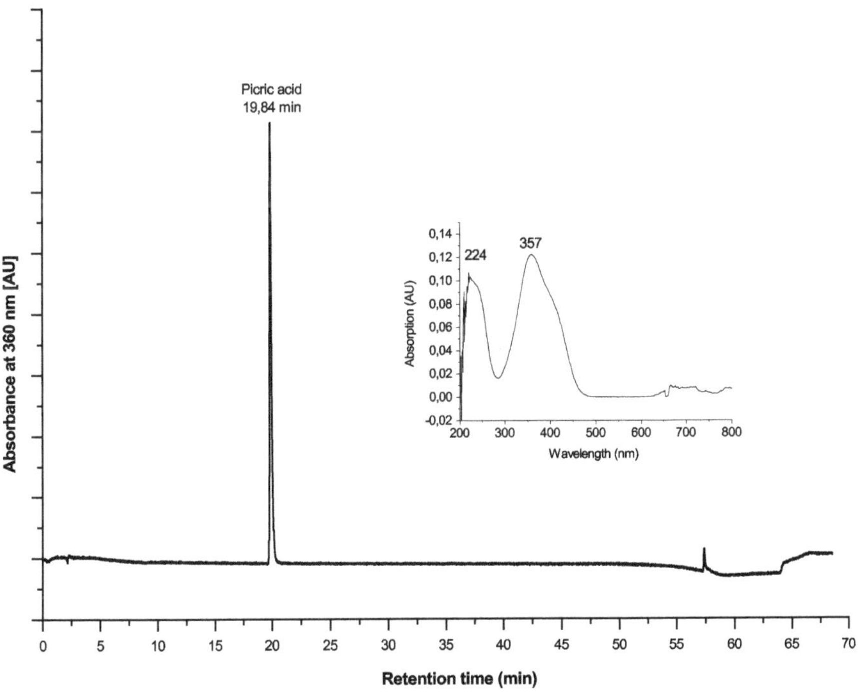


**Fig. 10.3** Chromatogram and spectrum of Picric acid (RCE no. 1657)

## *Other Applications*

By 1867, considerable quantities of Picric acid were produced, 80–100,000 lb in England and France, for the dye industry; the use in explosives had not yet started. Before that time chemists assumed that only the salts of Picric acid were explosive, not the acid itself. The German-British chemist Hermann Johann Philip Sprengel (1834–1906) proved the opposite and discovered the highly explosive nature of Picric acid in 1871. He was born at Schillerslage near Hanover, earned his degree at the University of Heidelberg in 1858 and moved one year later to England, where he became a naturalised Englishman. In March 1871, he detonated Picric acid at the gunpowder works of John Hall & Sons in Faversham in Kent, England. In the same year Sprengel filed two patents (B.P. 921 and B.P. 2.642, both filed in 1871) in Britain for 'safety explosives' (i.e. stable explosives). In the latter patent he proposed using Picric acid dissolved in nitric acid as an explosive [13]. In 1885, based on the research of Sprengel, the French chemist François Eugène Turpin (1848–1927) patented the use of Picric acid in blasting charges and artillery shells. In 1887, the French government adopted it under the name Melinite, a mixture of Picric acid with the addition of gun cotton. Lyddite is an explosive developed in 1888 in the United Kingdom, obtained using Picric acid with the addition of Vaseline. The name Lyddite was adopted from the town Lydd, in the district of Kent, England. Japan followed with an improved formula known as Schimose. In 1889, a similar material, a mixture of ammonium cresylate with trinitrocresol (2,4,6-trinitro-*m*-cresol), or an ammonium salt of trinitrocresol, started to be manufactured under the name Ecrasite in Austria-Hungary. By 1894, Russia was manufacturing artillery shells filled with Picric acid. Ammonium picrate, known as Dunnite or Explosive D, the ammonium salt of Picric acid, was used by the United States of America beginning in 1906. As an explosive, Picric acid was used in the Battle of Omdurman (1898), the Second Boer War (1899–1902), the Russo-Japanese War (1904–1905) and the First World War (1914–1918). In the twentieth century, Picric acid was largely replaced by trinitrotoluene (TNT).

Due to its explosive nature, Picric acid is known to have played a role in several tragic explosions. On the 22nd of June 1887, an explosion wrecked the Roberts, Dale & Company Chemicals Works in Manchester. One life was lost. The works produced, among other chemicals, Picric acid, lead nitrate, nitric acid, hydrochloric acid, Aurin, Manchester brown, Manchester yellow, lakes for staining paper and Emerald green; lead nitrate and its precursor litharge were present in a large amount. A fire started at the stove used for drying the Picric acid. It spread quickly, and in a few minutes the first, moderate, explosion followed. After hardly a minute, a second, devastating explosion ruined the works and caused great damage in the neighbourhood. The fire was thought to have been caused by a careless smoker, and the second explosion was probably due to the molten Picric acid reacting with the lead compounds which were placed nearby. The works at Manchester soon closed [14–16].

Another huge explosion took place on the 25th of April 1901 at the Chemische Fabrik Griesheim near Frankfurt am Main in Germany. Within an environment of 5 km all windows were broken, 26 people were killed and around 100 persons were injured [17].

A maritime disaster, known as the Halifax Explosion, happened on the morning of the 6th December 1917 in Halifax, Nova Scotia, Canada. The Norwegian vessel

*SS Imo* collided with the *SS Mont-Blanc*, a French cargo ship. The *SS Mont-Blanc* was fully loaded with the explosives TNT and Picric acid, the highly flammable fuel benzole, and guncotton. Approximately 2,000 people were killed by the blast, debris, fires or collapsed buildings, and an estimated 9,000 others were injured.

## 10.3 Martius Yellow

Colour Index name: Acid Yellow 24; Colour Index number: 10315
IUPAC name: 2,4-Dinitronaphthalen-1-ol
Discoverer: Carl Alexander Martius, 1864
Contemporary terminology: Manchester yellow (En), Jaune naphthol (Fr), Martinsgelb (Ge), Manchestergeel (NL)
Obsolete terminology: 2,4-Dinitronaphthol (En), Jaune d'or (Fr), Manchestergelb, Naphtolgelb, Naphtylamingelb, Naphtalingelb, Ganahlgelb (Ge)
Chemical formula: $C_{10}H_6N_2O_5$ (Fig. 10.4)

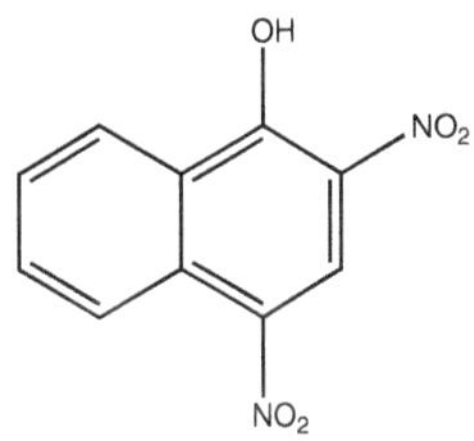


**Fig. 10.4** Structural formula of Martius yellow

### *History*

In the 1850s, a great number of German chemists went to Britain because German universities were educating more scientists than society needed and jobs for chemists were scarcer than qualified applicants. They were employed at various British dye firms. Two of them, Carl Alexander Martius (1838–1920) and Heinrich Caro (1834–1910), were employed by Roberts, Dale & Company in Manchester. In 1864, Martius obtained a yellow dye by dinitration of 1-naphthol, which was named Manchester yellow or Martius yellow [18–24]. The patent for Martius Yellow was filed under the joint names of Dale, Caro and Martius (B.P. 2.785, filed in 1864) (see Table 10.2). In 1864, the Roberts, Dale & Company of Manchester and Bayer in Barmen, Germany, already produced the yellow acid dye commercially.

Although Martius is known as the discoverer of Martius yellow, it is possible that he was not the first to synthesise this component. In the 1856 *Annalen der Chemie und Pharmacie* L. Chiozza reports that R. Ganahl experimented with naphthylamine and nitrous vapours (fumes). By continuous action of nitrous acid, a resin-like residue was formed together with a crystalline powder and dissolved in alcohol

**Table 10.2** Discoverers and patents of Martius yellow

| Name<br>C.I. name<br>C.I. number | Discoverers | Year of discovery | Patents |
|---|---|---|---|
| Martius yellow<br>Acid Yellow 24<br>10315 | R. Ganahl | 1856 | – |
| | C.A. Martius | 1864 | B.P. 2.785/1864 |
| | J. Dale, H. Caro and C.A. Martius | – | B.P. 89/1869 |

it gave a lemon-yellow colour. It is likely that this was the component later known as Martius yellow [25].

## *Trade Names*

Martius yellow was put on the market by AGFA and Beyer & Kegel. Other trade products were put on the market, such as: Naphtolgelb and Jaune Naphthol by Sévoz-Boasson; Naphtylamingelb by Kalle and Bayer; Jaune d'or by Durand-Huguenin, Manufacture Lyonnaise and Sandoz; Naphtalingelb by Cassella, Leonhardt, Weiler-ter Meer, Levinstein and Wülfing-Dahl; and Manchestergelb by Read Holliday, Levinstein and Roberts, Dale & Co.

## *Applications*

The dye was used for dyeing silk and wool, and became a competitor to Picric acid. In a dye bath acidified with acetic acid, Martius yellow produced a beautiful golden-yellow colour, unlike Picric acid, which provides a more greenish-yellow tone. The yellow dye is often used on its own or to produce mixed colours. One kilogram of Martius yellow dyes 200 kg of wool [18]. Martius yellow cannot be used for cotton. Martius yellow does not have the bitter taste associated with the other nitro dyes and therefore in the past had a special application for colouring soap and food, especially noodles and macaroni.

## *Properties*

Martius yellow in its dry state has the appearance of an orange-yellow powder. In his book Lehne wrote:[2] 'After 14 days of light exposure in July, the colour has almost completely faded'. From the Colour Index, no fastness properties are known, but as mentioned, the light fastness according to Lehne is poor.

[2] Martiusgelb: Echtheit: 'Nach 14 tägiger Einwirkung der Witterung (Juli) war die Färbung nahezu vollständig verblasst'.

## *Recipes*

The recipe described by Lehne is similar to that for Picric acid; however, the concentration of dye required was lower, typically 0.75–1%. It is noted that the silk is more vividly and evenly coloured after dyeing using acetic acid or tartaric acid, but excess of sulfuric acid should be avoided.

## *Identification Using UHPLC-PDA*

Analysis of a reference sample of Martius yellow (Aldrich CAS 605-69-6) using UHPLC-PDA results in a clear, single peak as presented in Fig. 10.5.

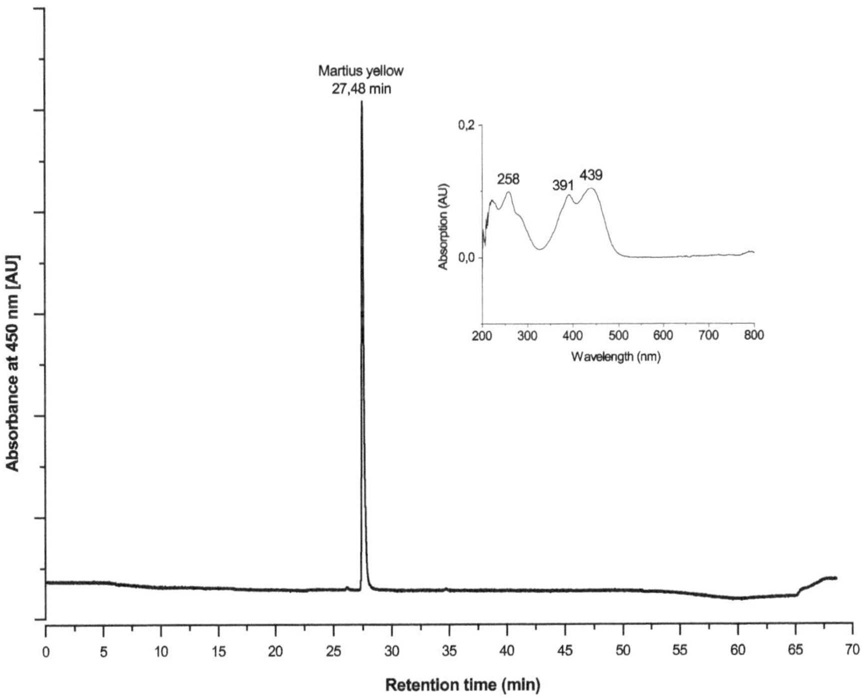


**Fig. 10.5** Chromatogram and spectrum of Martius yellow (Aldrich CAS 605-69-6)

## 10.4 Naphthol Yellow S

Colour Index name: Acid Yellow 1; Colour Index number: 10316
IUPAC name: Disodium 5,7-dinitro-8-oxidonaphthalene-2-sulfonate
Discoverer: Heinrich Caro, 1879
Contemporary terminology: Naphthol yellow S (En), Jaune acide (Fr), Naphtolgelb S (Ge), Naftolgeel S (NL)
Obsolete terminology: Acid yellow S, Amacid yellow S, Sodium flaviante, Flavianic acid (En), Citronin, Säuregelb, Echtgelb, Anilingelb, Schwefelgelb, Naphtolgelb (Ge)
Chemical formula: $C_{10}H_4N_2Na_2O_8S$ (Fig. 10.6)

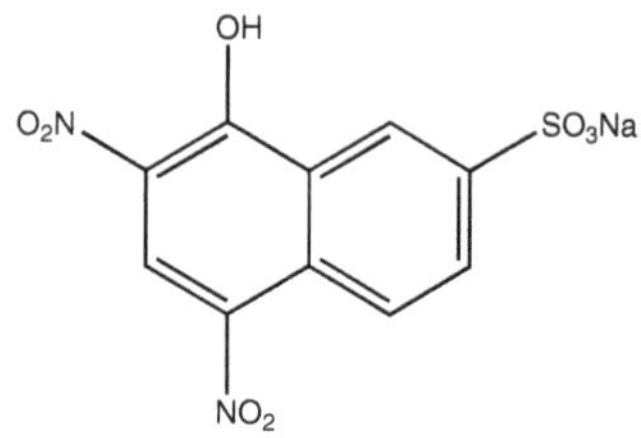


**Fig. 10.6** Structural formula of Naphthol yellow S

### *History*

The most widely known and significant useful nitro dye is Naphthol yellow S. Nearly all the German scientists involved in the dye industry returned from England to Germany after 1865. With their knowledge obtained in England, they contributed to the rise of new firms in Germany. Martius was co-founder of the predecessor of the dye company AGFA in Berlin and Caro became research director of BASF in Ludwigshafen am Rhein in 1868. In the following years, Caro was repeatedly disappointed by attempts to obtain German patents for the production of synthetic Alizarin and benzene. The reason was that a patent protection was not available in Germany until unification and formation of the German Empire in 1871 and the creation of an all-German patent system in 1877. This meant that anyone could enter the synthetic dye business and could imitate the products of existing producers, both national and international. After the introduction of a German patent system, in 1879 Caro discovered Naphthol yellow S and this was patented by BASF in Germany (G.P. 10.785 and B.P. 5.305, both filed in 1879; U.S.P. 225.108 and F.P. 134.632, both filed in 1880) (see Table 10.3). Naphthol yellow S can be obtained by the action of nitric acid on 1-naphthol 2,4,7-trisulfonic acid, when the $-SO_3H$ groups are replaced by $-NO_2$ groups [26–28]. Already in 1879 BASF produced Naphthol yellow S commercially.

Naphthol yellow S was the subject of some patent infringements in the 1880s. The German-British chemist Ivan Levinstein (1845–1916) manufactured Naphthol yellow S, without using a license agreement, the dye having been patented by BASF

**Table 10.3** Discoverer and patents of Naphthol yellow S

| Name<br>C.I. name<br>C.I. number | Discoverer | Year of discovery | Patents |
|---|---|---|---|
| Naphthol yellow S<br>Acid Yellow 1<br>10316 | H. Caro | 1879 | G.P. 10.785/1879<br>B.P. 5.305/1879<br>USP. 225.108/1880<br>F.P. 134.632/1880<br>G.P. 20.176/1882<br>B.P. 5.692/1882<br>USP. 289.543/1883<br>B.P. 11.318/1887 |

following its discovery by Caro. After his chemistry studies at Berlin University, Levinstein had moved to Britain where he founded the dye factory Levinstein Limited in Blackley near Manchester in 1865. Between 1866 and 1875 he did not file patents; in the period 1876–1883 the firm received eight British patents and afterwards Levinstein pursued the strategy of patenting as much as possible, especially azo dyes, just as the German firms did. Due to this strategy, the firm obtained 67 British patents between 1888 and 1914. Levinstein's chemists undertook improvements to the Naphthol yellow S production process which were patented in England and the United States of America towards the end of 1882 (B.P. 5.692, filed on the 30th of November 1882; U.S.P. 289.543, filed on the 4th of December 1883). Two years later, BASF considered that these patents were infringements upon their patents and made preparations for a lawsuit against Levinstein. However, both parties were aware of the risks of a lawsuit: a patent dispute does not necessarily lead to a total monopoly for one party and a total loss for the other. The uncertainty about the outcome of lengthy lawsuits fostered the willingness to compromise, to license and to form cartels and as result the number of conventions increased in the 1880s.

The German chemist Heinrich August Bernthsen (1855–1931), who worked at BASF, claimed infringement of BASF's American Naphthol yellow S patent by Levinstein during 1886–1888. In addition, Leonhardt (B.P. 11.318, filed in 1887) produced this yellow dye, which was a serious threat to BASF. To prevent there being any influence on the negotiations in the United States of America and England, in 1886 Caro asked the Russian-Swiss-German chemist Otto N. Witt (1853–1915) not to publish his works on Naphthol yellow S. In the end the tactic of avoiding this publication was successful and in 1888 BASF reached an agreement with Ivan Levinstein for mutual sharing of the British and U.S. patents. When BASF reached an agreement with Levinstein on Naphthol yellow S, Levinstein started an action against Dan Dawson Brothers of Milnsbridge, near Huddersfield. Dawson produced Naphthol yellow S and was supported by the firm Read Holliday & Sons, also located in Huddersfield. Several firms were involved in the negotiations. The outcome was to the satisfaction of BASF, namely the upholding of its patent, followed by the granting of a license to Dawson. Levinstein was also happy as it provided him with an opportunity to draw the British Naphthol yellow S manufacturers Dawson and Read Holliday, his own firm and the Manchester agent of BASF (Schott, Segner & Co.) into a cartel arrangement [9].

## Trade Names

Naphtholgelb S was put on the market by BASF, AGFA, Farbwerke ML&B, Cassella, American Aniline Co., Bulls Ferry, Bayer, Beyer & Kegel, Claus & Rée, Durand-Huguenin, CIBA, Griesheim-Elektron, Ruch, Sandoz and Leonhardt. Other trade names were also used to sell the dye, including: Naphtholgelb S extra konz. by Weiler-ter Meer; Naphtholgelb by Wülfing-Dahl; Naphtholgelb FY AS by Levinstein; Citronin A by Leonhardt; Schwefelgelb S by Kalle and Wiescher; Jaune acide by Durand-Huguenin and Picard; Jaune acide C by Manufacture Lyonnaise; and Jaune OS by Soc. Anon. St. Denis.

## Applications

The sodium salt of Naphthol yellow S is readily soluble in water, extensively used for dyeing wool and silk and is much faster to washing than the other nitro dyes. It was often used in combination with other acidic dyes for mixing colours to obtain, for example, the colour green. Naphthol yellow S gave margarine a beautiful yellow colour. It is clear that because of its toxicity this was not really a suitable additive.

## Properties

Naphthol yellow S is an orange-yellow powder and is the most widely known and useful dye of the nitro group. The colour of Naphthol yellow S is not fugitive and does not discolour after steaming. Lehne wrote in 1893:[3] 'After 5 days of weather exposure in May the colour had changed quite a bit, had become a brownish tint and after 18 days it had nearly completely faded'. The fastness to washing is good and much better than Picric acid and Martius yellow. Compared to the other nitro dyes, Naphthol yellow S possesses a greater resistance and is used for pure yellow dyeing or for mixed colours. However, the light fastness of Naphthol yellow S is very poor according to the ISO standard, ISO 1.

## Recipes

The recipe described by Lehne is similar to that for Picric acid; however, the concentration of dye required is lower, typically 0.75–1%. Silk is dyed in a dye bath acidified with sulfuric acid. Wool is dyed with a little sulfuric acid and tin chloride increases the vividness of the colour.

[3] Naphtolgelb S: Echtheit: 'Nach 5 tägiger Einwirkung der Witterung (Mai) war die Färbung ziemlich stark verändert, bräunlich geworden, nach 18 Tagen war sie nahezu vollständig verblasst'.

### *Identification Using UHPLC-PDA*

Analysis of a reference sample of Naphthol yellow S (Sigma-Aldrich CAS 846-70-8) using HPLC-PDA results in a clear, single peak as presented in Fig. 10.7.

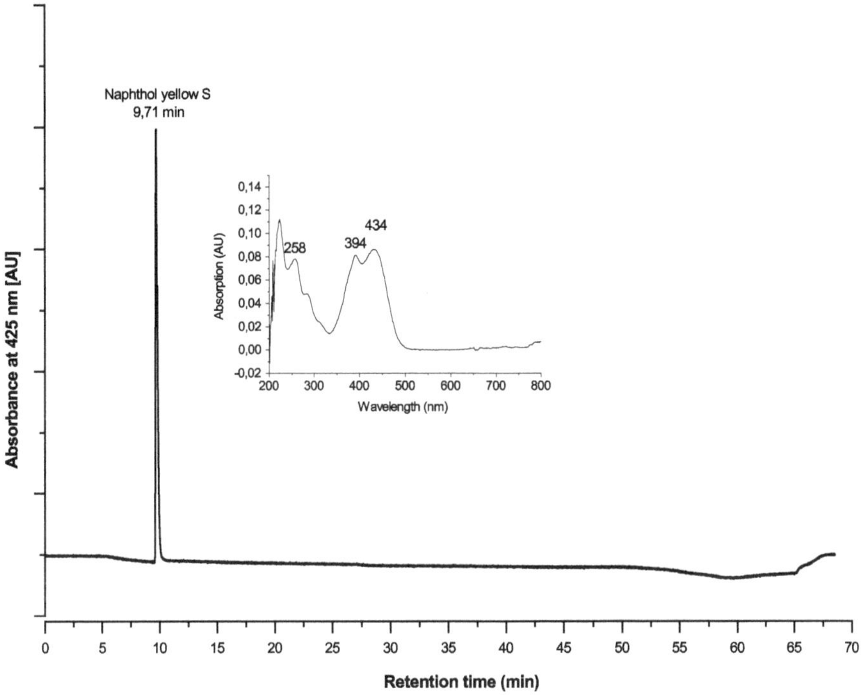


**Fig. 10.7** Chromatogram and spectrum of Naphthol yellow S (Sigma-Aldrich CAS 846-70-8)

## 10.5 Case Studies

It seemed that Picric acid and Naphthol yellow S were popular dyes for a long time. Émile Bernard used both dyes in his embroideries *Breton Women, Fire-Screen* (1892), *Plants and Flowers on a Grey Background* (1892–1904), *Hérodias* (1903) and *Flowers on an Orange Background* (1927), all dated between 1892 and 1927. These case studies, currently owned by the Van Gogh Museum, Amsterdam, are described in Chap. 13.

## *Vincent Van Gogh's Chinese Lacquer Box with Coloured Balls of Wool, 1887*

The Van Gogh Museum is the owner of a Chinese lacquer box filled with coloured balls of wool belonging to Vincent van Gogh (1853–1890), dated 1887. The red lacquered box with a calligraphic decoration was originally a Chinese tea caddy, see Fig. 10.8.

How much of an interest one artist had in colour harmonies and in the phenomena of colour contrast is demonstrated by the contents of this small box. Why did he need these balls of wool near the end of his lifetime? He used them for the development of his colour theory, which shows the relation between certain colour combinations and colour compositions in his paintings [29]. This allowed him to combine differently coloured yarns of wool to examine the perception of colour contrasts through the combined effect of the yarns, for instance yellow and violet yarns or red, yellow and green yarns. This can be seen in some of his paintings, such as *Courtesan (after Eisen)* (F0373, JH 1298), *Red Cabbages and Garlic* (F0374, JH 1338) and *Pink Peach Tree in Blossom (Reminiscence of Mauve)* (F0394, JH1379).

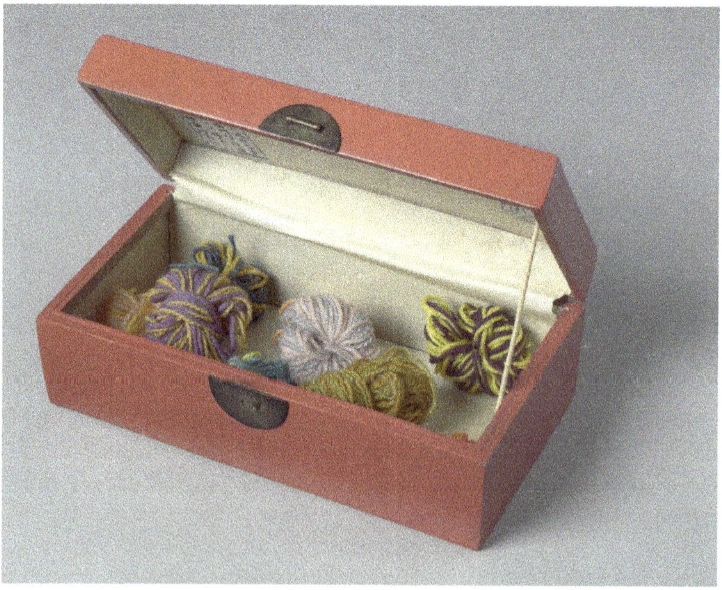

**Fig. 10.8** Replica of the red lacquer box made by Ruud Daems (1964) (10.3 cm × 29.8 cm × 15.9 cm, vgd00298) of Vincent van Gogh's Chinese lacquer box with 16 balls of wool (original), Amsterdam, Van Gogh Museum (Vincent van Gogh Foundation)

Analyses show that the woolen balls were dyed with synthetic organic dyes (see Table 10.4), including Picric acid, Martius yellow, Naphthol yellow S, Fuchsine, Methyl violet, Water blue, Orange ll, Eosin, Malachite green, Victoria blue B and Auramine. Half of the analysed synthetics belong to the di- and tri-arylmethanes. The three nitro dyes Picric acid, Martius yellow and Naphthol yellow S were identified in the balls of wool together with the di-arylmethane dye Auramine. Victoria blue B and Auramine, discovered in 1883 and 1884, were already used for dyeing wool 4 years after their discovery showing how rapidly synthetic dyes came on the market. It should be noted that some dyes remain unidentified; although their spectra resemble those of known reference materials, their retention times do not. These dyes are labelled as equivalents and could be side products formed during the synthesis of the main dye. The exception is the Picric acid equivalent found in sample 1B, where it is the main compound. This could be either an unknown dye, or a degradation product. An arguable identification is the presence of Patent blue equivalents. Patent blue was discovered in 1888, while the woolen balls are dated 1887. Probably, although the spectra in the samples resemble that of Patent blue, another blue or maybe green dye was used.

It is striking that many samples contained mixtures of dyes, probably to achieve a particular shade or hue. Several dyes were present at low concentrations; these could be either side products or cross-contamination, that is, the sample was contaminated with some fibres from another woolen yarn.

**Table 10.4** Overview of the dyes detected in the woolen yarns in the Chinese lacquer box from Vincent van Gogh

| Sample | Colour | Dyes detected | Trace or cross-contamination |
|---|---|---|---|
| 1A | Yellow | Picric acid | |
| 1B | Pink | Picric acid equivalent<br>Picric acid<br>Eosin | |
| 2 | Yellow | Picric acid | |
| 3A | Orange | Naphthol yellow S<br>Orange II | |
| 3B | Red | Ponceau G | Methyl violet (trace) |
| 4A | Orange | Naphthol yellow S<br>Orange II | |
| 4B | Light blue | Water blue<br>Naphthol yellow S | Orange II (trace) |
| 5A | Light blue | Water blue | |
| 5B | Pink | Eosin<br>Picric acid equivalent | |
| 6A | Light purple | Methyl violet<br>Patent blue equivalent<br>Picric acid equivalent | Auramine (trace) |

(continued)

**Table 10.4** (continued)

| Sample | Colour | Dyes detected | Trace or cross-contamination |
|---|---|---|---|
| 6B | Dark purple | Methyl violet<br>Fuchsine<br>Auramine | |
| 7A | Purple | Methyl violet<br>Fuchsine | Naphthol yellow S<br>Picric acid |
| 7B | Yellow | Naphthol yellow S<br>Orange II | Cochineal (natural dye) |
| 8A | Yellow | Picric acid | |
| 8B | Ochre | Naphthol yellow S<br>Picric acid | Orange II (trace)<br>Martius yellow |
| 8X | Green fibres from 8B | Naphthol yellow S<br>Picric acid<br>Diamond green equivalent | Martius yellow (trace) |
| 9 | Ochre | Naphthol yellow S<br>Picric acid | Orange II (trace)<br>Methyl violet (trace) |
| 10A | Yellow | Naphthol yellow S | Orange II (trace) |
| 10B | Dark purple | Methyl violet<br>Methyl violet equivalent<br>Auramine | |
| 11A | Yellow | Picric acid<br>Martius yellow | |
| 11B | Dark purple | Methyl violet<br>Fuchsine | Auramine (trace)<br>Martius yellow (trace) |
| 12A | Yellow | Naphthol yellow S<br>Picric acid | Martius yellow (trace)<br>Patent blue equivalent (trace) |
| 12B | Blue-green | Fuchsine<br>Methyl violet<br>Victoria blue B | Auramine (trace)<br>Martius yellow (trace) |
| 13A | Dark yellow | Naphthol yellow S<br>Picric acid<br>Martius yellow | Patent blue (trace) |
| 13B | Blue | Water blue<br>Picric acid<br>Auramine | Naphthol yellow S (trace)<br>Martius yellow (trace) |
| 14 | Green | Patent blue or Diamond green equivalent | |
| 15A | Yellow | Naphthol yellow S<br>Picric acid | Orange II (trace) |
| 15B | Yellow-green | Naphthol yellow S<br>Picric acid<br>Diamond green B<br>Diamond green B equivalents | Orange II (trace)<br>Martius yellow (trace) |
| 15X | Purple fibre from 15A | Fuchsine<br>Methyl violet | |

### *The Interior Upholstery of the Train Carriage Hannibal*

In 1832 the railway track between Budweis (České Budějovice, Bohemia, Czech Republic) and Linz (Upper Austria, Austria) was laid. The main purpose of the horse-drawn railway was to transport salt from Upper Austria to Bohemia, but passenger carriages were also added to the rolling stock at the beginning of the 1840s. Hannibal, built in 1841, is unique because it is the last surviving carriage that belonged to this railway, and it was in service until 1872 (see Fig. 10.9).

Hannibal was exhibited at the Viennese World Fair at the Prater in 1873. In 1885 the Imperial-Royal Directorate General of the Austrian State Railways issued instructions to source and collect objects for a planned museum. The first exhibits were displayed in the Main Building of the Western Railway in 1886. No rolling stock was on show. Two years later, in 1888, Hannibal was exhibited at the Jubilee Trade Exhibition at the Prater, after which it was dismantled for storage in the attic of the Western Railway station. The exhibits moved to the Administration Building of the Directorate-General of the Imperial-Royal State Railways and in 1893 the Museum of the History of the Imperial-Royal State Railways opened to the general public. Hannibal is not mentioned in the railway museum's catalogue, but photographic evidence dated between 1893 and 1902 showing Hannibal exhibited at the Administration Building site exists. In 1902 Hannibal is mentioned in the museum's catalogue with the sequential position No. 55.

The carriage is now part of the collection of the Vienna Museum of Science and Technology and has been conserved to prepare it for re-display in the permanent

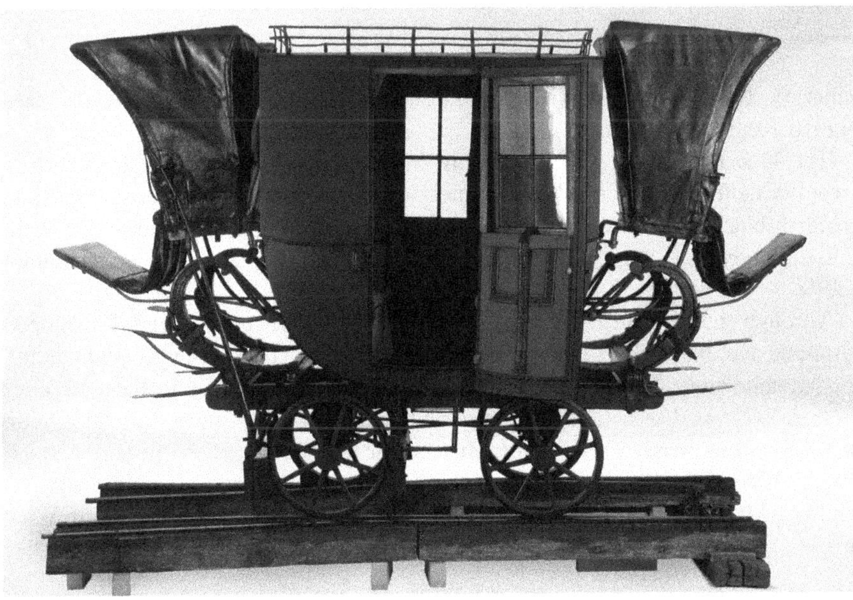

**Fig. 10.9** Hannibal, the last surviving carriage of a horse-drawn railway. (© Vienna Museum of Science and Technology, Vienna, Austria. Photograph: Peter Sedlaczek)

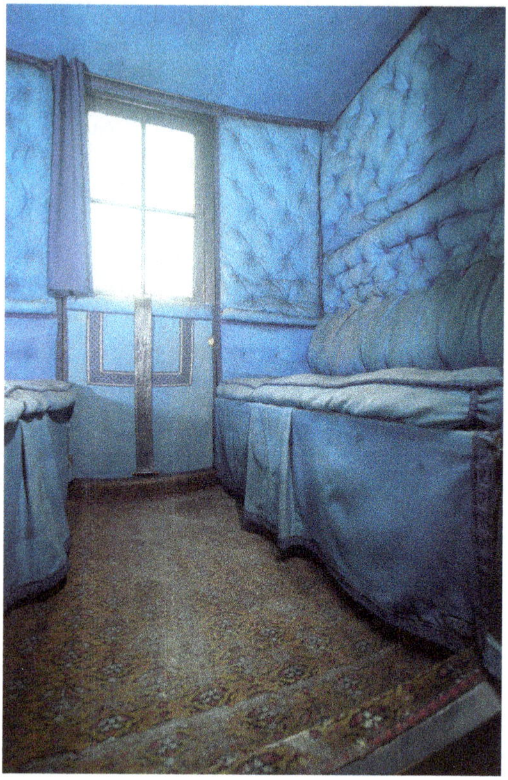

**Fig. 10.10** Interior of Hannibal, the blue upholstery. (© Vienna Museum of Science and Technology, Vienna, Austria. Photograph: Peter Sedlaczek)

galleries. In order to establish the chronology of the interior upholstery (see Fig. 10.10), the dyes were analysed.

The blue upholstery was dyed with Water blue and shows the characteristic greenish fading as described by Lehne (see Fig. 10.11).[4] The bluish-black and greenish-black bands used in the upholstery were dyed with the natural dye Weld (*Reseda luteola* L.), Indigo carmine, Water blue, Methyl violet and Naphthol yellow S.

Through the identification of Naphthol yellow S, the most recently developed synthetic dye, patented in 1879, it is possible to assign a terminus post quem dating for the upholstery. The most probable dates for installing the upholstery are after the carriage was dismantled for storage in the attic of the Western Railway station in 1888 and its reassembly for the Museum of the History of the Imperial-Royal State Railways in the last decade of the nineteenth century. As an alternative, it could also have been redecorated for the opening of the Technisches Museum Wien in 1916.

[4] Wasserblau. Echtheit: 'Nach 5 tägiger Einwirkung der Witterung (Juni) Wollfärbung etwas matter, nach 18 Tagen heller und grünlicher.

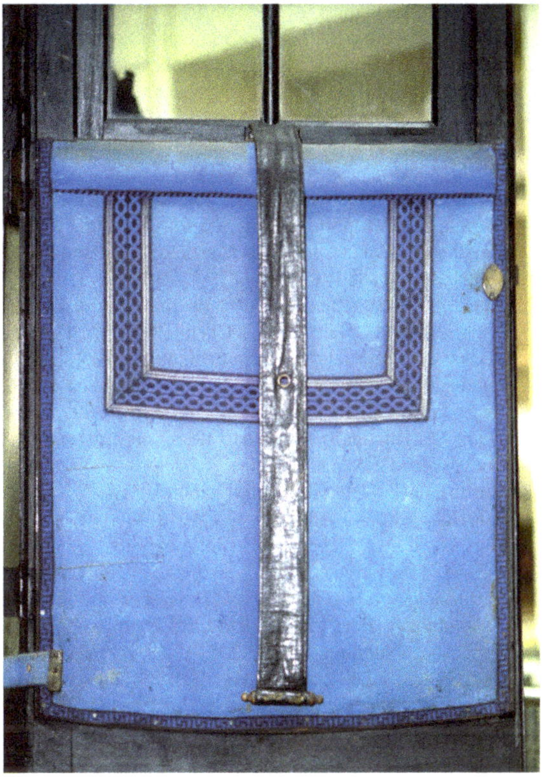

**Fig. 10.11** Covering on the inside of the door. (© Vienna Museum of Science and Technology, Vienna, Austria. Photograph: Peter Sedlaczek)

# References


1. Woulfe, P.: A method of dying wool and silk, of a yellow colour, with indigo; and also with several other blue and red colouring substances and Receipt for making the yellow dye. Philosophical Transactions of the Royal Society. 61, 127–130 (1771)
2. Welter, J.J.: Notice sur quelques matières particulières trouvées dans les substances animales, traitées par l'acide nitrique. Annales de Chimie et de Physique. 29, 301–305 (1799)
3. Dumas, J.-B.: Quatrième mémoire sur le types chimiques. Annales de Chimie et de Physique. Imprimerie De Bachelier, Paris. 3, 204–232 (1841)
4. Laurent, A.: Ueber das Phenyl und die daraus abgeleiteten Verbindungen. Annalen der Chemie und Pharmacie. Academische Verlagsbuchhandlung von C.F. Winter, Heidelberg. 43, 200–223 (1842)
5. Laurent, A.: Sur le phényle et ses dérivés. Annales de Chimie et de Physique, Troisième Série. 221–228 (1841)
6. Knecht, E., Rawson, C., Loewenthal, R.: Manual of Dyeing: For the Use of Practical Dyers, Manufactures, Students and All Interested in the Art of Dyeing. , London (1925)
7. Farrar, W.V.: Synthetic dyes before 1860. Endeavour. 33, 149–155 (1974)
8. Wagner, D.: Ernst Sell, der Begründer der Teerdestillation in Deutschland - Zum 150jährigen Bestehen der ersten Teerdestillation Deutschlands. Mitteilungen, Gesellschaft Deutscher Chemiker/Fachgruppe Geschichte der Chemie, Frankfurt/Main. 9, 42 (1933)

9. Reinhardt, C., Travis, A.S.: Heinrich Caro and the Creation of Modern Chemical Industry. Kluwer Academic Publishers, Dordrecht/Boston/London (2000)
10. Schmitt, R.E., Glutz, L.: Ueber Diazophenole. Berichte der deutschen chemischen Gesellschaft zu Berlin. Ferd. Dümmler's Verlagsbuchhandlung, Berlin. 2, 51–53 (1869)
11. Rzepa, H.: Chinese Green. Committee RSC Historical Group Newsletter August. 10–14 (2010)
12. Cooksey, C., Dronsfield, A.: Pre-Perkin Synthetic Routes to Purple. Journal of Dyes in History and Archaeology. 19, 119–122 (2003)
13. Sprengel, H.: On a new class of explosives which are non-explosive during their manufacture, storage, and transport. J Chem Soc. 26, 796–808 (1873)
14. Manchester Police newspaper article 'The Inquest': News article 'Fatal Explosion at Manchester', (1887)
15. Majendie KCB, C.V.D.: Report on the Circumstances Attending a Fire and Explosion at Messrs. Roberts, Dale & Co.'s Chemical Works, Cornbrook, Manchester. Gov. Report No. 1. Journal of the Society of Chemical Industry. 835 (1887)
16. Roberts, D.: Obituary of John Dale, F.C.S. and co-founder of Roberts, Dale & Co. Journal of the Society of Chemical Industry. 528–530 (1889)
17. Moos, G.: Rundgang durch das alte Frankfurt-Höchst. Wartberg Verlag GmbH & Co. KG., Gudensberg-Gleichen (2001)
18. Martius, C.A.: Ueber das Binitronaphtol. Zeitschrift für Chemie. Verlag von Quandt & Hände, Leipzig. 11, 80–83 (1868)
19. Martius, C.A.: Ueber das Binitronaphtol. Journal für die praktische Chemie. 102, 442–447 (1867)
20. Darmstädter, L., Wichelhaus, H.: Ueber einige Abkömmlinge des Naphtols und ein Naphtobioxyl. Berichte der deutschen chemischen Gesellschaft zu Berlin. Ferd. Dümmler's Verlagsbuchhandlung, Berlin. 2, 113–115 (1869)
21. Darmstädter, L., Wichelhaus, H.: Ueber Abkömmlinge des Naphtalins. Annalen der Chemie und Pharmacie. C.F. Winter'sche Verlagshandlung, Leipzig und Heidelberg, Band CLl. 298–311 (1869)
22. Liebermann, C., Dittler, Aug.: Studien in der Naphtalingruppe; von Demselben. Justus Liebigs Annalen der Chemie. C.F. Winter'sche Verlagshandlung, Leipzig und Heidelberg. 183, 249 (1876)
23. Ballo, M.: Ueber die Darstellung des Naphtylamins. Berichte der deutschen chemischen Gesellschaft zu Berlin. Ferd. Dümmler's Verlagsbuchhandlung, Berlin. 3, 673–676 (1870)
24. Bender, F.: Ueber die aus alpha-Naphtol entstehenden Sulfosäuren. Berichte der deutschen chemischen Gesellschaft. Commissionsverlag von R. Friedländer & Sohn, Berlin. 22, 993–1000 (1889)
25. Chiozza, L.: Notizen aus dem chemischen Laboratorium zu Mailand: Einwirkung der salpeterigen Säure auf Naphtylamin. Annalen der Chemie und Pharmacie. C.F. Winter'sche Verlagshandlung, Leipzig und Heidelberg, Band XCIX. 240–244 (1856)
26. Lauterbach, P.: Ueber die Sulfosäure des Binitronaphthols. Berichte der deutschen chemischen Gesellschaft. Commissionsverlag von R. Friedländer & Sohn, Berlin. 14, 2028–2031 (1881)
27. Graebe, C.: Ueber β-Sulfophtalsäure. Berichte der deutschen chemischen Gesellschaft. Commissionsverlag von R. Friedländer & Sohn, Berlin. 18, 1126–1130 (1885)
28. Knecht, E., Hibbert, E.: Zur Kenntnis des Naphtolgelb S. Berichte der deutschen chemischen Gesellschaft. Commissionsverlag von R. Friedländer & Sohn, Berlin. 37, 3475–3477 (1904)
29. Coyle, L.: Strands interlacing: colour theory, education and play in the work of Vincent van Gogh. Van Gogh Museum Journal, Waanders Publishers, Zwolle. 119–132 (1996)

# Chapter 11
# Mauveine: The First Synthetic Dye

## 11.1 Introduction

In the period 1856–1865, four developments led to the foundation of fundamental research, resulting in hundreds of synthetic dyes. The start of the synthetic dye industry can be dated very precisely. From a historical point of view, the violet dye Mauveine is interesting, since it was the first artificial aniline or coal tar dye discovered. In 1856, the British student William Henry Perkin worked as an assistant to August Wilhelm Hofmann at the Royal College of Chemistry in London. Actually, Perkin had the task of finding a way to synthesise quinine. He tried to synthesise an artificial drug for the treatment of malaria, and instead discovered the first synthetic violet dye Mauveine.

By the end of the 1850s, purple was a favourite colour for clothing in fashion, stimulated by important people such as Queen Victoria and Empress Eugénie of France. In 1858, Queen Victoria wore a purple dress for her daughter's wedding and she appeared with a long mauve gown at the Royal Exhibition in London in 1862. Empress Eugénie of France, the wife of Napoleon lll, also enjoyed wearing dresses of the same colour. But the enthusiasm was only temporary and ebbed away after a few years. This was enough time for Perkin to get very wealthy, however: by the age of 21, he was already a millionaire and in 1873, at the age of 36, he sold his company and retired to devote himself exclusively to research in organic chemistry.

## 11.2 Mauveine

Colour Index name: Basic dye; Colour Index number: 50245
IUPAC name: 3-methyl-8-N-(4-methylphenyl)-10-phenylphenazin-10-ium-2,8-diamine (Mauveine A); 3,6-dimethyl-8-N-(4-methylphenyl)-10-phenylphenazin-10-ium-2,8-diamine (Mauveine B)



M. R. van Bommel, M. de Keijzer, *Bright Colours from the Past*, Cultural Heritage Science, https://doi.org/10.1007/978-3-031-88664-5_11

Discoverer: William Henry Perkin, 1856
Contemporary terminology: Mauveine (En), Mauve (Fr), Mauveïn (Ge), Mauveine (NL)
Obsolete terminology: A variety of equivalent names are known, including Perkin's violet, Perkin's mallow, Perkin's mauve, Mauve (Malve), Mauve dye, Tyrian Purple, Aniline violet (En), Rosolane, Violet au bichromate, Violet pâte (Fr), Anileïn, Anilinviolett, Anilinpurper, Violeïn, Perkins Violett, Indisin, Phenamin, Phenameïn, Purpurin, Tyrian, Lydin, Harmelin, Tyralin, Roseïn, Chromviolett (Ge) [1]
Chemical formula: $C_{27}H_{25}N_4$ (Fig. 11.1)

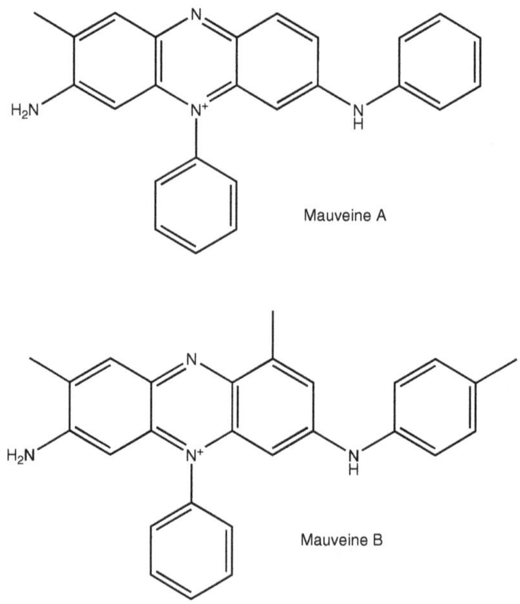


**Fig. 11.1** Structural formula of Mauveine

## 11.3 History

The German chemist August Wilhelm Hofmann (1818–1892) became the director of the Royal College of Chemistry in London (now part of Imperial College, London) in 1845. His students and assistants worked on reactions leading to a variety of amino compounds, as well as on the analysis of natural substances, including dyes. In 1853, the 15-year-old William Henry Perkin (1838–1907) became one of

Hofmann's students. In 1856, he worked as an assistant to Hofmann at the Royal College of Chemistry and was commissioned to synthesise quinine. Quinine was at that time an expensive and limited natural substance extracted from the bark of the cinchona tree (*Cinchona officinalis L.*), which only grew in the tropical forests of the Andes (Ecuador, Colombia, Bolivia and Peru). It was then the only effective drug against malaria. The need for quinine was very high, as many people suffered from this disease [2–4].

During the Easter holidays of 1856, Perkin was working in the laboratory he had set up in a room at his parent's home on Cable Street in East London. He was trying to synthesise quinine from impure crude aniline, a colourless aromatic oil derived from coal tar, by oxidising it with potassium dichromate. The aniline employed by Perkin was impure and consisted of aniline and small amounts of *o*- and *p*-toluidines. After reacting impure aniline with potassium dichromate in a cold dilute sulfuric acid solution, he did not succeed in synthesising quinine, but obtained a black precipitate. He did not throw it away, as was the custom at the time, but by extracting it with alcohol he observed that he obtained a solution with a purple colour [5]. With this discovery, the violet dye Mauveine, William Henry Perkin entered the world of synthetic dye chemistry.

The next step was that Perkin decided to dye a piece of silk with this purple solution and noticed that the colour gave a satisfactory result. He found that the dye had good light and wash fastness. With his elder brother Thomas Dix Perkin (1831–1891) and his friend and British chemist Arthur Herbert Church (1834–1915), he carried out further experiments with the dye. For a detailed assessment, he sent samples of Mauveine to the Scotsman Robert Pullar (1828–1912) of the renowned textile dyeing company J. Pullar and Sons Ltd. in Mill Street, Perth, Scotland. This company was founded in 1824 and had connections throughout Europe. In mid-June 1856, Perkin received an encouraging response for his novel material: 'If your discovery does not make the [dyed] goods too expensive, then it is certainly one of the most valuable in a long time', was the verdict from Scotland [2, 6].

Perkin showed Pullar's letter to Church, who advised him to patent his discovery immediately. However, at 18, Perkin was not yet of age, but he sought legal advice and learned that age should not matter as the patents were granted by the Crown [2]. On the 26th of August 1856, Perkin filed a patent for his invention (B.P. 1.984, Dyeing Fabrics). Perkin describes in his patent that he obtained the violet dye by the action of potassium dichromate on aniline sulfate: 'I take a cold solution of sulphate of aniline, or a cold solution of sulphate of toluidine, or a cold solution of sulphate of xylidene, or a cold solution of sulphate of cumidene, or a mixture of any one of such solution with any others or other of them, and as much of a cold solution of a soluble bichromate as contains base enough to convert the sulphuric acid in any of the above-mentioned solutions into a neutral sulphate.' A few months later, in October, he left the Royal College of Chemistry.

In the beginning, Mauveine had many difficulties and the British textile industry did not accept the dye. In the early winter of 1856, Perkin travelled to Perth, where he met Pullar, conducted experiments and visited Scottish dye works. The results were disappointing. Mauveine dyed well on silk, but had a limited application since the dye adhered poorly to cotton without a mordant. Unlike the vegetable dyes available, Mauveine was not acidic and all the mordants Pullar had at his disposal failed. On cotton, Mauveine faded noticeably after washing [2]. The result was that the product was difficult to sell because of the difficulties of attachment to cotton. Consequently, there was little enthusiasm from the calico printers and dyers in Lancashire and Scotland [4].

It was in January 1857 that an East London silk dyer in Bethnal Green named Thomas Keith, who had the largest silk dye works in London at that time, did some experiments with Mauveine himself and Pullar reported that the new shades obtained 'are certainly the best I ever saw' [2]. Shortly afterwards, Pullar and Perkin independently discovered that a tannin mordant on cotton improved the wash and light fastness of Mauveine dyeing.

In 1857, with the support of his father George Fowler Perkin (1802–1865) and brother Thomas, Perkin founded the factory of Perkin & Sons at Greenford Green near Harrow on the banks of the Grand Junction Canal in Middlesex, north-west of London, to manufacture Aniline purple. Perkin's father sold his construction business at the end of June 1857 and started building the dye factory, which took 6 months [5]. For the production of Mauveine aniline is required, which was only present in small quantities in coal tar and was not available on a large scale. Therefore, Perkin soon started to produce aniline himself by nitration and subsequent reduction of benzene distilled from coal tar. The benzene was of such a poor quality, however, that Perkin found he had to purify it before use. It can be assumed that the benzene used by Perkin as the starting material would also have been contaminated with toluene. In addition, the nitric acid available to prepare nitrobenzene was not strong enough [7].

The first batch of Mauveine was delivered to Keith in December 1857 and in the same month, the first silk fabrics were dyed in his dye works. It was probably well into 1858 before production took place on a regular basis [8, 9].

Perkin refused to license his patented process to other companies in Britain; however, he could not obtain the French patent for Mauveine because he had not applied for it within 6 months of the English patent. This meant the French were free to produce Mauve. He soon discovered that at least one reputable dye factory in Lyon was copying his manufacturing process and producing a very similar dye. It seems that some French manufacturers had easier access to supplies of nitrobenzene and more practised industrial techniques to produce the colour before Perkin did [2, 8]. Lyon dyers specialising in silk soon experimented with the aniline reaction. They, together with innovative firms such as Depouilly Frères and Laurent et Castelhaz, produced Mauveine on a large scale from the end of 1858, as did one or two German firms [4]. Still, it is unlikely that much Mauveine was produced in France, both because of the inherent difficulties of the process and because something much better was soon available, namely the synthetic dye Fuchsine [8].

As mentioned before, in those days the colour purple was popular, but alternatives using natural dyes, including lichens and the semi-synthetic dye Murexide, derived from guano, were not fast to light. There was one exception: the introduction of a fast and brilliant natural lichen dye called French purple or Pourpre française, an improved form of orchil produced by the firm of Guinon, Marnas et Bonnet, formerly Guinon, Marnas & Co. of Lyon, posed a threat to Mauveine.

After developing a suitable mordant process for dyeing cotton by first treating Mauveine with tannic acid, an improvement achieved by Perkin and Pullar independently, Mauveine became the leading fashion colour in Paris in the second half of 1857 and reached London one year later. The decisive impulse seems to have come from the Empress Eugénie of France (1826–1920), wife of the French Emperor Napoleon III, who appeared in public in silk dresses dyed with Mauveine. At the end of 1857, the *Illustrated London News* mentioned Eugénie's affection for Mauve [2, 8]. This opened the door of desirability and in the years 1859–1861, Mauve mania dominated in British and French society and the colour was seen throughout Europe [10, 11].

The violet dye was first named Tyrian purple, after the iconic purple used in the ancient world, and afterwards Aniline purple, from the starting material used to produce it [4]. Later in 1859, the French chemist Alcide Poirrier (1832–1917) from Saint-Denis coined the name 'Mauve' from the colour of the mallow flower (genus *Malva* of the Malvaceae family). Mauve appealed more to the public in both countries than 'Aniline purple' and chemists named it Mauveine [8]. In Britain, Aniline purple was renamed Mauve during the spring of 1859.

## 11.4 Trade Names

The purple dye was put on the market as Mauveine by Perkin & Sons. The French names Rosolane and Violet au bichromate were used by Soc. Anon. St. Denis, and Violet pâte by John Casthelaz, Bruère & Co.

## 11.5 Applications

Mauveine has been used in the printing and textile industries, especially for silk and cotton. In 1893, Adolf Lehne wrote that Mauveine was still rarely used. The printing of the mauve English 6d postage stamps between 1867 and 1880 is considered the last commercial application [12].

## 11.6 Properties

Lehne studied the fastness of the dye and wrote:[1] 'After 8 days of exposure to the weather almost completely faded. Not changed by sulfuric acid of 4.75 %. The silk dyeings are fast in water. In hot water slightly soluble'. Mauveine is insoluble in cold and sparingly soluble in hot water, producing a red-violet colour. In ethanol, it is soluble giving a purple colour. Mauveine dyes silk reddish-violet and cotton with a violet tone. No fastness properties of Mauveine are given in the Colour Index, but it is known for its poor light fastness.

## 11.7 Recipes

Dyeing silk with Mauveine occurs in a neutral or in a weakly acetic acid dye bath. Lehne provides a recipe_ for dyeing silk, V7, using 0.2% of dye.

## 11.8 Identification Using UHPLC-PDA

The molecular structure of Mauveine proved to be difficult to determine. The definitive structures of the chromophores were identified for the first time by mass spectroscopy and NMR spectrometry in 1994. The study led to the determination of a mixture of two major chromophores in the commercial dye, named Mauveine A and Mauveine B [13]. Studies conducted in 2007 and 2008 revealed the presence of four main chromophores: the already known Mauveine A and B, and Mauveine B2, an isomer of Mauveine B with methyl on a different aryl group, and Mauveine C, which has one more *p*-methyl group than Mauveine A over a mixture of at least 13 methyl derivatives [14, 15]. In 2015 a crystal structure was reported for the first time [16].

UHPLC-PDA analysis of Mauveine (RCE no. 2454) showed the presence of both Mauveine A and B (see Fig. 11.2).

---

[1] Mauveïn: 'Nach 8 tägiger Einwirkung der Witterung fast völlig verblasst. Durch schweflige Säure von 4½ °Bé nicht verändert. Die Seidefärbungen sind wasserecht. In heissem Wasser schwer löslich'.

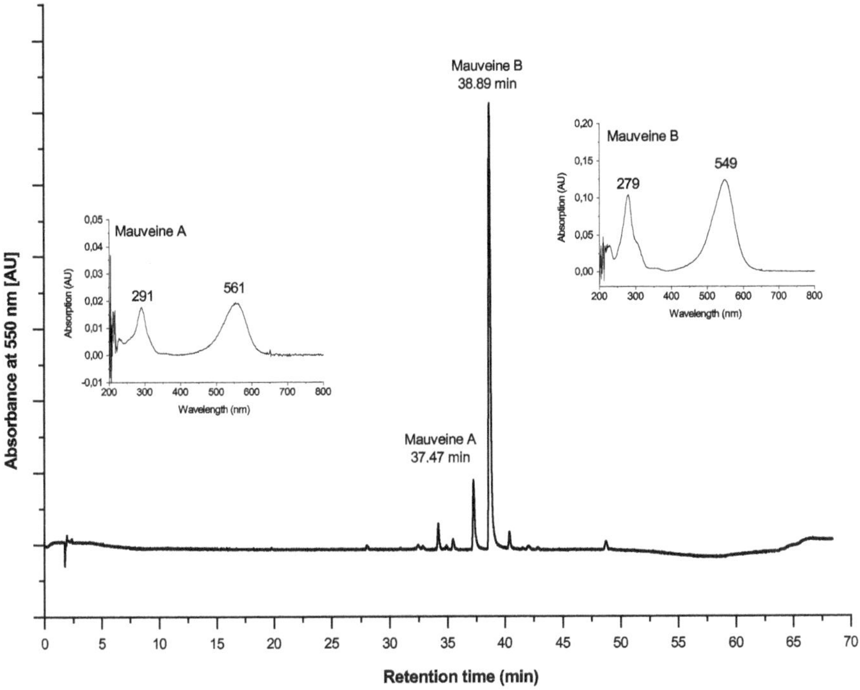


Fig. 11.2 The chromatogram and spectrum of Mauveine (RCE no. 2454)

## 11.9 Case Study: Carpet and Wall-Hangings of Huis Willet-Holthuysen, Amsterdam, the Netherlands

The canal-house Herengracht 605 in Amsterdam was built at the end of the seventeenth century. In the middle of the nineteenth century, the house was owned by the Willet-Holthuysen family and in 1865–1866 they redecorated the living room on the first floor (see Fig. 11.3). Thirty years later in 1895, a year after the death of Mrs. Sandrina Louisa Geertruida Willet, born Holthuysen (1824–1895), the museum Willet-Holthuysen opened its doors (nowadays Huis Willet-Holthuysen).

Some years ago, research was undertaken in cooperation with the Amsterdam Historisch Museum (nowadays Amsterdam Museum) on the interior of the building to determine the original appearance of the room in those days. Questions arose about the original colours of the velouté wallpaper and the carpet. The velouté wallpaper now appears a greyish-green shade, but where it has been protected from light it is clear that it has faded from a reddish-violet colour (see Fig. 11.4).

This discolouration could well point to the use of synthetic dyes and indeed, HPLC analysis of the unfaded material showed that the wallpaper was coloured using Fuchsine and Mauveine. It is therefore likely that under these lighting

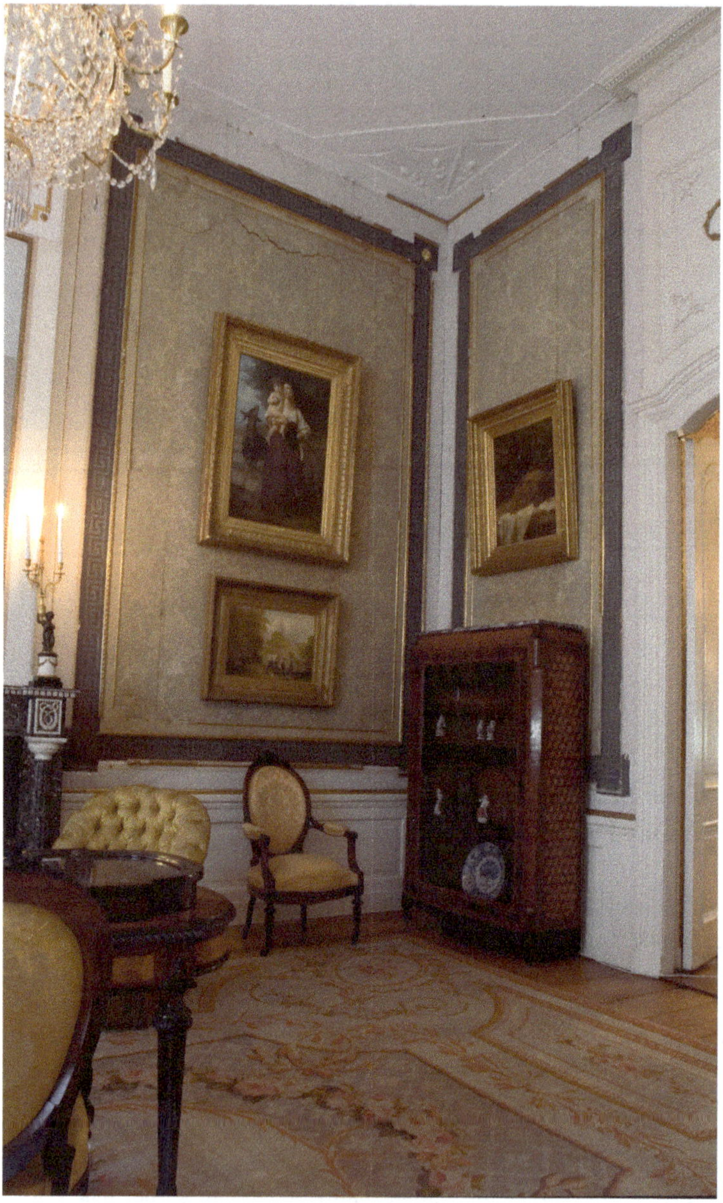

**Fig. 11.3** Huis Willet-Holthuysen, small salon (test arrangement), 2007, Amsterdam Museum. Photograph: Monique Vermeulen

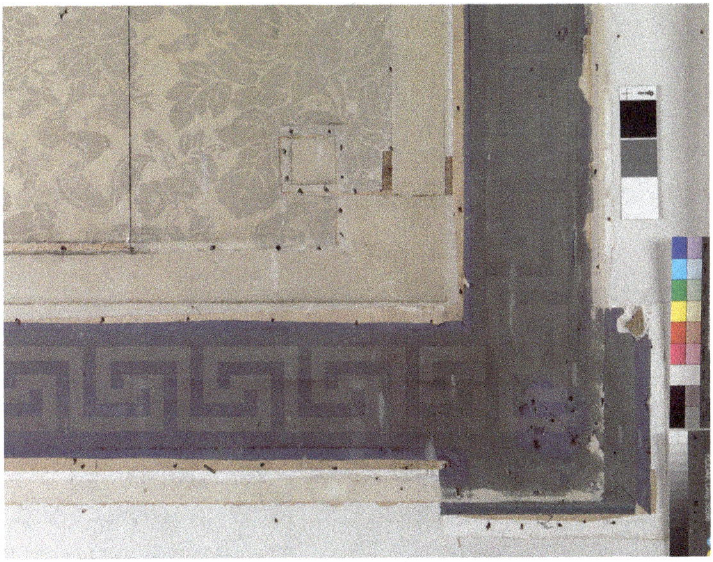

**Fig. 11.4** Detail of velouté wallpaper. Areas protected from light show a reddish-violet colour, whereas the part exposed to light is now greyish-green. Huis Willet-Holthuysen, small salon, wallpaper left of Porte Brisee/double doors, 2007, Amsterdam Museum. Photograph: Monique Vermeulen

conditions, the Fuchsine in the wallpaper has been discoloured to green [17]. The front of the carpet has discoloured to a greyish-blue (see Fig. 11.5). In the areas protected from light, Indigo carmine and Mauveine were identified by HPLC.

In the archives of the Manufacture de Tapis et Tapissiers Braquenié & Co. in Paris, an order, number 1071, dated February 13, 1865, was found in which 'Monsieur Willet, 243 Herengracht près de l'Amstel à Amsterdam' ordered a carpet and other textile objects [18]. The order number and the identification of the three synthetic dyes, Indigo carmine, Mauveine and Fuchsine, which were industrially produced in the period 1865–1866, are indications that these objects were bought for redecorating the living room. Almost certainly this is the carpet at present in the house today.

## References


1. Schultz, G.: Farbstofftabellen. Weidmannische Buchhandlung, Berlin (1914)
2. Garfield, S.: Chapter 7: The terrible glare. In: Mauve: How one Man invented a Colour that changed the World. pp. 78, 83–84. Faber and Faber Ltd., London (2000)

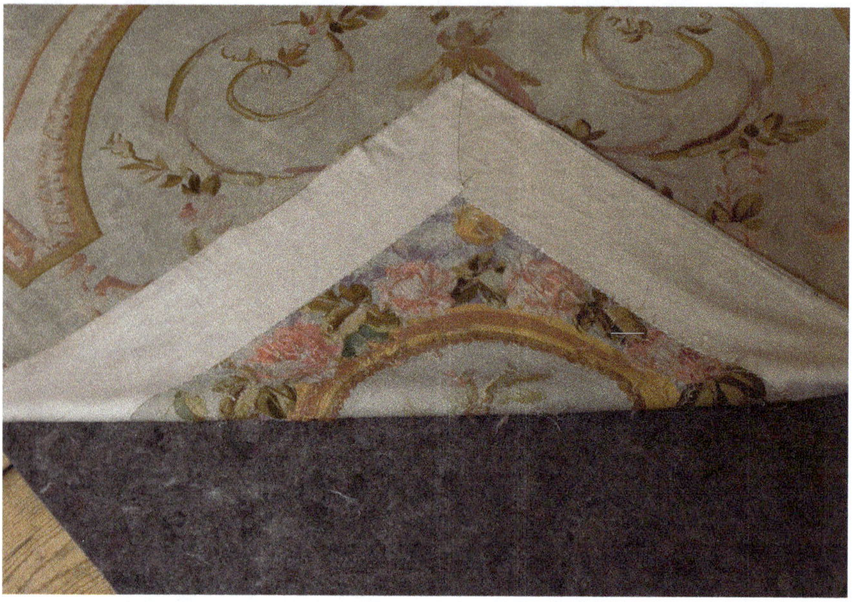

**Fig. 11.5** Detail of the carpet. Areas protected from light show still vibrant colours, whereas the part exposed to light is severely faded. Huis Willet Holthuysen, small salon, carpet, front (top) and back (middle), 2007, Amsterdam Museum. Photograph: Monique Vermeulen


3. Travis, A.S.: August Wilhelm Hofmann (1818–1892). Endeavour. 16, 59–65 (1992)
4. Morris, P.J.T., Travis, A.S.: A history of the international dyestuff industry. American Dyestuff Reporter. 81, 3–6 (1992)
5. Bäumler, E.: Die Rotfabriker, Familiengeschichte eines Weltunternehmens. R. Piper GmbH & Co. KG, München (1988)
6. Edelstein, S.M.: Sir William Henry Perkin. American Dyestuff Reporter. 45, 598–608 (1956)
7. Perkin, W.H.: Hofmann Memorial Lecture. The Origin of the Coal-Tar Colour Industry, and the Contributions of Hofmann and his Pupils. J Chem Soc. 596–625 (1896)
8. Farrar, W.V.: Synthetic dyes before 1860. Endeavour. 33, 149–155 (1974)
9. Voegelin, W.: Sir William Henry Perkin, der Mensch. CIBA-Rundschau. 8–14 (1956)
10. Kay-Williams, S.: The Story of Colour in Textiles, imperial purple to benim blue. Bloomsbury Publishing Plc, London (2013)
11. Plater, M.J., Raab, A.: Liquid chromatography–mass spectrometry analysis of mauveine extracted from silk fabrics: A Victorian dress in Gunnersbury Park and Museum and a modern Charles Rees bow tie. J Chem Res. 46, 648 651 (2022). https://doi.org/10.1177/17475198221103999
12. Da, M., Oliveira, C., Dias, A., Douglas, P., Sørgio Seixas De Melo, J.: Perkin's and Caro's Mauveine in Queen Victoria's Lilac Postage Stamps: A Chemical Analysis. Chem. Eur. J. 20, 1808–1812 (2014). https://doi.org/10.1002/chem.201303232
13. Meth-Cohn, O., Smith, M.: What did W. H. Perkin actually make when he oxidised aniline to obtain mauveine? J Chem Soc Perkin 1. 5–7 (1994). https://doi.org/10.1039/p19940000005
14. Seixas de Melo, J., Takato, S., Sousa, M., Melo, M.J., Parola, A.J.: Revisiting Perkin's dye(s): the spectroscopy and photophysics of two new mauveine compounds (B2 and C). Chemical Communications. 2624–2626 (2007). https://doi.org/10.1039/b618926a

15. Sousa, M.M., Melo, M.J., Parola, A.J., Morris, P.J.T., Rzepa, H.S., de Melo, J.S.S.: A Study in Mauve: Unveiling Perkin's Dye in Historic Samples. Chemistry – A European Journal. 14, 8507–8513 (2008). https://doi.org/10.1002/chem.200800718
16. Plater, M.J., Harrison, W.T.A., Rzepa, H.S.: Syntheses and Structures of Pseudo-Mauveine Picrate and 3-Phenylamino-5-(2-Methylphenyl)-7-Amino-8-Methylphenazinium Picrate Ethanol Mono-Solvate: The First Crystal Structures of a Mauveine Chromophore and a Synthetic Derivative. J Chem Res. 39, 711–718 (2015). https://doi.org/10.3184/174751915X14474318419130
17. Internal ICN report, project no. 2003-084: Kleuronderzoek aan de velouté behangsels en het vloertapijt uit de linker voorzaal (0.4) van de bel-étage van het Museum Willet-Holthuysen, Herengracht 605, te Amsterdam
18. Smit, H.: Interieurtextiel in het museum Willet-Holthuysen te Amsterdam, De periode Willet, circa 1857–1895. Textiel Historische Bijdragen. 77–95 (1997)

# Chapter 12
# Fuchsine and Methyl Violet

## 12.1 Introduction

The tri-arylmethanes belong to one of the most important and oldest synthetic organic dye classes. Its members include species with brilliant yellow, red, violet, blue and green hues. The tri-arylmethanes can be divided in four main groups: the hydroxys, the aminohydroxys, the diamines (the so-called Malachite series) and the triamines (the so-called Rosaniline series). Only the dyes of the diamine and triamine classes are of real importance in the textile industry. The basic structure consists of three phenyl groups linked by a central carbon atom (see Fig. 12.1).

The hydroxy compounds include the dye Aurin, already discovered by the German chemist Friedlieb Ferdinand Runge (1794–1867) in 1834 and 25 years later rediscovered by the chemists Hugo Müller, Adolph Wilhelm Hermann Kolbe, Rudolf Schmitt and Jules Persoz [1–3]. The dyes belonging to the aminohydroxy class were invented at the end of the nineteenth and the beginning of the twentieth

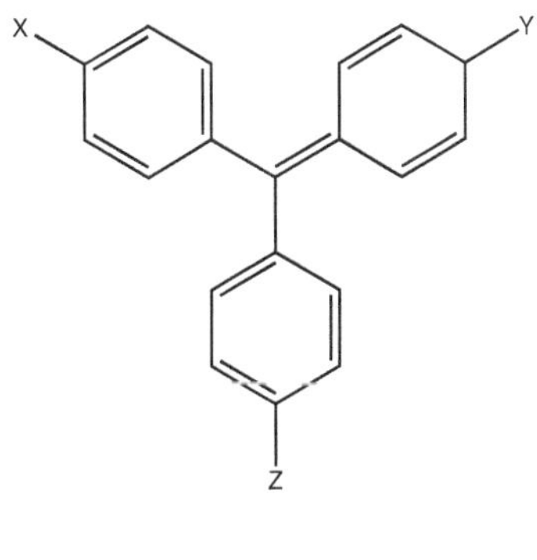


**Fig. 12.1** Basic structure of the tri-arylmethanes, X, Y, Z indicates the typical position of nitrogen


M. R. van Bommel, M. de Keijzer, *Bright Colours from the Past*, Cultural Heritage Science, https://doi.org/10.1007/978-3-031-88664-5_12

century; a typical example is Resorcine violet. The diamines include the successful dyes Malachite green, discovered in 1877 by Otto Philipp Fischer and independently in 1878 by Oscar Doebner, and Brilliant green, invented in 1879 by the factory Bindschedler & Busch. The last group, the triamino class, contains, among others, the famous dyes Fuchsine, Methyl violet, Regina purple, Hofmann's violet, Crystal violet, Alkali blue, Aniline blue, Aldehyde green, Iodine green and Methyl green. In the 1888, 1891 and 1902 editions of *Tabellarische Übersicht der künstlichen organischen Farbstoffe*, by the German chemists Gustav Schultz and Paul Julius, and in the 1914 book *Farbstofftabellen*, by Gustav Schultz, hundreds of synthetic dyes are listed. The number of dyes shows the importance of a dye class: the azo dyes are the largest group with nearly 400 members, followed by the tri-arylmethanes with almost 80 being described as of commercial importance [4–7]. In this chapter the two most famous members of the Rosaniline series, namely Fuchsine and Methyl violet, will be discussed. Several other tri-arylmethane dyes are presented in Chap. 13.

## 12.2 Fuchsine

Colour Index name: Basic Violet 14; Colour Index number: 42510
IUPAC name: 4-[(4-aminophenyl)-(4-imino-3-methylcyclohexa-2,5-dien-1-ylidene) methyl]aniline
Discoverer: Jakub Natanson, 1856
Contemporary terminology: Aniline red, Magenta (En), Fuchsine (Fr), Fuchsin (Ge), Fuchsine (NL)
Obsolete terminology: Marron (Fr), Solferino, Azalein, Rubin, Rosein, Erythrobenzin, Rubianit, Hermelin Brillantfuchsin, Fuchsiacin, Nitrobenzolfuchsin, Arsenfuchsin, Rosanilin, Juchtenrot (Ge)
Chemical formula: $C_{20}H_{20}N_3Cl$ (Fig. 12.2)

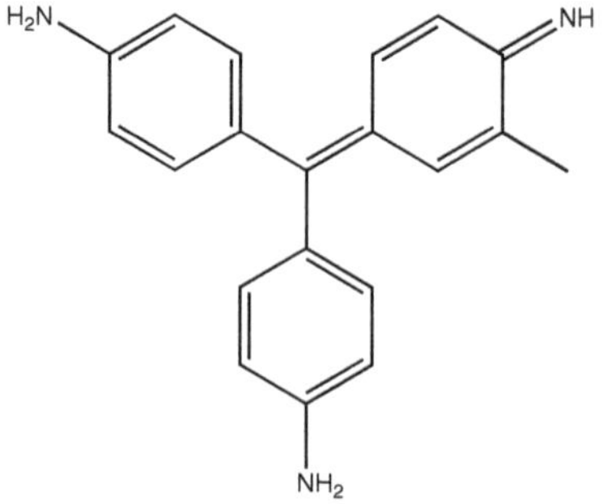


**Fig. 12.2** Structural formula of Fuchsine. Fuchsine is usually available as a mixture of four compounds that differ in their degrees of methylation. They are also known as Magenta 0 (no methyl group), Magenta I (structure below), Magenta II (two methyl groups and Magenta III (three methyl groups)

## History

Just months after the discovery of Mauveine by the British chemist William Henry Perkin (1838–1907) in 1856, the Polish chemist Jakub Natanson (1832–1884), working on his doctoral studies at the University of Dorpat (nowadays known as Tartu, Estonia), invented another and more successful synthetic dye, later named Fuchsine, by heating aniline (phenylamine) and ethylene chloride (1,2-dichloroethane) in a sealed tube at 200 °C. The colourless mixture turned into a rich blood-red colour during the heating process [8, 9].

Two years later, in 1858, the German chemist August Wilhelm Hofmann (1818–1892) obtained the same aniline-derived product by reacting still impure aniline with carbon tetrachloride ($CCl_4$). He achieved a small quantity of material as a secondary product, which he described as 'a very soluble substance of a magnificent crimson colour'. [9–11] In 1862 Hofmann produced the dye in a pure form and named it Rosaniline [12–15]. Curiously, neither chemist recognised the potential of their product as a textile dye. In 1860, the French chemist Antoine Béchamp (1816–1908) wrote that Fuchsine dissolves in water, alcohol and ether, resulting in a red colour, but he too did not mention the potential of Fuchsine as a textile dye [16].

In April 1859, the Frenchman François-Emmanuel Verguin (1806–1864), an industrial chemist in Lyon and professor at the Collège of Lyon, developed another oxidative method for preparing Fuchsine by heating crude aniline and tin(IV) chloride ($SnCl_4$) near to the boiling point of aniline, but he did not exploit this invention because he was ill at that time. Verguin had directed the Picric acid factory of Louis Raffard et Cie at Givray, near Péage de Roussillon before his employment by Renard Frères but suffered from poor health. As he was thus unable to benefit from his invention, he sold it for a low price to Francisque Renard (1811–1900) and Joseph Renard (1822–1882). In this way the production process came into the hands of Renard Frères, proprietors of a silk-dyeing firm in Lyon, who filed the French patent no. 40.635 on the 8th of April 1859 with six additions on the 1st of October, 19th of November, 26th of November, 15th of December 1859, 14th of February and 3rd of July 1860 and three British patents under the name of A.R. Brooman four days later in England (filed on the 12th of April, 27th of October and 20th of November 1859) [9, 17–19]. They named the dye Fuchsine and introduced it on the market in May 1859. The French patent describes the production of Fuchsine and its use to dye silk, wool, cotton and for the dyeing of skins and feathers [9]. Later that year (before November 1859) the firm Renard Frères et Franc was formed. Émile Hyppolite Franc was a French dyer, who patented the formation of a red dye obtained from from toluidine and cumidine on the 5th of July 1859 [18].

Tin chloride was expensive, so the production of Fuchsine led to the investigation of other, and cheaper, oxidising agents that would work upon aniline and this was achieved with other metallic salts, such as those of mercury, iron and silver. The result was that between 1859 and 1861 a number of patents for the production of Fuchsine were filed, which led to several patent struggles. On the 29th of October 1859, the French chemist Jean (Johann) Gerber-Keller (1809–1884) with his son

Armand in Mulhouse (Alsace) patented a better process in France where tin chloride was replaced by mercury(I) nitrate ($Hg_2(NO_3)_2$) (F.P. 42.621). During the reaction, metallic mercury was formed due to reduction, which was easily recovered for later (re)conversion to the nitrate salt. Their dye had a good quality and was sold under the trade name Azaléine [9]. Another dye firm, Depouilly Frères, also developed and patented processes for manufacturing the red dye from aniline; their product was called Aniléine [18]. When first introduced, Fuchsine prepared by this method was not purified, but put on the market in a crude state so that before using it the dyer had to extract it with water. In the preparation of Fuchsine by this process, all the mercury from the nitrate of mercury employed was recovered in the metallic form, and although this process may possess some advantages, the use of mercury salts was most undesirable on account of their fearfully deleterious influence upon the workmen.

However, the mercury salt was soon replaced by arsenic acid ($H_3AsO_4$ or $AsO(OH)_3$) as an oxidant, which was reduced to arsenous acid ($H_3AsO_3$ or $As(OH)_3$). On the 18th of January 1860, the British chemist Henry Medlock (c. 1840–1875), a consultant to a firm of dyers in Coventry, patented a better and more efficient commercial process for the production of Fuchsine, which he named Magenta, by heating aniline with arsenic acid (B.P. 126, *Red and Purple Dyes*). Eight days later, on the 26th of January, 1860, the British chemist and dye manufacturer Edward Chambers Nicholson (1827–1890) filed a similar patent (B.P. 184) but did not proceed with it when he learnt of Medlock's intentions [9, 13, 15, 17]. Medlock's patent is notorious for the amount of litigation it generated owing to the occurrence in it of the words 'dry arsenic acid'. Two major cases kept the London High Court occupied for several months, both revolving around the precise interpretation of this description. In one case, the judge informed the jury he was relieved that they, and not he, had to decide on the outcome of the case, as there had been so much bewildering and conflicting expert evidence [9, 20]. Finally, in 1865, this dispute was resolved and Medlock's patent was used by many English firms who took out a licence to produce it, including Simpson, Maule & Nicholson who produced the dye under the name Roseine [9, 12, 13].

From 1855 onwards, the French chemical assistant Adam Charles Girard (1837–1918) worked with another assistant, Georges de Laire (1836–1908), in the laboratory of Théophile-Jules Pelouze (1807–1867). On May 1, 1860, they also filed a patent for producing Fuchsine in France by using aniline and arsenic acid (F.P. 44.958). On May 26, 1860, both French chemists protected this process in England (B.P. 1.300) [9, 12, 13]. The research and development had actually been done in Pelouze's laboratory in Paris. In the early 1850s, Pelouze owned a benzene factory in Champerret, near Paris, and started a nitrobenzene production factory around 1859 in Brentford, near London. Patent affairs in France made the production of Fuchsine there complicated; as a result, the arsenic acid production was introduced at the factory in Brentford, managed by De Laire, in 1860. In 1892, Girard wrote that he and De Laire had been forced by the Renard factory's monopoly on Fuchsine to cede their arsenic acid patent to the Lyonese firm by forming an association, acquiring a certain percentage claim of the financial benefits of this firm, so-called '*intéressés*'. Thus in 1861, Girard and De Laire transferred the arsenic acid process patent and the Aniline blue patents to the firm Renard Frères, Franc et Cie [18]. The use of arsenic acid soon replaced the older methods and the yield of Fuchsine rose from about 2% to 20% as a result.

At the beginning of the 1860s, the firm Renard Frères et Franc had initially successfully defended its fuchsine patent against French dye firms in Mulhouse, Paris and Lyon. The Renards' firm attempted to control the French markets with monopolies based on their patents. They claimed precedence for their fuchsine process and instituted lawsuits for infringement of their Fuchsine patent against competitors and against any person who sold or used their competitors' products. However, the French patent law of 1844 allowed only products to be patented, not their production processes. So, producing an already patented product using another method would not result in a new patent. The patent cases dragged on from January 1860 until March 1863 and ended in a legal victory for the Renards.

After the patent quarrels, the firm Renard Frères et Franc entered into an alliance with the participation of the recently formed Crédit Lyonnais bank to set up the monopoly concern Société La Fuchsine. Société La Fuchsine convinced the courts to grant it a master patent for Fuchsine, stating that it was not only a product, but also a key intermediate for other dyes and thereby Société La Fuchsine was able to use the French police to shut down many rival producers. The result was that many chemists and at least seven dye firms closed down production and relocated to Switzerland, which had no patent law at that time. They settled in the textile and dye city of Basel. Some of the firms that remained in France reverted to the production of natural and other dyes of minor importance.

Around 1862, several persons who worked and lived in the neighbourhood of the fuchsine plant in Pierre-Bénite, south of Lyon, were poisoned by arsenic or arsenic products. The poisoning caused frequent and prolonged interruptions in production and no obvious economic solution seemed possible. In 1866, there was no regular production of the dye for more than one year and production was even brought to a complete stop in 1867. In February 1868 steps were taken to resume work at the plant, but the local residents protested strongly. An economic factor also played a role. By the late 1860s, German companies had become very strong and exported their Fuchsine product to France; the price of Fuchsine was reduced as a result and, in the end, the Société La Fuchsine sold it at cost price. Société La Fuchsine, which evidently had no incentive to be innovative, went into financial decline and was declared bankrupt in 1868, although it continued its legal existence until 1875 [18]. After the demise of the Société La Fuchsine, the patents and licenses were transferred to the Paris firm of Poirrier & Chappat fils in 1868, who already produced Mauveine in 1861 and Methyl violet, sold as Violet de Paris in 1866.

Société La Fuchsine's patent position was not unassailable and skilled innovators were able to find ways of circumventing the monopoly. The French chemist Jean Théodore Coupier (1820–1908), an aniline producer in Poissy near Paris, improved the separation of hydrocarbons present in coal tar in the early 1860s. The starting material for commercial aniline, benzene, was actually a mixture of benzene, toluene, xylene, cumene and other substances. In the French patent, filed on the 4th of April 1863, Coupier devised a special distillation apparatus to separate and remove the lower boiling hydrocarbons and thus produced purer aniline, toluidine and other substances. He discovered that he could produce a red dye from pure toluidine alone, without using aniline, which he called Rouge de Toluidine (Toluidine red). The commercial exploitation was obstructed by Société La Fuchsine, who claimed priority of the 1859 French patent for the production of Fuchsine from toluidine and cumidine. In 1861, the French chemist Charles Lauth (1836–1913) observed the formation of Fuchsine from nitrobenzene, aniline and tin chloride. In the same year, the firm Laurent & Castelhaz succeeded in forming Fuchsine by the reaction using nitrobenzene, iron and hydrochloric acid (F.P. 52.223, filed on the 10th of December 1861). In 1866, this process was further developed and patented by Coupier (filed on the 5th of April 1866), based on the reaction between aniline and nitrobenzene in the presence of iron chloride, the latter being prepared from hydrochloric acid and iron powder. This so-called nitrobenzene process ultimately replaced the old toxic arsenic acid method. The French chemist Paul Schützenberger (1829–1897) praised the method as giving a product of greater purity and yield than the arsenic acid method. Commercial manufacturers, however, complained that the hydrochloric acid corroded their iron retorts [14, 21].

At the Paris International Exhibition in 1867 the owners of the firm Farbwerke ML&B were informed about Coupier's red and the nitrobenzene process, and in 1869 they visited him to buy his nitrobenzene process. The negotiations had no result. In the meantime, on the 24th of November 1868, the German industrialist and founder of the BASF Friedrich Engelhorn (1821–1902) signed a contract with Coupier, which enabled BASF to deploy the nitrobenzene process he had developed. Then, on January 13th, 1873, the German chemist and industrialist Johann Adolf Brüning (1837–1884) announced that in October 1872 Farbwerke ML&B was successful in producing Fuchsine from nitrobenzene and aniline by the nitrobenzene method, also avoiding arsenic acid. An in-house report on Fuchsine reveals that the manufacture in Frankfurt-Hoechst was profitable for the first time in 1876; in the first years, the yield of Fuchsine had been poor [22]. Nevertheless, at the end of the 1870s, the majority of factories still produced Fuchsine by the arsenic acid method. Only large companies, such as Farbwerke ML&B, BASF and AGFA, which could produce their own nitrobenzene, were able to produce Fuchsine by this process [17, 21–25] (Table 12.1).

## Trade Names

The common trade names are Fuchsine and Magenta, further identified by adding letters or numbers, and were manufactured by many international dye companies. The large number of trade names and producers showed the importance of this dye.

**Table 12.1** Discoverers and patents of Fuchsine

| Name<br>C.I. name<br>C.I. number | Year of discovery | Discoverers | Patents |
|---|---|---|---|
| Fuchsine,<br>Basic Violet 14<br>42510 | 1856 | Jakub Natanson | — |
| | 1858 | August Wilhelm Hofmann | — |
| | 1859 | François-Emmanuel Verguin | — |
| | 1859 | Renard Frères et Franc | F.P. 40.635/1859 and six additions/1859 and 1860 and B.P. 921/1859 and two additions |
| | 1859 | Gerber and Keller | F.P. 42.621/1859 |
| | 1860 | Henry Medlock | B.P. 126/1860 |
| | 1860 | Edward Chambers Nicholson | B.P. 184/1860 |
| | 1860 | Adam Charles Girard and Georges de Laire | B.P. 1.300/1860<br>F.P. 44.958/1860 |
| | 1861 | Laurent & Casthelaz | F.P. 52.223/1861<br>(10 December 1861) |
| | 1866 | Jean Théodore Coupier | 5 April 1866 |

Depending on the synthesis procedure, other names, such as Cerise, Grenadin, Marron and Cardinal, are mentioned, with the addition of letters, especially by German dye factories.

## *The Elucidation of the Constitution of Fuchsine*

From the time of its discovery, many chemists were engaged in elucidating the constitution of Fuchsine, and also its method of formation. Among these were Hofmann and the German chemists Emil Hermann Fischer (1852–1919) and Otto Philipp Fischer (1852–1932) (not twin brothers). The first important scientific research on this subject was done by Hofmann. In 1863, he discovered that, in fact, pure aniline or pure toluidine (2-methyl phenylamine) did not yield Fuchsine with the usual reagents: a mixture of aniline and *o*- and *p*-toluidines was required. Distillation of coal tar provided a benzene fraction that contained significant amounts of toluene (methyl benzene). This fraction was nitrated and the nitro compounds were reduced to give aniline (derived from benzene) and toluidines (derived from toluene). However, the exact chemical structure of Fuchsine remained unclear. Five years later, in 1868, the German chemist Friedrich August Kekulé (1829–1896) suggested that the methyl group of the toluidine held three phenyl groups together. Many other chemists made contributions, but Emil Hermann Fischer and Otto Philipp Fischer succeeded in 1878 in converting pararosaniline (one of the components of Fuchsine) into triphenylmethane and solved the problem [15, 26–34].

## *Explanation of the Names of Fuchsine, Magenta and Solferino*

There are over 30 synonyms for Fuchsine, including Magenta and Solferino. The common name of the dye is described as Fuchsine, named after the colour of the flowers of *Fuchsia*—*Fuchsia* is a genus of the family Onagraceae—so named in honour of the German physician and botanist Leonhart Fuchs (1501–1566). Fuchs saw the importance of practical experience and offered botanical field trips for students, where he demonstrated the medicinal plants in situ [9, 20]. The plant *Fuchsia* was discovered in Santo Domingo, in the Caribbean, in 1696–1697 by the French scientist Dom Charles Plumier (1646–1704), who published the first description of a species of *Fuchsia* under the name of 'Fuchsia triphylla, flore coccineo' in his *Nova plantarum Americanarum genera* [35, 36].

The other two names, Magenta and Solferino, are related to famous battles. The first name is after the Battle of Magenta which was fought on the 4th of June 1859, during the Franco-Sardinian war against Austria, resulting in a French victory, an important step towards Italian independence. Twenty days later, on the 24th of June 1859, the Battle of Solferino took place, which again resulted in the victory of the allied French and Sardinian armies against the Austrians. It was a particularly gruelling battle, lasting over nine hours and resulting in more than 40,000 soldiers killed, wounded, missing or captured. The colour of blood spilt as a consequence of these battles was related to the colour of Fuchsine, thus the synonyms Magenta and Solferino were used [9, 13, 15, 17]. In passing, the Battle of Solferino would have an effect on the future conduct of military actions. The Swiss humanitarian and businessman Jean-Henri Dunant (1828–1910), who witnessed the conflict in person, was motivated by the horrific suffering of wounded soldiers left on the battlefield, to begin a campaign that would result in the Geneva Conventions and the establishment of the International Red Cross in 1864 [37, 38].

## *Applications*

The application of Fuchsine is mainly for dyeing silk and wool, which are dyed in neutral baths. Cotton is rarely dyed with Fuchsine. The advent of the early dyes, however, with these intense colours available at a reasonable cost, offered new opportunities to manufacturers of processed foods and drinks for artificially colouring their products. Pure Fuchsine, as well as Fuchsine that has not been produced with arsenic acid, is not toxic. Fuchsine was popular, even though the toxic nature of arsenic had been recognised for years. The product could still contain 6% of arsenic, however, and numerous poisonings observed after ingestion of Fuchsine could be traced back to its arsenic content. In the nineteenth century Fuchsine was used to kill flies and rodents. Fuchsine was being added for colouring wines within a year of its discovery in 1859 [39, 40]. Fuchsine was also used in sweetmeats and sausages and to colour liqueurs, jams, syrups, fruit juices, ice cream and red wine in the late nineteenth century, but in Germany its use as a food dye was prohibited from 1883.

## Properties

The dyes on textiles and the pigments made from Fuchsine generally have a poor light fastness, which is well documented in literature of the period. For Fuchsine, Lehne wrote in his 1893 book[1]: 'After 5 days of light exposure, in June, the colour of fuchsine is more bluish and duller and in 18 days the colour has been very strongly faded'. Artificial light ageing tests presented in the Colour Index showed that Fuchsine dyed on silk has a light fastness of ISO 1. The wash fastness is not very good, ISO 2.

The German chemist Heinrich Caro (1834–1910), one of the famous colour chemists, reported in 1892 on the discovery of Fuchsine as a textile dye and its light fastness[2]: 'The success of Fuchsine is much greater than Perkin's Mauveine. Such a dye one never had seen before. How dull appears the colour of the pink of Cochineal! Its tempting brightness displaces all doubts concerning its lack of fastness; with it the industry follows the dangerous pathway of the fugitive colours. Fashion wants it. The ladies' world likes this fugitive way, when it is new and dresses (trendily). Thereby the prompt mass production' [9].

## Recipes

According to Lehne, wool can be dyed using recipe V1, silk using recipe V7, cotton or linen following recipe V11 and jute applying recipe V15. The concentrations of dye used are typically between 0.5 to 1.5%.

## Identification Using UHPLC-PDA

Fuchsine (Merck CAS 632-99-5) consists of a mixture of components, of which four are usually present in a significant ratio. These components can be well separated using liquid chromatography as shown in Fig. 12.3.

As can be seen from the spectra, the four components show very similar UV/Vis absorption behaviour which suggests a strong chemical similarity. Fuchsine is a mixture of four compounds that differ in their degree of methylation: not methylated (1st peak, known as pararosaniline and as Magenta 0); singly methylated (2nd peak, known as Fuchsine and as Magenta I); doubly methylated (3rd peak, known as Magenta II) and triply methylated (4th peak, known as New fuchsine or Magenta III). It must be noted that nowadays Magenta 0 and Magenta III can be obtained as pure compounds.

[1] Fuchsin: Echtheit: 'Nach 5 tägiger Einwirkung der Witterung (Juni) bläulicher und stumpfer, nach 18 Tagen sehr stark verblasst'.

[2] 'Der Erfolg des Fuchsins ist viel größer als der des Perkin 'schen Violets (Mauvein). Einen solchen Farbstoff hatte man nie zuvor gesehen. Wie matt erscheint dagegen das Rosa der Cochenille! Sein verführerischer Glanz drängt alle Bedenken gegen seine mangelnde Echtheit zurück, geräth auch mit ihm die Industrie auf die gefährliche Bahn der unechten Farben. Die Mode will es. Die Damenwelt liebt auch das Flüchtige, wenn es nur neu ist und kleidet. Dabei die sofortige Massenproduktion'.

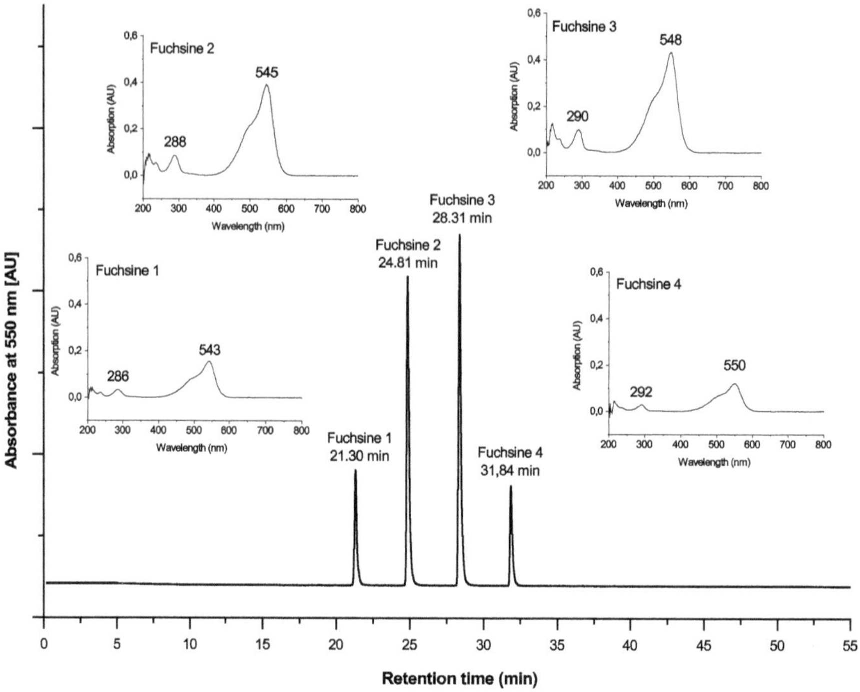


**Fig. 12.3** Chromatogram and spectra of the four components present in Fuchsine (Merck CAS 632-99-5)

## 12.3 Case Studies

### *The Tassel from the Church Sankt Leopold in Vienna Donaufeld, Austria*

Around 1880 the annual worldwide production of Fuchsine amounted to 750 tons. However, the importance of Fuchsine was decreasing in the twentieth century because new and better red-violet dyes had been discovered. But the decline was slow, and its popularity survived well into the twentieth century: the German I.G. Farbenindustrie AG was still producing about 450 tons of Fuchsine in 1930. The church Sankt Leopold in Vienna Donaufeld was built between 1905 and 1914. Above the altar hangs a lamp with a red-violet tassel (see Fig. 12.4). The investigation of the tassel (see Fig. 12.5), shows the continuing popularity of Fuchsine. Many yarns of the tassel were exposed to light, are discoloured and show now a greyish colour. However, the yarns on the inside of the tassel still show some reddish colour, since this part was protected from light. HPLC analysis reveals that the tassel was dyed with Fuchsine [41, 42].

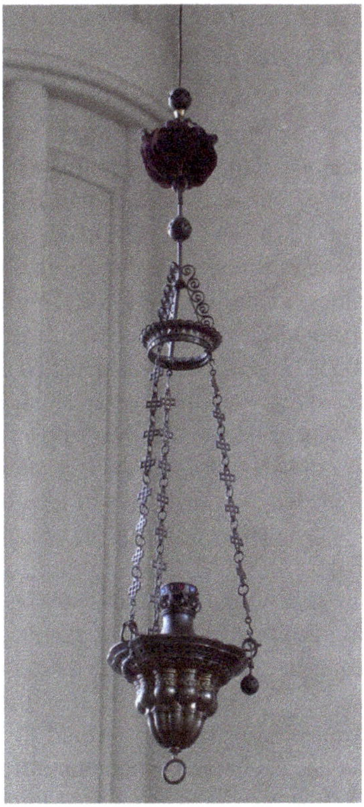

**Fig. 12.4** Lamp with tassel above the altar. (© Institute of Conservation, University of Applied Arts Vienna, Austria. Photograph: Martina Griesser-Stermscheg)

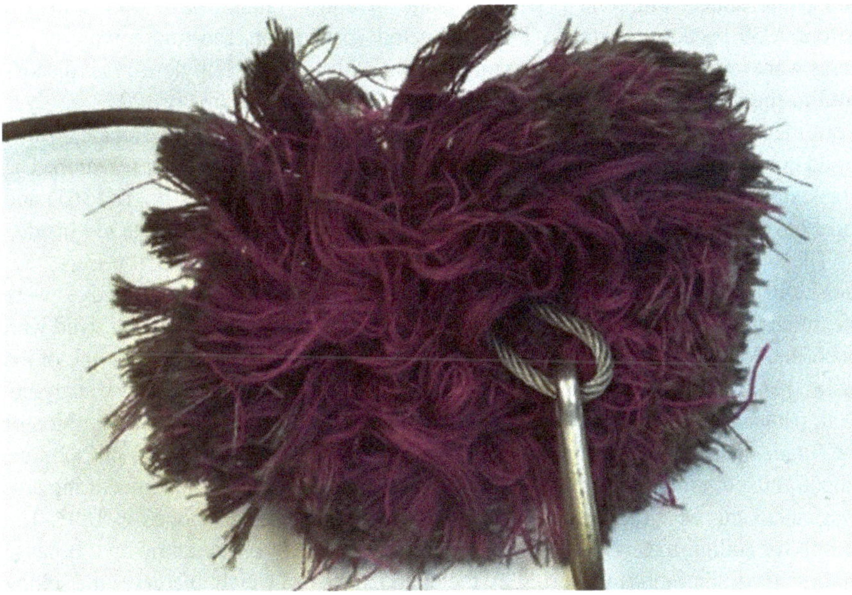

**Fig. 12.5** Tassel from the lamp above the altar. Inside of the tassel. (© Institute of Conservation, University of Applied Arts Vienna, Austria. Photograph: Martina Griesser-Stermscheg)

## *Parasols and Umbrellas from the vorarlberg Museum, Bregenz, Austria*

The vorarlberg museum in Bregenz has a collection of umbrellas and parasols, illustrating a wide spectrum of types and styles from the early nineteenth century to the 1950s. The research was focused on the dyes of several parasols and umbrellas (see Fig. 12.6) [43]. The results are presented in Table 12.2.

One aim of the dye analysis was to contribute to the dating of the umbrellas. The oldest umbrellas (Tr 1118, Tr 1220) date from after 1820 and are made of cotton dyed with natural dyes, which are not suitable for dating. In two parasols (Tr 0421, Tr 0128), Fuchsine was detected as the main dye and the presence of this dye proves that they were made after 1856. In the violet silk of the ladies' parasol (Tr 1224) Fuchsine occurs together with unknown violet and blue dyes, maybe comparable to Water blue (1862). In the loop of the violet silken ladies' parasol (Tr 1830), Fuchsine was identified with dyes of later origin.

An interesting case is the checked parasol with pinkish-purple stripes (Tr 0128, see Figs. 12.7 and 12.8). According to the museum's inventory book, this umbrella was dated between 1815 and 1848, the Biedermeier period. However, the use of a metal frame indicates that it was manufactured after 1870, and the identification of Fuchsine confirms that the umbrella dates from the second half of the nineteenth century.

Other synthetic dyes identified were Mauveine (1856), Safranine T (1858), Methyl violet (1861), Fast red AV (1876), Ponceau G (1878), Amaranth (1878), Azo flavine 3R (1880), Rhodamine B (1887) and Patent blue V (1888).

Mauveine was also found in the black cotton of the umbrella for men (Tr 1227) which suggests that it was made after 1856. The dark-blue decorative bow of the blue and white ladies' umbrella (Tr 0751) made of cotton cannot have been attached before 1859 because Safranine T was detected. In addition, four unknown components were found in the bow, which might indicate the use of Methylene blue, discovered in the year 1876. In the black umbrellas (Tr 0752, Tr 1062, Tr 1227), no dyes were identified. It is possible that the fabrics were dyed with Aniline black, which could not be analysed with the techniques used, but this requires further research. The black silken ladies' umbrella has a red lining made of cotton and silk (Tr 1365) and must have been produced after 1878, as Fast red AV was identified in the red inside.

The silks of the violet ladies' parasol and the violet loop (Tr 1830, see Figs. 12.9 and 12.10) are dyed differently with multiple dyes. In the parasol, three acid dyes were identified: Fast red AV, Azo flavine 3R and Patent blue V, while the loop was dyed with Fuchsine, Methyl violet, Ponceau G, Fast red AV and Amaranth. In the pink silk of the ladies' parasol (Tr 1867) Rhodamine B and an unknown red synthetic dye were analysed. Both parasols showed showed a similar phenomenon of degradation. The condition of the fragile silk threads was investigated by SEM-EDX. The elements tin, phosphorus, silicon and aluminium were detected. They correspond with a textile manufacturing process called silk weighting according to the Neuhaus process, developed in 1893. The reason for adding metals to silk was to increase the weight of the fabric and, because in those days silk fabrics were sold by the pound, the extra weight increases the selling price of the fabric. The disadvantage of weighted silk is that it loses much of its strength and durability [44]. It can be concluded that neither parasol was made before 1893.

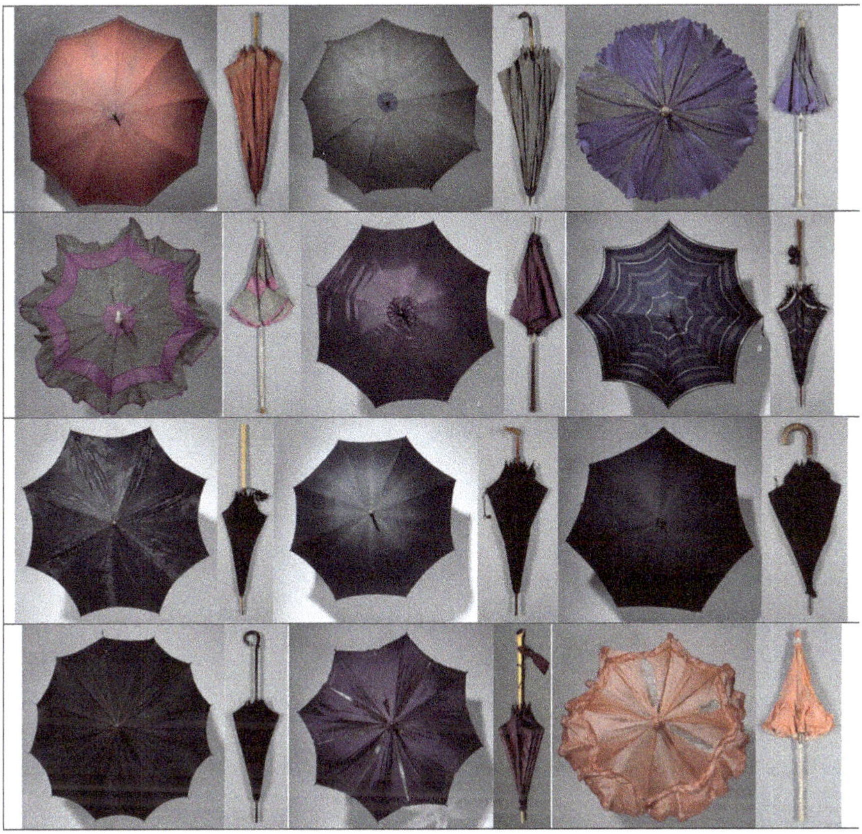

**Fig. 12.6** Parasols and umbrellas from the collection of the vorarlberg museum, Bregenz, Austria, first row: red umbrella, cotton (Tr 1118), greenish-blue umbrella, cotton (Tr 1220), violet ladies' parasol, silk (Tr 1224); second row: checked parasol with pinkish-purple stripes, silk (Tr 0128), violet ladies' parasol, silk (Tr 0421), blue and white ladies' umbrella, cotton (Tr 0751); third row: black ladies' umbrella, silk (Tr 0752), black umbrellas for men, cotton (Tr 1062, Tr 1227); fourth row: black silken ladies' umbrella with red lining of cotton and silk (Tr 1365), violet ladies' parasol with loop, silk (Tr 1830), pink ladies' parasol, silk (Tr 1867). (© Institute of Conservation, University of Applied Arts Vienna, Austria. Photograph: Leonie Tscherner and Stefan Oláh)

**Table 12.2** Dye analysis of the parasols and umbrellas

| Object (object number) | Analysis results | Conclusion | Dating (museum)<br>Dating (based on dye analysis) |
|---|---|---|---|
| Red umbrella (Tr 1118), cotton | Alizarin<br>Purpurin<br>Two Alizarin equivalents<br>Purpurin equivalent | Madder (*Rubia tinctorum* L.) | After 1820<br>Not possible, only natural dyes |
| Greenish-blue umbrella (Tr 1220), cotton | Indigotin<br>Indirubin<br>Isatin<br>Traces of alizarin and purpurin | Indigo (*Indigofera tinctoria* L.)<br>Madder, probably added to the dye bath to enhance fermentation | After 1820<br>Not possible, only natural dyes |

(continued)

**Table 12.2** (continued)

| Object (object number) | Analysis results | Conclusion | Dating (museum)<br>Dating (based on dye analysis) |
|---|---|---|---|
| Violet ladies' parasol (Tr 1224), silk | Fuchsine<br>Unknown violet and blue dyes, maybe comparable to Water blue (1862) | Fuchsine (1856)<br>Unknown violet or blue synthetic dye | After 1835<br>After 1856 |
| Checked parasol with pinkish-purple stripes (Tr 0128), silk | Fuchsine<br>Two unknown violet components<br>Unknown yellow components, probably degradation products | Fuchsine (1856) | After 1870<br>After 1856 |
| Violet ladies' parasol (Tr 0421), silk | Fuchsine<br>Unknown violet and yellow components | Fuchsine (1856) | c. 1900<br>After 1856 |
| Dark-blue bow of a blue and white ladies' umbrella (Tr 0751), cotton | Safranine T<br>Four unknown components, poor match with Methylene blue (1876) | Safranine T (1859)<br>Unknown blue synthetic dye | 1900–1920<br>After 1859 |
| Passementerie from a black ladies' umbrella (Tr 0752), silk | Black dye not identified | Maybe Aniline black | 1900–1920<br>Unknown |
| Black umbrella for men (Tr 1062), cotton | Black dye not identified | Maybe Aniline black | After 1870<br>Anknown |
| Black umbrella for men (Tr 1227), cotton | Mauveine<br>Mauveine equivalents<br>Unknown orange components<br>Black dye not identified | Mauveine (1856<br>Unknown synthetic orange dye<br>Maybe Aniline black | After 1870<br>After 1856 |
| Red inside of a black silken ladies' umbrella (Tr 1365), cotton and silk | Fast red AV<br>Three fast red equivalents<br>Unknown yellow or orange dye | Fast red AV (1878)<br>Unknown synthetic yellow or orange dye | 1890–1910<br>After 1878 |
| Violet ladies' parasol with a violet loop (Tr 1830), silk | *Parasol*: Patent blue V<br>Azo flavine 3R<br>Fast red AV<br>Three Fast red AV equivalents<br>Neuhaus silk weighting process (1893) | Fast red AV (1878)<br>Azo flavine 3R (1880)<br>Patent blue V (1888) | 1900–1920<br>After 1893 |
| | *Loop*: Fuchsine<br>Methyl violet<br>Ponceau G<br>Fast red AV<br>Amaranth | Fuchsine (1856)<br>Methyl violet (1861)<br>Ponceau G (1878)<br>Fast red AV (1878)<br>Amaranth (1878) | 1900–1920<br>After 1878 |
| Pink ladies' parasol (Tr 1867), silk | Rhodamine B<br>Two Rhodamine B equivalents<br>Azo flavine equivalent<br>Unknown acid red dye<br>Neuhaus silk weighting process (1893) | Rhodamine B (1887)<br>Unknown synthetic red dye | After 1870<br>After 1893 |

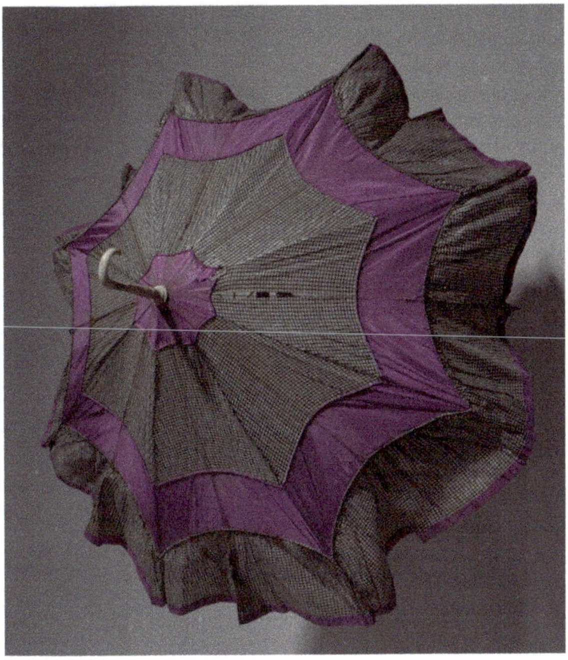

**Fig. 12.7** Checked silken parasol with pinkish-purple stripes, dyed with Fuchsine (Tr 0128). vorarlberg museum, Bregenz, Austria. (© Institute of Conservation, University of Applied Arts Vienna, Austria. Photograph: Stefan Oláh)

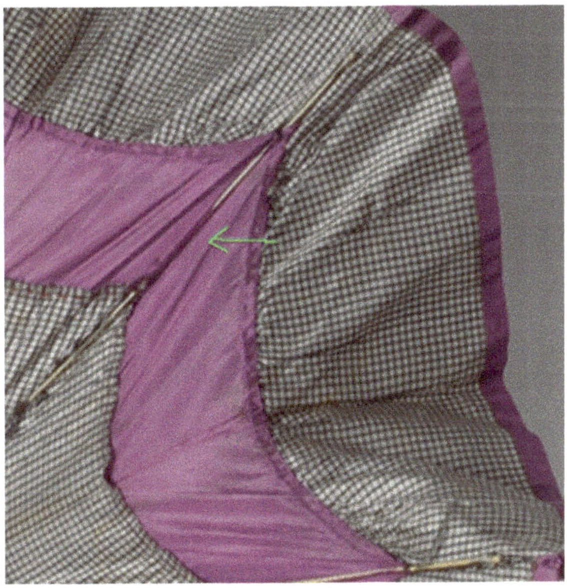

**Fig. 12.8** Detail of the checked silken parasol with pinkish-purple stripes (Tr 0128). vorarlberg museum, Bregenz, Austria. (© Institute of Conservation, University of Applied Arts Vienna, Austria. Photograph: Leonie Tscherner)

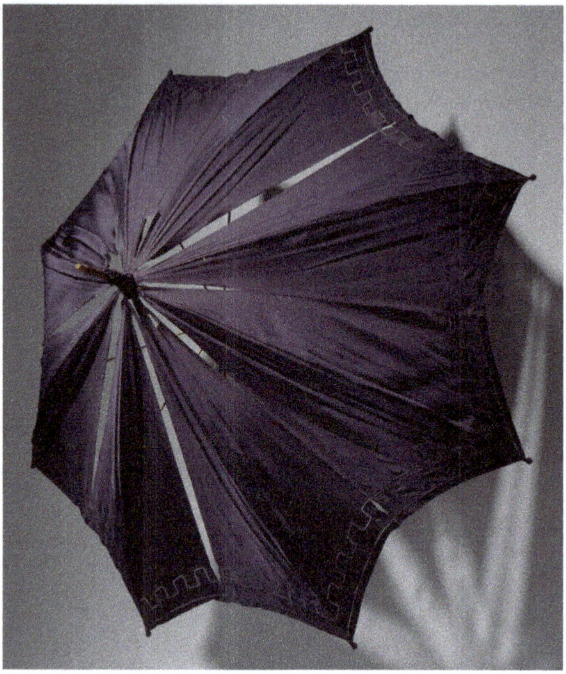

**Fig. 12.9** Violet silken ladies' parasol with loop (Tr 1830) (before conservation), vorarlberg museum, Bergenz, Austria. (© Institute of Conservation, University of Applied Arts Vienna, Austria. Photograph: Stefan Oláh)

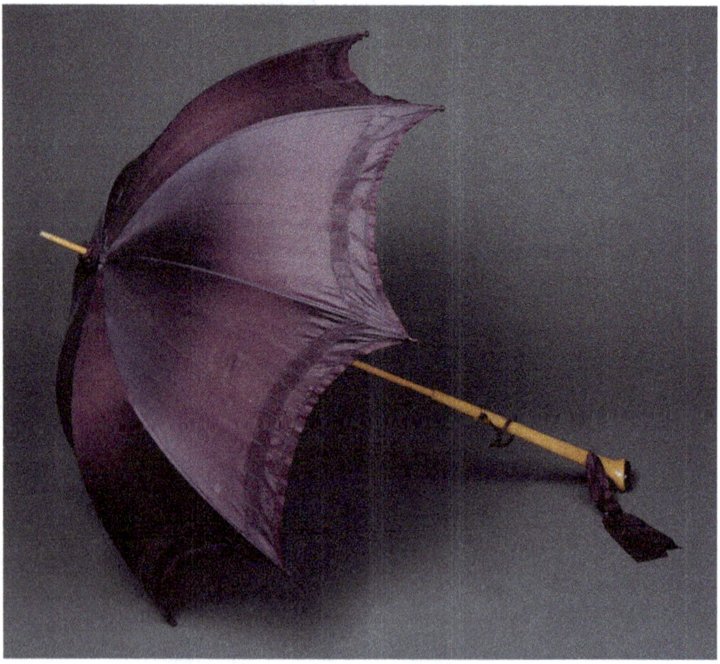

**Fig. 12.10** Violet silken ladies' parasol with loop (Tr 1830) (after conservation), vorarlberg museum, Bergenz, Austria. (© Institute of Conservation, University of Applied Arts Vienna, Austria. Photograph: Georg Oberlechner)

### Other Case Studies

Fuchsine was found in the wall-hangings of the Museum Willet-Holthuijsen (see Chap. 11). In addition, Fuchsine was also detected in three embroideries designed by Émile Bernard, which are described in Chap. 13, and in the coloured balls of wool in Vincent van Gogh's Chinese lacquer box presented in Chap. 10. Furthermore, Fuchsine was identified in the Hofdamask wall-hangings in Schloß Schönbrunn and the Hofburg in Vienna, Austria, discussed in Chap. 15.

## 12.4 Methyl Violet

Another important dye from the tri-arylmethane class is Methyl violet. Since its history is closely related to Hofmann's violet, which is actually a name for a group of dyes, these dyes will be discussed together.

Colour Index name: Basic Violet 1; Colour Index number: 42535
IUPAC name: N-(4-(bis(4-(dimethylamino)phenyl)methylene)cyclohexa-2,5-dien-1-ylidene)methanaminium chloride (this is the pentamethylated form)
Discoverer: Charles Lauth, 1861
Contemporary terminology: Methyl violet (En), Violet de Paris (Fr), Methylviolett (Ge), Methylviolet (NL)
Chemical formula: $C_{24}H_{28}N_3Cl$ (Fig. 12.11)

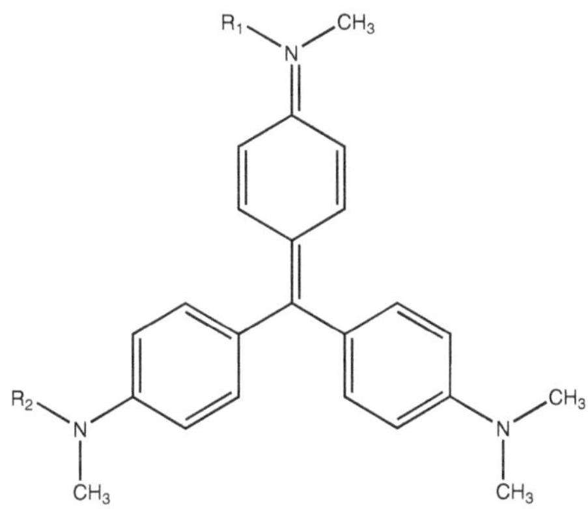


**Fig. 12.11** Structural formula of Methyl violet. The side groups can be respectively $R_1 = R_2 = H$ (tetramethylated), $R_1 = H$, $R_2 = CH_3$ (pentamethylated), $R_1 = R_2 = CH_3$ (hexamethylated also known as Crystal violet). Methyl violet consists of a mixture of these components; often other methylated forms are present as well

## *Hofmann's Violet*

Colour Index name: Basic dye; Colour Index number: 42530
IUPAC name: N-ethyl-4-[[4-(ethylamino)phenyl]-(4-ethylimino-3-methylcyclohexa-2, 5-dien-1-ylidene)methyl]aniline;hydrochloride
Discoverers: August Wilhelm Hofmann and Geyger, 1863
Comtemporary terminology: Hofmann's violet (En), Violet Hofmann (Fr), Hofmann's Violett (Ge), Hofmann's violet (NL)
Obsolete terminology: Primula, Jodviolett, Dahlia, Patentviolett, Alexandra 5R (Ge)
Chemical formula: $C_{26}H_{32}N_3Cl$ (Fig. 12.12)

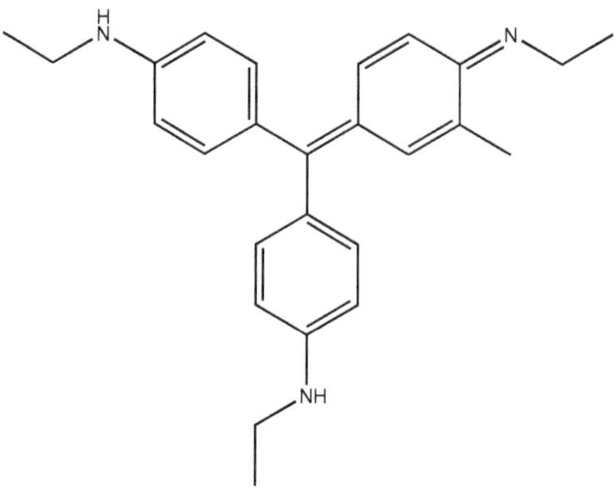


**Fig. 12.12** Structural formula of Hofmann's violet. Note that, as in the case of Methyl violet, several additional methyl- and ethyl groups can be attached to nitrogen

## *History*

At the end of the 1850s, the tri-arylmethane dye class was extended notably when violet dyes, such as Regina purple, were discovered. Soon new materials were prepared, and one of the most important of these was dimethylaniline. In 1861, the French chemist Charles Lauth (1836–1913) began to study the reaction of mono- and dimethylaniline. He performed some experiments on the product Hofmann had obtained by heating aniline with methyl iodide; the product he described as methylaniline. Research showed, although this was not immediately recognised, that this process yields a mixture of methylaniline and dimethylaniline (mainly dimethylaniline), and by oxidising this mixture with copper(II) chloride ($CuCl_2$), sand and nitric acid, Methyl violet was prepared. Lauth obtained a beautiful violet dye, but he did not consider the continuation of his experiments worthwhile since the light fastness of his product was less compared to that of the natural dyes used at that time [9, 14, 15, 18, 26, 45, 46]. So he stopped his experiments for a time.

In 1863, after Lauth's discovery of Methyl violet, August Wilhelm Hofmann showed that different alkyl groups could be introduced into the fuchsine molecule to produce dyes of various purple or violet colours. Several alkylated fuchsine dyes (trimethylrosaniline and triethylrosaniline), named Hofmann's violet, were made by heating Fuchsine in an alcoholic solution with an alkyl halide, either methyl iodide ($CH_3I$) or ethyl iodide ($C_2H_5I$), and caustic soda. The colour depends on the amount of iodine; for a reddish-violet this is lower than for a violet. A maximum of three methyl or ethyl groups can be introduced by this method. The colour could be varied by adjusting the extent of alkylation and has a spectrum of compositions depending on the production method [40]. The ethyl derivative is redder than the methyl product. The dyes are sold as chlorine hydrates and produce a very pure reddish violet. The methyl- and ethylrosanilines led to the production of various forms of Hofmann's violet, which is explained by the English chemist Frederick Crace-Calvert (1819–1873). 'By varying the circumstances of experiment, instead of three of the hydrogen being replaced by ethyl, dyes may be obtained having two, or only one, replaced by ethyl; moreover, by substituting methyl iodide for ethyl iodide, corresponding methyl compounds may be prepared. In this way the Hofmann violets are obtained of different shades, varying from RRR, the very red, which is principally a salt of monomethylated rosaniline to BBB, the bluest shade' [47].

In the same year, 1863, Hofmann patented his discovery (B.P. 1.291 and F.P. 59.309) and published an article in which he described several violet dyes which he named the Hofmann's violets. Société La Fuchsine acquired the French rights of this dye in May 1863. From 1864, the Hofmann's violets displaced Perkin's Mauveine. Maybe Hofmann's article and the introduction of these violets were reasons for Lauth to start again with his experiments. Eventually, a successful *N,N*-dimethylaniline process was discovered by Charles Bardy, a French chemist to the firm Poirrier & Chappat fils, which involved heating a mixture of aniline with hydrochloric acid and methyl alcohol to 280–300 °C in a closed vessel (F.P. 71.970, filed on the 16th of June 1866, and Lauth's British patent no 3.195, filed in 1866). Lauth noted that Methyl violet is produced from pure aniline, free from toluidine. Aniline containing more than 5% of toluidine is not suitable for the production of Methyl violet. By 1866, large quantities were available from this process and *N,N*-dimethylaniline was used in the production of Methyl violet. For a long time, Methyl violet was only manufactured by the firm Poirrier & Chappat fils in Saint-Denis near Paris. A block, weighing about 150 kilogrammes and of a fine golden-green colour, which they named Violet de Paris, was exhibited in the Paris International

Exhibition of 1867 [9, 14, 15, 18, 26, 45, 46]. The introduction of Methyl violet soon almost displaced the Hofmann's violets.

The actual structure of Methyl violet remained a mystery for several years and was not solved until 1873. The question arose as to whether Methyl violet is identical to or an isomer of Hofmann's violet. Lauth considered that it was isomeric. Hofmann, interested in the relationship between these colouring matters, investigated the subject and published his results in 1873. He studied the conditions under which the colouring matter could be formed, showing that Methyl violet could be produced from pure dimethylaniline [26].

## The Letters R and B

Red or blue shades of various alkyl substitution products of pararosaniline are indicated by adding the letters R, ranging from R up to 5R, and B, ranging from B up to 10B, and so forth to the name Methyl violet. Methyl violet is a mixture of the hydrochlorides of tetramethylpararosaniline, pentamethylpararosaniline and hexamethylpararosaniline, and gives bluer shades than Hofmann's violet; these correspond to the letters B and 2B. Hofmann's violet (trimethylrosaniline and triethylrosaniline) is a mixture of the lowest substitution products of pararosaniline and gives redder shades, corresponding to the letters R, 2R and 3R. The six hydrogen atoms of the rosaniline base, which are in the amino groups, can be replaced by a methyl group, an ethyl group or a benzyl group. The most commonly known compounds of the rosaniline series, containing 4, 5 and 6 methyl groups in these positions, are known as the tetramethyl-, pentamethyl- and hexamethylpararosanilines respectively. It must be noted that Methyl violet consists of a mixture of the three dyes, but nowadays they are also available in their pure form. When a component contains a larger number of methyl groups, it becomes bluer in colour. As an example, the tetramethyl form is known as Methyl violet 2B; the pentamethyl is known as Methyl violet 6B and is darker (in dye form) than 2B. Compounds designated as Methyl violet 5B, 6B, and 7B are still bluer in shade and contain pentamethylpararosaniline, in which a benzyl group has been introduced into one of the amino groups. The only textile dye in the series which has a definite chemical formula is Crystal violet, the hexamethylated form, also known as Methyl violet 10B, prepared by the phosgene method. This dye is a deep-blue violet, darker than Methyl violet 2B and Methyl violet 6B [48, 49].

## Trade Names of Methyl Violet

The firm Poirrier & Chappat fils in Saint-Denis was for a long time the only manufacturer of Methyl violet under their trade name Violet de Paris. Methyl violet was put on the market by AGFA, BASF, Bayer, Cassella, Ruch, Beyer & Kegel, Jäger,

**Table 12.3** Discoverers and patents of Methyl violet and Hofmann's violet

| Name<br>C.I. name<br>C.I. number | Discoverers | Year of discovery | Patents |
|---|---|---|---|
| Methyl violet<br>Basic Violet 11 | Charles Lauth | 1861 | B.P. 3.195/1866 |
| 42535 | Poirrier & Chappat,<br>Violet de Paris | 1866 | F.P. 71.970/1866 |
| Hofmann's violet<br>Basic dye<br>42530 | August Wilhelm Hofmann and Geyger | 1863 | B.P. 1.291/1863<br>F.P. 59.309/1863 |

Weiler-ter Meer, Farbwerke ML&B, Griesheim-Elektron, Read Holliday, Roberts, Dale & Co., Wülfing-Dahl, Claus & Rée, Griesheim G.m.b.H. and Heller & Merz. Other trade names are Methylviolett BBN by Sandoz; Methylviolett BO by Cassella and Leonhardt; Methylviolett V3 by Kalle; Methylviolett bläulich, rötlich by Jäger; Dahlia B by Wülfing-Dahl; Violet au Méthyl by Manufacture Lyonnaise; Violet 3B extra by Soc. Chim. Us. Rhône and Violet Méthyle 2B, B extra konz. by Wiescher. Methyl violet is also produced under many different trade names, such as Direct Violet, Violet 3B, Gentian Violet, Best Violet, Brilliant, Indian dye, Fine Violet, Genuine Violet, Superfine Violet, Superior Violet, Malbery Blue, Violet de Methylaniline, Violet non plus ultra, Violet soluble, Pariser Violett, Violett B, Violettbase 2B, Violett B, 3R, various brands esp. B, 2B, AA, 4B conc., 2B cryst., 2B lumps, 4B extra, 5B, 4BN, NE, NP, R. conc. pdr, and conc. pencil special.

### *Trade Names of Hofmann's Violet*

Although, after the discovery of Methyl violet, the Hofmann's violets were almost completely replaced, several dye companies still manufactured this violet dye under various trade names. The dye was put on the market under the trade name Rotviolett 5R extra by BASF. Other trade names were Primula R wasserlöslich by Farbwerke ML&B; Violett 4RN by CIBA; Violett R and RR by Soc. Chim. Us. Rhône; Violet Hofmann by Soc. Anon. St. Denis; Rotviolett and Hofmann's Violet by Weiler-ter Meer (Table 12.3).

### *Applications*

The main use of Methyl violet is as a purple dye for textiles, especially silk and cotton, and to give deep violet colours in paint and ink [39].

## Properties

Methyl violet shows a green metallic lustre as a solid and is easily soluble in water and alcohol. Methyl violet is sold as a chlorine derivative or as a zinc double salt. The latter consists of small crystals, while the former is usually in a powder form or forms irregular pieces.

The dyes on textiles and the pigments of Methyl violet generally have a poor light fastness, which is well documented in literature of the period. In 1867, Charles Lauth wrote on Methyl violet, which he called Violet de Paris: 'The violets obtained from methyl-aniline possess a richness and purity which leave nothing to be desired.... Nevertheless, they were not adopted by manufacturers, who, indeed, at the time above-mentioned [1861], attached less importance to the beauty of a colour than to its permanence. In this latter respect, the methyl-aniline violets do not excel, and consequently, dyers would have nothing to do with them. Gradually, however, people have become accustomed to colours which fade on exposure to the sun. Indeed, the public taste at the present day, in colours as in everything else, inclines rather to tinsel than to solid excellence'. [15].

For Methyl violet and Hofmann's violet, Adolf Lehne wrote[3]: 'After 5 days of light exposure, in June, the colour of Methyl violet and Hofmann's violet is duller and in 18 days the colour has been strongly faded'. Artificial light ageing tests presented in the Colour Index showed that Methyl violet, dyed on wool, has a light fastness of 1 according to the German norms developed in 1951, so quite poor. The wash fastness is 4.

## Recipes

According to Lehne, wool can be dyed using recipe V1, silk using recipe V7, cotton or linen following recipe V11 and jute applying recipe V15. The concentrations of dye used are typically between 0.5 to 1.5%. When used as a printing ink, recipes V19 for wool and V23A for cotton are given. Lehne also mentioned that it can be used to colour paper.

## Identification Using UHPLC-PDA

Like Fuchsine, Methyl violet (Sigma-Aldrich CAS 8004-87-3) is composed of a mixture of structurally related components. In case studies, four components are found that differ in methylation. In the chromatogram shown (see Fig. 12.13), the

[3] Methyl Violett and Hofmann's Violett: Echtheit: 'Nach 5 tägiger Einwirkung der Witterung (Juni) viel matter, nach 18 Tagen sehr stark verblasst'.

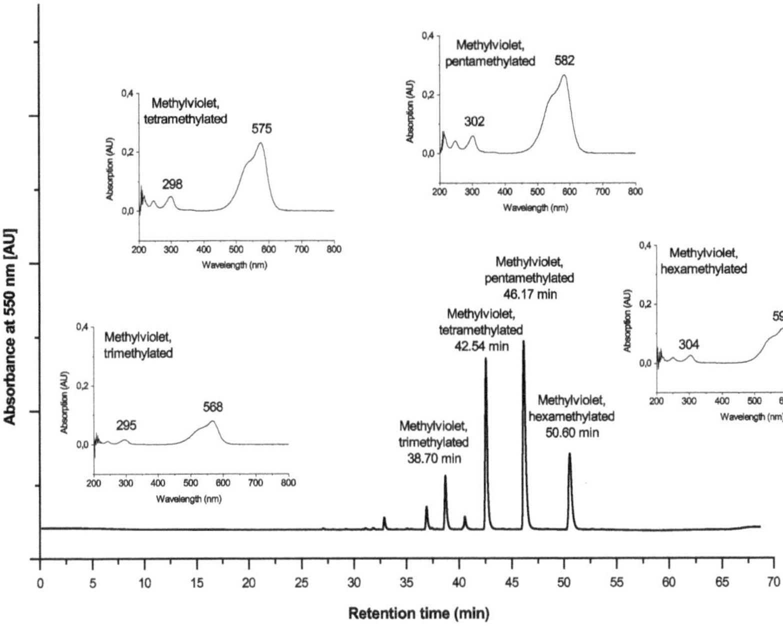


**Fig. 12.13** Chromatogram and spectra of the components present in Methyl violet (Sigma-Aldrich CAS 8004-87-3)

trimethylated form elutes at 38.495 min., the tetra form at 42.312 min., the pentamethylated one at 45.985 min. and finally at 50.417 min. the hexamethylated component is found: this is also known as Crystal violet. It is usually these four components which are found in historical objects, with approximately the ratio observed in Fig. 12.13. Sometimes they are accompanied by side-products which could be due to the synthesis process or result from degradation.

## 12.5 Case Studies

### *Fuchsine and Methyl Violet as Artists' Paints*

The main problem for the use of early synthetic organic dyes and pigments is that they were insufficiently permanent for the purpose of art. Seduced by the brilliance of the colours, artists used them with disastrous results. Even as late as 1930, in the book *The Permanent Palette* by the German-North American physician and artist Martin Fischer (1879–1962), artists are warned: 'It is a good rule for the painter to eliminate everything from the manufacturer's catalogue which bears a name unfamiliar to him or the content of which is not wholly disclosed to him' [50]. In his chapter on the natural and synthetic lakes, he condemns many pigments of the triarylmethane series, including Fuchsine, Rose aniline, Magenta red, Aniline red,

Solferino rose, Methyl violet, Malachite green, Methyl green, Brilliant green and Aniline blue. Their use, he says, should be forbidden in works of art that aspire to permanence [50].

## *Paint Analysis on Paintings of the Paris Period of Vincent van Gogh*

Vincent van Gogh used lakes of several varieties in his paintings. During his stay in Paris, from 1886 to 1888, he mainly painted his reddish hues with a variety of natural organic pigments, such as lakes made from Kopp's purpurin, which is purpurin-rich dye extracted from Madder using diluted sulfuric acid, Cochineal and Redwood, or mixtures of these. Interestingly, in the painting *Glass with Yellow Roses*, (F0218, Van Gogh Museum, Amsterdam (Vincent van Gogh Foundation)), painted late June–mid-July 1886, a synthetic organic violet pigment was found. Under a translucent layer of Kopp's purpurin-rich madder lake, there is a paint layer consisting of the pigments vermilion, lead white, zinc white, red ochre, synthetic ultramarine blue and a violet material identified as Methyl violet [51]. This proves that Methyl violet, put on the market in 1866 by the firm Poirrier & Chappat fils in Saint-Denis, was used by Vincent van Gogh twenty years later as an artist's paint.

## *Bureau de ministre by Pieter Lodewijk Kramer*

The Dutch architect, sculptor, interior and furniture designer Piet Kramer (1881–1961) was one of the founders of the Amsterdam School. Kramer was known to use strong brilliant colours in furniture designed by him: he used bright green, bright orange, bright violet and deep black colours. But he used synthetic dyes and these tints sadly have now completely faded. In 1915, he designed a writing desk (*Bureau de ministre*) (Inv. AB2155, RCE) shown in Fig. 12.14. During a restoration treatment, a violet colour was observed on the areas where it was protected from light (see Fig. 12.15). HPLC analysis indicates the presence of a mixture of Methyl violet and Nigrosine [52].

## *Synthetic Inks*

Synthetic dyes are not only examined on painted objects, but research on synthetic inks is also performed. The study of ink recipes showed that Methyl violet was a popular dye. Unfortunately, a Methyl violet ink, although very beautiful, is not particularly durable, as can be observed in the letters of Vincent van Gogh and in his drawings, such as *Montmajour* (F1423, Van Gogh Museum, Amsterdam (Vincent van Gogh Foundation)). The drawing is completely faded, but the edge was covered by a passepartout, which protected the ink from light. As a result, traces of the violet colour are still present [53].

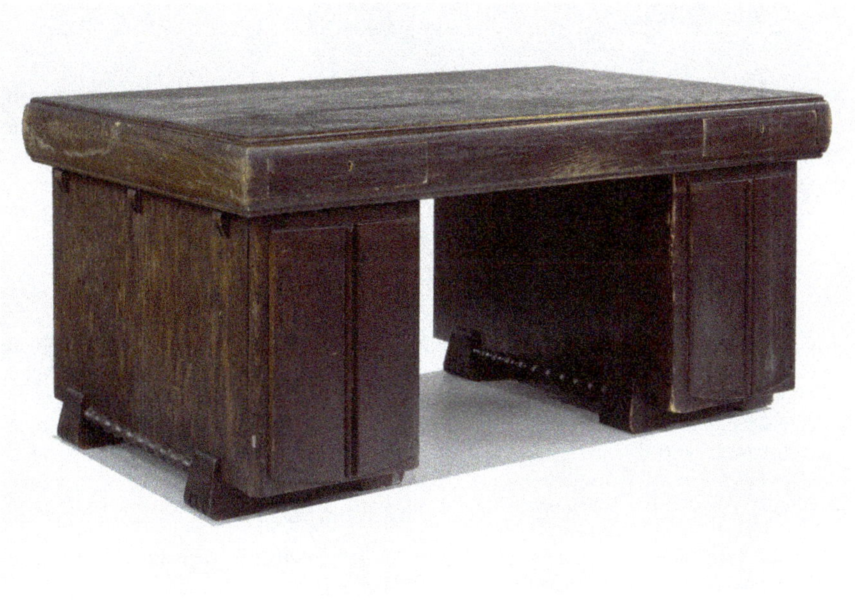

**Fig. 12.14** Piet Kramer, *Bureau de ministre*, 1915, Ash veneer on plywood, solid elm & spruce, 75 × 154 × 110 cm (H × W × D), inventory number AB2155. (Collection Cultural Heritage Agency of the Netherlands. Photograph: Cultural Heritage Agency of the Netherlands)

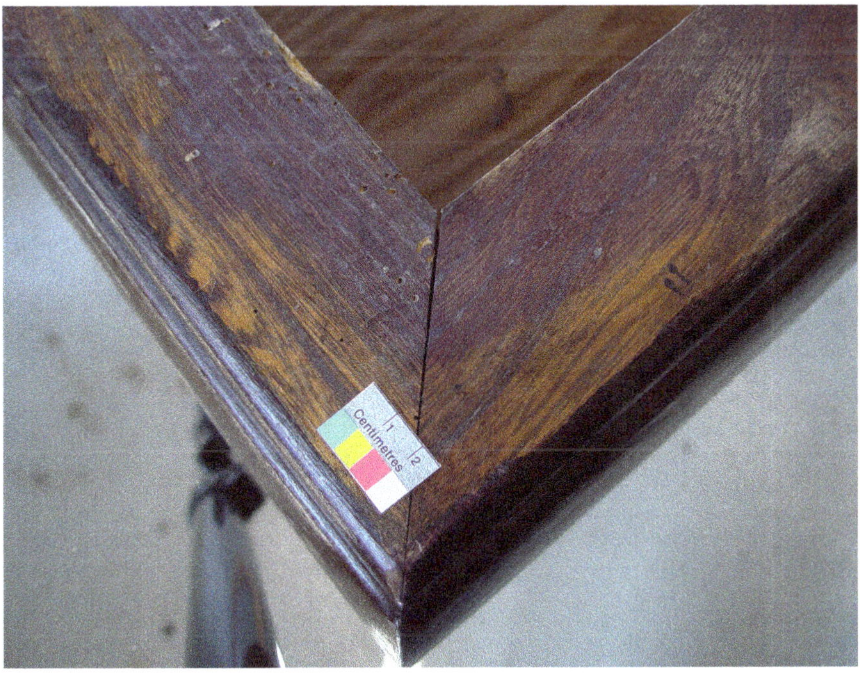


**Fig. 12.15** Detail of *Bureau de ministre* (Inv. AB2155), corner top backside showing the different colours. Top: original layer, middle: faded layer, bottom: a latter applied finish. Photograph: Ron Kievits

## Other Case Studies

Methyl violet was identified in the embroidery *Breton Women* designed by Émile Bernard and is presented in Chap. 13, and in the Hofdamask wall-hangings from the Hofburg and Schloß Schönbrunn in Vienna, Austria, discussed in Chap. 15.

## References

1. Schützenberger, P.: Traité des matières colorantes, comprenant leurs applications à la teinture et à l'impression et des notices sur les fibres textiles les épaississants et les mordants, Vol 2. Victor Masson et fils, Paris (1867)
2. Schützenberger, P.: Traité des matières colorantes, comprenant leurs applications à la teinture et à l'impression et des notices sur les fibres textiles les épaississants et les mordants, Vol 1. Victor Masson et fils, Paris (1867)
3. Bolley, P., Kopp, E.: Traité des matières colorantes artificielles dérivées du goudron de l'houille. Librairie F. Savy, Paris (1874)
4. Schultz, G., Julius, P.: Tabellarische Übersicht der künstlichen organischen Farbstoffe. R. Gaertner's Verlagsbuchhandlung, Berlin (1902)
5. Schultz, G., Julius, P.: Tabellarische Übersicht der künstlichen organischen Farbstoffe. R. Gaertner's Verlagsbuchhandlung, Berlin (1891)
6. Schultz, G., Julius, P.: Tabellarische Übersicht der künstlichen organischen Farbstoffe. R. Gaertner's Verlagsbuchhandlung, Berlin (1888)
7. Schultz, G.: Farbstofftabellen. Weidmannische Buchhandlung, Berlin (1914)
8. Natanson, J.: Mittheilungen aus dem Universitätslaboratorium in Dorpat, 2) Ueber das Acetylamin. Justus Liebig's Annalen der Chemie und Pharmacie. C.F. Winter'sche Verlagshandlung; Leipzig und Heidelberg. 98, 291–301 (1856)
9. Caro, H.: Referate, "Ueber die Entwicklung der Theerfarben-Industrie." Berichte der deutschen chemischen Gesellschaft zu Berlin. Commissionsverlag von R. Friedländer & Sohn, Berlin. 25, 955–1105 (1892)
10. Will, P.: Organische Chemie, Organische Basen. Jahresbericht über die Fortschritte der reinen, pharmaceutischen und technischen Chemie. Physik, Mineralogie und Geologie für 1858, J. Ricker'sche Buchhandlung, Giessen. 348–355 (1859)
11. Hofmann, A.W.: Ueber Aethylen-Anilin. Journal für die praktische Chemie. Verlag von Johann Ambrosius Barth, Leipzig. 77, 188–190 (1859)
12. Hofmann, A.W.: Ueber die aus Anilin entstehenden Farbstoffe. Journal für die praktische Chemie. Verlag von Johann Ambrosius Barth, Leipzig. 87, 227 (1892)
13. Mühlhäuser, O.: Die Fabrikation des Arsensäurefuchsins. Dingler's Polytechnisches Journal. Verlag der J.G. COTTA'schen Buchhandlung, Stuttgart. 266, 458–547 (1887)
14. Friedlaender, P.: Fortschritte der Theerfarbenfabrikation und verwandter Industriezweige 1877-1887. Verlag von Julius Springer. 30–33 (1888)
15. Perkin, W.H.: Hofmann Memorial Lecture. The Origin of the Coal-Tar Colour Industry, and the Contributions of Hofmann and his Pupils. J Chem Soc. 596–625 (1896)
16. Béchamp, A.: Ueber die Erzeugung des Fuchsins. Zeitschrift für Chemie und Pharmacie. Verlag von Ferdinand Enke, Erlangen. 3, 460–461 (1860)
17. Voigtländer-Tetzner, Dr.: Die Triphenylmethanfarbstoffe. Firmengeschichte BASF. 45–141 (1956)
18. Van Den Belt, H.: Why monopoly failed: The rise and fall of Société La Fuchsine. The British Journal for the History of Science. 25, 45–63 (1992). https://doi.org/10.1017/S0007087400045325

19. Chastrette, M.: La découverte de la fuchsine. Actualite Chimique. 48–53 (2009)
20. Garfield, S.: Chapter 7: The terrible glare. In: Mauve: How one Man invented a Colour that changed the World. pp. 78, 83–84. Faber and Faber Ltd., London (2000)
21. Wagner, J.R.: Vl Gruppe, Technologie der Gespinnstfasern, Theerfarbstoffe. Jahres-Bericht über die Leistungen der Chemischen Technologie für 1869. Verlag von Otto Wigand, Leipzig. 568–570 (1870)
22. Epting, Dr.: Bericht über die Fuchsinfabrikation im Jahr 1902, Neunzig Jahre Fuchsin in Hoechst. Farbwerke Hoechst Aktiengesellschaft vormals Meister Lucius & Brüning. Dokumente aus Hoechster Archiven, Beiträge zur Geschichte der Chemie. 34 (1965)
23. Brüning, A.: Darstellung des Fuchsins. Berichte der deutschen chemischen Gesellschaft zu Berlin. Ferd. Dümmler's Verlagsbuchhandlung, Berlin. 25, 25–26 (1873)
24. Coupier: Ueber die neue Methode zur Darstellung des Fuchsins, nach Brüning. Berichte der deutschen chemischen Gesellschaft zu Berlin. Ferd. Dümmler's Verlagsbuchhandlung, Berlin. 6, 423 (1873). https://doi.org/10.1002/cber.187300601140
25. Brüning, A.: Antwort auf Coupiers Bemerkungen über die Darstellung des Fuchsin ohne Arsensäure. Berichte der deutschen chemischen Gesellschaft zu Berlin. Ferd. Dümmler's Verlagsbuchhandlung, Berlin. 6, 1072 (1873). https://doi.org/10.1002/cber.18730060259
26. Hofmann, A.W.: Ueber die violetten Farbabkömmlinge der Methylaniline. Berichte der deutschen chemischen Gesellschaft zu Berlin. Ferd. Dümmler's Verlagsbuchhandlung, Berlin. 6, 352–368 (1873). https://doi.org/10.1002/cber.187300601122
27. Fischer, E., Fischer, O.: Ueber Farbstoffe der Rosanilingruppe. Berichte der deutschen chemischen Gesellschaft zu Berlin. Ferd. Dümmler's Verlagsbuchhandlung, Berlin. 12, 2344–2353 (1879)
28. Fischer, E., Fischer, O.: Darstellung des Triphenylmethans. Berichte der deutschen chemischen Gesellschaft zu Berlin. Ferd. Dümmler's Verlagsbuchhandlung, Berlin. 14, 1943–1944 (1881)
29. Fischer, E., Fischer, O.: Ueber Triphenylmethan und Rosanilin. Justus Liebig's Annalen der Chemie. C.F. Winter'sche Verlagshandlung, Leipzig und Heidelberg. 194, 295 (1878)
30. Fischer, E., Fischer, O.: Ueber einige Farbstoffe der Rosanilingruppe. Berichte der deutschen chemischen Gesellschaft zu Berlin. Ferd. Dümmler's Verlagsbuchhandlung, Berlin. 11, 2095–2099 (1878). https://doi.org/10.1002/cber.187801102240
31. Fischer, O., German, L.: Zur Kenntniss der violetten Derivate des Triphenylmethans. Berichte der deutschen chemischen Gesellschaft zu Berlin. Ferd. Dümmler's Verlagsbuchhandlung, Berlin. 16, 706–710 (1883)
32. Fischer, O., Körner, G.: Zur Kenntniss der violetten Derivate des Triphenylmethans. Berichte der deutschen chemischen Gesellschaft zu Berlin. Ferd. Dümmler's Verlagsbuchhandlung, Berlin. 16, 2904–2911 (1883). https://doi.org/10.1002/cber.188301602250
33. Fischer, O., Koerner, G.: Ueber die violetten Abkömmlinge des Triphenylmethans. Berichte der deutschen chemischen Gesellschaft zu Berlin. Ferd. Dümmler's Verlagsbuchhandlung, Berlin. 17, 98–101 (1884). https://doi.org/10.1002/cber.18840170121
34. Wichelhaus, H.: Zur Kenntniss des Farbstoffs aus Dimethylanilin und Chloranil. Berichte der deutschen chemischen Gesellschaft zu Berlin. Ferd. Dümmler's Verlagsbuchhandlung, Berlin. 16, 2005–2007 (1883). https://doi.org/10.1002/cber.18830160297
35. Leonhart Fuchs - Kräuterbuchautor Biografie, https://www.garten-literatur.de/Leselaube/persoenl/fuchs_p.htm
36. Impressum. Mycoses. 24, 120–120 (2009). https://doi.org/10.1111/j.1439-0507.1981.tb01848.x
37. Magenta - Wikipedia, https://en.wikipedia.org/wiki/Magenta
38. Solferino - Wikipedia, https://en.wikipedia.org/wiki/Solferino
39. Davies, B.H.: Dyes (and Pigments) as Food Additives (and Adulterants). Dyes in History and Archaeology. 20, 175–181 (2005)
40. Benedikt, R.: Allgemeine Waarenkunde und Rohstofflehre, Die künstlichen Farbstoffe. Verlag von Theodor Fischer, Kassel. II, 103–117 (1883)

41. Hofmann-de Keijzer, R., van Bommel, M.R., de Keijzer, M.: Materialtechnische Untersuchung einer violettroten Quaste aus der Donaufelder Kirche, Wien, Floridsdorf. Universität für angewandte Kunst Wien, Abteilung Archäometrie / Instituut Collectie Nederland, Afdeling Onderzoek. (2004)
42. Griesser-Stermscheg, M.: Buntmetall und elektrische Glühbirnen. Die Kirchenausstattung der Donaufelder Kirche im Zeichen des Wiener Sezessionismus. Doctoral thesis. University of Applied Arts Vienna, Institute of Conservation., (2005)
43. Tscherner, L.: Von Sonnendächern und Zweiflern. Die Schirmsammlung des Vorarlberger Landesmuseums. Diploma thesis. University of Applied Arts Vienna, Institute of Conservation, (2011)
44. Hacke, M.: Weighted silk: history, analysis and conservation. Studies in Conservation, Issue sup2: Reviews in Conservation, Volume 9. 53, 3–15 (2008)
45. Lauth, Ch.: On the New Aniline Dye "Violet de Paris." Laboratory. 1, 138–139 (1867)
46. Mühlhäuser, O.: Die Fabrikation des Methylviolett. Dingler's Polytechnisches Journal. Verlag der J.G. COTTA'schen Buchhandlung, Stuttgart. 264, 37 (1887)
47. Grace-Calvert, F., Stenhouse, John., Groves, C.E.: Dyeing and Calico Printing. Palmer & Howe, Manchester (1878)
48. Conn, H.J.: an Investigation of American Gentian Violets Report of Committee on Bacteriological Technic. J Bacteriol. 7, 529–536 (1922). https://doi.org/10.1128/jb.7.5.529-536.1922
49. Creighton, H.J.M.: A method for making methyl violet. Proceedings and Transactions of the Nova Scotian Institute of Science. 15, 57–61 (1919)
50. Fischer, M.: The Permanent Palette. National Publishing Society, New York (1930)
51. van Bommel, M.R., Geldof, M., Hendriks, E.: An investigation of organic red pigments used in paintings by Vincent van Gogh (November 1885 to February 1888). Art Matters, Vol. 3, 111–137 (2005)
52. van Bommel, M.R.: Internal ICN report. The restoration treatment was performed by Ron Kievits at the Instituut Collectie Nederland. (2005)
53. Neevel, H.: Logwood Writing Inks: History, Production, Forensics, and Use. Restaurator. 42, 169–191 (2021). https://doi.org/10.1515/res-2021-0015

# Chapter 13
# A Colour Boom: The Tri-Arylmethane Dyes

## 13.1 Introduction

After the discovery of Fuchsine, the first tri-arylmethane dye of the Rosaniline series, in 1856, it turned out that Fuchsine could not only be used as a dye but was also a key intermediate for other synthetic dyes of this dye class. In the beginning of the 1860s, the number of the tri-arylmethane dyes was increasing, especially violets and blues, for example Imperial violet (1859), Regina purple (1860), Hofmann's violet (1863), Britannia violet (1864), Methyl violet (1866), Aniline blue (1860), and a series of sulfonated, water-soluble blue dyes (1862).

In 1860, Adam Charles Girard and Georges de Laire obtained Aniline blue (Bleu de Lyon or Lyon blue), the first synthetic blue, by heating Fuchsine with an excess of aniline. Two years later, in 1862, Edward Chambers Nicholson found that the water-insoluble basic dye Aniline blue, when treated with fuming sulfuric acid, could be converted into the water-soluble acid dye Alkali blue (Nicholson blue). This discovery led to the preparation of a number of water-soluble mono-, di- and tri-sulfonic blue acid dyes, such as Methyl blue, Water blue, Cotton blue, Navy blue and Soluble blue. Subsequently, fuming sulfuric acid was used not only for water-insoluble dyes, but also for water-soluble dyes, such as Fuchsine. This process was patented by the BASF in 1877, resulting in the water-soluble acid dyes Fuchsine S, Red violet 4RS and Red violet 5RS.

With new intermediates, such as dimethylaniline prepared from methyl iodide and aniline and by oxidising with copper chloride, Charles Lauth obtained Methyl violet in 1861, which soon replaced the Hofmann's violets.

By using Fuchsine and Methyl violet as starting materials, the green dyes Aldehyde green (1863), Perkin's green (1864), Iodine green (1866) and Methyl green (1871) were obtained. Iodine green was prepared by treating Fuchsine with an excess of methyl iodide, but this was soon replaced by Methyl green, prepared from Methyl violet and methyl chloride.



M. R. van Bommel, M. de Keijzer, *Bright Colours from the Past*, Cultural Heritage Science, https://doi.org/10.1007/978-3-031-88664-5_13

At the end of the 1870s, the basic green dyes of the so-called Malachite series, such as Malachite green (1878), Brilliant green (1879) and Victoria green 3B (1883), were discovered. In 1877, Otto Philipp Fischer prepared a bluish-green colour by condensing dimethylaniline with benzaldehyde in the presence of sulfuric acid, which he named Bittermandelölgrün. Simultaneously, but independently, Oscar Doebner used dimethylaniline with benzotrichloride and prepared the same dye that Fischer had developed earlier. In collaboration with Doebner, AGFA filed a patent in 1878 to protect the process and the dye was sold as Malachite green.

The basic dyes of the Malachite series can also be treated with fuming sulfuric acid, resulting in the water-soluble green acid dyes, including Helvetia green (1878), Light green (1879) and Guinea green B (1883), and the Patent blues (1888).

In 1883, a new episode of the tri-arylmethanes commenced, when Alfred Kern from the aniline factory Bindschedler & Busch employed the toxic gas phosgene in dye synthesis. To overcome the technical and scientific difficulties associated with phosgene, the company joined forces with BASF and for a short period Kern and Heinrich Caro worked together. The process involved the reaction of dimethylaniline with phosgene to give the intermediate Michler's ketone, which reacted with additional dimethylaniline in the presence of phosphorus oxychloride and hydrochloric acid, resulting in Crystal violet. The co-operation was so successful that in 1883 a series of violet and blue tri-arylmethanes, including Ethyl violet and the naphthylamino derivatives Victoria blue B, Victoria blue 4R and Night blue, were invented. These were followed by several sulfonated acid violets and alkali violets, such as Acid violet 7B, and Alkali violet, developed between 1884–1886 and 1891 by Carl L. Müller of BASF. In 1884, the collaboration between the two companies also opened the way to a new class of dyes, the di-arylmethanes, of which yellow Auramine was the most successful.

In the 1902 book *Tabellarisch Übersicht der im Handel befindlichen künstlichen organischen Farbstoffe* by the German chemists Gustav Schultz and Paul Julius, 681 synthetic dyes are listed [1]. The number of dyes shows the importance of a dye class: the azo dyes are the greatest group with nearly 400 members, followed by the tri-arylmethanes with almost 70.

The main problem for the use of tri-arylmethane dyes is their poor lightfastness. In Adolf Lehne's 1893 book, *Tabellarische Übersicht über die künstlichen organischen Farbstoffe und ihre Anwendung in Färberei und Zeugdruck*, more than 40 tri-arylmethane dyes are listed and the light fastness of each dye is mentioned. In most cases, he writes that, after five days of light exposure in June, the colour has strongly faded or very strongly faded and after 18 days, the colour has very strongly faded or almost completely faded. For example, on Malachite green he wrote the following text[1]: 'After 7 days exposed to the elements (July) very strongly faded'. On Patent blue Lehne wrote[2]: 'After 5 days exposed to the elements (June) changed

[1] Malachitgrün: Echtheit: 'Nach 7 tägiger Einwirkung der Witterung (Juli) sehr stark verblasst'.

[2] Patentblau: Echtheit:'Nach 5 tägiger Einwirkung der Witterung (Juni) sehr wenig verändert, etwas matter, nach 18 Tagen ziemlich stark verändert'.

very little, somewhat duller, but severely changed after 18 days'. After 1875 there is a shift towards the use of synthetic dyes with an increased light fastness. The early tri-arylmethanes were replaced by better, more light fast synthetic dyes, and it seems that sulfonated dyes have a rather good light fastness.

As a consequence of their poor fastness properties, some of the tri-arylmethane dyes developed were only on the market for a few years or decades. As a result, several were not commercially available at the end of the century and, for example, not recorded in the Colour Index. In Table 13.1, an overview of the most relevant dyes developed are presented. Several of them will be discussed in more detail below.

**Table 13.1** Important tri-arylmethanes

| Name | C.I. name | C.I. number | Discoverers | Year of discovery |
|---|---|---|---|---|
| Fuchsine | Basic Violet 14 | 42510 | Natanson | 1856 |
| Violet Imperial<br>Regina purple | Basic dye | 42515 | Girard and De Laire<br>Nicholson | 1859/1860<br>1862 |
| Methyl violet | Basic Violet 1 | 42535 | Lauth | 1861/1866 |
| Hofmann's violet | Basic dye | 42530 | Hofmann and Geyer | 1863 |
| Britannia violet | n.a. | n.a. | Perkin | 1864 |
| Aniline blue | Solvent Blue 3 | 42775 | Girard and De Laire | 1861 |
| Alkali blue | Acid Blue 119 | 42765 | Nicholson | 1862 |
| Nicholson blue | Acid Blue 110 | 42750 | Nicholson<br>Gilbee | 1862<br>1862 |
| Methyl blue | Acid Blue 22 | 42755 | Nicholson | 1862 |
| Water blue IN | Acid Blue 93 | 42780 | Hofmann<br>Nicholson | 1858<br>1862 |
| Methyl blue<br>Water-soluble | Acid Blue 48 | 42770 | Nicholson | 1862 |
| Opal blue SS | Solvent Blue 23 | 42760 | Girard and De Laire | 1866 |
| Aldehyde green | n.a. | n.a. | Cherpin and Usèbe<br>Lucius | 1862<br>1863 |
| Iodine green | Basic dye | 42556 | Keisser | 1866 |
| Perkin's green | n.a. | n.a. | Perkin | 1864 |
| Para Rosaniline<br>Para Magenta | Basic Red 9 | 42500 | Coupier | 1869 |
| Methyl green | Basic Blue 20 | 42585 | Lauth and Baubigny | 1871 |
| Malachite green | Basic Green 4 | 42000 | O. Fischer | 1878 |
| Brilliant green | Basic Green 1 | 42040 | Bindschedler & Busch | 1879 |
| Victoria green 3B | Basic dye | 42030 | Bindschedler & Busch | 1883 |
| Helvetia green | n.a. | n.a. | AGFA | 1878 |
| Light green SF, yellowish | Acid Green 5 | 42095 | Köhler | 1879 |
| Light green SF, bluish | Acid Green 6 | 42075 | Köhler | 1879 |

(continued)

**Table 13.1** (continued)

| Name | C.I. name | C.I. number | Discoverers | Year of discovery |
|---|---|---|---|---|
| Guinea green B | Acid green 3 | 42085 | Schultz and Streng | 1883 |
| Patent blue V | Acid Blue 3 | 42051 | Hermann | 1888 |
| Crystal violet | Basic Violet 3 | 42555 | Kern and Caro | 1883 |
| Ethyl violet | Basic Violet 4 | 42600 | Kern, Caro (independently) | 1883 |
| Victoria blue B | Basic Blue 26 | 44045 | Kern and Caro | 1883 |
| Victoria blue 4R | Basic Blue 8 | 42563 | Kern and Caro | 1883 |
| Night blue | Basic Blue 15 | 44085 | Kern and Caro | 1883 |
| Auramine yellow | Basic Yellow 1 | 41000 | Kern and Caro | 1884 |
| Acid Magenta (Fuchsine S) | Acid Violet 19 | 42685 | Caro | 1877 |
| Redviolet 4RS | Acid dye | 42690 | Caro | 1877 |
| Redviolet 5RS | Acid dye | 42695 | Caro | 1877 |
| Alkali violet | Acid dye | 42630 | C. L. Müller | 1886 |
| Acid violet 6B | Acid Violet 49 | 42640 | Schultz and Zierold | 1889 |
| Acid violet 6B | Acid Violet 17 | 42650 | Sandmeyer<br>Weinberg | 1890<br>1890 |
| Victoria blue R | Basic Blue 11 | 44040 | Nastvogel and Reingruber | 1892 |

## 13.2 The Violet Dyes

Fuchsine and Methyl violet are discussed in Chap. 12 due to their importance of use; however, both dyes also served as a starting material for many other dyes. The most relevant ones will be described in more detail below, but many more were developed.

### *Regina Purple*

Colour Index name: Basic dye; Colour Index number: 42515
IUPAC name: 4-((4-aminophenyl)(4-(o-tolylamino)phenyl)methylene)cyclohexa-2, 5-dien-1-iminium
Discoverers: Adam Charles Girard and Georges de Laire, 1860; Edward Chambers Nicholson, 1862
Contemporary terminology: Regina purple (En), Violet Imperial rouge (Fr), Reginaviolett (Ge), Reginaviolet (NL)
Obsolete terminology: Regina violet (En), Violet phenylique (Fr)
Chemical formula: $C_{20}H_{24}N_3^+.CH_3COO^-$ (Fig. 13.1)

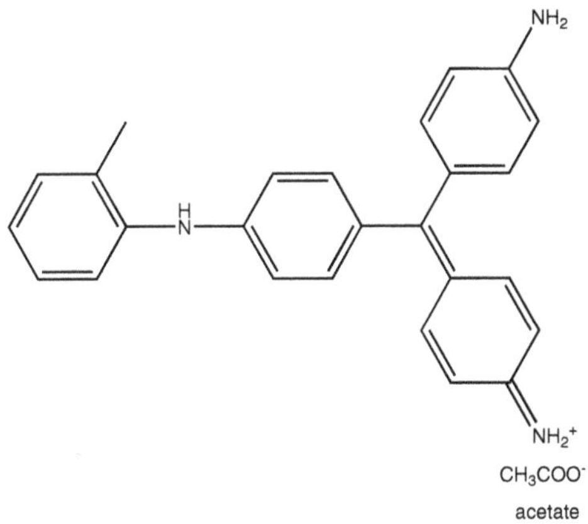


**Fig. 13.1** Structural formula of Regina purple. The dye usually consists of a mixture of derivatives of *N*-phenyl-pararosanilines

## History

In 1860, the French chemists Adam Charles Girard (1837–1918) and Georges de Laire (1836–1908) discovered that Fuchsine, when heated with aniline, gave a violet dye, which they named Violet Imperial; later it becomes known as Regina purple. Both chemists patented their process (F.P. 45.826, filed on the 6th of July 1860; French patent (number unknown), filed on the 2nd of January 1861, and B.P. 97, filed on the 12th of January 1861). Regina purple was prepared by heating a mixture of equal weights of Fuchsine and aniline to 160 °C for several hours. The purple mass is ready for use after washing with water and hydrochloric acid to remove unreacted aniline or Fuchsine [2]. A few years later, Nicholson found that on heating rosaniline acetate up to 200–215 °C, ammonia was released, and he obtained the same product, which he called Regina purple [2, 3]. He patented this process on the 20th of January 1862. This substance was found to be a monophenylrosaniline. Imperial violet, also known as Regina purple, often consists of a mixture of monophenyl- and mono-tolylrosaniline acetates. Different shades can be obtained by the production of more or less phenylated pararosanilines.

The dye company Simpson, Maule & Nicholson, who had been engaged in the manufacture of Rosaniline for some time, began the production of Girard and De Laire's Imperial violet (together with Bleu de Lyon), obtained by heating a salt of rosaniline with aniline (British patent filed on the 12th of January 1861). The German aniline factory Heinrich Tillmann added the phenyl derivatives of Fuchsine discovered by Girard and De Laire to its range and launched dyes under the name Regina purple as early as 1863 [4]. Very large quantities of the dye were produced and used, but the consumption declined rapidly with the introduction of new violets. However, it is remarkable that Regina purple is described in Gustav Schultz's 1901 book, *Die Chemie des Steinkohlentheers mit besonderer Berücksichtigung der künstlichen organischen Farbstoffe—die Farbstoffe*, as still being commercially available [5].

### Applications

The dye is known to be used to dye silk and wool.

### Properties

The dye is easily soluble in water, alcohol and acetic acid. The light fastness of Regina purple is described by Lehne as follows[3]: 'After 5 days exposed to the elements (June) duller, after 18 days strongly faded'. In the Colour Index, no properties are given for Regina purple.

### Recipes

Silk can be dyed following recipe V8, while for dyeing wool recipe V3A can be used. For wool, a concentration of 1.5% dye is given.

## *Other Violet Dyes*

In addition to Fuchsine, Methyl violet and Regina purple, several other violet dyes were discovered in the first decade after the invention of Mauveine. In 1863 the German chemist August Wilhelm Hofmann (1818–1892) patented the Hofmann's violets, which were prepared by alkylating Fuchsine with methyl and ethyl iodides, chlorides or bromides in methyl alcohol. However, the introduction of Methyl violet soon displaced the Hofmann's violets, which are merely of historical interest.

Purple and violet derivatives were also obtained from Fuchsine using brominated turpentine. These were known as Britannia violets and were manufactured in various shades. In 1864, the British chemist William Henry Perkin (1838–1907) patented Britannia violet, which was made by alkylating Fuchsine in methylated spirit

[3] Reginaviolett: Echtheit: 'Nach 5 tägiger Einwirkung der Witterung (Juni) matter, nach 18 Tagen stark verblasst'.

or methyl alcohol under pressure with aqueous hydrogen bromide and heating in a closed vessel for 8 h at 150 °C (B.P. 2.181, filed on the 6th of September 1864). Perkin was unable to purchase a source of aqueous hydrogen bromide, so it was made in-house by treating turpentine with bromine [3]. Britannia violet is very soluble in water, made in large quantities and was extensively employed for dyeing and printing. It could be produced in any shade, from purple to a blue violet. However, before the end of the century it was no longer commercially available; as a result, the dye is not mentioned by Lehne or in the Colour Index.

## 13.3 The Blue Dyes

### *Aniline Blue*

Colour Index name: Solvent Blue 3; Colour Index number: 42775
IUPAC name: 4-((E)-(4-amino-3-methylphenyl)((E)-4-(phenylimino)cyclohexa-2,5-dien-1-ylidene)methyl)-N-(cyclohexa-1,5-dien-1-yl)aniline
Discoverers: Adam Charles Girard and Georges de Laire, 1860
Contemporary terminology: Aniline blue, (En), Bleu de Lyon (Fr), Anilinblau (Ge), Anilineblauw (NL)
Obsolete terminology: Rosaniline blue, Opal blue, Spirit blue, Lyon blue (En), Bleu d'aniline, Bleu de nuit, Bleu de Paris (Fr), Anilinblau spritlöslich, Feinblau, Nachtblau (Ge)
Chemical formula: $C_{32}H_{27}N_3$ (Fig. 13.2)

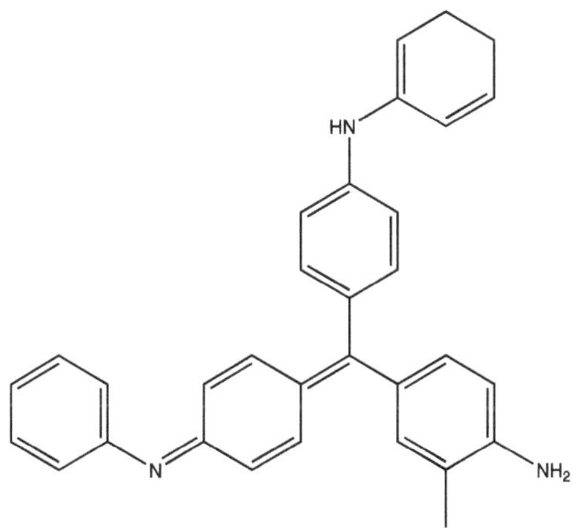


**Fig. 13.2** Structural formula of Aniline blue

## History

When pure aniline is treated with a salt of Fuchsine, a remarkable change takes place: at first the colour gradually becomes purple (as described above and resulting in Regina purple), but afterwards it turns blue. This curious reaction was observed by Girard and De Laire in 1860, and they found that this change in colour was due to the development of a new dye, which they called Bleu de Lyon. This first synthetic blue dye, which became known as Aniline blue, was obtained by heating Fuchsine and sodium acetate to 180 °C with an excess of pure aniline and a little benzoic acid. The aniline used to prepare Aniline blue must be as pure as possible and free from toluidine; technically it was known as 'Aniline oil for blue'. Aniline blue was manufactured in enamelled iron pots heated by an oil bath [6].

At first, the red colour of the mixture changed slowly, but then rapidly and it was considered to be ready when a good blue colour had been obtained. The cheaper grades of the dye were obtained by simply treating the crude product several times with hydrochloric acid, which removed all the free aniline and most of the red and purple impurities. For the production of better qualities, a process was used in which the raw product was mixed with methylated spirits of wine, that is, denatured alcohol, poured into water acidified with hydrochloric acid and then washed with water. The final product was a very intense blue in a pure state, the hydrochloride of triphenylrosaniline. Both chemists patented this process in January 1861 and their Aniline blue patents were soon acquired by the Société Renard Frères Franc et Cie in Lyon. The Renards were closely connected with the firm Simpson, Maule & Nicholson, which in 1862 was granted a license to manufacture Aniline blue in Britain [7, 8]. The main drawback of Aniline blue is its insolubility in water; it was therefore of little importance as a dye, but it served for the production of important sulfonic acid dyes.

A few months after the discovery of Aniline blue, another colouring matter, called Bleu de Paris, was obtained by the French chemists Jean-François Persoz (1805–1868), Victor-Joseph-Hippolyte de Luynes (1828–1904), and Louis-Alphonse Salvétat (1820–1882). These chemists found that when aniline was heated with tin chloride in a sealed tube to 180 °C for 30 h, neither a red nor a violet, but a very pure blue, was produced [9, 10]. Bleu de Paris was difficult to prepare in large quantities, however, and had never been introduced commercially [3].

## Trade Names

The dye was put on the market under the commonly used trade name Aniline blue by BASF, Leonhardt, Weiler-ter Meer, Claus & Rée, Read Holliday, Wülfing-Dahl, Roberts, Dale & Co. and CIBA. Other trade names were used, such as: Spritblau SFC by Kalle; Blau ll spritlöslich by Farbwerke MLB; Opalblau by Cassella; Bleu opale by Manufacture Lyonnaise; Lichtblau by Weiler-ter Meer; Bleu lumière by Soc. Anon. St. Denis; Gentianblau 6B by AGFA; Anilinblau spritlöslich, Feinblau, Nachtblau, Bleu de nuit, Bleu de Paris and Bleu de Lyon.

### Properties

Aniline blue is a dark red-brown powder that is insoluble in water, but dissolves in ethanol with a splendid blue colour (spirit-soluble Aniline blue). Aniline blue is supplied to the consumers as a coarse powder having a coppery lustre, or in alcoholic solutions. The light fastness is given by Lehne[4]: 'After 5 days exposed to the elements (June) duller, after 18 days strongly faded'. In the Colour Index, no information can be found about this colourant used as a textile dye.

### Recipes

Although the dye is not water-soluble, it can still be used to dye silk and wool if the dye is first dissolved in alcohol. Silk can be dyed using recipe V8, by adding the dye solution gradually to the almost boiling soap bath containing crushed alkaline soap. Wool can be dyed in the same way according to recipe V3A. For wool a concentration of 1.5% is mentioned.

## *Opal Blue SS*

Colour Index name: Solvent Blue 23; Colour Index number: 42760
IUPAC name: N-(4-(bis(4-(phenylamino)phenyl)methylene)cyclohexa-2,5-dien-1-ylidene)benzenaminium chloride
Discoverers: August Wilhelm Hofmann, 1864; Adam Charles Girard and Georges de Laire, 1866
Contemporary terminology: Diphenylamine blue (En), Bleu de ciel à l'alkool (Fr), Bayrischblau spritlöslich (Ge), Bavariablauw (NL)
Obsolete terminology: Spirit blue, Bavaria blue, Blue direct (En), Diphenylaminblau spritlöslich, Lichtblau superfein spritlöslich (Ge)
Chemical formula: $C_{37}H_{30}ClN_3$ (Fig. 13.3)

[4]Anilinblau spritlöslich: Echtheit: 'Nach 5-tägiger Einwirkung der Witterung (Juni) matter, nach 18 Tagen stark verblasst'.

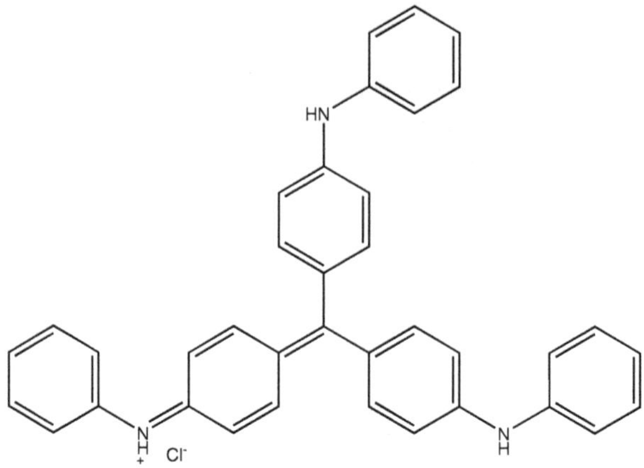


**Fig. 13.3** Structural formula of Opal blue SS

**History**

Hofmann was the first to notice in 1864 that a blue colour was produced by the reaction of various reagents with diphenylamine [11, 12]. Diphenylamine is prepared by heating a mixture of aniline and dry aniline hydrochloride under pressure. Next, diphenylamine is treated with oxalic acid at a high temperature, resulting in Opal blue SS [3]. Two years later, in 1866, Girard and De Laire, again working in Lyon, discovered, developed and patented Opal blue SS suitable for use in dyeing [13] (B.P. 1.093 and B.P. 2.686, both filed in 1866; F.P. 70.876, filed on the 21st of May 1866; F.P. 75.101 and F.P. 75.168, both filed in 1867). Opal blue SS, a dark-blue powder, is not soluble in water, but can be dissolved in alcohol giving a splendid blue colour. The main difference compared to Aniline blue is that a third phenyl group is attached to the pararosaniline structure. Opal blue SS was a competitor for Aniline blue, with similar properties and applications. It also served as a building block for acid blue dyes, as will be discussed below.

## *Alkali Blue*

Colour Index name: Acid Blue 110; Colour Index number: 42750
IUPAC name: Sodium 4-((4-((4-amino-3-methylphenyl)(4-(cyclohexa-1,5-dien-1-ylamino)phenyl)(hydroxy)methyl)cyclohexa-2,5-dien-1-ylidene)amino) benzenesulfonate
Discoverers: Edward Chambers Nicholson, 1862; Gilbee, 1862
Contemporary terminology: Alkali blue (En), Bleu d'alcalin (Fr), Alkaliblau (Ge), Alkaliblauw (NL)
Obsolete terminology: Nicholson blue, Blue soluble (En), Nicholsonblau, Lösliches Anilinblau (Ge)
Chemical formula: $C_{32}H_{27}N_3O_4SNa$ (Fig. 13.4)

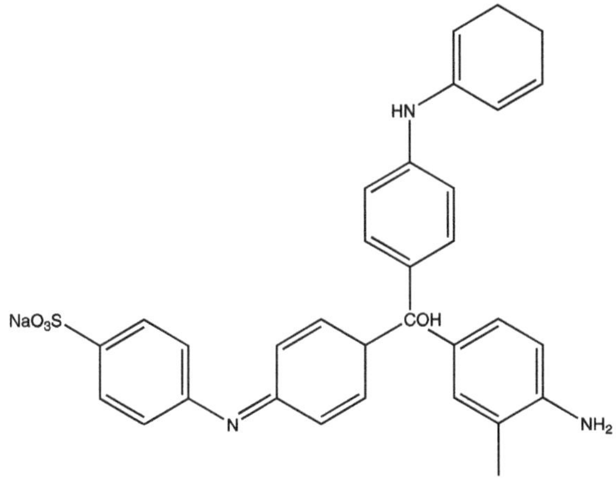


**Fig. 13.4** Structural formula of Alkali blue. Note that the exact position of the sulfonic acid group on the ring is not clear

## History

In 1862, Nicholson found that the water-insoluble basic dye Aniline blue could be converted into the water-soluble monosulfonic acid dye Alkali blue, also known as Nicholson blue, when treated it with fuming sulfuric acid (B.P. 1.857, filed on the 1st of June 1862 by Nicholson and B.P. 1.939, filed on the 3rd of July 1862 by Gilbee) [3]. He found that sulfonation gave Aniline blue the important property of solubility: a process similar to that forming Indigo carmine, obtained by treating natural Indigo with concentrated sulfuric acid. As a sodium salt, Alkali blue was widely used as an acid dye for the dyeing of wool and silk. However, the name Alkali blue leads to confusion. According to the Colour Index, Alkali blue is used both for Acid Blue 110 and Acid Blue 119. These are considered to be the monosulfonated versions of solvent blue dyes: Acid Blue 110 is derived from Aniline blue, i.e. Solvent Blue 3, while Acid Blue 119 used Opal blue SS, i.e. Solvent Blue 23, as a starting material. They differ in the number of phenyl groups attached to pararosaniline, two for Aniline blue and three for Solvent Blue 23. It could be that in the early days of the synthesis of dyes, components were not very pure. As such, it is likely that mixtures of di- and tri-phenyl pararosanilines were used and as a result, mixtures of the sulfonated derivatives should also have been formed. What complicates the matter is that both dyes were discovered in 1862 by Nicholson, Acid Blue 110 was simultaneously discovered by Gilbee, while Acid Blue 119 was rediscovered by Girard and De Laire in 1866.

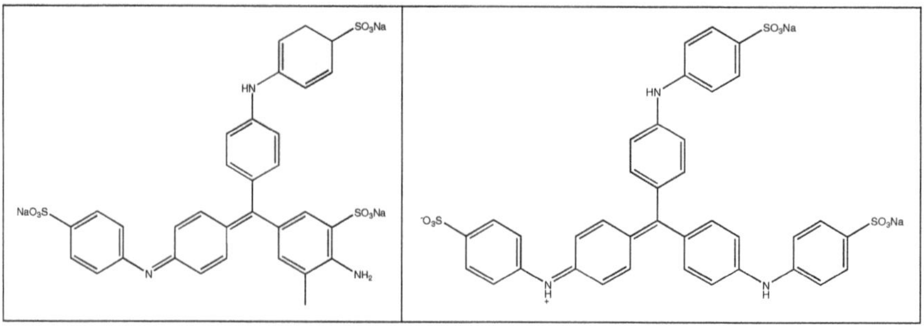

Fig. 13.5 Structural formula of Methyl blue (left) and Water blue (right). Note that the exact position of the sulfonic acid groups on the ring is not clear

Nicholson obtained two sulfonic acid varieties—a mono and a tri—the first being known as Alkali blue (Nicholson's blue) and the latter as Soluble blue. The discovery led to the preparation of a series of water-soluble mono-, di- and tri-sulfonic blue acid dyes derived from Aniline blue and Opal blue SS. Methyl blue for silk is derived from Aniline blue and contains one or two sulfonic acid groups; it was sold under trade names such as Seidenblau, Marineblau, Bleu Marin and Soluble Blue. Methyl blue, used to dye cotton, containing two or three sulfonic acid groups, is also obtained from Aniline blue, with trade names such as Baumwollblau, Methylwasserblau, Methylbaumwollblau, Bleu Méthyl, Wool blue and Helvetiablau. Water blue is derived from Opal blue SS, which contains two or three sulfonic acid groups and appears under trade names including Cotton blue and similar names to Methyl blue [14].

Another example is Methyl blue water-soluble, also called Soluble blue 6B, which can be converted into a pigment called Reflex blue and is derived from Opal blue SS with two sulfonic acid groups attached. One, two, three or even four hydrogen atoms in the phenyl groups are replaced by sulfonic groups. The stronger the acid and the higher the temperature, the more hydrogen atoms are replaced. The dyes are mainly mixtures of the ammonium or sodium salts. By the sulfonation of the water-insoluble basic dye Aniline blue, numerous water-soluble acid blues were obtained, which are far better adapted for the dyeing of wool than the basic dyes from which they are derived (Fig. 13.5).

## Trade Names

The dye was put on the market under the commonly used trade name Alkali blue by the addition of different letters and numbers by AGFA, BASF, Bayer, Farbwerke ML&B, Cassella, Geigy, Leonhardt, Kalle, Oehler, Claus & Rée, Durand-Huguenin,

Hudson River, Heller & Merz, Griesheim G.m.b.H., Levinstein, CIBA, Wülfing-Dahl, Schoellkopf and Weiler-ter Meer. Nicholson blue was manufactured by the firm Simpson, Maule & Nicholson and was extensively employed in Great Britain for wool-dyeing. Other trade names were used, such as: Bleu Nicholson by Soc. Anon. St. Denis and Bleu alcalin by Wiescher and Manufacture Lyonnaise.

#### Applications

Alkali blue is mainly used for dyeing or wool and silk; it can also be used to print wool.

#### Properties

Lehne reported about the fastness of both dyes, which seems to be similar[5]: 'After 5 days of exposure to the elements (June) unchanged, after 18 days quite strongly faded, much duller in appearance'. According to the Colour Index, the light fastness is ISO 2, while the wash fastness is ISO 2–3.

#### Recipes

In the publication by Lehne the same recipes are used for Acid Blue 110 and 119, for dyeing wool and silk. The recipe given reads[6]: 'For dyeing wool (yarn, piece goods): 10 kg wool, 250 g sodium carbonate (or 400 g borax), dye solution. Add the textile at 50 °C, heat and boil gently for about 1/2 hour until the sample immersed in diluted hot sulfuric acid shows the desired colour. The goods are then rinsed and the colour is developed in a hot acid bath containing about 150 g sulfuric acid. Silk is also dyed using borax or Marseille soap'. The sample present in the book is dyed with a concentration of 1% of dye.

---

[5] Alkaliblau: Echtheit: 'Nach 5 tägiger Einwirkung der Witterung (Juni) unverändert, nach 18 Tagen ziemlich stark verblasst, viel matter im Schein'.

[6] Zum Färben von Wolle (Garn, Stückwaare): 10 kg Wolle. 250 g Krystallsoda (oder 400 g Borax), Farblösung. Eingehen bei 50 °C. zum Kochen erhitzen, schwach kochen lassen, etwa 1/2 Stunde, bis eine in verdünnte heisse Schwefelsäure eingetauchte Probe die gewünschte Nüance zeigt. Sodann wird die Waare gespült und in einem heissen Säurebad, welches etwa 150 g Schwefelsäure enthält, die Farbe entwickelt, gespült. Seide wird ebenso unter Anwendung von Borax oder Marseiller Seife gefärbt.

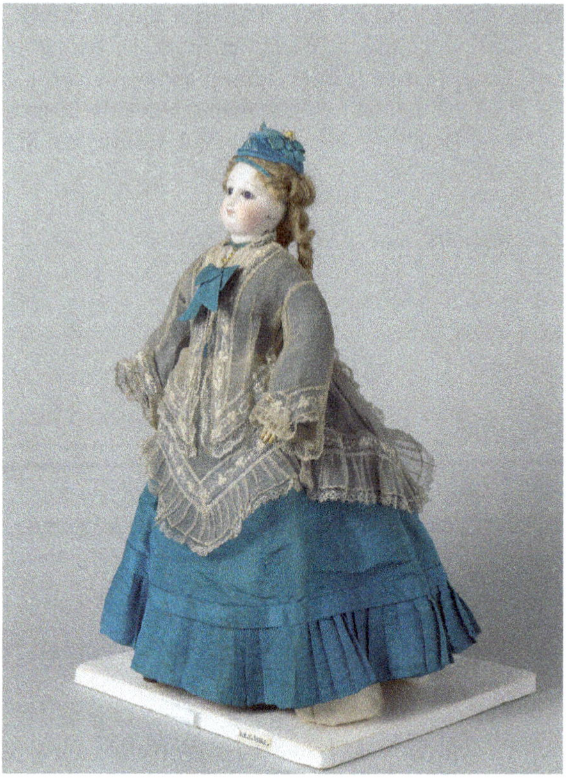

**Fig. 13.6** Doll with fashion gown, silk, 1875 RBK- 15162 Collection Rijksmuseum, Amsterdam the Netherlands. Photograph: Rik Klein Gotink

## 13.4 Case Studies: Majestic and Heavenly

In co-operation with the Rijksmuseum Amsterdam and the University of Amsterdam, the Cultural Heritage Agency of the Netherlands (RCE) started research on a series of blue-dyed textiles, dated between 1857 and 1890. One of the aims of the research was to determine which colourants were used. Some dresses contained Prussian blue, which are described in Chap. 8. One interesting dress was present on a porcelain doll with blond hair (see Fig. 13.6). She wears a blue silk gown, with a very bright light blue colour. In this dress we identified the tri-sulfonic acid dye Water blue. The dress of the doll shows the characteristic greenish fading for Water blue, as described by Lehne.[7]

Water blue was also found in the interior upholstery of the train carriage Hannibal, which is described in Chap. 10.

[7] Wasserblau. Echtheit: 'Nach 5 tägiger Einwirkung der Witterung (Juni) Willfärbung etwas matter, nach 18 Tagen heller und grünlicher.—Baumwollfärbung nach 5 Tagen viel heller, nach 18 Tagen sehr stark verblasst'.

## 13.5 The Green Dyes

### *Aldehyde Green*

No information on the Colour Index name and Colour Index number.
IUPAC name: n.a.
Discoverers: Pierre Cherpin, C.J. Usèbe, 1862; Eugen Lucius, 1863
Contemporary terminology: Aldehyde green (En), Vert d'Usèbe (Fr), Aldehydgrün (Ge), Aldehydegroen (NL)
Obsolete terminology: Aniline green (En), Usèbegrün, Anilingrün (Ge)
Chemical formula: $C_{22}H_{27}N_3S_2O$ (according to August Wilhelm Hofmann) [2]

#### History

The earliest synthetic greens developed in the 1860s had good colours, but were not lightfast. By using Fuchsine and Methyl violet as intermediates, chemists obtained green dyes, such as Aldehyde green, Iodine green and Methyl green. One of the first green organic synthetic dyes was a deep green obtained from aniline, named Émeraldine, discovered and patented by Messrs. Calvert, Clift and Lowe in 1860. When this dye is passed through a solution of potassium dichromate, the colour changes to a deep blue; the dye obtained is called Azurine. Both colours are dull and were of little importance.

The names Aldehyde green, Light green and Aniline green, which were often used synonymously at the time, covered two variants based on the reaction product of Fuchsine with acetaldehyde. The history of the basic dye Aldehyde green starts in France. Towards the end of 1861, the French chemist Charles Lauth (1836–1913) prepared a blue colouring matter, named Aldehyde blue, by heating acetaldehyde with a solution of Fuchsine and sulfuric acid [3, 15]. Although this blue produced pure and vivid shades on silk, wool and cotton, it was unusable due to the great difficulty in fixing it and to its poor light fastness.

One year later, in 1862, the dyer, foreman and chemist Pierre Cherpin, of the dyeing firm Usèbe near Saint Ouen in Paris, told a friend, who was a photographer, how difficult it was to fix the blue dye. The man advised him to use sodium hyposulphite, which he used for fixing his photographs. Unfortunately, Cherpin did not succeed in fixing the blue colourant; however, he discovered Aldehyde green by converting the unstable blue into a stable and beautiful green. This was done by treating Aldehyde blue with an aqueous solution of sodium thiosulfate. After boiling for a short time, the product, which contains Aldehyde green, was filtered off (French patent (number unknown) filed on the 26th of June 1861 and B.P. 254, filed in 1863). This green could not be produced in a solid form and was only available in solution. C.J. Usèbe obtained a French patent for this discovery in 1862

(F.P. 56.109, filed on the 28th of October 1862). A note for the preparation of Vert d'Usèbe (Aldehyde green) was given in an anonymously published article; however, Usèbe claimed the anonymously published preparation process as his patented property [3, 16–18].

The second variant was discovered by the German chemist Eugen Lucius (1834–1903) of the company Farbwerke ML&B and protected by English patents at the beginning of 1864. (B.P. 200, filed on the 23th of January 1864 and B.P. 301, filed on the 5th of February 1864). In 1863, Lucius wanted to dye a piece of leather with a solution of Aldehyde green. Interestingly, he noticed the formation of a dark-green precipitate of Aldehyde green caused by the tanning material sumach in the leather: instead of dyeing the leather, the dye precipitated and settled as a paste. Thus, Lucius discovered a production process by which Aldehyde green could be manufactured and sold as a paste [19]. The two processes, discovered by Cherpin and Lucius, were originally in vogue for the production of Aldehyde green. The Lucius process consists of the reaction of aldehyde, with hydrogen sulfide and then sulfurous acid with Fuchsine, while Usèbe used sodium thiosulfate following the process developed by Cherpin. The green dyes obtained in these processes differ in composition. The Aldehyde green obtained in Lucius's process contains two atoms of sulfur, while that of Cherpin only contains one. Aldehyde green was sold in both forms. The German aniline factory Heinrich Tillmann started producing Aldehyde green in 1864.

In 1874, Lucius stated that in order to produce Farbwerke ML&B's Aldehyde green in 1867, the Aldehyde blue was treated with sodium thiosulfate and the dye was precipitated with the tannin material sumach and sodium acetate. Here he described the original sulfurising agent in combination with modified precipitation and not the alternative sulfurisation method he had patented [4].

In 1870, Hofmann scientifically investigated the composition and chemical structure and showed that Aldehyde green contained sulfur and assigned an empirical formula: $C_{22}H_{27}N_3S_2O$ [18]. In 1893, Lehne reported that Aldehyde green was no longer on the market [20].

## Trade Names

This green was much used and marketed under the names Vert d'Usèbe, Aldehyde green or Aniline green and formed the greater part of the sales by the company Meister, Lucius & Co. (later Farbwerke ML&B).

## Applications

Aldehyde green dyes silk and wool a splendid green colour; unfortunately, it cannot be used on cotton. Aldehyde green is principally employed in silk dyeing. It possesses a magnificent colour, is very brilliant both by day and by artificial light and does not look blue in gaslight as several other green dyes do.

#### Properties

According to Lehne,[8] the dye is almost insoluble in water. It can be dissolved by adding ethanol to the dye (ratio dye:solvent is 1:1) and next it can be diluted in hot water. Since the dye is not listed in the Colour Index, no recent fastness properties can be found. However, Lehne reported on Aldehyde green[9]: 'After 8 days exposure to the elements (April) the colour is strongly faded. The silk dyeings are washfast'.

#### Recipes

Silk can be dyed following recipe V8, using a dye concentration of 18%. Lehne mentioned that the more acidic the dye bath, the more yellowish the green colour obtained.

## 13.6 Case Studies

### *The Aldehyde Green-Dyed Silk Robe of Empress Eugenie of France*

As mentioned above, Eugen Lucius of Farbwerke ML&B dyed a piece of leather with a solution of Aldehyde green and noticed that a dark-green precipitate of Aldehyde green was formed, caused by the tanning material sumach in the leather. In December 1863, the Belgian August de Ridder (1837–1911), the commercial manager of the company, travelled to Lyon by train with ten cans of Aldehyde green. When he arrived, he immediately visited the main important silk factory Renard & Villet and handed over one of the cans. One day later, Renard signed an exclusive contract with De Ridder, buying the entire annual production, including the nine remaining cans, and thus preventing the supply of Aldehyde green to other French companies. Renard also had a second demand: De Ridder had to leave Lyon immediately without visiting any other companies in Lyon; Renard even came to the platform to be sure that De Ridder left. Silk dyed with Aldehyde green was used for a robe, donated by the city of Lyon, to Empress Eugenie (1826–1920), the wife of the French Emperor Napoleon III (1808–1873). When she wore this superb green evening gown at the Paris Opera, she caused a sensation. Green festive clothing soon became a fashion, as Aldehyde green retained its luminous green colour in gas light. The clever merchant De Ridder concludes his report with the laconic sentence[10]: 'Renard did a great deal and so did we'. [21].

[8] In Wasser nahezu unlöslich.—Den Farbstoff mit der gleichen Gewichtsmenge Sprit anfeuchten, sodann heisses Wasser zufügen.

[9] Aldehydgrün: Echtheit: 'Nach 8 tägiger Einwirkung der Witterung (April) Farbe stark verblasst. Die Seidefärbungen sind wasserecht'.

[10] Renard machte ein grosses Geschäft und wir auch.

## *Iodine Green*

Colour Index number: Basic dye; Colour Index number: 42556
IUPAC name: [4-[(E)-[4-(dimethylamino)phenyl]-(4-dimethylazaniumylidene-3-methylcyclohexa-2,5-dien-1-ylidene)methyl]phenyl]-trimethylazanium; di-iodide or dichloride
Discoverer: J. Keisser, 1866
Contemporary terminology: Iodine green (En), Vert iode (Fr), Jodgrün (Ge), Joodgroen (NL)
Obsolete terminology: Vert lumière, Vert en pâte (Fr), Nachtgrün, Metternichsgrün (Ge)
Chemical formula: $C_{27}H_{35}N_3I_2$ (or $Cl_2$) (Fig. 13.7)

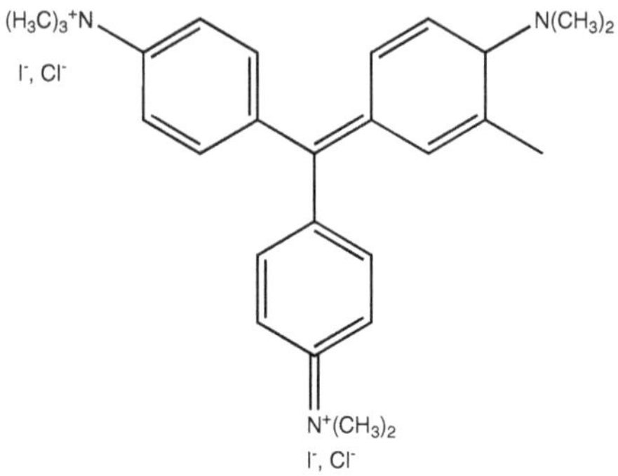


**Fig. 13.7** Structural formula of Iodine green

### History

Not every dye from the company Farbwerke ML&B enjoyed imperial publicity. Although Aldehyde green was produced in considerable quantities for some years, it did not remain a consistent sales favourite and was fairly soon replaced by the important dye Iodine green in 1867. The production of Iodine green was based on the treatment of Hofmann's violet, obtained from Fuchsine, using an excess of ethyl or methyl iodide. A bluish-green product remained in solution which could be precipitated with Picric acid to form a poorly soluble picrate.

To determine the exact discovery date of Iodine green is difficult. The German aniline factory Heinrich Tillmann in Crefeld, Meister, Lucius & Brüning (later Farbwerke ML&B) and Poirrier & Chappat fils all seem to have manufactured this green dye in 1865 [2]. A process was published by the British chemist James Alfred Wanklyn (1834–1906), professor of chemistry at the London Institution from 1863 to 1870, which was included in a patent filed on the 6th of November 1865. Although this process was of relatively little value from an industrial point of view, it served as the starting point for the production of Iodine green. Tillmann introduced the colourant to the market in 1866 as a competitor for Aldehyde green. The product earned him a special citation in the exhibition reports at the Paris World Exhibition in 1867 and the award of the gold medal [4].

It proved to be a valuable green dye and was eventually supplied in small quantities to dyers under the name of Iodine green. Some later, it was found by J. Keisser of Lyon (F.P. 71.625, filed on the 18th of April 1866) that Iodine green could be produced in much larger quantities by methylating Fuchsine, dissolved in methyl alcohol, with an excess of methyl iodide, at a relatively low temperature [3, 22]. On the Continent, Iodine green was generally prepared as a paste precipitated with Picric acid; in England, it was sold in an alcoholic solution. The main disadvantages of Iodine green were that it was too costly, due to the use of expensive iodine and produced in limited quantity; furthermore it had a fugitive character and was easily altered by heat, so that it could not be boiled without changing to violet. When heated, it is converted into violet methyl fuchsine, with the loss of methyl iodide. As a consequence, Iodine green was also not long-lived and was soon replaced by Methyl green.

## Applications

The green dye was much used for cotton and silk dyeing. Its colour is bluer than that of Aldehyde green and is therefore more useful, as it yields, with the addition of Picric acid, a greater variety of green shades.

## Properties

Iodine green is soluble in alcohol and in water, but insoluble in ether and in benzene. Regarding the light fastness of Iodine green, Lehne writes[11]: 'After 8 days exposed to the elements (April) strongly faded'. In the Colour Index, no information about the properties is given.

## Recipes

Although Lehne does not provide an exact recipe,[12] it is stated that it can be used to dye silk at 50–60 °C, brightening it with acetic acid. Cotton can be dyed following recipe V11. Wool only absorbs the dye if it has been pre-mordanted with zinc chlo-

[11] Jodgrün: Echtheit: 'Nach 8 tägiger Einwirkung der Witterung (April) stark verblasst'.

[12] Diente zum Färben der Seide; bei 50–60 ° C. ausfärben, mit Essigsäure aviviren. Baumwolle (V11). Wolle nimmt den Farbstoff erst auf, wenn sie vorher mit Chlorzink, Tannin oder Galläpfolabkochnng gebeizt ist. (G. Schultz.)—Zum Bedrucken der Baumwolle empfahl G. Schäffer, den Farbstoff in Essigsäure zu lösen, essigsaure Thonerde beizufügen und mit benzoësaurem Ammoniak einen Farblack zu bilden. Dieser wurde ausgewaschen und mit Albumin verdickt aufgedruckt. Der Druckfarbe wurde etwas Salmiak und soviel Weinsäure, dass das Eiweiss nicht gerann, zugesetzt.

ride, tannins or gallnut extract. For printing on cotton, G. Schäffer recommended dissolving the dye in acetic acid, adding aluminium acetate and forming a lake with benzoic acid ammonia. This should be washed and thickened with albumin to allow printing. A little ammonia and sufficient tartaric acid were added to the printing ink to prevent the protein from coagulating.

Other aniline blue-greens include Perkin's green. After the discovery of Mauveine, the British chemist William Henry Perkin (1838–1907) continued his active research in organic chemistry, inventing and marketing two other synthetic dyes: Britannia violet, obtained from brominated turpentine oil and methanol, and Perkin's green, obtained in 1864 by treating Britannia violet with acetyl chloride. The properties of Perkin's green resemble those of Iodine green more closely than those of Aldehyde green and it is precipitated by solutions of alkaline carbonates, such as sodium carbonate. Like Iodine green, it is precipitated by Picric acid, forming a picrate which crystallises from alcohol in small prisms with a golden reflection. This bluish-green dye is principally employed for calico printing and considerable quantities were prepared when Iodine green was too expensive. After some time, Perkin's green was displaced by Iodine green [3].

## *Methyl Green*

Colour Index name: Basic Blue 20; Colour Index number: 42585
IUPAC name: 4-((4-(dimethylamino)cyclohexa-2,5-dien-1-ylidene)(4-(dimethyliminio)-4 15-phenyl)methyl)-N,N,N-trimethylbenzenaminium, dichloride
Discoverers: Charles Lauth and Henry Baubigny, 1871; Wischin, 1873; Prosper Monnet and Frédéric Reverdin, 1874
Contemporary terminology: Methyl green (En), Vert methyl (Fr), Methylgrün (Ge), Methylgroen (NL)
Obsolete terminology: Vert lumière, Vert de methylaniline (Fr), Parisergrün (Ge)
Chemical formula: $C_{26}H_{33}N_3^{2+}$, chloride salt (Fig. 13.8)

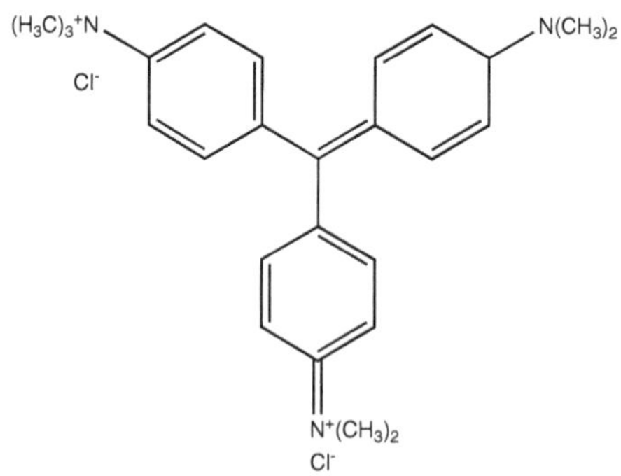


**Fig. 13.8** Structural formula of Methyl green

### History

Methyl green was obtained by the alkylation of Methyl violet by methyl nitrate and was discovered by the French chemists Charles Lauth and Henry Baubigny (1842–1913) in 1871. Two years later, in 1873, Methyl green was produced using methyl chloride instead of methyl nitrate by Wischin at the German company G. Siegle in Stuttgart, and some later, in 1874, by the French chemist Prosper Monnet (1834–1914) and the Swiss chemist Frédéric Reverdin (1849–1931) [23]. First, methyl alcohol and hydrochloric acid are heated to 100 °C under a pressure of 30 to 35 atmospheres. The methyl chloride obtained is passed into an alkaline solution of Methyl violet in methanol until the solution is saturated, after which it is heated to 95 °C under pressure. The resulting product is distilled with an alkali, and the solution containing Methyl green is filtered [24]. Methyl green was introduced by the German dye factory Tillmann in 1872 and replaced Iodine green the following year. Methyl green was the only green aniline dye produced in large quantities until the discovery of Malachite green in 1877 [25]. Finally, this green colourant was replaced by the cheaper Malachite green in combination with Brilliant green, sold by the German aniline factory Heinrich Tillmann from 1881 onwards [4].

### Trade Names

The dye was put on the market under the trade name Methylgrün by Bayer, Weiler-ter Meer, AGFA, Kalle and Roberts, Dale & Co. Other trade names were used, such as Vert Etincelle by Soc. Chim. Us. Rhône; Doppelgrün SF by Kalle; Lichtgrün by Weiler-ter Meer and Vert Méthyle by Soc. Anon. St. Denis.

### Applications

Methyl green can be used to dye silk, wool and cotton. Methyl green, dyed on cotton using a tannin mordant, was better suited for this purpose than Malachite green.

### Properties

The dye has a greenish-blue colour. The chloride salt dissolves well in water, but not in ethanol according to the Colour Index, while Lehne indicates that it can also be dissolved in ethanol. Lehne wrote about the fastness of Methyl green[13]: 'After 5 days exposure to the elements (June) strongly faded. – The silk dyeings are very resistant to water'. According to the Colour Index, the light fastness is ISO 1–2, and the wash fastness is ISO 2.

[13] Methylgrün: Echtheit: 'Nach 5 tägiger Einwirkung der Witterung (Juni) stark verblasst.—Die Seidefärbungen sind sehr wasserecht'.

**Recipes**

Lehne stated that silk can be dyed using recipe V7 at 35 °C. Wool dyeing requires more preparation.[14] 'To dye the wool, it must be prepared by heating it with 15% sodium hyposulphite and 12% hydrochloric acid from 50° to 80 °Bé,[15] for 1½ h and rinsing well after being left for 10 h. Dye at 50 to 60 °C with the addition of 1% acetic acid and a little zinc acetate. The mordant and dye baths are stored. Copper or lead vessels should be avoided. It is best to mordant and to dye in a wooden vat using rubber vapour tubes. To nuance the colour, Picric acid, Naphthol yellow S etc. can be used. Note that larger quantities of Picric acid results in a precipitation of the dye. Cotton can be dyed cold, using recipe V11'.

## 13.7 The Malachite Series

### *Malachite Green*

Colour Index name: Basic Green 4; Colour Index number: 42000
IUPAC name: N-(4-((4-(dimethylamino)phenyl)(phenyl)methylene)cyclohexa-2,5-dien-1-ylidene)-N-methylmethanaminium chloride
Discoverers: Otto Philipp Fischer, 1877; Oscar Doebner, 1878
Contemporary terminology: Malachite green, Diamond green B (En), Vert malachit (Fr), Malachitgrün (Ge), Malachietgroen (NL)
Obsolete terminology: Bittermandelölgrün, Echtgrün, Benzoylgrün (Ge)
Chemical formula: $C_{23}H_{25}ClN_2$ (Fig. 13.9)

[14] Zum Färben der Wolle muss diese vorbereitet werden, indem sie mit 15% unterschwefligsaurem Natron und 12% Salzsäure von 50° bis 80° Bé während 1½ Stunden erwärmt und nach 10-stündigem Liegen gut gespült wird. Färben bei 50 bis 60 °C. unter Zusatz von 1% Essigsäure und etwas essigsaurem Zink. Beiz- und Färbebad werden aufbewahrt. Kupferne oder bleierne Gefässe sind zu vermeiden. Am besten wird im Holzbottich unter Anwendung von Kautschuk-Dampfröhren gebeizt und gefärbt. Zum Nuanciren dient Pikrinsäure, Naphtolgelb u. m. Grössere Mengen von Pikrinsäure fällen den Farbstoff. (By.) Baumwolle (V11, kalt färben).

[15] In the literature, the concentration of hydrochloric acid can be found up to 24 °Bé, which equals 40%. Probably, 50–80° Bé is very concentrated hydrochloric acid although the exact concentration is currently unknown.

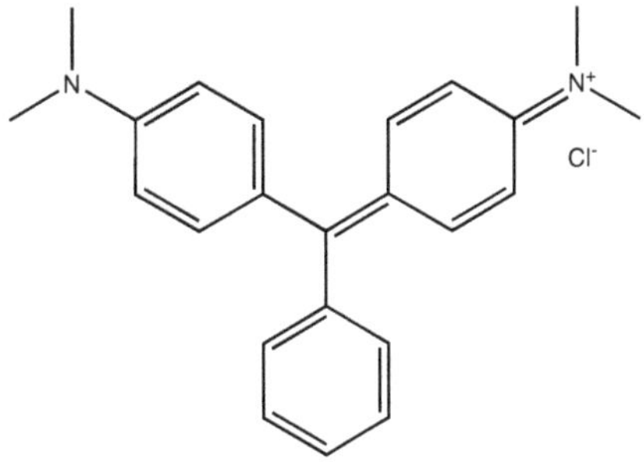


**Fig. 13.9** Structural formula of Malachite green

## History

At the end of the 1870s, the so-called Malachite series were invented. The older members of this group are green; these include Malachite green (1878), Brilliant green (1879) and Victoria green 3B (1883).

In 1877, the German chemist Otto Philipp Fischer (1852–1932), working at the Academy of Science in Munich, studied the reaction of dimethylaniline with benzaldehyde in the presence of sulfuric acid and obtained a white crystalline material. This compound, when exposed to the air, rapidly oxidised to a bluish-green colour, which Fischer named Bittermandelölgrün (Bitter almond oil green). The name referred to the natural product Bittermandelöl, the source for benzaldyde used in this process. Some time later, he showed that by treating this white crystalline base with oxidising agents, such as potassium permanganate, potassium dichromate, manganese dioxide or (even better) lead dioxide, good yields of this green dye could be produced more easily [26].

Shortly afterwards, in 1878, the German chemist Oscar Doebner (1850–1907), working at the University Laboratory in Berlin, was exploring the chemistry of benzotrichloride [23, 27–30]. After the reaction of dimethylaniline with benzotrichloride, he obtained the same dye that Fischer had developed earlier. In collaboration with Doebner, AGFA in Berlin filed a patent in February 1878 to protect the process and used this process for the manufacture of the dye (G.P. 4.322, filed on the 26th of February 1878; B.P. 828, filed in 1878; B.P. 4.762 and U.S.P. 222.257, both filed in 1879) (see Table 13.2) [31]. The company scaled up the manufacture using Doebner's procedure for only a short time and sold the dye under their name Malachitgrün. The industrial process is based on heating benzotrichloride, dimethylaniline and zinc chloride as a dehydrating agent. When the reaction is complete, the excess of the amine is removed by steam distillation. However, it appeared that the method developed by Fischer using benzaldehyde gave better results [32].

The main difficulty in the production of this dye was the preparation of benzaldehyde. The usual method of obtaining benzaldehyde in Fischer's method, from natural bitter almonds, was not practical. As an alternative, benzaldehyde was

**Table 13.2** Discoverers and patents of Malachite green and Brilliant green

| Name<br>C.I. name<br>C.I. number | Discoverers | Year of discovery | Patents |
|---|---|---|---|
| Malachite green<br>Basic green 4<br>42000 | O.P. Fischer<br>O. Doebner | 1877<br>1878 | –<br>G.P. 4.322/1878<br>B.P. 828/1878<br>B.P. 4.762/1879<br>U.S.P. 222.257/1879 |
| Brilliant green<br>Basic green 1<br>42040 | Bindschedler & Busch | 1879 | – |

prepared from pure toluene. The process was based on the reaction of benzaldehyde with dimethylaniline and the first experiments for the production of the green were successfully employed for dye production by BASF in March 1878. As such, the technical difficulties involved in producing benzaldehyde were successfully overcome by BASF and Doebner's benzotrichloride process, used by AGFA, was abandoned [3].

## Trade Names

By adding various letters and numbers, the dye was marketed under the common trade name Malachite green by AGFA, Kalle, Farbwerke ML&B, Küchler & Buff (later Weiler-ter Meer), Durand-Huguenin, Claus & Rée, Wiescher, Jäger and Read Holliday. Other trade names were used, such as: Malachitgrün B by BASF; Benzalgrün OO by Griesheim-Elektron; Neuvictoriagrün by BASF and Küchler & Buff (later Weiler-ter Meer); Diamantgrün by BASF and Soc. Chim. Us. Rhône; Neugrün by Bayer and Küchler & Buff (later Weiler-ter Meer); Solidgrün O by CIBA; Solidgrün cryst. by Cassella and Leonhardt; Vert solide cristaux O by Manufacture Lyonnaise; Diamantgrün by Soc. Chim. Us. Rhône; Diamantgrun B by BASF; Echtgrün by Griesheim G.m.b.H.; Bittermandelölgrün by Küchler & Buff (later Weiler-ter Meer); Green Crystals by Read Holliday; Vert solide B extra and LB extra by Soc. Chim. Us. Rhône. Export trade names for Malachite green were Superior New, Fine New, Imperial, Royal, Diamond, Nawa, China, Emerald and Green crystals.

## Applications

Malachite green has attained considerable importance and was used as an effective and brilliant dye for wool and silk. It has a much better tinctorial power than Methyl green and in addition possesses the advantages of dyeing wool easily and resisting the action of heat. It was also used to dye cotton and linen and as a printing ink.

## Properties

The dye is very easily soluble in water and ethanol, resulting in a bluish-green colour. The light fastness of Malachite green according to Lehne is rather poor[16]: 'After 7 days exposure to the elements (July) very faded'. The Colour Index states that the light fastness is ISO 2, while the wash fastness is ISO 4.

## Recipes

Lehne provided several recipes[17]: For dyeing cotton and linen, recipe V11 was used and a recipe for jute is given as well (V15); dye at 60–70 °C. The dye was often used for dyeing silk following recipe V7 (in a neutral bast soap bath, left over from silk degumming; when brightening with acetic acid, it can be shaded with acid dyes—e.g. with Naphthol yellow S). Wool can be dyed following recipe V1 (a concentration of 1.5% dye is given); hard water must be improved by adding 1–2 L of acetic acid per 1,000 litres. Cotton can be printed following recipe V23A, while for printing on wool, recipe V19 is given.

## Identification Using UHPLC-PDA

The analysis of a Malachite green (Sigma-Aldrich CAS 569-64-2), showed a clear peak as presented in Fig. 13.10.

---

[16] Malachitgrün: Echtheit: 'Nach 7 tägiger Einwirkung der Witterung (Juli) sehr stark verblasst'.

[17] Zum Färben von Baumwolle und Leinen (V11), Jute (V15); bei 60–70 °C. Viel benutzt zum Färben der Seide (V 7, neutrales Bastseifenbad; bei dem Aviviren mit Essigsäure kann mit Säurefarbstoffen z. B. mit Naphtolgelb S abgetönt werden). Wolle V1; hartes Wasser muss durch Zusatz von 1–2 Liter Essigsäure auf je 1000 Liter verbessert werden. Baumwolldruck V23A. Wolldruck V19.

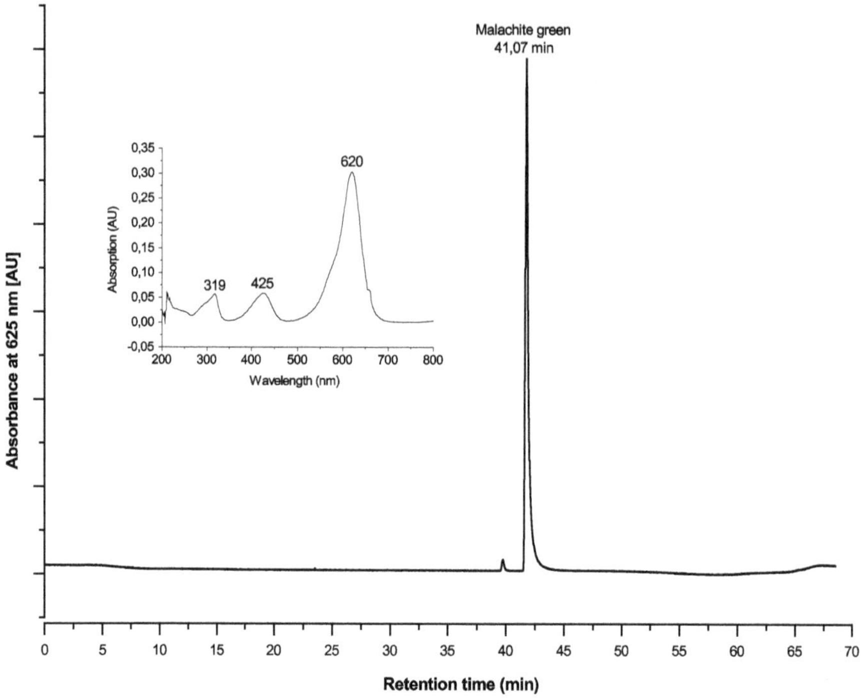


**Fig. 13.10** Chromatogram and spectrum of Malachite green (Sigma-Aldrich CAS 569-64-2)

## *Brilliant Green*

Colour Index name: Basic Green 1; Colour Index number: 42040
IUPAC name: N-(4-((4-(diethylamino)phenyl)(phenyl)methylene)cyclohexa-2,5-dien-1-ylidene)-N-ethylethanaminium hydrogen sulfate
Discoverer: dye company Bindschedler & Busch, 1879
Contemporary terminology: Brilliant green, Diamond green G (En), Vert brillant (Fr), Brillantgrün (Ge), Briljantgroen (NL)
Obsolete terminology: Neuvictoriagrün (Ge)
Chemical formula: $C_{27}H_{34}N_2O_4S$ (Fig. 13.11)

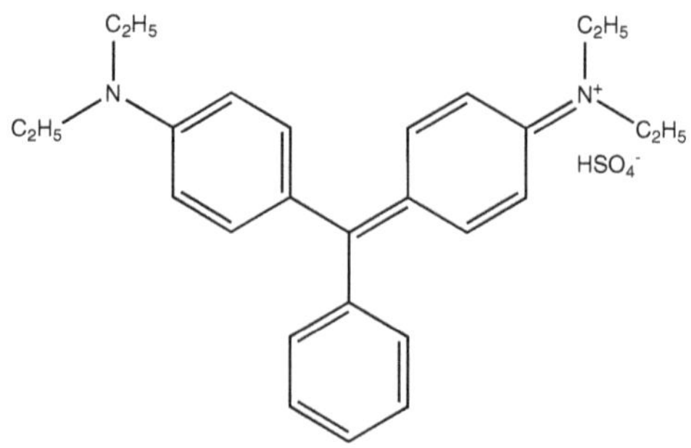


**Fig. 13.11** Structural formula of Brilliant green

### History

In 1879, one year after the successful production of Malachite green by BASF, a similar compound, the di-ethyl derivative of Malachite green, called Brilliant green, was invented and first reported by the dye company Bindschedler & Busch in Basel (see Table 13.2) [27, 33, 34]. It was prepared from di-ethylaniline and benzaldehyde and appeared on the market by converting the dye to the sulfate form.

### Trade Names

The dye was put on the market under the commonly used trade name Brillantgrün by BASF, Farbwerke ML&B, Cassella, Manufacture Lyonnaise, Weiler-ter Meer, Bayer, Kalle, Leonhardt, Claus & Rée, Durand-Huguenin, Griesheim G.m.b.H., Read Holliday, Wiescher, Jäger and Griesheim-Elektron. Other trade names were used, including: Malachitgrün G and Diamantgrün G by BASF; Aethylgrün by AGFA; Smaragdgrün by Bayer, CIBA and Küchler & Buff (later Weiler-ter Meer), Solidgrün J by Soc. Chim. Us. Rhône; Solidgrün JJO by CIBA; Vert ethyle extra by Soc. Chim. Us. Rhône and Vert brilliant by Soc. Anon. St. Denis.

### Applications, Properties and Recipes

Brilliant green has the same application range and properties as Malachite green; textiles can be dyed and printed using the same recipes as mentioned above.

### Identification Using UHPLC-PDA

The analysis of Brillant green (Sigma-Aldrich CAS 633–03-4) shows a clear peak as show in Fig. 13.12.

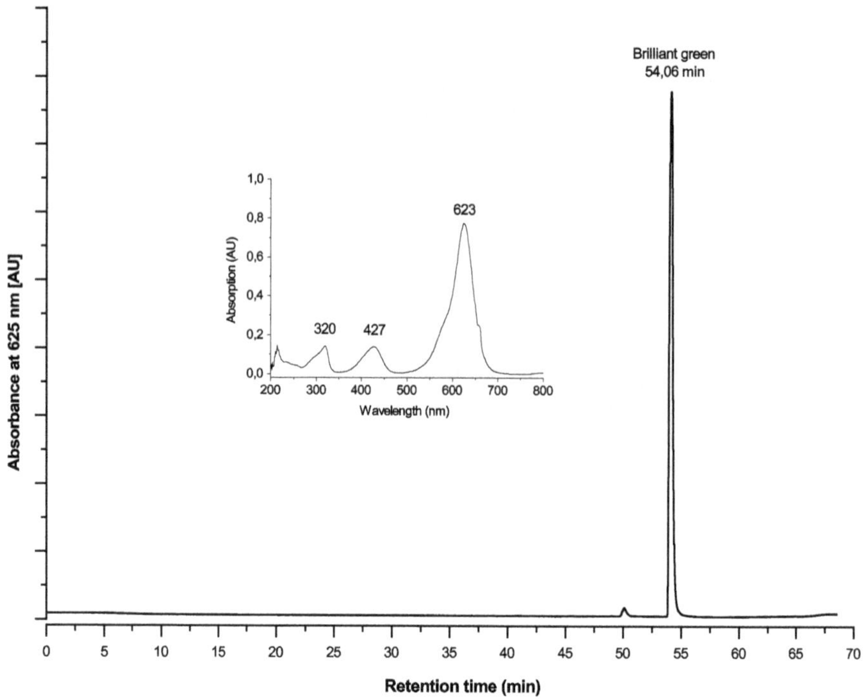


**Fig. 13.12** Chromatogram and spectrum of Brilliant green. (Sigma-Aldrich CAS 633-03-4)

After the introduction of Malachite green and Brilliant green, other green dyes were developed. Victoria green 3B is interesting because it shows the influence on the colour of a compound of this series by the introduction of chlorine atoms into the unsubstituted benzene ring. Fischer discovered and patented this green dye in 1883 (G.P. 25.827, filed on the 23th of June 1883). It is prepared from tri-chlorobenzaldehyde and dimethylaniline, with subsequent oxidation of the leuco-base formed. The shade of green of this dye is much bluer than that of Malachite green. The dye was put on the market under the trade name Victoriagrün 3B by BASF. Other trade names were used like: Neusolidgrün BB and 3B by Bindschedler & Busch (later CIBA).

## *Sulfonic Acids of the Malachite Green Series*

Sulfonation of the dyes from the Malachite series using concentrated sulfuric acid resulted in the water-soluble green acid dyes, such as Helvetia green (1878), Light green (1879), and Guinea green B (1883). It was not until 1888 that the sulfonated greenish-blue members, the Patent blues, were developed.

Helvetia green, the sodium salt of sulfonated Malachite green, was patented by AGFA in 1878 (G.P. 6.714, filed on the 27th of October 1878; B.P. 4.406, filed in 1878, and F.P. 127.298, no date known), and one year later, in 1879, by the Swiss aniline factory Bindschedler & Busch in Basel (G.P. 10.410, filed on the 10th of

June 1879; B.P. 2.509, filed in 1879; G.P. 14.944, filed on the 3rd of April 1880 (First additional patent for G.P. 10.410) and F.P. 131.325). Around 1883, Helvetia green was replaced by the dye Light green SF.

In 1883, Guinea green B was discovered by the German chemists Gustav Schultz and Streng, working for AGFA, but it was not patented until the late 1880s (G.P. 50.782, filed on the 9th of April 1889; B.P. 7.550 and F.P. 198.415, both filed in 1889). Other trade names are Säuregrün 2BG by Weiler-ter Meer; Neusäuregrün 3BX by Bayer and Vert sulfo BB by Soc. Anon. St. Denis. However, the most important representative of this group is Light green SF.

## *Light Green SF*

Colour Index name: Acid Green 5; Colour Index number: 42095
IUPAC name: Sodium 4-((ethyl((1Z,4Z)-4-((4-(ethyl(4-sulfonatobenzyl)amino) phenyl)(4-sulfonatophenyl)methylene)cyclohexa-2,5-dien-1-lidene)ammonio) methyl)benzenesulfonate
Discoverer: Köhler, 1879
Contemporary terminology: Light green SF (En), Vert clair SF (Fr), Lichtgrün SF gelblich, (Ge), Lichtgroen SF (NL)
Obsolete terminology: Säuregrün SOF, Lichtgrün S (Ge)
Chemical formula: $C_{37}H_{34}S_3O_9.Na_2$ (Fig. 13.13)

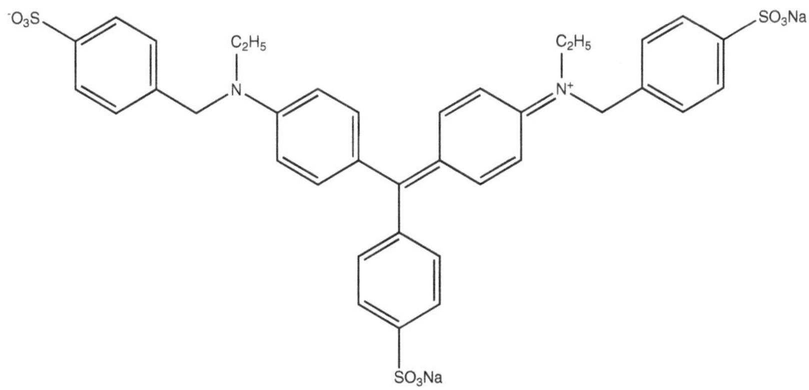


**Fig. 13.13** Structural formula of Light green SF. Note that the exact positions of the sulfonic acid groups are not clear

### History

Light green SF was discovered in 1879 by the German chemist Köhler, working for BASF. This green dye is prepared by sulfonation and oxidation of the condensation product of benzaldehyde with benzylethylaniline. The advantage of this dye is that sulfonation is much easier when a benzyl group is present. The advantages of Light green SF are a better solubility and a more even dyeing of textiles [35].

#### Trade Names

The dye was put on the market under the common trade name Lichtgrün SF by BASF. By adding various letters and numbers, the dye was mainly marketed under the name Säuregrün by BASF, Bayer, Leonhardt, Farbwerke ML&B, CIBA, Cassella, Griesheim-Elektron and Weiler-ter Meer; Vert acide JJ extra concentré by Manufacture Lyonnaise; Guineagrün 2G by AGFA and Vert sulfo J extra by Soc. Anon. St. Denis.

#### Applications

Light green SF was used on a very large scale and dyes wool and silk from an acid bath. It was also used to stain leather and dye jute. In addition, it is also used as a food colourant, known as C.I. Food Colorant 2.

#### Properties

The dye is very soluble in water, but almost insoluble in ethanol. Lehne wrote on Light green SF[18]: 'After 5 days exposure to the elements (June) very little faded, very faded after 18 days'. According to the Colour Index, the light fastness is ISO 2, while the wash fastness is ISO 1–2.

#### Recipes

Wool can be dyed following recipe V3A, which results in a green colour. It can be combined with Acid violet to replace Indigo carmine. Silk can be dyed using recipe V8, while for jute recipe V17 can be employed.

#### Identification Using UHPLC-PDA

The analysis of Light green SF (RCE no. 5641) showed a mixture of compounds with similar UV/Vis spectra, see Fig. 13.14.

[18] Lichtgrün SF: Echtheit: 'Nach 5 tägiger Einwirkung. der Witterung (Juni) sehr wenig verändert, nach 18 Tagen stark verblasst'.

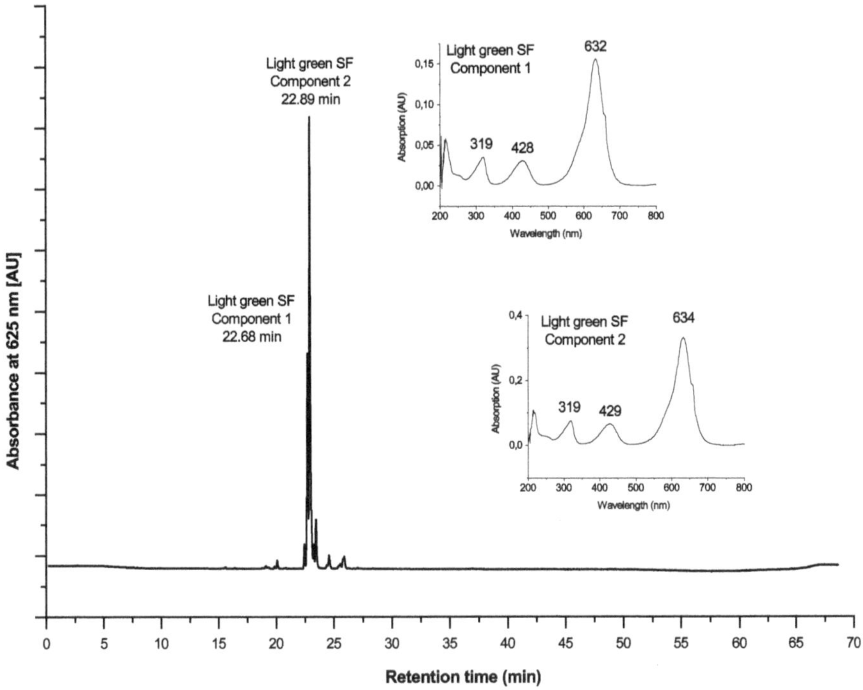


**Fig. 13.14** Chromatogram and spectra of Light green SF (RCE no. 5641)

## *Patent Blue V*

Colour Index name: Acid Blue 3; Colour Index number: 42051
IUPAC name: 4-{[4-(Diethylamino)phenyl](5-hydroxy-2,4-disulfophenyl)methylene}-N,N-diethyl-2,5-cyclohexadien-1-iminium
Discoverer: Albert Herrmann, 1888
Contemporary terminology: Patent blue (En), Bleu patenté (Fr), Patentblau (Ge), Patentblauw (NL)
Chemical formula: $C_{27}H_{32}N_2NaO_7S_2$ (Fig. 13.15)

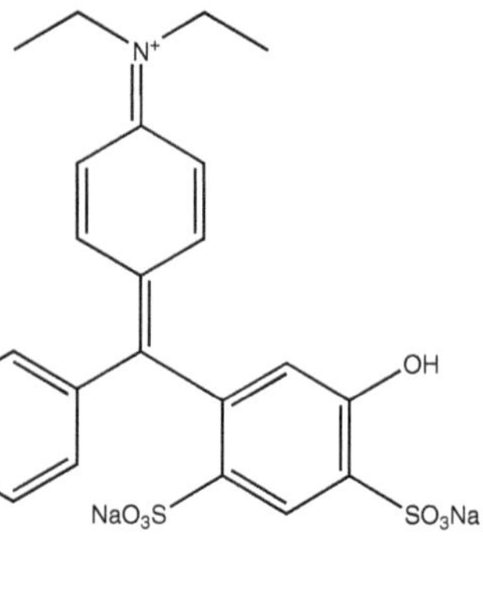


**Fig. 13.15** Structural formula of Patent blue V

**Table 13.3** Discoverer and patents of Patent blue V

| Name<br>C.I. name<br>C.I. number | Discoverer | Year of discovery | Patents |
|---|---|---|---|
| Patent blue V<br>Acid Blue 3<br>42051 | A. Herrmann | 1888 | G.P. 46.384/1888<br>G.P. 48.523/1888<br>B.P. 12.796/1888<br>G.P. 50.286/1889<br>G.P. 50.293/1889<br>G.P. 50.440/1889<br>G.P. 55.621/1890<br>U.S.P. 412.613/1889<br>U.S.P. 412.614/1889<br>U.S.P. 412.615/1889<br>G.P. 66.791/1892<br>G.P. 71.156/1892<br>G.P. 74.014/1893 |

## History

An important step forward was the discoveries of the Patent blues. In the 1890 article 'Patent blau', Ostersetzer describes that Farbwerke ML&B had introduced this important dye of the amido-oxy group on the market a year earlier [36]. In 1888, the German chemist Albert Herrmann (1859–1921), working for Farbwerke ML&B, discovered the sulfonated greenish-blue members, the Patent blues. Patent blue V was the first representative of this dye group, patented by G.P. 46.384 on the 18th of August 1888, G.P. 48.523 patented on the 13th of October 1888, and B.P. 12.796, filed in 1888, followed by many international patents (see Table 13.3) [37].

They are prepared by condensation of dimethylaniline, di-ethylaniline and ethylbenzylaniline with *m*-oxybenzaldehyde to form leuco compounds, which can be converted into valuable greenish-blue dyes by sulfonation and subsequent oxidation. In the same article, Ostersetzer writes that 'Unfortunately Patent blue is still too expensive today. Farbwerke ML&B has recognised the importance of their new product and have recently built an extensive plant for the production of Patent blue, so that this dye will soon be able to compete with other products, not only in terms of its excellent properties, but also in terms of price'.[19] The dyes of the Patent blues differ from Malachite green by only one hydroxyl group in their benzene nuclei. Depending on the position of this hydroxyl, in the *m*-position blue to green colours are produced, whereas in the *p*-position violet to bordeaux tones are created. The shades produced by the Patent blues are similar to those of Indigo and have the advantages of being water-soluble, and therefore easier to dye with, and a greater brilliancy. The Patent blues were successful in replacing Indigo carmine. However, one disadvantage of Patent blue V is its low light fastness.

### Trade Names

The dye was put on the market under the commonly used trade name Patentblau (V, N, L, superfein, konz. and extra) by Farbwerke ML&B and AGFA. Other trade names were used, such as: Neptunblau BG by BASF; Brillantsäureblau V by Bayer; Säureblau G by Weiler-ter Meer and Tetracyanol V and S by Cassella.

### Applications

The dye can be used to dye wool and silk, and was also used as a printing ink for wool.

### Properties

The dark-blue powder is easily soluble in water and alcohol. In an acidic solution, Patent blue V is greenish-yellow; in a basic solution it is blue. Lehne wrote about the light fastness[20]: 'After 5 days exposure to the elements (June) very little changed, slightly duller, after 18 days quite strongly changed'. According to the Colour Index, the light fastness is ISO 1–2, and the wash fastness ISO 2–3.

### Recipes

Wool can be dyed following recipe V3A ('short bath, avoid the use of copper vessels')[21]: It could be used on its own but was also used in combination with other acid dyes such as Acid violet N. A typical concentration for dyeing wool is 1.5%.

[19] Leider ist das Patentblau heut noch etwas teuer; die Farbwerke vorm. Meister, Lucius & Brüning haben in Erkenntnis der Bedeutung ihres neuen Produkts in letzterer Zeit einen sehr ausgedehnten Betrieb zur Erzeugung von Patentblau eingerichtet, so dass dieser Farbstoff bei seiner bevorstehenden Preisniedrigung bals nicht nur durch seine hervorragenden Eigenschaften, sondern auch bezüglich des Preises mit allen anderen einschlägigen Produkten wird concurriren können.

[20] Patentblau: Echtheit: 'Nach 5 tägiger Einwirkung der Witterung (Juni) sehr wenig verändert, etwas matter, nach 18 Tagen ziemlich stark verändert'.

[21] Allein oder in Verbindung mit anderen Säurefarbstoffen, wie Säureviolett N u. s. w., in bedeutenden Mengen in der Wollfärberei als Ersatz des Indigocarmms benutzt (V3A, kurzes Bad; kupferne Kessel sind zu vermeiden). Zum Färben der Seide (V8, nicht zu viel Säure!).

Lehne mentioned explicitly that it was often used to replace Indigo carmine in wool dyeing. Silk can be dyed using recipe V8, Lehne mentioned that not too much acid should be used.

For printing on wool, the following recipe is given[22]: 'Rub 60 g of dye well with 500 g of cold water, add 500 g of gum water 1:1, and 60 g of powdered aluminium sulfate. (For Patent blue superfine and N. use 400 g water and 600 g gum water). – Do not dry too hot and steam fairly moist for 1 hour'.

## Identification Using UHPLC-PDA

The analysis of Patent blue (Sigma-Aldrich CAS 3536-49-0) shows a clear peak. It must be noted that in many historical samples, several compounds are found with similar absorption spectra, likely to be side products of the synthesis (Fig. 13.16).

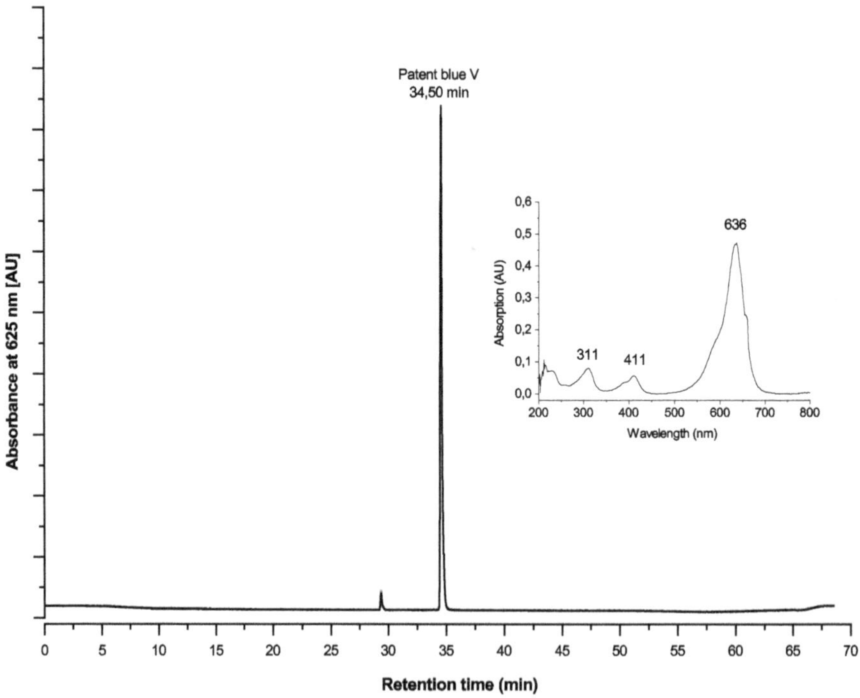


**Fig. 13.16** Chromatogram and spectrum of Patent blue. (Sigma-Aldrich CAS 3536-49-0)

[22]Wolldruck: 60 g Farbstoff mit 500 - kaltem Wasser gut verreiben, dazu 500 - Gummiwasser 1:1, ferner 60 - gepulverte schwefelsaure Thonerde. (Bei Patentblau superfein und N. nehme man 400 g Wasser und 600 g Gummiwasser.) - Nicht zu scharf trocknen und 1 Stunde ziemlich feucht dämpfen.

## 13.8 New Episode of the Tri-Arylmethane Dyes

### *Crystal Violet*

Colour Index name: Basic Violet 3; Colour Index number: 42555
IUPAC name: [4-[bis[4-(dimethylamino)phenyl]methylidene]cyclohexa-2,5-dien-1-ylidene]-dimethylazanium; chloride
Discoverers: Alfred Kern and Heinrich Caro, 1883
Contemporary terminology: Crystal violet, Gentian violet (En), Violet cristal (Fr), Kristallviolett (Ge), Kristalviolet (NL)
Chemical formula: $C_{25}H_{30}N_3.Cl$ (Fig. 13.17)

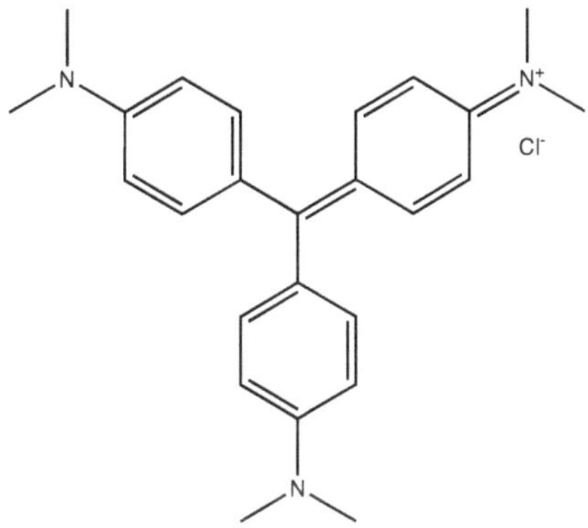


**Fig. 13.17** Structural formula of Crystal violet

#### History

In 1876, the German chemist Wilhelm Michler (1846–1889) synthesised bis(dimethylamino)benzophenone, also known as Michler's ketone by the reaction of phosgene with *N,N*-dimethylaniline [38]. Michler became professor at the Zurich Federal Polytechnic (Eidgenossische Technische Hochschule Zürich) in 1878. He left Europe for a journey to study the natural products of South America in August 1881. He reached Brazil, where he became professor at the Escola Politécnica in Rio de Janeiro and he conducted his research there between 1882 and his death in 1889. He died when he was gassed by one of his colleagues while working in a closed laboratory room with phosgene, with motives unclear. But his discovery of Michler's ketone reaction laid the foundation of the synthesis of Crystal violet, shown at the first Swiss national exhibition in Zurich in 1883.

Early in 1883, the plant manager of Bindschedler & Busch (later CIBA), the Swiss chemist Alfred Kern (1850–1893) employed the highly toxic gas phosgene in the synthesis of aniline dyes. The process was complicated and not fully understood. So, in the company's view, it was too risky to develop it fully on its own to the point of industrially profitable production. To overcome the technical and scientific problems, the company joined forces with BASF and for a short time during 1883 and 1884, the two companies worked together [39]. There were two reasons for the Swiss involvement: firstly, their need for inexpensive raw materials and intermediates, available only from the German dye industry; secondly, the Swiss had little expertise in patent affairs and BASF could provide this knowledge. The result was an agreement dated the 11th of May 1883, which regulated patents, prices for products and awarded a 10% share for the Swiss firm for the dyes developed together. The contract was tied to the duration of the patents. This allowed the German chemist Heinrich Caro (1834–1910), the research-director of the BASF, to improve Kern's synthesis (see Table 13.4) [40]. The new process involved the reaction of *N,N*-dimethylaniline with phosgene to give the intermediate Michler's ketone, which reacted with additional dimethylaniline in the presence of phosphorus oxychloride and hydrochloric acid, yielding Crystal violet. He named the dye Crystal violet because of its crystallinity and brilliant shades.

Caro realised the importance of the new method, as is indicated in a letter to William Perkin written in April 1884, in which he declared: 'An endless variety of Rosaniline-analogues may now be built up with the same certainty and ease as Azo-compounds'.

**Table 13.4** Discoverers and patents of Crystal violet

| Name<br>C.I. name<br>C.I. number | Discoverers | Year of discovery | Patents |
|---|---|---|---|
| Crystal violet<br>Basic Violet 3<br>42555 | A Kern and H. Caro | 1883 | G.P. 26.016/1883<br>G.P. 27.032/1883<br>G.P. 27.789/1883<br>U.S.P. 290.856/1883<br>U.S.P. 290.891/1883<br>U.S.P. 290.892/1883<br>B.P. 4.428/1883<br>B.P. 5.450/1883<br>F.P. 157.430/1883<br>F.P. 158.438/1883<br>B.P. 4.850/1884<br>B.P. 5.038/1884 |

### Trade Names

The dye was put on the market under the commonly used trade name Kristallviolett by BASF, Sandoz, Farbwerke ML&B, Bayer, CIBA, Weiler-ter Meer, Cassella and AGFA. Other trade names were used, such as: Violet G and Violet C by Soc. Anon. St. Denis and Violett 7B extra by Soc. Chim. Us. Rhône.

### Applications

Crystal violet is typically used for the dyeing of wool, silk, cotton or linen.

### Properties

The dye Crystal violet is present as greenish crystals, which dissolves easily in water and ethanol. According to Lehne,[23] the fastness to light is poor: After 5 days exposure to the elements (June) much duller, very faded after 18 days, grey.' The Colour Index also indicates that the light fastness of Crystal violet on wool is poor, ISO 1 to ISO 2, while on silk only slightly better ISO 2 to ISO 3. The wash fastness varies from ISO 2 to 3.

### Recipes

Wool is dyed following recipe V1, using 1% of dyestuff, while silk is dyed with recipe V7. Cotton or linen can be dyed according to recpes V11 and a recipe for jute (V15) is present in the book by Lehne as well. The colourant can also be used to print on wool (recipe V19) and cotton (recipe V23A).

---

[23] Krystallviolett: Echtheit: 'Nach 5 tägiger Einwirkung der Witterung (Juni) viel matter, nach 18 Tagen sehr stark verblasst, grau'.

## Identification Using UHPLC-PDA

Analysis of Crystal violet (Sigma-Aldrich CAS 548–62-9) showed a clear main peak and a compound also found in Methyl violet, which is the penta-methylated form (Fig. 13.18).

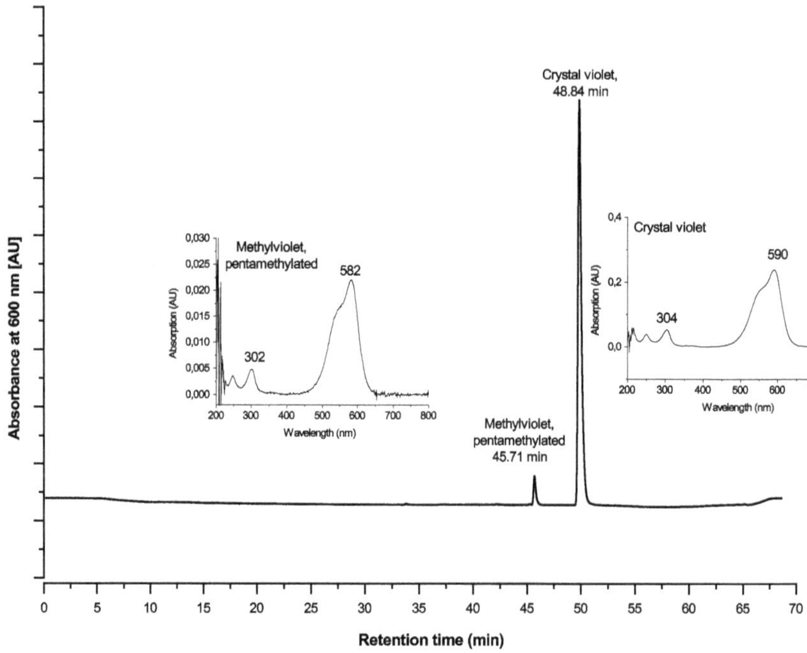


**Fig. 13.18** Chromatogram and spectra of Crystal violet (Sigma-Aldrich CAS 548–62-9)

## The Fruitful Collaboration Between Kern and Caro

The co-operation between both chemists and firms was so successful that in 1883 a series of valuable violet and blue tri-arylmethanes, including Ethyl violet, the naphthylamino derivatives Victoria blue B (see Fig. 13.19), Victoria blue 4R and Night blue were invented and manufactured using tetraalkyddiamidobenzophenones as starting material [41]. One year later, in 1884, the collaboration between both companies also opened the way to an entirely new dye class, the di-arylmethane dyes, of which the yellow Auramines were the most important products. This was followed by several sulfonated Acid violets such as Acid violet 7B (1884), Acid violet 6B (1891) and Alkaline violets (Alkali violet, 1886) developed between 1884 and 1891 by the German chemist Carl L. Müller of BASF.

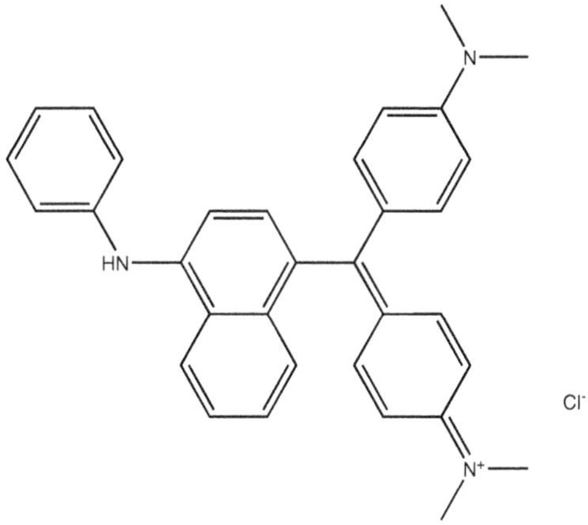


**Fig. 13.19** Structural formula of Victoria blue B

**Table 13.5** Discoverers and patents of Victoria blue B, Victoria blue 4R and Night blue

| Name<br>C.I. name<br>C.I. number | Discoverers | Year of discovery | Patents |
|---|---|---|---|
| Victoria Blue B<br>Basic Blue 26<br>44045 | H. Caro and A Kern | 1883 | G.P. 27.789/1883<br>B.P. 5.038/1884<br>B.P. 11.159/1884<br>U.S.P. 297.414/1884<br>F.P. 160.090/1884<br>B.P. 12.022/1886<br>B.P. 19.062A/1891 |
| Victoria blue 4R<br>Basic Blue 8<br>42563 | H. Caro and A Kern | 1883 | G.P. 27.789/1883<br>G.P. 29.962/1884<br>U.S.P. 308.748/1884<br>B.P. 5.038/1884<br>F.P. 160.090/1884<br>B.P. 12.022/1886 |
| Night blue<br>Basic Blue 15<br>44085 | H. Caro and A. Kern | 1883 | GP 27.789/1883<br>GP 29.962/1884<br>BP 5.038/1884<br>BP 12022/1886 |

By condensing Michler's ketone with *N*-phenyl-1-naphthylamine in the presence of phosphorus oxychloride, Caro and Kern synthesised the very clear and pure Victoria blue B in 1883 [41, 42]. Victoria blue B was sold by BASF, CIBA, Farbwerke ML&B, AGFA, Bayer, Weiler-ter Meer, Read Holliday and Schoellkopf. It was also sold under the trade name Bleu Victoria by Wiescher.

By varying the reaction, still in 1883, Caro and Kern synthesised other dyes, such as Victoria blue 4R and Night blue (Nachtblau). Victoria blue 4R was manufactured by BASF, CIBA, Farbwerke ML&B, AGFA, Weiler-ter Meer and Sandoz [43]. Night blue is prepared from tetraethyldiamidobenzophenone and *p*-tolyl α-naphthylamine and sold by BASF, CIBA and Sandoz using the tradename Nachtblau (see Table 13.5). In 1892, the German chemists Oscar Nastvogel (1863–1910) and Friedrich Reingruber, assignors to the Farbenfabriken vormals Friedr. Bayer & Co., developed Victoria blue R [44, 45].

## Identification Using UHPLC-PDA

Analysis of Victoria blue B (Sigma CAS 2580-56-5) showed clearly separated peaks, together with some minor compounds. Although the absorption maximum of both compounds is similar, 612 nm, the shape of the UV/Vis spectrum is clearly different and the compounds are well separated (Fig. 13.20).

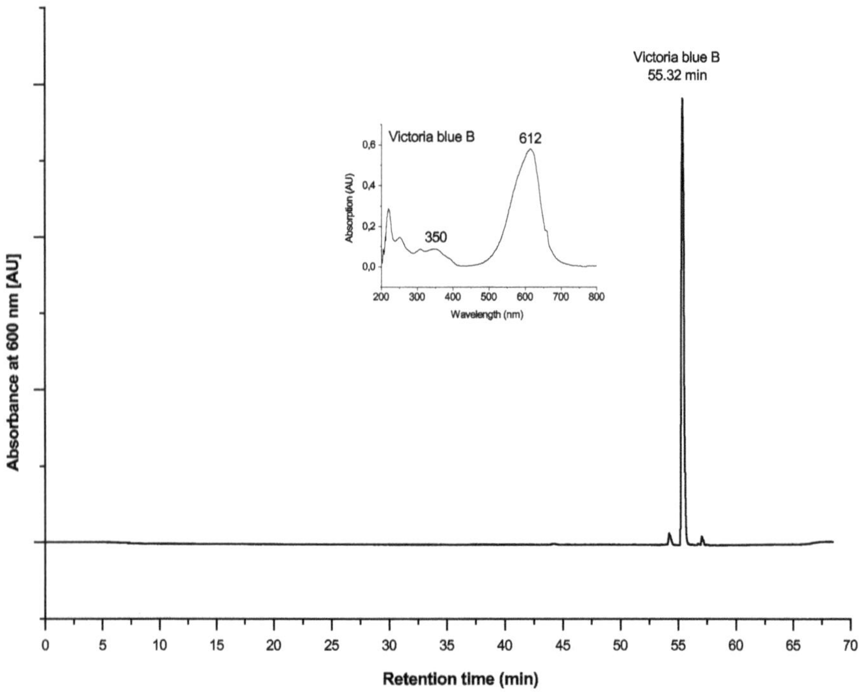


**Fig. 13.20** Chromatograms and spectrum of Victoria blue B. (Sigma CAS 2580-56-5)

## *Fuchsine S*

Colour Index name: Acid Violet 19; Colour Index number: 42685
IUPAC name: Disodium;(3Z)-3-[(4-amino-3-sulfonatophenyl)-(4-amino-3-sulfophenyl)methylidene]-6-imino-5-methylcyclohexa-1,4-diene-1-sulfonate
Discoverer: Heinrich Caro, 1877
Contemporary terminology: Acid magenta, Acid fuchsine, Fuchsine acid (En), Fuchsine acide (Fr), Fuchsin S (Ge), Fuchsine S (NL)
Chemical formula: $C_{20}H_{17}N_3Na_2O_9S_3$ (Fig. 13.21)

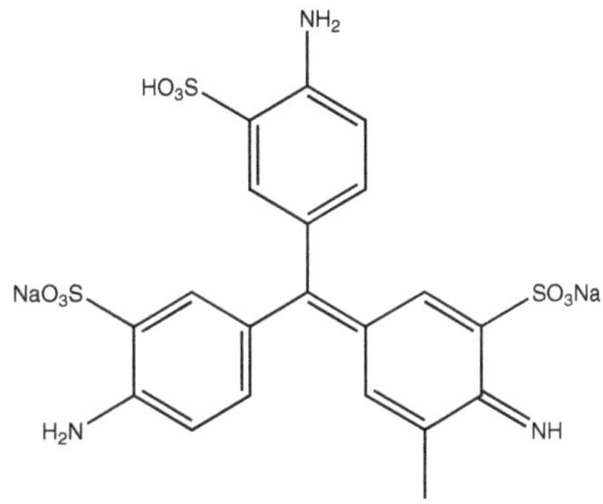


**Fig. 13.21** Structural formula of Fuchsine S

### History

In order to obtain durable acid dyes that could be dyed with Indigo carmine, at the same time in an acid bath, sulfonation, using concentrated sulfuric acid, was not only used for water-insoluble basic dyes, but also for water-soluble basic dyes such as Fuchsine, Methyl violet and Malachite green, as mentioned above. Heinrich Caro of the BASF patented this process in 1877 resulting in, among others, the water-soluble acid dye Fuchsine S (G.P. 2.096, filed on the 16th of December 1877; U.S.P. 250.201 and B.P. 3.731, both filed in 1877; B.P. 2.828, filed in 1879 and F.P. 122.721) [23]. This dye class was extended by the acid dyes Red violet 4RS and Red violet 5RS. Fuchsine S is usually the acid solution of a mixture of mono- and disulfonic acids obtained by sulfonation of Fuchsine.

### Trade Names

The dye was put on the market under the common trade name Fuchsin S by BASF with the addition of different letters. Other trade names were used, such as: Säurefuchsin, with various letters, by Farbwerke ML&B, Kalle, Leonhardt, AGFA, Weiler-ter Meer, Cassella; Fuchsine alcide by Soc. Anon. St. Denis, Durand-Huguenin and Ruch; Acid Magenta by Read Holliday; and Acid Magenta O by Schoellkopf.

### Applications

The dyes were used to dye silk and wool.

### Properties

The dye is soluble in hot and cold water, but not in ethanol. Lehne reported about Fuchsin S[24]: 'After 5 days exposure to the elements (June) more bluish and duller, after 18 days very faded'. The Colour Index indicated a light- and wash fastness of ISO 1.

### Recipes

Wool can be dyed following recipe V3A and silk using recipe V8. For wool, a concentration of 1.5% is mentioned by Lehne.

## 13.9 Case Studies: Embroideries Designed by Émile Bernard

The Van Gogh Museum in Amsterdam owns five embroideries designed by Émile Bernard (1868–1941), a well-known painter and good friend of Vincent van Gogh. These objects are part of the former collection of Andries Bonger. Bonger (1861–1934) was the brother-in-law of Theo van Gogh (1857–1891), Vincent's younger brother. When Bonger lived in Paris, Theo introduced him to important avant-garde artists, such as Émile Bernard and Odilon Redon (1840–1916). After returning to the Netherlands, Bonger started to collect art, in particular by these artists.

In 2004, we started the research into the embroideries in collaboration with the Van Gogh Museum. Although the embroideries were found to be in a generally good condition and only little affected by insects, it was clear that several colours of the woollen yarns were discoloured due to exposure to daylight. This is particularly evident in the comparison of the front and the back of the objects. Long yarn ends present at the back of all embroideries showed colours not exposed to light and allowed an easy removal of different coloured samples without disturbing the integrity of the objects. During the research we analysed yarns of four of these embroideries presented in Table 13.6 and Figs. 13.22, 13.23, 13.24 and 13.25.

---

[24] Fuchsin S: Echtheit: 'Nach 5 tägiger Einwirkung der Witterung (Juni) bläulicher und stumpfer, nach 18 Tagen sehr stark verblasst'.

**Table 13.6** The investigated embroideries of Émile Bernard[a]

| Title | Abbreviation | Date | Inventory number | Size |
|---|---|---|---|---|
| *Breton Women, Fire Screen* | BW | 1892 | v0109N1996 | 79 cm × 61.5 cm |
| *Hérodias* | Hé | 1903 | v0110N1996 | 49 cm × 75.5 cm |
| *Plants and Flowers on a Grey Background* | PFGF | 1906 | v0111N1996 | 41 cm × 103 cm |
| *Plants and Flowers on an Orange Background* | PFOF | 1927 | v0112N1996 | 89 cm × 74.5 cm |

[a]Information obtained from https://www.vangoghmuseum.nl/nl/collectie, accessed 08-12-2024. Note that the embroideries are framed, the size given without the frames

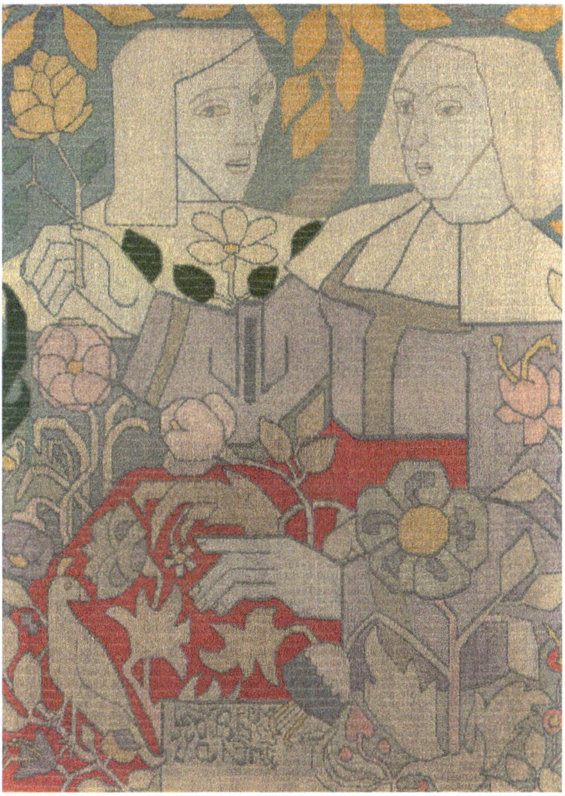

**Fig. 13.22** Émile Bernard (1868–1941), *Breton Women, Fire Screen*, c. 1892, embroidery and polychromed carvings 79 cm × 61.5 cm, v0109N1996, Van Gogh Museum, Amsterdam, the Netherlands

The front of *Breton Women, Fire Screen* is now extremely faded and is dominated by a greenish-grey hue, whereas the unexposed parts, the back, of the textile, still show a wide range of bright colours. The date and the extreme fading suggest that early synthetic dyes were present in this embroidery and in the others. An overview of the dyes found is presented in Table 13.7.

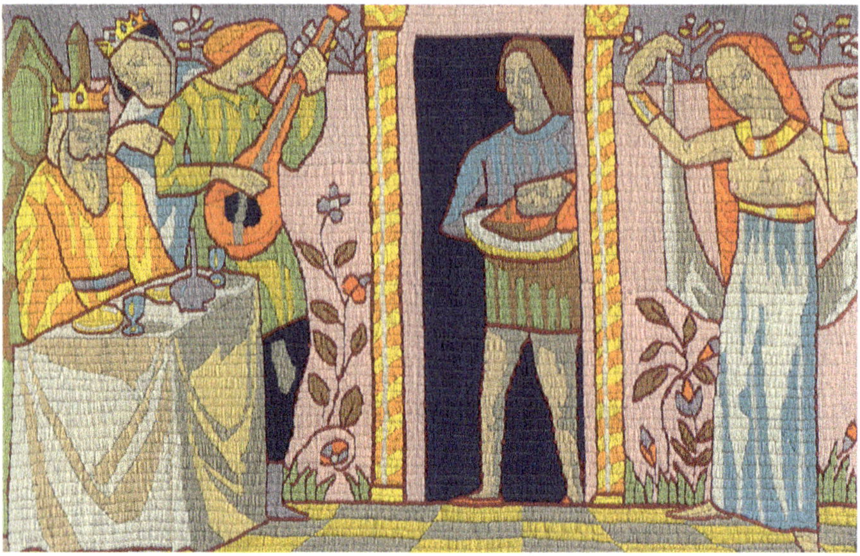

**Fig. 13.23** Émile Bernard (1868–1941), *Hérodias*, 1903, embroidery, 49 cm × 75.5 cm, v0110N1996, Van Gogh Museum, Amsterdam, the Netherlands

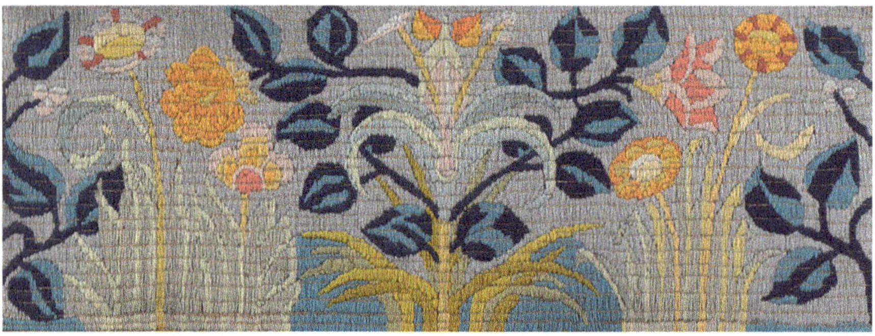

**Fig. 13.24** Émile Bernard (1868–1941), *Plants and Flowers on a Grey Background*, 1906, embroidery, 41 cm × 103 cm, v0111N1996, Van Gogh Museum, Amsterdam, the Netherlands

One of the earliest discovered dyes was Indigo carmine, which was found in *Breton Women*, *Hérodias* and *Plants and Flowers on an Orange Background*, surprisingly since the dye was known not to be very light fast. In 1903 and 1927, faster blue dyes were widely available, yet Indigo carmine was still used. Picric acid was also found in three of the embroideries, the most recent dating from 1906. In *Breton Women*, green was obtained by mixing Indigo carmine and Picric acid. Like Indigo carmine, Picric acid did not have a good light fastness. In the same three objects, Fuchsine and Water blue IN were found, which also had a poor light fastness. It seems that Bernard was not fully aware of the poor stability of these dyes. Methyl violet was only detected in *Breton Women*. Two dyes were detected in all four embroideries, Naphthol yellow S and Cochineal red A.

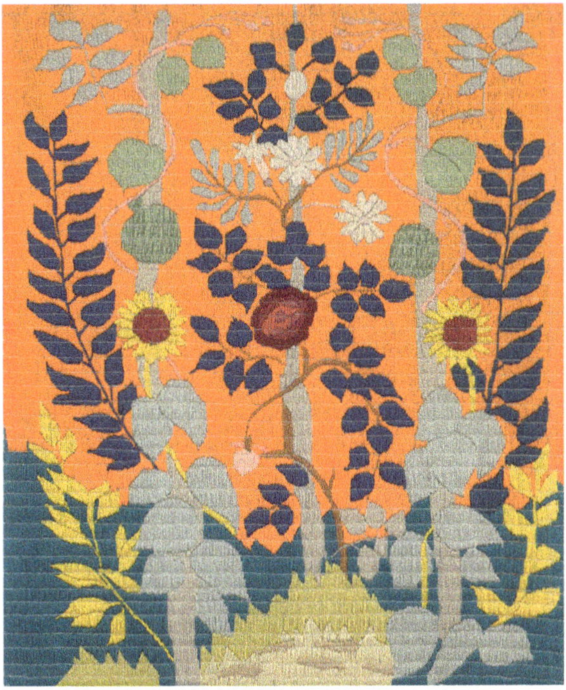

**Fig. 13.25** Émile Bernard (1868–1941), *Plants and Flowers on an Orange Background*, 1927, embroidery, 89 cm × 74.5 cm, v0112N1996, Van Gogh Museum, Amsterdam, the Netherlands

Orange I, Quinoline yellow, Rhodamine B and Patent blue V were detected in the embroideries *Hérodias*, *Plants and Flowers on a Grey Background* and *Plants and Flowers on an Orange Background*, but were not found in *Breton Women*. Auramine has been identified in *Breton Women* and *Plants and Flowers on an Orange Background* designed more than 35 years apart.

Ponceau 2R was present in three of the four objects, but in the embroidery *Hérodias*, Fast red A was detected, which was not found in the other textile objects. The synthetic dyes Amido naphthol red, Rhodamine B, Fast acid magenta and Victoria blue R were only found in *Plants and Flowers on an Orange Background*.

In all embroideries, synthetic colourants were detected which could not be identified due to lack of reference materials. Their UV-Vis spectra resemble those of known colourants, but the retention times were different from those of the reference materials analysed. As such, these were labelled equivalents. In addition, in all four embroideries natural dyes were found as well, showing that although the natural dye industry suffered heavily from the introduction of synthetic colourants, they were sometimes still applied.

Most of the dyes in *Plants and Flowers on an Orange Background* (1927) were developed after 1875, while in the first three embroideries the dyes discovered before 1875 predominate. Research led to the production of better synthetic dyes after 1875. Comparing the four textile objects, there is a shift towards the use of dyes with an increased light fastness.

**Table 13.7** Dyes detected in the four embroideries of Émile Bernard

| Synthetic dyes | Date of discovery | BW (1892) | Hé (1903) | PFGF (1906) | PFOF (1927) |
|---|---|---|---|---|---|
| Indigo carmine | 1743 | X | X | | X |
| Picric acid | 1771 | X | X | X | |
| Fuchsine | 1856 | X | X | X | |
| Methyl violet | 1861 | X | | | |
| Water blue IN | 1862 | X | X | X | |
| Naphthol yellow S | 1875 | X | X | X | X |
| Orange II | 1876 | | X | X | X |
| Amido naphthol red | 1876 | | | | X |
| Fast red A | 1877 | | X | | |
| Cochineal red A | 1878 | X | X | X | X |
| Ponceau 2R | 1878 | X | | X | X |
| Quinoline yellow | 1882 | | X | X | X |
| Auramine | 1883 | X | | | X |
| Tartrazine yellow | 1884 | | | | X |
| Rhodamine B | 1887 | | X | X | X |
| Patent blue V | 1888 | | X | X | X |
| Fast acid magenta | 1890 | | | | X |
| Victoria blue R | 1892 | | | | X |
| Equivalents | | | | | |
| Basic violet | | X | | | |
| Unknown acid red | | X | X | | |
| Patent blue | | | X | | X |
| Unknown blue or violet | | | X | | |
| Diamond green | | | | X | |
| Azo fuchsine 6B | | | | | X |
| Azo flavine 3R | | | | | X |
| Tartrazine yellow | | | | | X |
| Mauveine | | | | | X |
| Natural dyes | | | | | |
| Cochineal | | X | | X | |
| Weld | | X | | | |
| Possibly Persian berries | | X | | | |
| Young fustic | | | | X | |
| Tannins | | | | X | |
| Henna | | | | | X |
| Possibly Old fustic | | | | | X |

## *Other Case Studies*

Tri-aryl methane dyes are found in many other objects. In addition to Fuchsine and Methyl violet, Malachite green, Water blue and Victoria blue B were found in the woollen yarns used by van Gogh as described in Chap. 10. Furthermore, Water blue was detected in the interior upholstery in the train carriage Hannibal, as described in Chap. 10. Water blue and Patent blue were also found in some parasols which are discussed in Chap. 12. Patent blue was identified in samples from billiard cloths, these results are presented in Chap. 9.

## References


1. Schultz, G., Julius, P.: Tabellarische Übersicht der im Handel befindlichen künstlichen organischen Farbstoffe, Vierte Auflage. Verlag Gaertner, Berlin (1902)
2. Crookes, W.: A Practical Handbook of Dyeing and Calico-Printing. Longmans, Green & Co., London (1874)
3. Gardner, W.M.: The British Coal-Tar Industry, its orgin, development, and decline. J.P. Lippincott Company / Williams and Norgate, Philadelphia / London (1915)
4. Scheinert, W.: Triphenylmethanfarbstoffe bei den Chemischen Fabriken vorm. Weiler-ter Meer und ihren Vorläuferfirmen in Krefeld und Uerdingen. Mitteilungen, Gesellschaft Deutscher Chemiker / Fachgruppe Geschichte der Chemie, (Frankfurt/Main). 20, 45–48 (2009)
5. Schultz, G.: Die Chemie des Steinkohlenteers mit besonderer Berücksichtigung der künstlichen organischen Farbstoffe. Zweiter Band, Die Farbstoffe, Druck und Verlag von Friedrich Vieweg und Sohn, Braunschweig. Dritte Auflage. (1901)
6. Perkin, W.H.: On the aniline or coal tar colours (Cantor lecture 2, December 14th, 1868). Journal of the Society of Arts. 842, 109–114 (1869)
7. Morris, P.J.T., Travis, A.S.: A history of the international dyestuff industry. American Dyestuff Reporter. 81, 3–6 (1992)
8. van den Belt, H.: Why monopoly failed: The rise and fall of Société La Fuchsine. The British Journal for the History of Science. 25, 45–63 (1992). https://doi.org/10.1017/S0007087400045325
9. Persoz, J.-F., Luynes, V. de, Salvétat, L.-A.: Nouvelles observations sur le bleu de Paris. C R Hebd Seances Acad Sci. 52, 700–702 (1861)
10. Persoz, J.-F., Luynes, V. de, Salvétat, L.-A.: Faits pour servir à l'histoire de l'aniline; production d'une nouvelle matière tinctoriale bleue, le bleu de Paris. C R Hebd Seances Acad Sci 52, 448–450 (1861)
11. Hofmann, A.W.: Researches on the Colouring-matters derived from Coal-tar. - III. Diphenylamine. Proceedings of the Royal Society of London 13, 341–346 (1863)
12. Hofmann, A.W.: Researches on the Colouring Matters Derived from Coal-Tar. On Aniline-Blue. Proceedings of the Royal Society of London. 13, 9–14 (1863)
13. Girard, C., Laire, G. de: Préparation du bleu de dlphénylamine. Bulletin de la Société chimique de Paris. 7, 363 (1867)
14. Schultz, G.: Farbstofftabellen, Fünfte Auflage. Weidmannische Buchhandlung, Berlin (1914)
15. Perkin, W.H.: On the aniline or coal tar colours (Cantor lecture 3, December 21st, 1868). Journal of the Society of Arts. 843, 121–122 (1869)
16. LII. 'Bereitung des grünen Anilins', (Excerpt from Moniteur scientif. 176, 1863, 362 / Excerpt from Moniteur scientif. 177, 1864, 432). Journal für praktische Chemie. Verlag von Johann Ambrosius Barth, Leipzig. 432, 22–23 (1864)

17. Gattermann, L., Wichmann, G.: Ueber Aldehydblau. Berichte der deutschen chemischen Gesellschaft zu Berlin. Commissionsverlag von Friedländer & Sohn, Berlin. 22, 227–236 (1889)
18. Hofmann, A.W.: Beobachtungen vermischten Inhalts, 1. Zur Kenntnis des Aldehydgrüns. Berichte der deutschen chemischen Gesellschaft zu Berlin. Ferd. Dümmler's Verlagsbuchhandlung, Berlin. 3, 761–762 (1870)
19. Lucius, E.: Jahres-Bericht über die Fortschritte und Leistungen der chemischen Technologie und technischen Chemie. Verlag von Otto Wigand, Leipzig. 10, 537–538 (1864)
20. Lehne, A.: Tabellarische Uebersicht über die künstlichen organischen Farbstoffe und ihre Anwendung in Färberei und Zeugdruck. Verlag von Julius Springer, Berlin (1893)
21. Farbwerke vorm. Meister Lucius & Brüning 1863–1913. Hausdruckerei der Farbwerke vorm. Meister Lucius & Brüning in Höchst am Main
22. Hofmann, A.W., Girard, C.: Ueber die chemische Natur des Anilingrüns. Berichte der deutschen chemischen Gesellschaft zu Berlin. Ferd. Dümmler's Verlagsbuchhandlung, Berlin. 2, 440–452 (1869)
23. Caro, H.: Referate, "Ueber die Entwicklung der Theerfarben-Industrie." Berichte der deutschen chemischen Gesellschaft zu Berlin. Commissionsverlag von R. Friedländer & Sohn, Berlin. 25, 955–1105 (1892)
24. Monnet., P., Reverdin, F.: Preparation of Alcoholic Chlorides and their Application in the Maufacture of Colouring Matters. Bulletin de la Sociétè chimique de Paris. XXIX, 106–108
25. Martius, C.A.: Zur Geschichte der grünen Anilinfarbstoffe. Die Chemische Industrie. Verlag von Julius Springer, Berlin. 2, 81–82 (1879)
26. Fischer, O.: Ueber Condensationsprodukte tertiärer aromatischer Basen. Berichte der deutschen chemischen Gesellschaft zu Berlin. Ferd. Dümmler's Verlagsbuchhandlung, Berlin. 10, 1623–1626 (1877)
27. Doebner, O.: Ueber die Bildung von Farbstoffen durch Einwirkung von Benzotrichlorid auf Phenole und tertiäre aromatische Basen. Berichte der deutschen chemischen Gesellschaft zu Berlin. Ferd. Dümmler's Verlagsbuchhandlung, Berlin. 11, 1236–1241 (1878)
28. Doebner, O.: Zur Kenntnis des Malachitsgrün. Berichte der deutschen chemischen Gesellschaft zu Berlin. Ferd. Dümmler's Verlagsbuchhandlung, Berlin. 11, 2274 (1878)
29. Fischer, E., Fischer, O.: Bemerkungen zu der Abhandlung des Hrn. O. Doebner Zur Kenntnis des Malachitsgrüns. Berichte der deutschen chemischen Gesellschaft zu Berlin. Ferd. Dümmler's Verlagsbuchhandlung, Berlin. 12, 791–803 (1879)
30. Doebner, O.: Erwiderung auf die Bemerkungen der HH. E. Fischer und O. Fischer. Berichte der deutschen chemischen Gesellschaft zu Berlin. Ferd. Dümmler's Verlagsbuchhandlung, Berlin. 12, 1010 (1879)
31. AGFA: G.P. 18.959 (1878), G.P. 23.775 (1882), F.P. 123.187, F.P. 144.169; Bayer & Co.: 1.976 (1878); Farbwerke ML&B: G.P. 11.412 (1879), B.P. 476 (1879), F.P. 133.704; Badische Co.: G.P. 27.789 (1883), B.P. 4.850 (1884), B.P. 5.038 (1884), F.P. 160.090
32. Cooksey, C., Dronsfield, A.: It happened by chance: the discovery of Malachite green. Dyes in History and Archaeology. 20, 165–174 (2005)
33. Fischer, O.: Ueber Condensationsprodukte tertiärer aromatischer Basen. Berichte der deutschen chemischen Gesellschaft zu Berlin. Ferd. Dümmler's Verlagsbuchhandlung, Berlin. 14, 2521 (1881)
34. Mühlhäuser, O.: Industrie der Bittermandelölgrün. Dingler's Polytechnisches Journal. Verlag der J.G. COTTA'schen Buchhandlung, Stuttgart. 263, 250 (1887)
35. Friedlander, P.: Kleinere Mittheilungen. Berichte der deutschen chemischen Gesellschaft zu Berlin. Ferd. Dümmler's Verlagsbuchhandlung, Berlin. 22, 588 (1889)
36. Ostersetzer, O.: Ueber Patentblau. Mittheilungen des K.K. Technologischen Gewerbe-Museums in Wien. Verlag von Carl Graeser, Wien. 4, 109–113 (1890)
37. Farbwerke ML&B: G.P. 50.286 (1889), G.P. 50.293 (1889), G.P. 50.440 (1889), B.P. 2.796 (1888), B.P. 14.822 (1888), G.P. 55.621 (1890), U.S.P. 412.613 (1889), U.S.P. 412.614 (1889), U.S.P. 412.615 (1889), G.P. 63.026, G.P. 66.791 (1892), G.P. 71.156 (1892), G.P. 74.014 (1893)

38. Michler, W.: Synthese aromatischer Ketone mittelst Chlorkohlenoxyd. Berichte der deutschen chemischen Gesellschaft zu Berlin. Ferd. Dümmler's Verlagsbuchhandlung, Berlin. 9, 716–718 (1876)
39. Reinhardt, C., Travis, A.S.: Heinrich Caro and the Creation of Modern Chemical Industry. Kluwer Academic Publishers, Dordrecht/Boston/London (2000)
40. BASF: G.P. 26.016 (1883), G.P. 27.032 (1883), G.P. 27.789 (1883), U.S.P. 290.856 (1883), U.S.P. 290.891 (1883), U.S.P. 290.892 (1883), B.P. 4.428 (1883), B.P. 5.450 (1883), B.P. 4.850 (1884), B.P. 5.038 (1884), F.P. 157.430 (1883) and F.P. 158.438 (1883)
41. BASF: G.P. 27.789 (1883), G.P. 29.962 (1884), U.S.P. 297.414 (1884), B.P. 5.038 (1884), B.P. 11.159 (1884), B.P. 12.022 (1886), B.P. 19.062A (1891) and F.P. 160.090 (1884)
42. Nathansohn, M., Müller, P.: Ueber Derivate und Reactionen des Tetramethyldiamidobenzophenons. Berichte der deutschen chemischen Gesellschaft zu Berlin. Commissionsverlag von R. Friedländer & Sohn, Berlin. 22, 1888–1892 (1889)
43. BASF: G.P. 27.032 (1883), G.P. 27.789 (1883), G.P. 29.962 (1883), B.P. 5.038 (1884), B.P. 11.159 (1884), B.P. 12.022 (1886), U.S.P. 308.748 and F.P. 160.090
44. Noelting, E., Philipp, K.: Zur Kenntnis der Triphenylmethanfarbbasen. Berichte der deutschen chemischen Gesellschaft zu Berlin. Ferd. Dümmler's Verlagsbuchhandlung, Berlin. 41, 583–584 (1908)
45. Bayer & Co.: B.P. 19.062A (1891), U.S.P. 517.473 (1894) and F.P. 217.020

# Chapter 14
# The Early Azo Dyes: Aniline Yellow, Bismarck Brown, Chrysoidine, Orangés, Fast Red A

## 14.1 Introduction

The first discovered synthetic dyes, Mauveine and Fuchsine, both named after the colour of flowers, were produced by British and French manufacturers. These new products were among the great sensations of the Great Exhibition in London in 1862. Two years after these discoveries, the azo dyes were introduced by Johann Peter Griess. Griess worked at the chemistry department at the German University of Marburg, led by German professor Adolph Wilhelm Hermann Kolbe (1818–1884). In 1858, he discovered the diazo coupling reaction. He found that when an aromatic amine is treated with nitrous acid at low temperatures, an unstable diazonium salt is obtained. Next, the diazonium salt reacts with aromatic amines, phenols, naphthols or pyrazolones to give coloured azo products. Azo dyes are characterised by the presence of a double nitrogen (diazo) group (an –N=N– bond) and represent the largest synthetic dye class. This discovery convinced August Wilhelm Hofmann and, on Kolbe's recommendation, Griess was offered a position at the Royal College of Chemistry in London. In 1864, Griess suggested the prefix 'azo' to indicate that the compound contained a nitrogen atom instead of a hydrogen atom and this definition is now generally accepted [1].

The development of azo dyes started slowly with the so-called ABC dyes. The first azo dye was Aniline yellow, discovered by Griess in 1859, which was made by the reaction of nitrous acid with aniline. The first commercially successful azo dye was Bismarck brown, also known as Manchester brown, discovered by Carl Alexander Martius in 1863, which was brought to the market in 1865. Between the discovery of Bismarck brown and the next useful azo dye lies a period of 12 years. In December 1875, Heinrich Caro, who worked in Germany, discovered the first basic azo dye for wool, Chrysoidine. The same dye was discovered independently by Otto Nikolaus Witt, working in England, in January 1876.



M. R. van Bommel, M. de Keijzer, *Bright Colours from the Past*, Cultural Heritage Science, https://doi.org/10.1007/978-3-031-88664-5_14

The breakthrough came in 1875 and led to the systematic start of the production of azo dyes. Griess found that he could couple sulfanilic acid with amines and phenols resulting in the dyes now known as Orangés but he did not publish his results until 1878. The same dye was discovered by François-Zacharie Roussin. In the spring of 1877, the dye firm of Alcide Poirrier in Paris introduced the name Orangés for the dyes developed by Roussin, and brought Orange 1 and Orange 11 onto the market.

The dye chemists feared that the range of the azo dyes was limited to yellow and orange colours, but the substitution of sulfanilic acid in Orange II by naphthionic acid led to the first useful red azo dye, called Fast red A, discovered by Caro in 1877. Soon a series of analogue compounds of this dye class was produced.

One year later, in 1878, the bright red and bluish-red azo dye, called Ponceau dyes, were discovered by Heinrich Baum, who worked at the factory Farbwerke ML&B in Frankfurt-Hoechst. The Ponceaus, which are further described in Chap. 15, are named after the colour of flowers, namely that of the wild corn poppy.

In 1879 Biebrich Scarlet, the first diazo dye, created by Rudolf Hugo Nietzki, was introduced, although this was not used as a textile dye. A significant advance was made by Paul Böttiger in 1884 with the direct tetra-azo dye Congo red, prepared by the interaction of benzidine with 1-naphthionic acid. Carl Duisberg discovered Benzopurpurine 4B, prepared from 1-naphthionic acid and *o*-tolidine in 1885. Both dyes have the property of dyeing unmordanted cotton directly, hence are named direct dyes or substantive dyes. These dyes are discussed in Chap. 18.

The discovery of Congo red marked the beginning of modern azo dye chemistry and history has shown that the azo dyes would become by far the most important chemical class of dyes and pigments (Table 14.1).

**Table 14.1** Discoverers, year of discovery, constitution and patents of the early azo dyes

| Name<br>C.I. name<br>C.I. number | Discoverers | Year | Constitution | Patents |
|---|---|---|---|---|
| Aniline yellow<br>Solvent Yellow 1<br>11000 | Charles Mène<br>John Dale and<br>Heinrich Caro | 1861<br>1863 | Aniline with nitrous acid | –<br>B.P. 3.307/1863 |
| Bismarck brown<br>Basic Brown 1<br>21000 | Carl Alexander<br>Martius<br>John Dale and<br>Heinrich Caro | 1863<br><br>1863 | *m*-phenylene-diamine with nitrous acid | –<br><br>B.P. 3.307/1863 |
| Chrysoidine<br>Basic Orange 2<br>11270 | Heinrich Caro<br>Otto Nikolaus Witt | 1875<br>1876 | Aniline with<br>*m*-phenylene-diamine | –<br>– |

(continued)

**Table 14.1** (continued)

| Name<br>C.I. name<br>C.I. number | Discoverers | Year | Constitution | Patents |
|---|---|---|---|---|
| Orange I<br>Acid Orange 20<br>14600 | Johan Peter Griess<br>François-Zacharie Roussin | 1876<br>1876 | Sulfanilic acid with α-naphthol | –<br>U.S.P. 210.054/1878 |
| Orange II<br>Acid Orange 7<br>15510 | Otto Nikolaus Witt<br>François-Zacharie Roussin | 1876<br>1876 | Sulfanilic acid with β-naphthol | –<br>U.S.P. 210.054/1878 |
| Orange III<br>Acid Orange 52<br>13025 | Johann Peter Griess<br>Otto Nikolaus Witt | 1875<br>1876 | Sulfanilic acid with dimethyl-aniline | –<br>– |
| Orange IV<br>Acid Orange 5<br>13080 | Otto Nikolaus Witt<br>François-Zacharie Roussin | 1876<br>1877 | Sulfanilic acid with diphenyl-amine | –<br>U.S.P. 210.054/1878 |
| Fast red A<br>Acid Red 88<br>15620 | Heinrich Caro<br>François-Zacharie Roussin | 1877<br>1878 | Naphthionic acid with β-naphthol | G.P. 5.411/1878<br>B.P. 786/1878<br>U.S.P. 204.799/1878<br>F.P. 123.148 |
| Crocein Orange G<br>Acid Orange 12<br>15970 | Johann Peter Griess | 1878 | | – |
| Ponceaus<br>Acid Reds | Heinrich Baum | 1878 | | G.P. 3.229/1878 |

## 14.2 Aniline Yellow

Colour Index name: Solvent Yellow 1; Colour Index number: 11000
IUPAC name: 4-(Phenyldiazenyl)aniline
Discoverers: Johann Peter Griess, 1859; Charles Mène, 1861
Contemporary terminology: Aniline yellow (En), Jaune d'aniline (Fr), Anilingelb (Ge), Anilinegeel (NL)
Obsolete terminology: Amidoazobenzol (Ge)
Chemical formula: $C_{12}H_{11}N_3$ (Fig. 14.1)

**Fig. 14.1** Structural formula of Aniline yellow

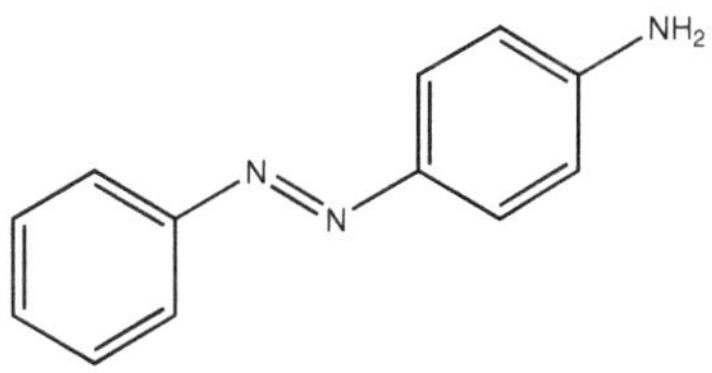

## History

The first azo dye, Aniline yellow, already discovered by the German chemist Johann Peter Griess (1829–1888) in 1859, was made by reacting cold nitrous acid with aniline. This dye was independently discovered by the French chemist Charles Mène in 1861 [2–4]. In 1862, he reported that nitrous fumes ($NO_2$) would convert aniline into a yellow-brown (then unnamed) product 'which dyes cotton and silk perfectly'. It was afterwards, in 1863, patented by the British chemist John Dale (1815–1889), the co-founder of Roberts, Dale & Company in Manchester, and the German chemist Heinrich Caro (1834–1910), working at that time in Britain (B.P. 3.307).

By 1863, the London dye factory Simpson, Maule and Nicholson manufactured this yellow dye and put it on the market under the name Aniline yellow, the first commercial azo dye. Its constitution was established by the German chemists Carl Alexander Martius (1839–1920) and Griess in 1866. Unfortunately, as a dye it was less than satisfactory because it proved to have quite poor dyeing properties, despite its golden-yellow, strongly glittering, leaflike crystals.

## Trade Names

The dye was put on the market under the trade name Amidoazobenzol by AGFA, Farbwerke ML&B, Wülfing-Dahl and Soc. Anon. St. Denis. Other trade names were used like: Spritgelb by Leonhardt and Weiler-ter Meer; Spritgelb G by Kalle; Anilingelb and Jaune d'aniline by Durand-Huguenin.

## Applications

As mentioned above, Aniline yellow was not satisfactory due to its poor dyeing properties. Aniline yellow gained some niche uses, but was not utilised by the garment industry.

## Properties

Aniline yellow has the appearance of a yellowish-brown powder and dissolves very well in oil and ethanol. The drawbacks are a very poor light fastness and the fact that it does not dissolve in water, making it unsuitable for use in dyeing. As a consequence, it is not described by Lehne. Dyers suggested this was due to its insolubility in water, but although its solubility is low, it dissolves sufficiently to dye yarn and fabric a fairly yellow colour. The reason why Aniline yellow failed as a fabric dye was that its affinity for cloth is poor because the single amine group on the molecule does not have sufficient interaction with the textiles. It coloured the fabric well enough, but it leached out on washing and as such was inferior to existing natural dyes (especially Weld, *Reseda luteola* L.) that produced the same colour [5]. The oxalate of Aniline yellow was commonly used as a yellow dye, but because it was volatile at a very low temperature, it quickly evaporated from the dyed fabric. Aniline yellow is no longer used because it has a very poor light fastness and easily volatilises when steamed. According to the Colour Index, the light fastness is good in mineral oil and the fastness to water is good as well. However, it is not mentioned as a textile dye but in spirit lacquers and varnishes, for wax products, oils and more recently in styrene resins.

## *Recipes*

In acidic solutions of Aniline yellow, the colour on silk turns red. Washing with water releases the acid and the free base remains on the fibre, which looks yellow [6].

## *Identification Using UHPLC-PDA*

The analysis of a reference sample (RCE no. 7760) showed the presence of two compounds which elute rather late. This indicates their hydrophobic nature (Fig. 14.2).

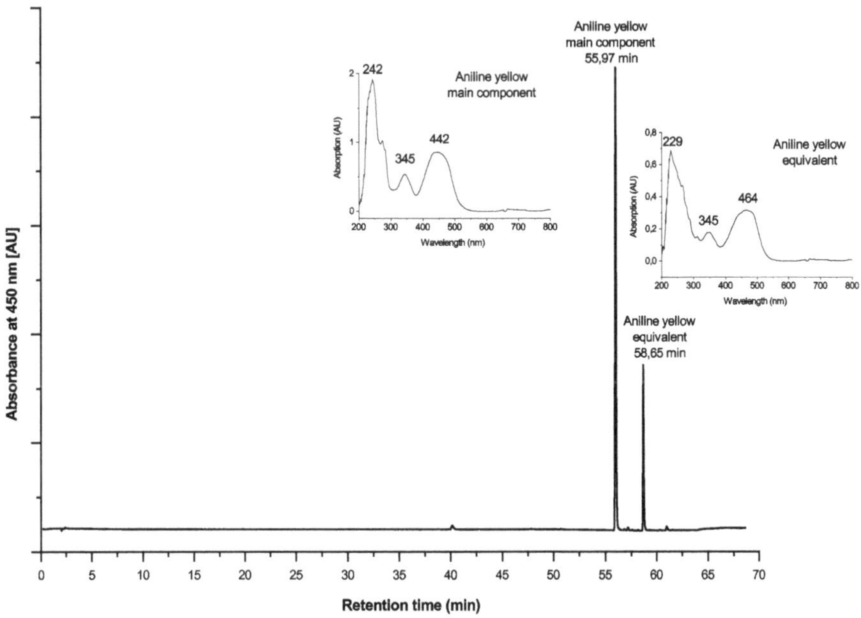


**Fig. 14.2** Chromatogram and spectra of Aniline yellow (RCE no. 7760)

## 14.3 Bismarck Brown

Colour Index name: Basic Brown 1; Colour Index number: 21000
IUPAC name: 4-[[3-[(2,4-diamino-5-methylphenyl)diazenyl]-4-methylphenyl] diazenyl]-6-methylbenzene-1,3-diamine;dihydrochloride
Discoverer: Carl Alexander Martius, 1863
Contemporary terminology: Bismarck brown, Vesuvine, Manchester brown (En), Brun Bismarck (Fr), Bismarck braun (Ge), Bismarckbruin (NL)
Obsolete terminology: Anilinbraun, Zimmetbraun, Canelle, Englischbraun, Goldbraun (Ge)
Chemical formula: $C_{18}H_{20}N_8$ (Fig. 14.3)

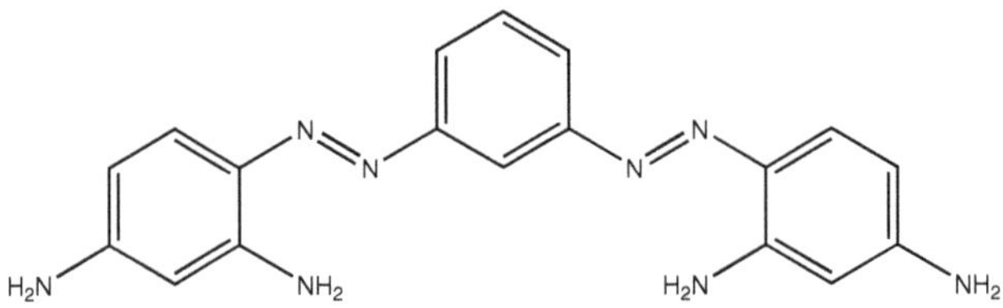


**Fig. 14.3** Structural formula of Bismarck brown

### *History*

The first commercially successful azo dye was Bismarck brown, also named Manchester brown, Phenylene brown or Vesuvine, discovered by Martius in 1863. Martius worked at that time at the firm Roberts, Dale & Company in Manchester. Bismarck brown was patented by Dale and Caro, who also stayed in England between 1850 and 1866 [7]. It was brought onto the market by this firm in 1866. Bismarck brown is produced by the reaction of nitrous acid with *m*-phenylenediamine and coupling to non-diazotised *m*-phenylenediamine. As in the case of Aniline yellow, the diamine is available in excess and is therefore available to couple with the initially formed tetrazo salt. The dye, which coloured fabrics brown with a tint of orange, consists of a mixture of three different bases, two of which are only present in very small quantities. According to the investigations of Caro and Griess, Bismarck brown has to be interpreted as triaminoazobenzene [8]. However, this dye is a diazo compound which actually contains four amino groups; these can diazotise and thus couple further. This remained unclear and even at the end of the nineteenth century, the constitution of Bismarck brown was examined by several German chemists [9–12].

## Trade Names

Bismarck brown was put on the market by AGFA, Bayer, Leonhardt, Kalle, William Brothers, Ruch, Durand-Huguenin and Schoellkopf. Other companies sold the dye, using different names such as Manchester brown by Levinstein, Roberts, Dale & Co. and Cassella; Brun Bismarck by Manufacture Lyonnaise; Vesuvin by Farbwerke ML&B and BASF; Phenylenbraun by Weiler-ter Meer and Lederbraun by Cassella.

## Applications

Bismarck brown is reasonably water-soluble and transfers well to silk, linen and leather. It is a basic dye, dyeing wool a red-brown colour. It can also dye cotton when this is mordanted with tannins. The dye does not leach out to any extent on washing and achieves rich brown tones. The brown dye still finds considerable application in the dyeing and leather industry.

## Properties

Bismarck brown dissolves in water and alcohol with a brown colour. Lehne wrote on the fastness:[1] 'After 8 days exposed to the elements, in July strongly, and after 4 weeks completely changed. Not changed by chlorination (1 part of a calcium hypochlorite solution of 5 °Bé[2] and 10 parts of water)'. According to the Colour Index, its light fastness is poor, between 1 and 2 out of a scale up to 5 following the German norms. For the wash fastness, different values are given by the Colour Index. According to the German norms of 1951, the wash fastness is categorised as class 1, the AATCC norm indicates only a slight change (class 4).

## Recipes

Lehne gives recipes for dyeing and printing cotton (V11 and V23A), dyeing jute (V15) and leather (all without a mordant). Silk can be dyed at 50 °C with the addition of about 1% Marseille soap and then dyed with acetic acid; a concentration of 1% is typically used. For the dyeing of cotton, it must be mordanted with tannic acid, Catechu (Cutch), etc. to fix the dye.

---

[1] Bismarck braun: Echtheit: 'Nach 8 tägiger Einwirkung der Witterung (Juli) stark, nach 4 Wochen vollständig verblasst. Durch Chloren (1 Theil Chlorkalklösung von 5 °Bé und 10 Theile Wasser) nicht verändert'.

[2] Degree Baumé, the exact concentration remains unknown.

## *Identification Using UHPLC-PDA*

The analysis of a reference sample (RCE no. 8511) showed the presence of several compounds, with similar UV-Vis absorption spectra. The compound eluting at a retention time of 38.21 min is the main compound; the others are likely to be side products formed during the synthesis of Bismarck brown (Fig. 14.4).

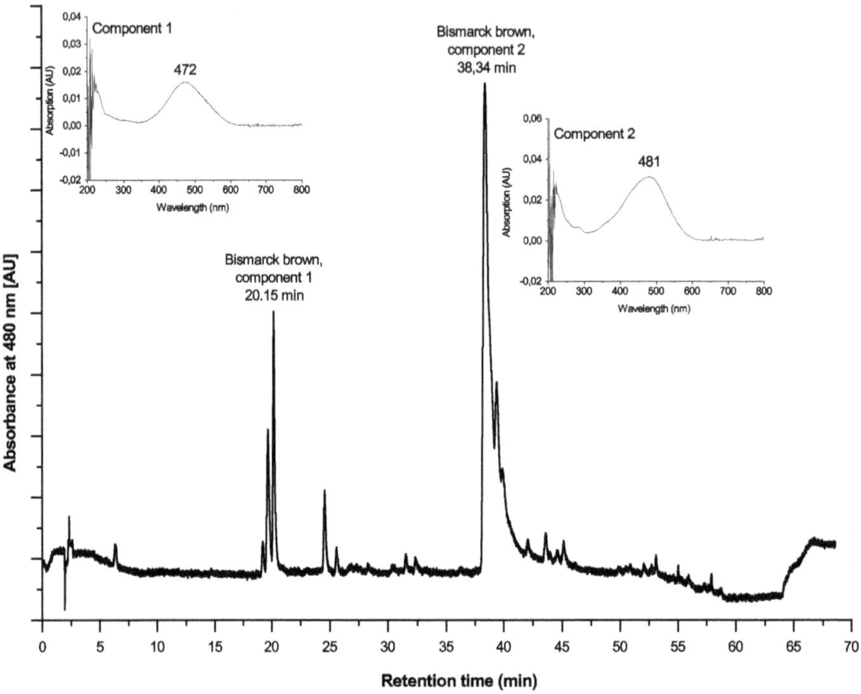


**Fig. 14.4** Chromatogram and spectra of Bismarck brown (RCE no. 8511)

## 14.4 Chrysoidine

Colour Index name: Basic Orange 2; Colour Index number: 11270
IUPAC name: 4-Phenyldiazenylbenzene-1,3-diamine;hydrochloride
Discoverers: Heinrich Caro, 1875; Otto Nikolaus Witt, 1876
Contemporary terminology: Chrysoidine (En), Chrysoidine (Fr), Chrysoidin (Ge), Chrysoidine (NL)
Chemical formula: $C_{12}H_{12}N_4$ (Fig. 14.5)

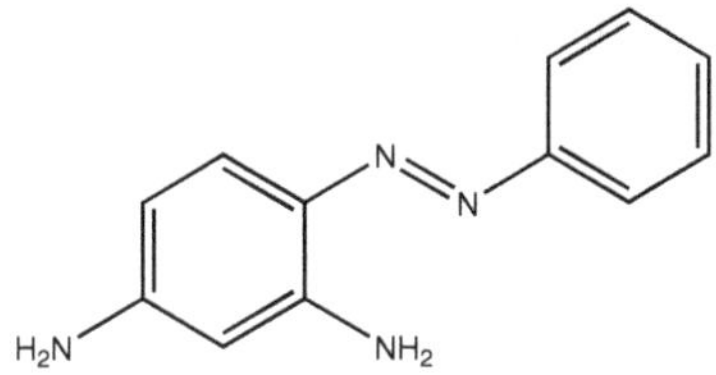


**Fig. 14.5** Structural formula of Chrysoidine

## *History*

Between the discovery of Bismarck brown and the next commercial azo dye lies a period of 12 years. It took a decade to fully understand the azo reaction and exploit it for the production of numerous azo and diazo dyes, starting with Chrysoidine, the first basic azo dye for wool. Chrysoidine was discovered in December 1875 by Caro. Independently from Caro, the dye was also discovered by the chemist Otto Nikolaus Witt (1853–1915) in England, in January 1876. Witt had a Russian, Swiss and German nationality. The dye was commercially produced by the firm Williams, Thomas & Dower at Brentford near London and was already introduced to the market in April 1876. This golden yellow, slightly tending towards orange, was derived from the reaction of diazotised aniline with *m*-phenylenediamine [13–15]. Hofmann determined the chemical composition of Chrysoidine and stated that it was the hydrochloride of diamino azobenzene [16].

## *Trade Names*

Different from many other dyes, no synonyms were used and the trade name chrysoidine was used by famous international dye companies in Germany, France, England and Switzerland. These companies were AGFA, BASF, Leonhardt, Cassella, Kalle, Farbwerke ML&B, Soc. Chim. Us. Rhône, Roberts, Dale & Co., Wülfing-Dahl, Williams Brothers, Bayer, Durand-Huguenin, Griesheim-Elektron, Weiler-ter Meer, CIBA, Soc. Anon. St. Denis, Read Holliday, Schoellkopf and Levinstein.

## *Applications*

Chrysoidine dyes wool, silk and mordanted cotton golden yellow, slightly tending to orange. However, according to Lehne, the dye is regularly used on tannin-mordanted cotton, often mixed with Fuchsine and Safranine (C.I. name: Basic Red 2; C.I. number: 50240).

## *Properties*

Chrysoidine is a yellowish-orange powder which dissolves well in water and ethanol. Lehne investigated the fastness of this dye; he wrote:[3] 'After 5 days exposed to the elements strongly changed. No fastness to chlorination'. According to the Colour Index, the light fastness (method B) when applied to cotton is between 1 and 2, depending on the colour intensity. The wash fastness is categorised as class 1 (method B).

## *Recipes*

Lehne gives a recipe for the dyeing of cotton (V11), using 0.15% dye to obtain a light colour and 1.5% dye to obtain a more intense colour.

## *Identification Using UHPLC-PDA*

The analysis of a reference sample (RCE no. 3289) showed the presence of one component, representing Chrysoidine (Fig. 14.6).

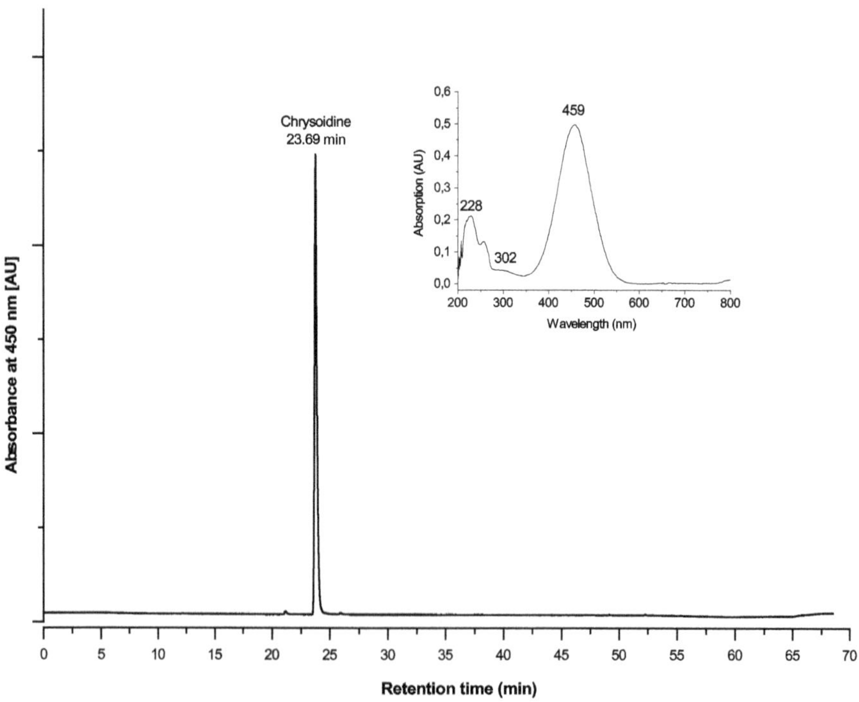


**Fig. 14.6** Chromatogram and spectrum of Chrysoidine (RCE no. 3289)

---

[3] Chrysoidin: Echtheit: 'Durch 5 tägige Einwirkung der Witterung stark verändert. Nicht chlorecht'.

## 14.5 The Orangés

### *History*

Soon after the discovery of Chrysoidine, a series of orange dyes were discovered, i.e. Orange 1, 11, III and IV known as 'the Orangés'. Several chemists worked on these dyes and developed them independently. The first was Griess, who in 1875 prepared the dye Hélianthine (C.I. name: Acid Orange 52; C.I. number: 13025) by coupling the diazo compound of sulfanilic acid with dimethylaniline [17]. He knew that the -$NH_2$ group of sulfanilic acid could be diazotised as a result of work published by the German chemist Rudolf Schmitt (1830–1898) in 1859. Griess did not publish his findings until 1878, when he applied for a patent (G.P. 3.224, filed on the 12th of March 1878). By then 'his' product was already being produced in Paris [18, 19]. He made the seminal discovery and was content to let others exploit the procedure and profit from it. He saw no commercial potential in his discovery.

In 1876, Witt developed the same dyes, and others, independently from Griess. He named the dyes Tropäolines. The dyes he worked on were Orange 1, III, IV and Chrysoine (C.I. name: Acid Orange 6; C.I. number: 14270), prepared using sulfanilic acid and resorcinol [20, 21]. In the spring of 1877, Alcide Poirrier's dye factory in Paris marketed the 'Orangés', which the Paris Exhibition of 1878 claimed to have been discovered independently of Griess's work by the French chemist François-Zacharie Roussin (1827–1894). Roussin was at first a pharmacist in a military hospital in Paris (1875–1879), but resigned from the post. He worked mainly on naphthalene derivatives and first used tin and hydrochloric acid in reducing nitro compounds. Orangés are prepared from sulfanilic acid and different naphthols, like α- and β-naphthol. Orange I, II, III and IV were prepared by reacting diazotised sulfanilic acid (4-aminobenzene-1-sulfonic acid) with 1-naphthol (α-naphthol), 2-naphthol (β-naphthol), *N,N*-dimethylaniline and diphenylamine respectively.

The most important ones were Orange II and Orange IV. These products show strong colours and the presence of the sulfuric acid group (-$SO_3H$, or -$SO_3Na$) made these dyes extremely water-soluble and also made the dyes adhere well to wool. In 1881, the German chemist Carl Theodor Liebermann (1842–1914) wrote that Orange 1 (α-Naphtholorange) was no longer on the market [22]. It is remarkable that the invention of the various Orangés was not patented and was only described and applied by Roussin in one American patent (U.S.P. 210.054, filed on October 22, 1878) [23] apparently in the belief that rivals would find it difficult to replicate the synthesis, an error that soon came to light. The dyes were investigated by Hofmann who concluded in 1877 that they were formed by coupling diazobenzene with naphtholsulfonic acid [24].

## *Orange I*

Colour Index name: Acid Orange 20; Colour Index number: 14600
IUPAC name: Sodium;4-[(4-hydroxynaphthalen-1-yl)diazenyl]benzenesulfonate
Discoverers: Johann Peter Griess, 1876; François-Zacharie Roussin, 1876; Otto Nikolaus Witt, 1876
Contemporary terminology: Orange 1 (En), Orangé 1 (Fr), Orange 1 (Ge), Oranje l (NL)
Chemical formula: $C_{16}H_{12}N_2O_4S$ (Fig. 14.7)

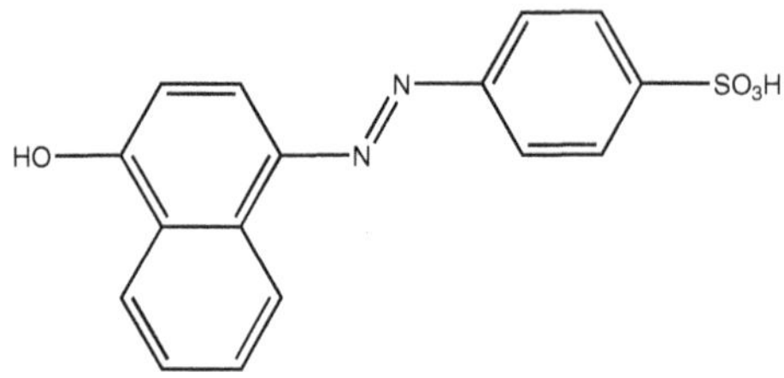


**Fig. 14.7** Structural formula of Orange l

### Trade Names

Orange l was put on the market by Jäger, Kalle, Weiler-ter Meer, Durand-Huguenin, Roberts, Dale & Co., Williams Brothers and Schoellkopf. Other trade names were used such as Orangé No. 1 by Poirrier & Chappat fils, who later on merged with Soc. Anon. St. Denis; Orange R extra by Soc. CHIM. Us. Rhône; Naphtholorange by AGFA and Beyer & Kegel; Tropaolin *G* by Sandoz, and Orange B by Leonhardt.

### Applications

Orange I was mainly used for the dyeing of wool although the Colour Index also reported the dyeing of silk.

## Properties

Orange I is a bright orange powder which dissolves well in water. Lehne researched the fastness of Orange I. He stated:[4] 'After 18 days exposed to the elements, in September and October, the colour has strongly faded. By sulfurous acid of 4.75% strongly changed'. No fastness properties are given in the Colour Index.

## Recipes

Adolf Lehne gives a recipe for the dyeing of wool (V3) using 1.25% of dye.

# *Orange II*

Colour Index name: Acid Orange 7; Colour Index number: 15510
IUPAC name: Sodium;4-[(2-hydroxynaphthalen-1-yl)diazenyl]benzenesulfonate
Discoverer: François-Zacharie Roussin, 1876
Contemporary terminology: Orange II (En), Orangé II (Fr), Orange II (Ge), Oranje II (NL)
Chemical formula: $C_{16}H_{12}N_2O_4S$ (Fig. 14.8)

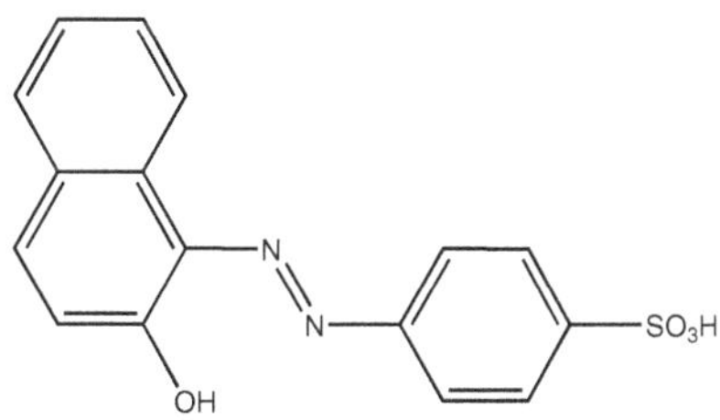


**Fig. 14.8** Structural formula of Orange II

## Trade Names

Orange II was put on the market by BASF, Beyer & Kegel, Durand-Huguenin, CIBA, Kalle, Levinstein, Manufacture Lyonnaise, Roberts, Dale & Co., Ruch, Williams Brothers, Griesheim-Elektron, Jäger, Weiler-ter Meer and Cassella. Other trade names used are Orange No. 2 by Farbwerke ML&B and Soc. Anon. St. Denis; Orange P by Griesheim-Elektron; Orange extra by Cassella; Orange A by Leonhardt

[4] Orange I: Echtheit: 'Nach 18 tägiger Einwirkung der Witterung (September und Oktober) stark verblasst. – Durch schweflige Säure von 4.5 °Bé sehr verändert'.

and Schoellkopf; Orange G by Beyer & Kegel and Read Holliday; Goldorange by Beyer & Kegel, Bayer, Wülfing-Dahl and Weiler-ter Meer; Mandarin or Mandarin G by AGFA, Beyer & Kegel and Bayer; Orange R extra by Soc. Chim. Us. Rhône; Orange II *B* by Bayer; Orange II P and Orange II *PL* by BASF. The number of trade names and producers showed the importance of this dye.

### Applications

Orange II can be used for dyeing wool, silk and for wool printing. In addition, the Colour Index reports that the dye can be also used to colour nylon.

### Properties

The orange powder is soluble in water, resulting in a reddish-yellow solution, and in ethanol where it appears orange. Lehne wrote about the fastness properties of Orange II the following:[5] 'After 18 days exposed to the elements, in September and October, the colour has hardly changed. By sulfurous acid of 4.75% strongly changed'. The Colour Index presents the light fastness as moderate, ISO 3–4. The wash fastness is categorised as ISO 1–2.

### Recipes

Lehne mentions recipes for the dyeing of wool (V3), using 1.5% dye, silk (V8) and wool printing (V21) for which a concentration of 30 g/L was used.

### Case Studies

Orange II was found in the coloured balls of wool in Vincent van Gogh's Chinese lacquer box, dated 1887, and this case study is described in Chap. 10. The dye was also identified in three out of the four embroideries designed by Émile Bernard, more information can be found in Chap. 13. The colourant was also detected in samples taken from the Hofdamask wall-hangings from the Hofburg and Schloß Schönbrunn in Vienna, Austria. This case study is presented in Chap. 15.

---

[5] Orange II: Echtheit: 'Nach 18 tägiger Einwirkung der Witterung (September und Oktober) sehr wenig verändert. – Durch schweflige Säure von 4,5 °Bé stark verändert (heller und gelber)'.

## *Orange III*

Colour Index name: Acid Orange 52; Colour Index number: 13025
IUPAC name: Sodium 4-{[4-(dimethylamino)phenyl]diazenyl}benzene-1-sulfonate
Discoverers: Johann Peter Griess, 1875; Otto Nikolaus Witt, 1876; François-Zacharie Roussin, 1876
Contemporary terminology: Orange III (En), Orangé III (Fr), Orange III (Ge), Oranje III (NL)
Chemical formula: $C_{14}H_{15}N_3O_3S$ (Fig. 14.9)

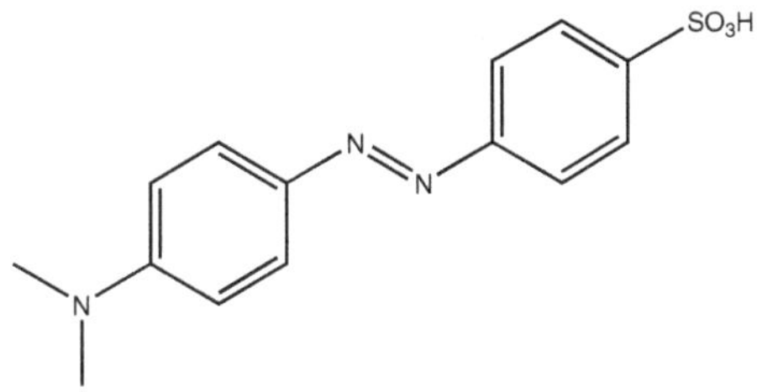


**Fig. 14.9** Structural formula of Orange III

### Trade Names

Orange III was put on the market by Soc. Anon. St. Denis, Weiler-ter Meer, Roberts, Dale & Co., Durand-Huguenin, Ruch and Williams Brothers. Other trade names used are Orange No. 3 by Soc. Anon. St. Denis; Helianthin by BASF; Methyl Orange MP and Goldorange MP by AGFA.

### Applications

Orange III was mainly used for the dyeing and printing of wool, although the Colour Index also mentioned the application on silk.

### Properties

Orange III forms an orange-yellow powder that dissolves easily in boiling water. According to the Colour Index it does not dissolve in alcohol. The dye gives a fiery orange to silk and wool. Lehne writes:[6] 'After 18 days exposed to the elements, in September and October, the colour has hardly changed. By sulfurous acid of 4.75% completely faded'. No fastness properties are given for Orange III in the Colour Index.

[6]Orange lll: Echtheit: 'Nach 18 tägiger Einwirkung der Witterung (September und Oktober) sehr wenig verändert. – Durch schweflige Säure von 4,5 °Bé stark vollständig verblasst'.

### Recipes

Lehne provides recipes for dyeing (V3) and printing (V18) of wool, for dyeing a concentration of 1.5% is mentioned. The addition of sulfuric acid should be limited as much as possible. So, dyeing is performed in weakly acidic solutions.

## *Orange IV*

Colour Index name: Acid Orange 5; Colour Index number: 13080
IUPAC name: Sodium;4-[(4-anilinophenyl)diazenyl]benzene-1-sulfonate
Discoverers: Otto Nikolaus Witt 1876; François-Zacharie Roussin, 1876
Contemporary terminology: Orange IV (En), Orangé IV (Fr), Orange IV (Ge), Oranje IV (NL)
Chemical formula: $C_{18}H_{15}N_3O_3S$ (Fig. 14.10)

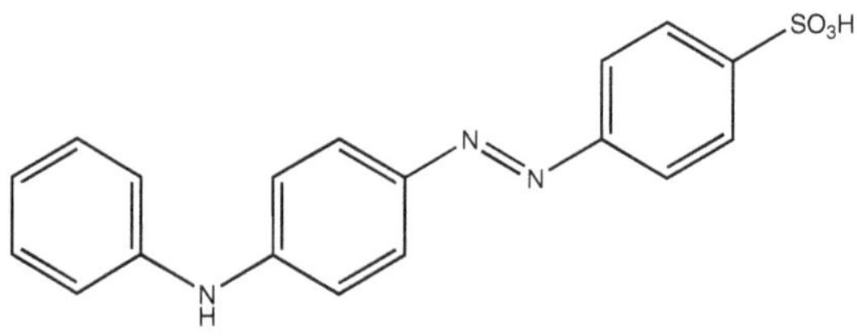


**Fig. 14.10** Structural formula of Orange IV

### Trade Names

Orange IV was put on the market by BASF, Beyer & Kegel, Kalle, Cassella, Durand-Huguenin, Picard, Manufacture Lyonnaise, Soc. Anon. St. Denis, Weiler-ter Meer, Bayer, Read Holliday, Roberts, Dale & Co., Ruch, Leonhardt, Williams Brothers, Schoellkopf and Wülfing-Dahl. The dye was sold using other trade names such as Orange No. 4 by Farbwerke ML&B and Soc. Anon. St. Denis; Orange N by BASF and CIBA; Orange GS by Griesheim-Elektron; Neugelb extra by Bayer; Säuregelb D extra by AGFA; Tropäolin OO and Säuregelb krist. by Cassella; Citronine Y double by Durand-Huguenin and Säuregelb DMP by AGFA.

## Applications

Orange IV is used to dye wool and silk.

## Properties

Orange IV is an orange-yellow crystalline powder and dissolves very easily in hot water. This dye gives a very beautiful golden yellow on silk and wool. Lehne writes about the fastness properties the following:[7] 'After 18 days exposed to the elements, in September and October, the colour has hardly changed. By sulfurous acid of 4.75% strongly changed'. According to the Colour Index the light fastness is moderate as expected based on the results from Lehne, ISO 3. The wash fastness is ISO 1–2.

## Recipes

Lehne reports recipes for dyeing of wool (V3), using a concentration of 0.75%, and silk (V9) with a concentration of 1.5%. For silk dyeing it is especially mentioned that it is often mixed with other dyes such as Fast red. The addition of sulfuric acid should be limited as much as possible.

## Identification Using UHPLC-PDA

A mixture of Orange I, II and IV was analysed (RCE no. 3016, 6928 and 6887 respectively). The compounds are well separated and each show their own unique UV/Vis absorption spectrum as shown in Fig. 14.11.

---

[7] Orange IV: Echtheit: 'Nach 14 tägiger Einwirkung der Witterung (Juli) wenig verändert. – Durch schweflige Säure von 4,5 °Bé stark vollständig verblasst'.

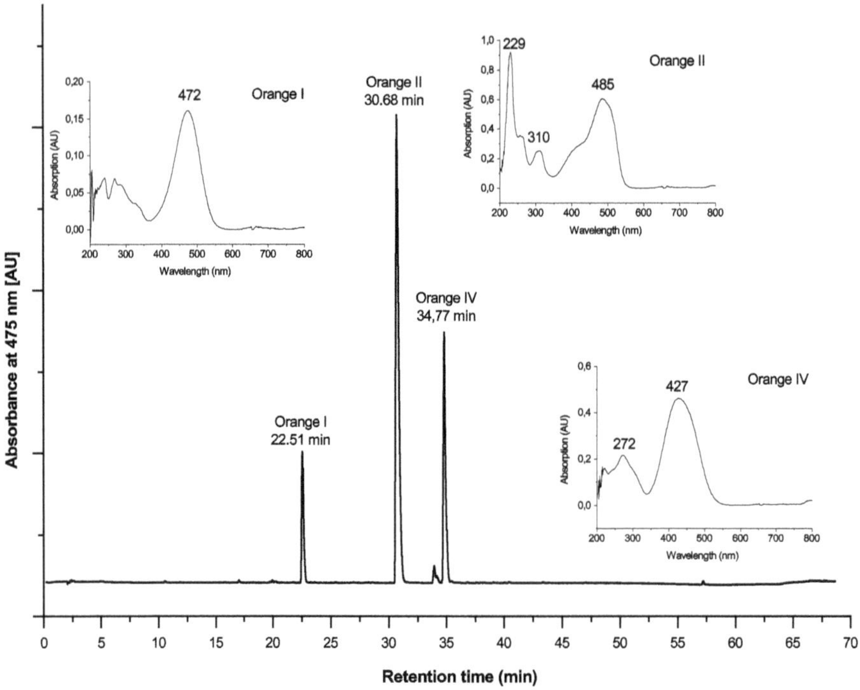


**Fig. 14.11** Chromatogram and spectra of a mixture of Orange I (RCE no. 3016), Orange II (RCE no. 6928) and Orange IV (RCE no. 6887)

### Case Study

Orange IV is one of the dyes found in the Hofdamask wall-hangings from the Hofburg and Schloß Schönbrunn in Vienna, Austria; this case study is discussed in Chap. 15.

## 14.6 Fast Red A

Colour Index name: Acid Red 88; Colour Index number: 15620
IUPAC name: Sodium;4-[(2-hydroxynaphthalen-1-yl)diazenyl]naphthalene-1-sulfonate
Discoverer: Heinrich Caro, 1877
Contemporary terminology: Fast red A (En), Roccelin (Fr), Echtrot A (Ge), Echtrood A (NL)
Obsolete terminology: Fast red AV (En), Orcellin, Rauracienne (Fr), Echtrot AV (Ge)
Chemical formula: $C_{20}H_{14}N_2O_4S$ (Fig. 14.12)

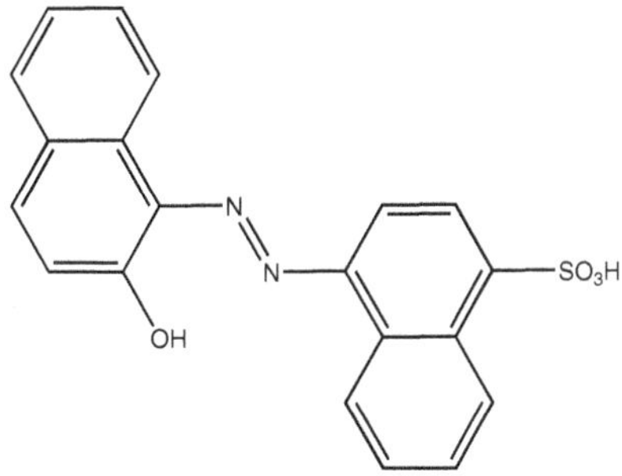


**Fig. 14.12** Structural formula of Fast red A

## *History*

The dye chemists feared that the range of the azo dyes was limited to yellow and orange colours. A technological breakthrough was the use of soluble sulfonic acid derivatives which led to the extension of the range to include red azo dyes. The use of naphthols, naphthylamines and naphthalene substitution products proved especially fruitful. In July 1877, Caro replaced sulfanilic acid in Orange II with naphthionic acid, which led to the first red azo dye Fast red A (Echtrot A or AV) (G.P. 5.411, filed on March 12, 1878) [7, 25–27]. Fast red A is obtained by diazotising naphthionic acid (4-amino-1-naphthalenesulfonic acid) and coupling it with 2-naphthol (β-naphthol). Already in the same month, BASF introduced this azo red dye under their trade name Echtrot AV [19]. Almost simultaneously, and independently from Caro, Fast red A seems to have been discovered by Roussin, in the dye factory of Alcide Poirrier in Paris. Fast red A became the first real threat to the natural insect dye American cochineal and the natural lichen dye Orseille. In a short time, a series of naphthalene-based red and brown analogues appeared, like Fast red E (C.I. name: Acid Red 13; C.I. number: 16045), Fast brown R (C.I. name: Acid Brown 6; C.I. number: 14625) and Fast brown 3B (C.I. name: Acid dye; C.I. number:14630), which are very suitable for dyeing wool. One year later, in 1878, the red and blue-red Ponceau dyes, which had much purer colours, were discovered.

## *Trade Names*

In Germany, the dye was sold under the name Echtrot A by AGFA, Beyer & Kegel, Bayer, Leonhardt, Levinstein, Kalle, Wülfing-Dahl, Jäger, Weiler-ter Meer and Griesheim-Elektron. Other trade names were used, such as Echtrot AV by BASF; Roccelin by Beyer & Kegel, Durand-Huguenin, Cassella, CIBA, Picard, Soc. Anon. St. Denis, Ruch, Sandoz, Weiler-ter Meer and Soc. Chim. Us. Rhône. Farbwerke ML&B sold the dye using the name Echtrot O; Schoellkopf used the name Fast red, and Durand-Huguenin, CIBA sold it as Cerasin and Rubidin was used by Beyer & Kegel and Read Holliday traded under the name Cardinal red.

## *Applications*

Fast red A was mainly used in wool and silk dyeing and for the printing of wool. When dyed on wool in an acidic solution, Fast red A produces vivid red nuances which, however, show a slight brownish tinge. The dye is used extensively in wool dyeing, especially for dyeing knitted articles and hosiery products, either on its own or mixed with other acid dyes. It is one of the most important and most widely used azo dyes.

## *Properties*

Due to its intense colour, solid samples appear almost black. The dye is slightly soluble in alcohol with a blood-red colour. It is also slightly soluble in cold water but easily soluble in hot water. Lehne studied the fastness properties of Fast red A; he stated:[8] 'After 30 days exposed to the elements, in May and June, the colour has hardly changed, a little brighter. By sulfurous acid of 4.75%, slightly brighter and more yellow'. According to the Colour Index, the light fastness is not very good, ISO 2–3. The wash fastness is categorised as ISO 1.

## *Recipes*

Recipes are present in the book from Lehne, for the dyeing of wool (V3) with a concentration of 2.5% and silk (V8) for which no concentration is given. According to Lehne, the dye was much used in wool and silk dyeing and wool printing.

## *Identification Using UHPLC-PDA*

UHPLC analysis of a reference sample of Fast red A (RCE no. 4989) showed a single component present (Fig. 14.13).

---

[8] Echtheit: 'Nach 30 tägiger Einwirkung der Witterung (Mai und Juni) sehr wenig verändert, etwas heller. – Durch schweflige Säure von 4,5 °Bé etwas heller und gelber'.

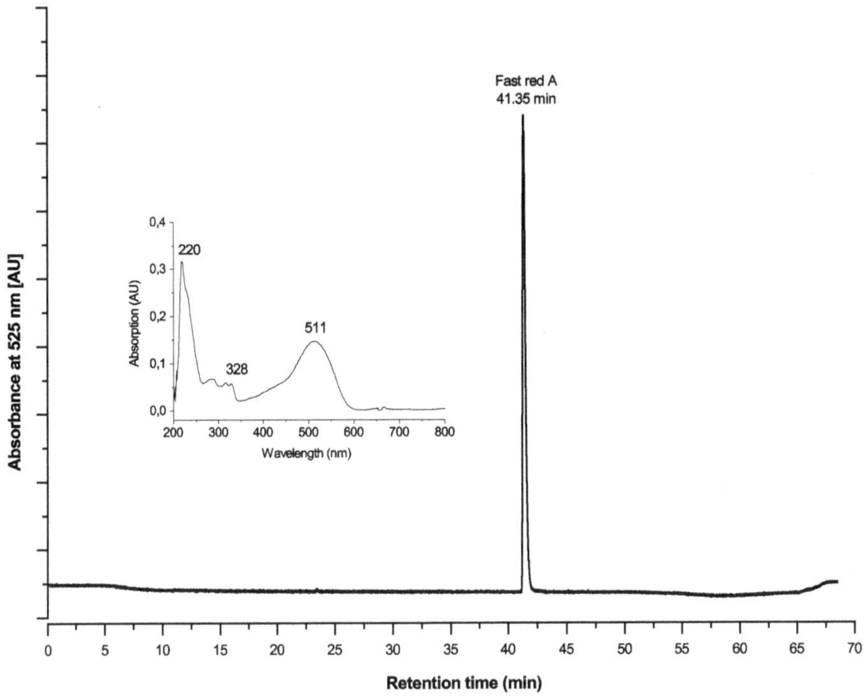


**Fig. 14.13** Chromatogram and spectrum of Fast red A (RCE no. 4989)

## 14.7 Case Study

### *The Chasuble from the St. Bavo Cathedral in Haarlem*

A chasuble from the St. Bavo Cathedral in Haarlem was investigated, presented in Fig. 14.14. Between 1895 and 1930 the Dutch architect Joseph Cuypers (1861–1949) built the Cathedral of St. Bavo in Haarlem. The chasuble was clearly designed for the Cathedral, given the names of the saints on it, including the Parish of St. Bavo. The chasuble is a Roman model, with armholes cut up at the front, the so-called 'fiddle-box model'. The orifice on the front is embroidered with a rose stem and lilies, and bands containing the names of the Saints Bavo, Willibrordus and Laurentius. On the reverse a cross is sewn, also embroidered with rose stems and lilies, with the names of the Saints Casper, Joseph and Martin and a central medallion with the image of Saint Mary. The maker and dating are unknown, but based on its form, style and embroidery technique it is dated before 1910. During the restoration of the St. Bavo Cathedral, between 2009 and 2013, the chasuble was temporarily stored folded in the attic. Due to leakage from the roof, the chasuble became wet and the red dyes of the lining leaked out.

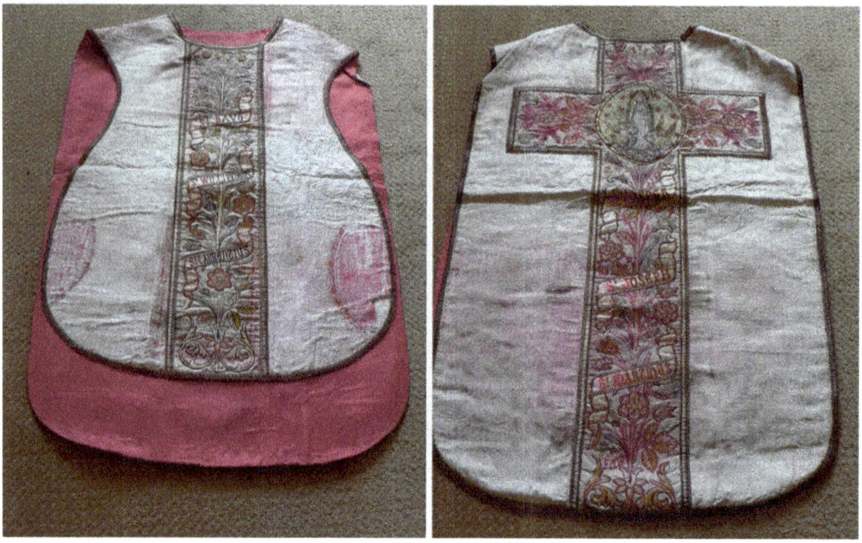

**Fig. 14.14** Chasuble, 1900–1925, Unknown maker, h. 140 cm, w. 68 cm, damask, natural silk, gold thread, galloon, St Bavo Cathedral Haarlem. Photograph: Museum Catharijneconvent, Utrecht, left front, right, back

The colour bleeding in the textile occurs in the red lining and in various red embroidery threads. The bleeding on the back of the chasuble shows a different pattern, namely the embroideries are surrounded by a red haze. This defect is caused by a dark-red embroidery thread worked as a contour around all the flowers and stems.

In the woolen samples of the chasuble Fast red A was the principal dye identified using UHPLC, together with Eosin, Orange ll, Rhodamine B and Alizarin. The dye analysis gave an unambiguous result. The objects are predominantly dyed using early synthetic acid and some basic red dyes known for their poor wash- and water resistance. The only non-bleeding dye occurs in the embroidery thread of the contours of plants and flowers on the front of the chasuble and consists of Alizarin, a mordant dye with excellent wash fastness properties. It is unclear if Alizarin was extracted from Madder root (*Rubia tinctorum* L.) or that it was synthetic Alizarin, which was available when the chasuble was made. The other dyes are known for inferior wash fastness. In particular, Fast red A has a poor wash fastness and is precisely the dye that was predominantly present in the chasuble.

Based on research, the chasuble can be dated. A terminus post quem of 1891 can be established for the chasuble due to the use of viscose in the lining and damask. The identification of Rhodamine B also provided a terminus post quem for the chasuble. This basic violet dye was discovered in 1887. These results support the dating around 1900 as previously suggested, based on form, style, material and technique [28].

## *Other Case Studies*

Fast red A was also found in the embroideries designed by Émile Bernard, these case studies are described in Chap. 13.

## References


1. Griess, P.: On a New Series of Bodies in Which Nitrogen is Substituted for Hydrogen. Philos Trans R Soc Lond. 154, 667–731 (1864)
2. Griess, J.P.: Ueber eine neue Klasse organischer Verbindungen, in denen Wasserstoff durch Stickstoff vertreten ist. Annalen der Chemie und Pharmazie. 121, 257–280 (1862)
3. Mène, C.: Jahres-Bericht über die Fortschritte und Leistungen der chemischen Technologie. Verlag von Otto Wigand, Leipzig. 7, 496 (1861)
4. Mène, C.: Note sur un nouveau de l'aniline'. Comptes rendus hebdomadaires des séanes de I. Academie des Sciences. 52, 311 (1861)
5. Cooksey, C.J., Dronsfield, A.: Peter Griess: a critical evaluation of his contribution to azo dye chemistry. The Diversity of Dyes in History and Archaeology. 354–357 (2017)
6. Benedikt, R.: Allgemeine Waarenkunde und Rohstoffleihe, Band ll, Die künstlichen Farbstoffe (Theerfarben), ihre Darstellung, Eigenschaften, Prüfung, Erkennung und Anwendung. Verlag von Theodor Fischer, Kassel. 198 (1883)
7. Roberts, D., Caro, H.: British patent 3307 of 31st of December 1863, (1863)
8. Caro, H., Griess, P.: Vorläufige Notiz über das Phenylenbraun. Zeitschrift für Chemie. Verlag von Quandt & Händel, Leipzig. 3, 278–280 (1867)
9. Täuber, E., Walder, F.: Ueber Bismarckbraun. ll. Berichte der deutschen chemischen Gesellschaft zu Berlin. Commissionsverlag von R. Friedländer & Sohn, Berlin. 30, 2899–2902 (1897)
10. Täuber, E., Walder, F.: Ueber Bismarckbraun. lll. Berichte der deutschen chemischen Gesellschaft zu Berlin. Commissionsverlag von R. Friedländer & Sohn, Berlin. 33, 2897–2899 (1900)
11. Täuber, E., Walder, F.: Ueber Bismarckbraun. Berichte der deutschen chemischen Gesellschaft zu Berlin. Commissionsverlag von R. Friedländer & Sohn, Berlin. 30, 2111–2117 (1897)
12. Möhlau, R., Meyer, L.: Ueber Bismarckbraun. Berichte der deutschen chemischen Gesellschaft zu Berlin. Commissionsverlag von R. Friedländer & Sohn, Berlin. 30, 2203–2206 (1897). https://doi.org/10.1002/cber.189703002203
13. Hofmann, A.W.: Zur Kenntnis des Chrysoïdins'. Berichte der deutschen chemischen Gesellschaft zu Berlin. Ferd. Dümmler's Verlagsbuchhandlung, Berlin. 10, 213–218 (1877)
14. Witt, O.N.: Das Chrysoïdin und seine Umsetzungen. Berichte der deutschen chemischen Gesellschaft zu Berlin. Ferd. Dümmler's Verlagsbuchhandlung, Berlin. 10, 654–662 (1877)
15. Witt, O.N.: Zur Geschichte des Chrysoïdins. Berichte der deutschen chemischen Gesellschaft zu Berlin. Ferd. Dümmler's Verlagsbuchhandlung, Berlin. 10, 350–351 (1877)
16. Hofmann, A.W.: Noch ein Wort über das Chrysoïdin. Berichte der deutschen chemischen Gesellschaft zu Berlin. Ferd. Dümmler's Verlagsbuchhandlung, Berlin. 10, 388–391 (1877)
17. Griess, J.P.: Neue Untersuchungen über Diazoverbindungen, lll. Mitteilung: Ueber Phenoldiazobenzol und diesem analoge Verbindungen. Berichte der deutschen chemischen Gesellschaft zu Berlin. Ferd. Dümmler's Verlagsbuchhandlung, Berlin. 9, 627–630 (1876)
18. Griess, J.P.: Verfahren zur Darstellung von Farbstoffen durch paarweise Verbindung von Diazophenolen mit Phenolen. G.P. 3.224 (1878)

19. Griess, J.P.: Neue Untersuchungen über Diazoverbindungen, VI. Mitteilung: Ueber die Einwirkung einiger Diazosulfosäuren auf Phenole. Berichte der deutschen chemischen Gesellschaft zu Berlin. Ferd. Dümmler's Verlagsbuchhandlung, Berlin. 11, 2191–2199 (1878)
20. Witt, O.N.: On Colouring Matters derived from Diazo-compounds. J Chem Soc. 35, 179–189 (1879)
21. Witt, O.N.: Ueber Azoderivate des Diphenylamins und Diphenylnitroamins. Berichte der deutschen chemischen Gesellschaft zu Berlin. Ferd. Dümmler's Verlagsbuchhandlung, Berlin. 12, 258–262 (1879)
22. Liebermann, C.: Ueber eine bequeme Darstellungsweise des α-Naphtochinons. Berichte der deutschen chemischen Gesellschaft zu Berlin. Ferd. Dümmler's Verlagsbuchhandlung, Berlin. 14, 1796 (1881)
23. Roussin, F.Z.: Improvement in coloring-matters derived from aniline and its homologues. U.S.P. 210.054 (1878), Assigned to Alcide François Poirrier
24. Hofmann, A.W.: Ueber einen neuen Farbstoff. Berichte der deutschen chemischen Gesellschaft zu Berlin. Ferd. Dümmler's Verlagsbuchhandlung, Berlin. 10, 1378–1381 (1877)
25. Reinhardt, C., Travis, A.S.: Heinrich Caro and the Creation of Modern Chemical Industry. Kluwer Academic Publishers, Dordrecht/Boston/London (2000)
26. BASF: Verfahren zur Darstellung rother und brauner Naphtalin-Farbstoffe (Sulfosäuren der Oxyazonaphtaline), G.P. 5.411 (1878)
27. Caro, H.: Improvement in Dye-Stuffs Produced From Naphthylamine, U.S.P. 204.799
28. de Bruijne, M.: BLOEDEND ROOD Uitlopende rode kleurstof in textiel rond 1900 (Bleeding Red. Red Dye Migration in Textiles around 1900), Master thesis, University of Amsterdam, (2015)

# Chapter 15
# The Ponceaus

## 15.1 Introduction

In the year 1878, Heinrich Baum, who worked at Farbwerke ML&B in Frankfurt-Hoechst, discovered the so-called Ponceau dyes [1–3]. The Ponceaus are named after the colour of flowers, namely those of the wild corn poppy (*Papaver rhoeas* L., family Papaveraceae), just like Mauveine and Fuchsine. They can be divided into two groups: dyes prepared from 2-naphthol-3,6-disulfonic acid (R-acid) (see Fig. 15.1, left), such as Ponceau 2G, Ponceau 2R, Bordeaux R and Amaranth; and dyes prepared from 2-naphthol-6,8-disulfonic acid (G-acid) (see Fig. 15.1, right), like Orange G, Crystal Ponceau 6R and Cochineal red A. Ponceau 6R is prepared from 2-naphthol-3,6,8-trisulfonic acid. The colours of the Ponceaus have a range from orange to scarlet and Bordeaux red or maroon, a deep, rather sombre red. The dyes prepared from R-acid show red to dark-red tints, while the dyes derived from G-acid have yellowish-red to red hues. The Ponceaus are used for dyeing animal fibres such as wool and silk.

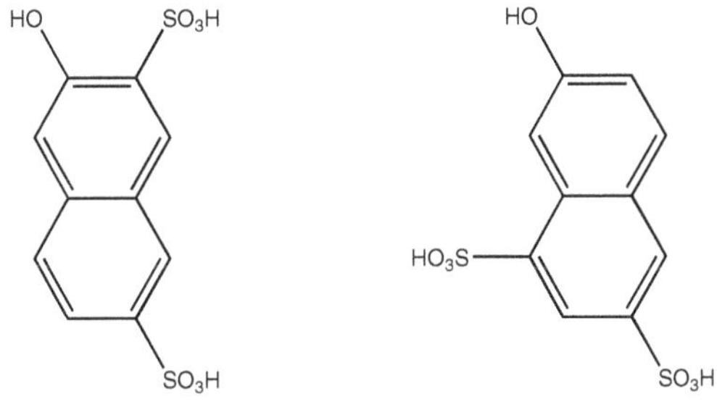


**Fig. 15.1** Left: 2-naphthol-3,6-disulfonic acid (R-acid); right: 2-naphthol-6,8-disulfonic acid (G-acid)



M. R. van Bommel, M. de Keijzer, *Bright Colours from the Past*, Cultural Heritage Science, https://doi.org/10.1007/978-3-031-88664-5_15

## 15.2 History

The firm Farbwerke ML&B at Frankfurt-Hoechst began the manufacture of the azo dyes only in 1878. In the years 1875–1876, the German chemist Heinrich Baum (1849–1923) started to investigate the azo dyes. He discovered two disulfo-acids derived from beta-naphthol-disulfonic acids, called R-acid (2-naphthol-3,6-disulfonic acid) and G-acid (2-naphthol-6,8-disulfonic acid) respectively, which in 1878 he coupled with diazotised amines. For the preparation of the Ponceaus, several components are important, including aniline and its derivatives toluidine, 2,4-xylidine, methyl- and ethylxylidine, 1-naphthylamine and naphthionic acid, which had to be diazotised. On the 24th of April 1878, Farbwerke ML&B received their first German Reichs Patent (G.P. 3.229), which protected the Ponceau dyes and is also one of the very first azo patents (Table 15.1 and 15.2). This patent forms the basis of the Ponceau dyes and claims first the production and separation of the beta-naphthol-disulfonic R- and G-acids. The second part of this patent describes the production of Ponceau and Bordeaux dyes by the coupling of diazotised compounds with R-acid or G-acid, resulting in dyes such as Ponceau R, Ponceau 2R, Bordeaux R, Orangegelb (Orange G), Bordeaux G and Ponceau G, which became of importance for wool dyeing [4].

Two other important inventions led to an extension of this dye class. Baum obtained a first supplement, G.P. 7.217, dated the 3rd of December 1878, in which

**Table 15.1** The Ponceau dyes derived from R-acid (2-naphthol-3,6-disulfonic acid)

| Name<br>C.I. name<br>C.I. number | Discoverer, year of discovery | Constitution | Patents |
|---|---|---|---|
| Ponceau 2G<br>Acid Orange 14<br>16100 | Heinrich Baum, 1878 | Diazotised aniline coupled with R-acid | G.P. 3.229/1878<br>B.P. 1.715/1878<br>U.S.P. 251.162/1881 |
| Ponceau 2R<br>Acid Red 26<br>16100 | Heinrich Baum, 1878 | Diazotised 2,4-xylidine coupled with R-acid | G.P. 3.229/1878<br>B.P. 1.715/1878<br>U.S.P. 210.233/1878<br>U.S.P. 251.162/1881<br>U.S.P. 251.164/1881<br>F.P. 124.811 |
| Bordeaux R<br>Fast red B<br>Acid Red 17<br>16180 | Heinrich Baum, 1878 | Diazotised 1-naphthyl-amine coupled with R-acid | G.P. 3.229/1878<br>G.P. 5.411/1878<br>B.P. 786/1878<br>B.P. 1.715/1878<br>U.S.P. 204.799/1878<br>U.S.P. 251.164/1881<br>F.P. 123.148<br>F.P. 124.811 |
| Amaranth<br>Acid Red 27<br>16185 | Heinrich Baum, 1878 | Diazotised naphthionic acid coupled with R-acid | G.P. 3.229/1878<br>G.P. 5.411/1878<br>B.P. 786/1878<br>B.P. 1.715/1878<br>U.S.P. 204.799/1881<br>F.P. 123.148<br>F.P. 124.811 |

**Table 15.2** The Ponceau dyes derived from G-acid (2-naphthol-6,8-disulfonic acid)

| Name<br>C.I. name<br>C.I. number | Discoverers, year of discovery | Constitution | Patents |
|---|---|---|---|
| Orange G<br>Acid Orange 10<br>16230 | Heinrich Baum, 1878 | Diazotised aniline coupled with G-acid | G.P. 3.229/1878<br>B.P. 1.715/1878<br>G.P. 36.491/1884<br>G.P. 33.916/1884<br>G.P. 35.019/1884<br>B.P. 816/1884<br>B.P. 7.097/1884<br>U.S.P. 314.938/1885<br>U.S.P. 331.059/1885<br>U.S.P. 351.056/1886<br>F.P. 124.811<br>F.P. 159.998<br>F.P. 161.840 |
| Crystal Ponceau 6R<br>Acid Red 44<br>16250 | Meinhard Hoffmann, 1883 | Diazotised 1-naphthyl-amine coupled with G-acid | G.P. 36.491/1884<br>B.P. 816/1884<br>U.S.P. 332.528/1885 |
| Cochineal red A<br>Acid Red 18<br>16255 | Heinrich Baum, 1878 | Diazotised naphthionic acid coupled with G-acid | G.P. 3.229/1878<br>G.P. 5.411/1878<br>B.P. 786/1878<br>B.P. 1.715/1878<br>U.S.P. 204.799/1878<br>G.P. 36.491/1884<br>B.P. 816/1884<br>U.S.P. 314.938/1885<br>F.P. 123.148<br>F.P. 124.811<br>F.P. 159.998 |

the coupling of diazotised compounds of phenols and naphthols with the disulfo-acids of beta-naphthol was described. A number of azo dyes are based on this addition, such as Anisidine Ponceau (Coccinine A, C.I. name: Acid dye; C.I. number: 16160), obtained by the coupling of alpha-anisidine with the R-acid, and Cresol red (Coccinine B, C.I. name: Acid dye; C.I. number: 16160), by coupling of amido-*p*-cresol methyl ether and R-acid.

The second supplement, G.P. 36.491, filed on the 1st of March 1884 by Farbwerke ML&B, describes the purification of the G-salt and the reaction of this purified G-acid (or gamma-acid) with naphthionic acid, which led to the Ponceau dyes Crystal Scarlet 6R (Crystal Ponceau 6R) and Brilliant Ponceau 4R, also known as Neucoccine O or Cochineal red A. The scarlet hues delighted the customers, not only because of their beauty, but also because of their low price. This patent proved to be valuable for Farbwerke ML&B because the dyes described in it surpassed all existing red wool dyes in terms of beauty and fastness. By using them, the natural Cochineal dye obtained from scale insects was gradually replaced in wool dyeing. One year later, in 1885, the German chemist Meinhard Hoffmann (1853–1936), who worked at the company Cassella, patented the preparation of 1-naphthylaniline and G-acid in an alkaline solution, which led to Crystal Ponceau 6R.

The dyes described in G.P. 3.229 were also patented in the United States of America under the U.S.P. 210.233 (combination of 2,4-xylidine with R and G acids), filed in 1878, and U.S.P. 251.162 (combination of aniline with R acid), U.S.P. 251.163 (combination of amidoethylxylidine with R acid) and U.S.P. 251.164 (combination of 1-naphthylamine with R- and G-acids), issued on November 2nd, 1881. In particular, the combinations of the xylidines with R-acid, sold under the trade names Ponceau R, Ponceau 2R and Ponceau 3R, attained great importance at that time, since these dyes were the first to be purely scarlet and were widely used for dyeing wool. The U.S. patents in Baum's name were all transferred to Farbwerke ML&B.

At the beginning of the 1880s, the Ponceau class was extended by Ponceau 6R and Crystal Ponceau 6R. In 1881 and 1882, the German chemist Carl Friedrich Leonhard Limpach (1852–1933) of Farbwerke ML&B patented the preparation of azo dyes such as Ponceau 6R from 2-naphthol-3,6,8-trisulfonic acid under alkaline conditions (G.P. 22.038, filed on the 26th of May 1882; F.P. 137.109, filed on the 27th of May 1882; B.P. 2.544, filed on the 30th of May 1882; U.S.P. 268.506, filed on the 22nd of June 1882, and F.P. 149.249).

## 15.3 Ponceau 2G

Colour Index name: Acid Orange 14; Colour Index number: 16100
IUPAC name: Sodium;3-hydroxy-4-phenyldiazenylnaphthalene-2,7-disulfonic acid
Discoverer: Heinrich Baum, 1878
Contemporary terminology: Ponceau 2G (En), Ponceau 2G (Fr), Ponceau 2G (Ge), Ponceau 2G (NL)
Chemical formula: $C_{16}H_{12}N_2O_7S_2$ (Fig. 15.2)

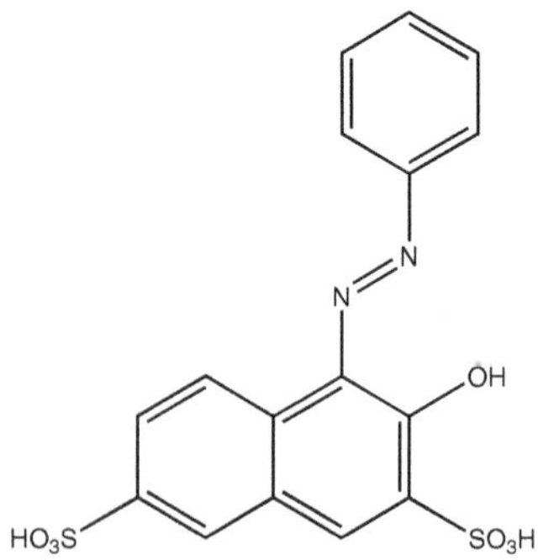


**Fig. 15.2** Structural formula of Ponceau 2G

### *Trade Names*

Ponceau 2G was put on the market by Farbwerke ML&B, BASF and Beyer & Kegel. Other trade names were also used, including Brilliant-Ponceau G by Cassella; Ponceau JJ by Manufacture Lyonnaise; and Orange R by Read Holliday.

## *Properties*

Lehne wrote in 1893 about the fastness of Ponceau 2G: ‘After 14 days exposed to the elements, in July, the colour is strongly changed, after 6 weeks the colour is nearly completely faded. Slightly yellower due to sulfuric acid of 4.75%’.[1] According to the Colour Index, the light fastness is ISO 4 and the wash fastness is ISO 2 (alteration).

## 15.4 Ponceau 2R

Colour Index name: Acid Red 26; Colour Index number: 16150
IUPAC name: Disodium;4-[(2,4-dimethylphenyl)diazenyl]-3-hydroxynaphthalene-2,7-disulfonate
Discoverer: Heinrich Baum, 1878
Contemporary terminology: Ponceau 2R (En), Ponceau 2R (Fr), Ponceau 2R (Ge), Ponceau 2R (NL)
Chemical formula: $C_{18}H_{16}N_2O_7S_2$ (Fig. 15.3)

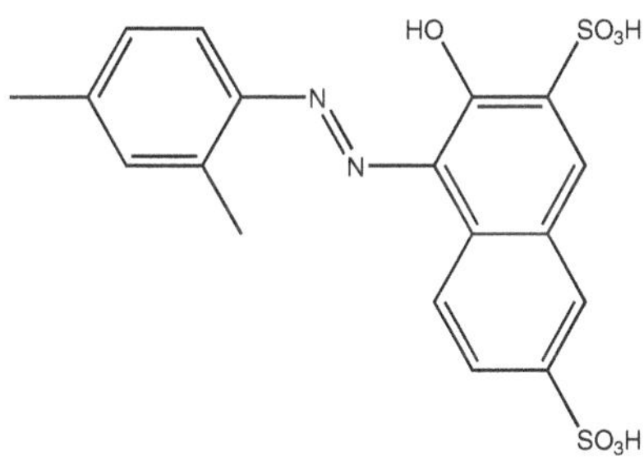


**Fig. 15.3** Structural formula of Ponceau 2R

## *Trade Names*

The dye was sold as Ponceau R and Ponceau 2R by AGFA, BASF, Farbwerke ML&B, Bayer, Beyer & Kegel, Jäger, Griesheim-Elektron, Levinstein, Soc. Anon. St. Denis, Ruch, Leonhardt, Sandoz and Weiler-ter Meer. The dye was also brought onto the market using other trade names, such as Ponceau G by Farbwerke ML&B, AGFA, BASF, Cassella, Beyer & Kegel, Levinstein and Weiler-ter Meer; Ponceau FR and FFR by Cassella; Brilliant ponceau R and RR by Weiler-ter Meer; Scarlet 2R by Read Holliday; Ponceau J by Manufacture Lyonnaise; and Ponceau GR by Farbwerke ML&B.

[1] Ponceau 2G: Echtheit: ‘Nach 14 tägiger Einwirkung der Witterung (Juli) stark verändert, nach 6 Wochen nahezu vollständig verblasst. – Durch schweflige Säure von 4.5 °Bé etwas gelber’.

## Properties

Lehne wrote about the fastness of Ponceau 2R: 'After 14 days exposed to the elements, in July, the colour is considerably paler, after 40 days the colour has strongly faded. Stable against sulfuric acid of 4.75%'.[2] According to the Colour Index, the light fastness is ISO 3 which is comparable with the findings of Lehne. The wash fastness is ISO 3–4.

## Identification Using UHPLC-PDA

The analysis of Ponceau 2R (RCE no. 5000) shows a strong signal, but several side products are present as well (Fig. 15.4).

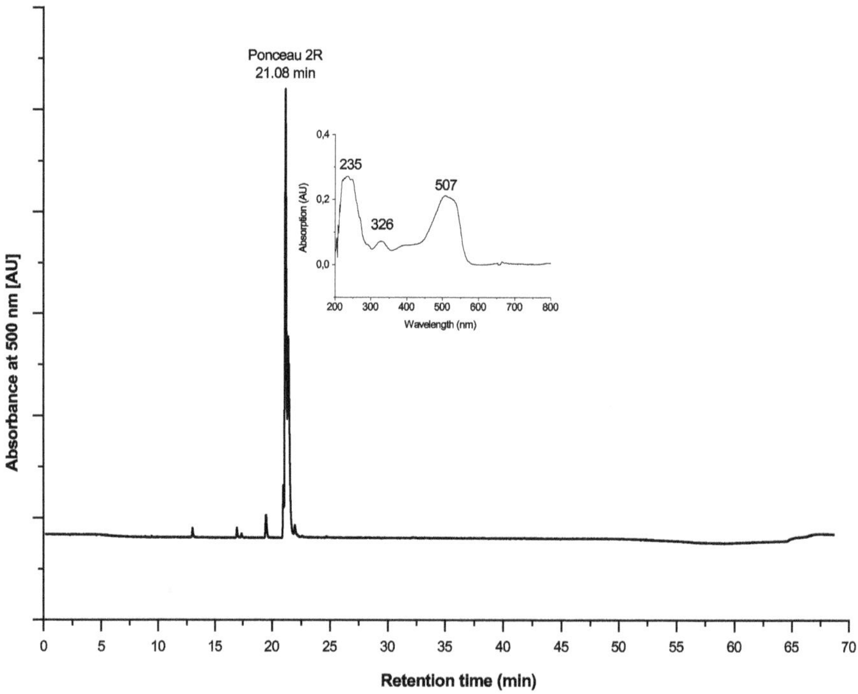


**Fig. 15.4** Chromatogram and spectrum of Ponceau 2R (RCE no. 5000)

[2] Ponceau 2R: Echtheit: 'Nach 14 tägiger Einwirkung der Witterung (Juli) erheblich heller, nach 40 Tagen stark verblasst. – Echt gegen schweflige Säure von 4.5 °Bé'.

## 15.5 Fast Red B

Colour Index name: Acid Red 17; Colour Index number: 16180
IUPAC name: Disodium;3-hydroxy-4-(naphthalen-1-yldiazenyl)naphthalene-2,7-disulfonate
Discoverer: Heinrich Baum, 1878
Contemporary terminology: Fast red B (En), Bordeaux R (Fr), Echtrot B (Ge), Echtrood B (NL)
Chemical formula: $C_{20}H_{14}N_2O_7S_2$ (Fig. 15.5)

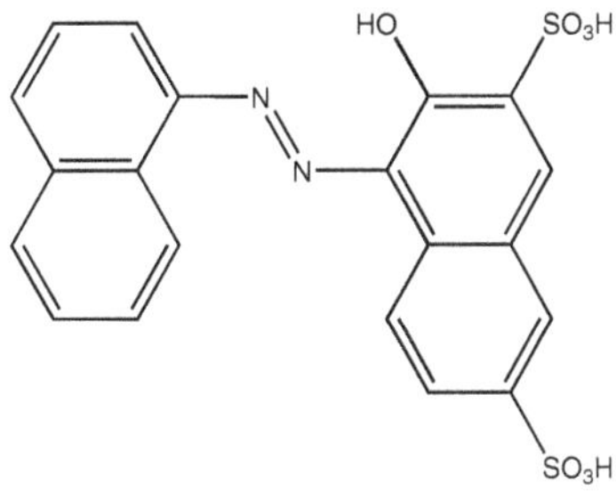


**Fig. 15.5** Structural formula of Fast red B

### *Trade Names*

Fast red B was sold as Echtrot B by BASF, Beyer & Kegel, Leonhardt and Jäger. The dye was introduced by other companies using different trade names such as Bordeaux B by AGFA, Beyer & Kegel, Picard, Farbwerke ML&B, Ruch and Read Holliday; Echtrot P extra by Bayer; Bordeaux BL by Cassella; Bordeaux G by Wülfing-Dahl; Bordeaux R extra by Farbwerke ML&B; Bordeaux R by Weiler-ter Meer; Cérasine by Soc. Anon. St. Denis; Cérasine R by Durand-Huguenin; Archelline 2B by Levinstein; Rouge B by Manufacture Lyonnaise; and Azo-Bordeaux by Schoellkopf.

## Properties

The fastness of Fast red B is described by Lehne under the name Bordeaux R: 'After 30 days exposed to the elements, in May and June, the colour is much duller, especially in appearance. Slightly yellower due to sulfuric acid of 4.75%'.[3] According to the Colour Index, the light fastness is ISO 3–4, the wash fastness is ISO 3.

## Identification Using UHPLC-PDA

The analysis of Fast red B (RCE no. 4995) shows a clear response, a side product is present at a low concentration (Fig. 15.6).

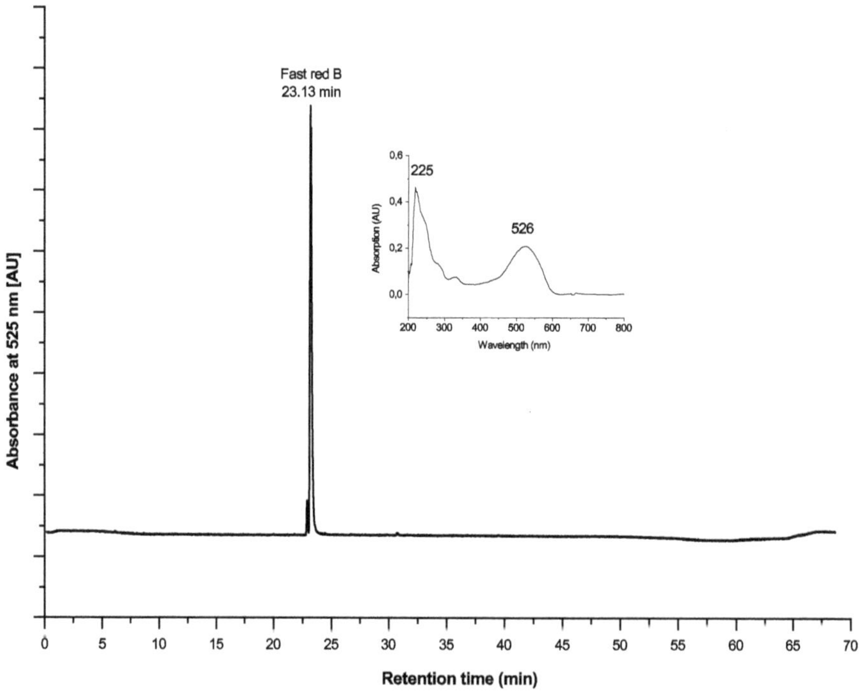


**Fig. 15.6** Chromatogram and spectrum of Fast red B (RCE no. 4995)

[3] Echtrot B: Echtheit: 'Nach 30 tägiger Einwirkung der Witterung (Mai und Juni) viel matter, besonders im Schein. – Durch schweflige Säure von 4.5 °Bé etwas gelber'.

## 15.6 Amaranth

Colour Index name: Acid Red 27; Colour Index number: 16185
IUPAC name: Trisodium;3-hydroxy-4-[(4-sulfonatonaphthalen-1-yl)diazenyl] naphthalene-2,7-disulfonate
Discoverer: Heinrich Baum, 1878
Contemporary terminology: Amaranth (En), Amaranth (Fr), Amaranth (Ge), Amarant (NL)
Chemical formula: $C_{20}H_{14}N_2O_{10}S_3$ (Fig. 15.7)

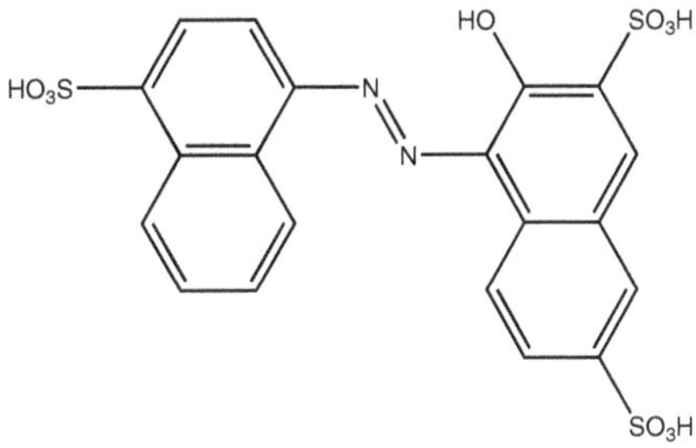


**Fig. 15.7** Structural formula of Amaranth

### *Trade Names*

Amaranth was put on the market by Cassella, Farbwerke ML&B, Beyer & Kegel, Soc. Anon. St. Denis, Manufacture Lyonnaise, Roberts, Dale & Co., CIBA, Durand-Huguenin, Weiler-ter Meer and Levinstein. Examples of trade names used by other companies are: Naphtolrot O by Farbwerke ML&B; Naphtolrot S by BASF and Beyer & Kegel; Naphtolrot C and EB by Cassella; Naphtylaminrot G by Bayer; Echtrot by Jäger; Echtrot D by BASF and Griesheim-Elektron; Echtrot NS by Bayer; Bordeaux SF by AGFA; Bordeaux S by AGFA, Ruch and Levinstein; Bordeaux DH by Durand-Huguenin; Viktoriarubin O by Farbwerke ML&B and Beyer & Kegel; Azorubin S by Sandoz; Azosäurerubin 2B by Beyer & Kegel, Wülfing-Dahl and Jäger; Tuchrot by Weiler-ter Meer; Acid Crimson by Read Holliday; Wollrot Extra by Kalle; Wool Red by Schoellkopf; and Pourpre by Picard.

## *Properties*

Lehne described the fastness of Amaranth as follows: 'After 30 days exposed to the elements, in May and June, the colour little changed.—Strongly changed due to sulfuric acid of 4.75%, yellower and lighter.—Silk dyeings are not fast to washing'.[4] According to the Colour Index, the light fastness is ISO 3–4 and seems to be less good than that given by Lehne. The wash fastness is ISO 3.

## *Identification Using UHPLC-PDA*

The analytical result of Amaranth (Sigma-Aldrich CAS 915-67-3) is shown in Fig. 15.8, Amaranth elutes rather early.

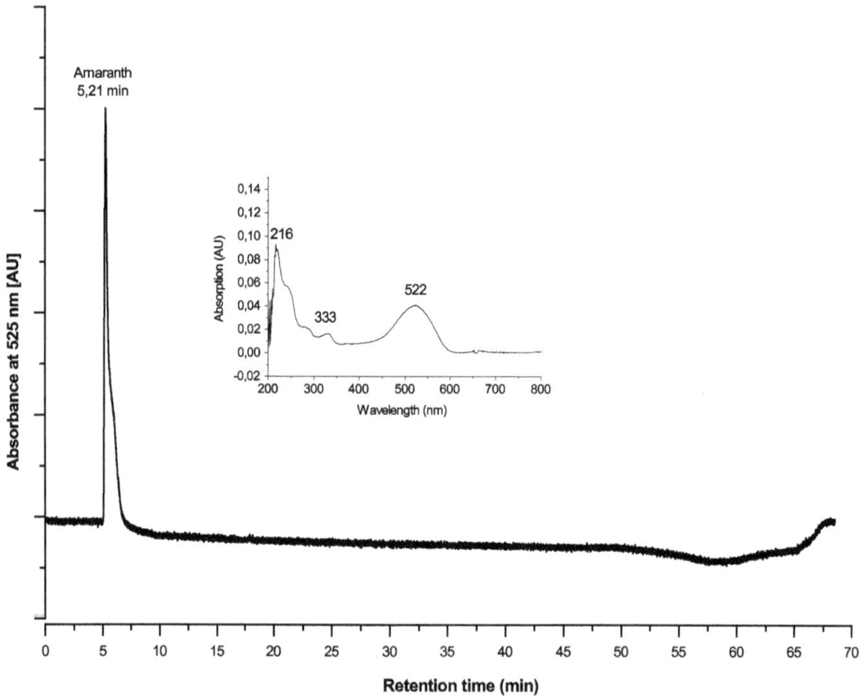


**Fig. 15.8** Chromatogram and spectrum of Amaranth (Sigma-Aldrich CAS 915-67-3)

[4]Amaranth: Echtheit: 'Nach 30 tägiger Einwirkung der Witterung (Mai und Juni) wenig verändert. – Durch schweflige Säure von 4,5 °Bé stark verändert, gelber und heller. – Seidenfärbung nicht wasserecht'.

## 15.7 Orange G

Colour Index name: Acid Orange 10; Colour Index number: 16230
IUPAC name: Disodium;7-hydroxy-8-phenyldiazenylnaphthalene-1,3-disulfonate
Discoverer: Heinrich Baum, 1878
Contemporary terminology: Orange G (En), Orangé G (Fr), Orange G (Ge), Oranje G (NL)
Chemical formula: $C_{16}H_{12}N_2O_7S_2$ (Fig. 15.9)

**Fig. 15.9** Structural formula of Orange G

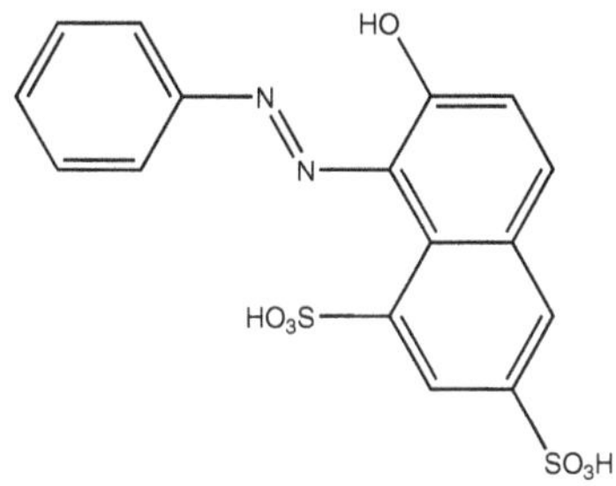


### *Trade Names*

Orange G was a popular dye, given many different trade names and put on the international market by many firms: AGFA, BASF, Cassella, Jäger, Kalle, Farbwerke ML&B, Weiler-ter Meer, CIBA and Griesheim-Elektron. Other companies introduced the dye on the market under different trade names, including Orange GG by Beyer & Kegel, Cassella and Wülfing-Dahl; Orange GMP by AGFA; Orange GG in Kristallen and Orange GR special by Cassella; Orangé en Cristaux GG by Manufacture Lyonnaise; Patentorange by AGFA; Kristallorange GG by Wülfing-Dahl; Echtlichtorange G by Bayer; and Orangé en cristaux by Manufacture Lyonnaise.

### *Properties*

The fastness of Orange G is given by Lehne as: 'After 30 days exposed to the elements, in May and June, the colour did not change. Fast to washing, also resistant to fairly light fulling. Yellower due to sulfuric acid of 4.75%'.[5] According to the Colour Index, the light fastness is ISO 4–5. For the wash fastness, ISO 1 is reported.

[5] Orange G: Echtheit: 'Nach 30 tägiger Einwirkung der Witterung (Mai und Juni) unverändert. – Waschecht, auch gegen leichtere Walke beständig. – Durch schweflige Säure von 4.5 °Bé gelber'.

## *Identification Using UHPLC-PDA*

The analysis of Orange G (RCE no. 3855) shows a clear signal as seen in Fig. 15.10.

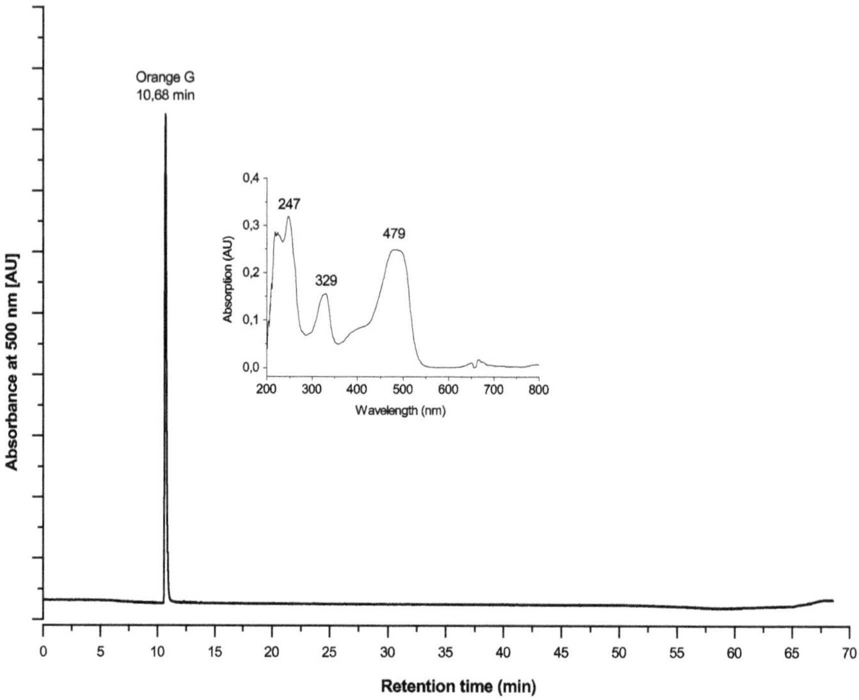


**Fig. 15.10** Chromatogram and spectrum of Orange G (RCE no. 3855)

## 15.8 Crystal Ponceau 6R

Colour Index name: Acid Red 44; Colour Index number: 16250
IUPAC name: Disodium;7-hydroxy-8-(naphthalen-1-yldiazenyl)naphthalene-1,3-disulfonate
Discoverer: Meinhard Hoffmann, 1884
Contemporary terminology: Crystal Ponceau 6R (En), Ponceau en cristaux 6R (Fr), Kristallponceau 6R (Ge), Kristalponceau 6R (NL)
Chemical formula: $C_{20}H_{14}N_2O_7S_2$ (Fig. 15.11)

Fig. 15.11 Structural formula of Crystal Ponceau 6R

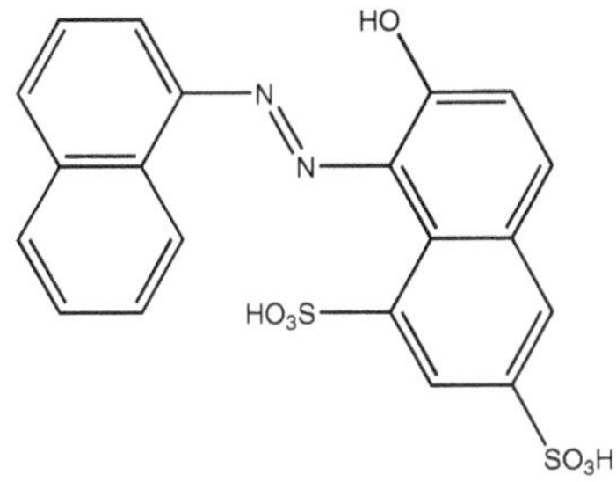


## *Trade Names*

The dye was put on the market using the name Kristallponceau 6R by Cassella, Farbwerke ML&B, Bayer and Beyer & Kegel. Other trade names used included Kristallponceau by AGFA, BASF, Kalle, Leonhardt and Wülfing-Dahl; Ponceau 6R by Weiler-ter Meer; Ponceau en cristaux 6R by Manufacture Lyonnaise; Ponceau cristallisé by Soc. Anon. St. Denis; Neucoccin R by AGFA; and Ponceau 6R krist. 88 N, Crystal Ponceau 6R and Brilliant Crystal Scarlet 6R.

## *Properties*

Lehne also studied the fastness of Crystal Ponceau 6R: 'After 5 days exposed to the elements in June, no changes, after 30 days strongly faded.—Slightly yellower due to sulfuric acid of 4.75%'.[6] According to the Colour Index the light fastness is ISO 3–4, which is in good agreement with the results obtained by Lehne. The wash fastness is ISO 2–3.

## 15.9 Cochineal Red A

Colour Index name: Acid Red 18; Colour Index number: 16255
IUPAC name: Trisodium;7-hydroxy-8-[(4-sulfonatonaphthalen-1-yl)diazenyl]naphthalene-1,3-disulfonate
Discoverer: Heinrich Baum, 1878
Contemporary terminology: Cochineal red A (En), Rouge de cochenille A (Fr), Cochenillerot A (Ge), Cochenillerood A (NL)
Chemical formula: $C_{20}H_{14}N_2O_{10}S_3$ (Fig. 15.12)

[6] Kristallponceau 6R: Echtheit: 'Nach 5 tägiger Einwirkung der Witterung (Juni) unverändert, nach 30 Tagen stark verblasst. – Durch schweflige Säure von 4.5 °Bé etwas gelber'.

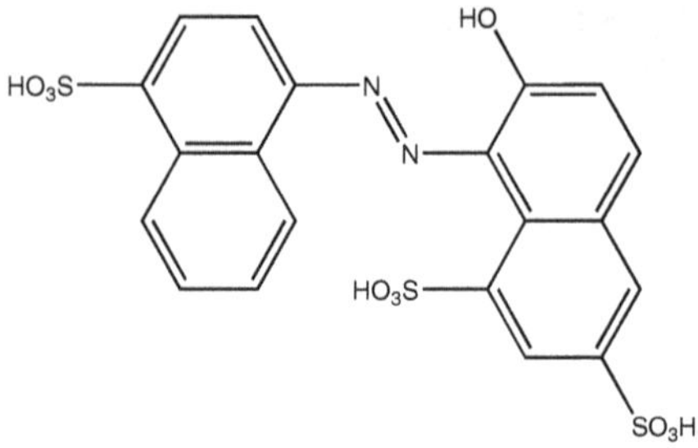


**Fig. 15.12** Structural formula of Cochineal red A

## *Trade Names*

Cochenillerot A was the trade name used by BASF. Other important trade names used were: Croceinescharlach 4BX by Kalle; Brilliant Scarlet by Levinstein and Cassella; Brilliantponceau 5R by Bayer and Wülfing-Dahl; Ponceau 4R by Beyer & Kegel; Brilliant Scarlet S by Schoellkopf; Ponceau brilliant 4R by Manufacture Lyonnaise; Neucoccin by AGFA and Ruch; Neucoccin O by Farbwerke ML&B; Scarlet ooooo by Read Holliday; Ponceau spécial by Soc, Anon. St. Denis, and Viktoriascharlach 4R extra by Weiler-ter Meer.

## *Properties*

Lehne stated the following about the fastness of Cochineal red A: 'After 40 days exposed to the elements, in July and August, the colour very little changed.—Strongly changed due to sulfuric acid of 4.75%, much yellower.—Produces very fiery colour shades, also used a lot in silk dyeing and in wool printing. Widely used for dyeing wool'.[7] According to the Colour Index the light fastness is ISO 4–5. The wash fastness is ISO 2.

## *Identification Using UHPLC-PDA*

The analysis of Cochineal red A (RCE no. 5311) is presented in Fig. 15.13.

[7] Cochenilleroth: Echtheit: 'Nach 40 tägiger Einwirkung der Witterung (Juli, August) sehr wenig unverändert. Durch schweflige Säure von 4.5 °Bé stark verändert, viel gelber. Liefert sehr feurige Farbtöne, viel gebraucht auch in der Seidenfärberei und im Wolldruck. Zum Färben von Wolle viel verwendet'.

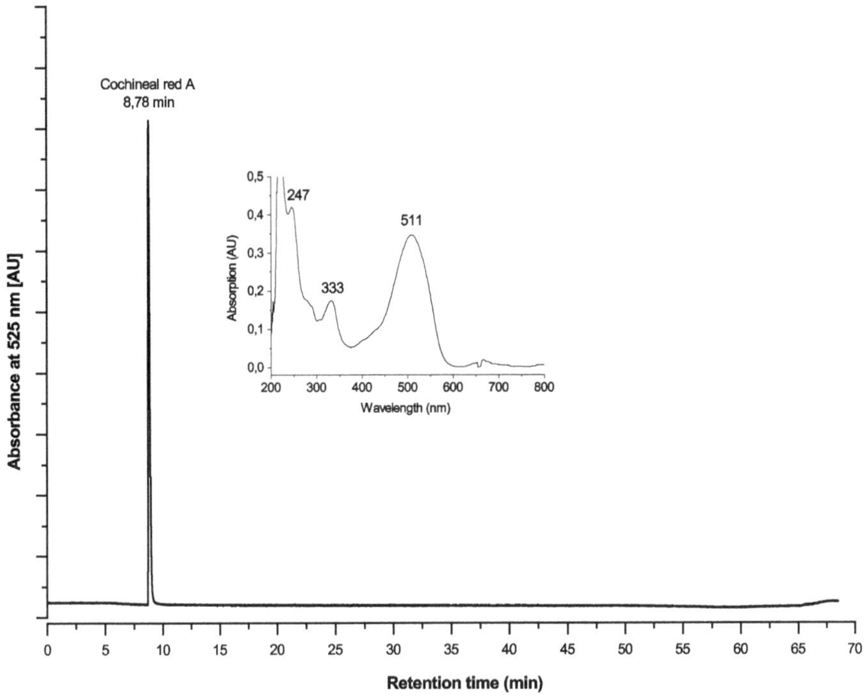


**Fig. 15.13** Chromatogram and spectrum of Cochineal red A (RCE no. 5311)

## 15.10 Ponceau 6R

Colour Index name: Acid Red 41; Colour Index number: 16290
IUPAC name: Tetrasodium;(8Z)-7-oxo-8-[(4-sulfonatonaphthalen-1-yl)hydrazinylidene]naphthalene-1,3,6-trisulfonate
Discoverer: Carl Friedrich Leonhard Limpach, 1881
Contemporary terminology: Ponceau 6R (En), Ponceau 6R (Fr), Ponceau 6R (Ge), Ponceau 6R (NL)
Chemical formula: $C_{20}H_{14}N_2O_{13}S_4$ (Fig. 15.14)

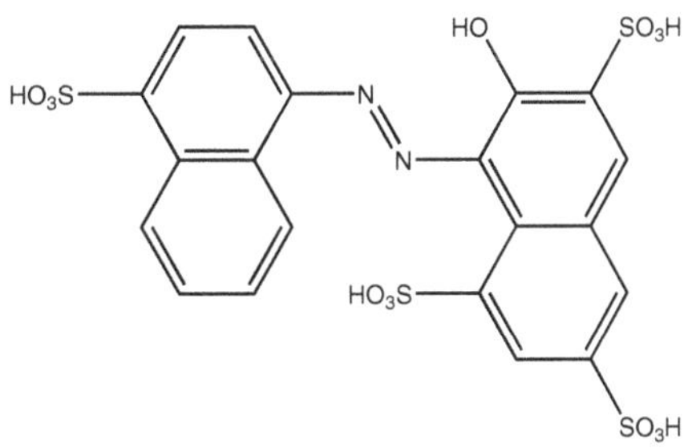


**Fig. 15.14** Structural formula of Ponceau 6R

**Table 15.3** The Ponceau dye derived from 2-naphthol-3,6,8-trisulfonic acid

| Name<br>C.I. name<br>C.I. number | Discoverer,<br>Year of discovery | Constitution | Patents |
|---|---|---|---|
| Ponceau 6R<br>Acid Red 41<br>16290 | Carl Friedrich Leopold Limpach, 1881 | Diazotised naphthionic acid coupled with 2-naphthol-3,6,8-trisulfonic acid | G.P. 22.038/1882<br>B.P. 2.544/1882<br>U.S.P. 268.506/1882<br>F.P. 137.109<br>F.P. 149.249 |

## *Trade Names*

The red dye Ponceau 6R was put on the market by Farbwerke ML&B and BASF. Scharlach 6R, another trade name, was also used to sell the dye. The patents are presented in Table 15.3.

## *Properties*

This dye was studied by Lehne in autumn: 'After 30 days exposed to the elements, in September and October, the colour did not change. Due to sulfuric acid of 4.75% the colour changed, orange. For dyeing wool and silk'.[8] According to the Colour Index, the light fastness is ISO 4–5 while for the wash fastness ISO 2–3 is given.

## *Identification Using UHPLC-PDA*

The analysis of Ponceau 6R (Sigma-Aldrich CAS 2766-77-0) shows a clear, single peak (Fig. 15.15).

[8] Ponceau 6R: Echtheit: 'Nach 30 tägiger Einwirkung der Witterung (September und Oktober) unverändert. – Durch schweflige Säure von 4.5 °Bé verändert, orange'. Zum Färben von Wolle und Seide'.

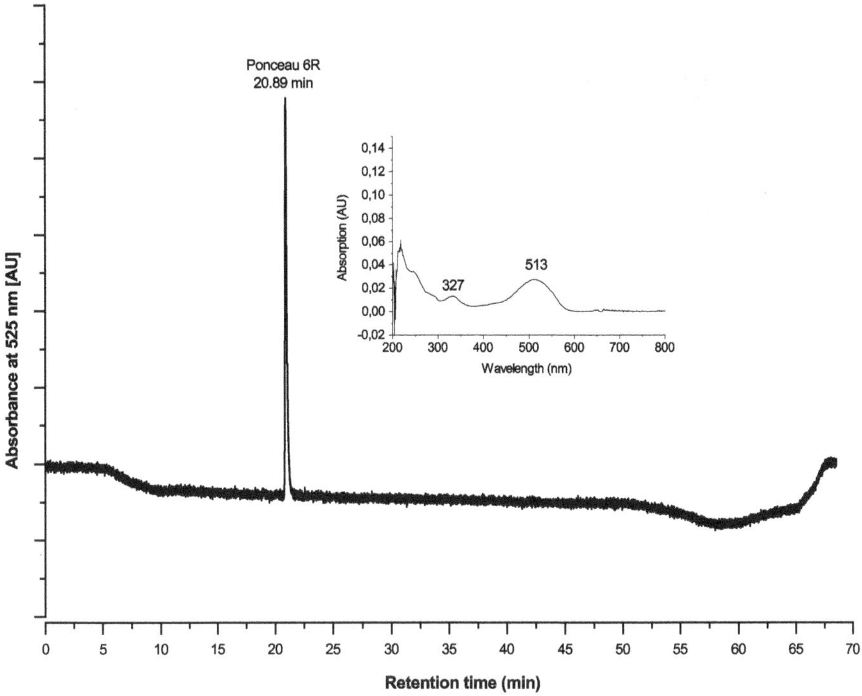


**Fig. 15.15** Chromatogram and spectrum of Ponceau 6R (Sigma-Aldrich CAS 2766-77-0)

## 15.11 General Information

The Ponceaus have been used for dyeing silk and wool in an acid bath and for printing on wool. The Ponceaus surpassed all existing red wool dyes in terms of beauty. They were widely used in practice and gradually replaced the natural dyes obtained from scale insects, Annatto (*Bixa orellana* L.) and Orchil (*Roccella tinctoria* DC) in wool dyeing and printing.

Orange G and Cochineal red A in particular were widely used for wool and silk. Orange G, Cochineal red A and Ponceau 6R were also used for non-textile purposes, like wallpaper, coloured paper, decorative paints and inks.

### *General Properties*

The colours of the Ponceaus range between orange and Bordeaux red. The dyes prepared from R-acid show red to dark-red tints, while the dyes derived from G-acid have a yellowish-red to deep-scarlet colour. The numbers 2, 3, etc. describe the increasing degree of darkness of the tones; for example, 3R is a darker red compared to 2R and 2B means a more blueish tone. In general, the Ponceaus are soluble in water, but are insoluble to very poorly soluble in alcohol. In general, the dyes show moderate light fastness.

Lehne wrote about the light fastness of these Ponceau dyes, as described above. The Ponceau dyes on textiles (wool and silk) show different colours and different light fastnesses. The dyes prepared from R-acid, such as Ponceau 2G, Ponceau 2R and Bordeaux R (Fast red B), have a poor light fastness; they fade strongly within a few weeks. The dyes made from G-acid, such as Orange G, Crystal Ponceau 6R and Cochineal red A, show a much better light fastness: they hardly change within several weeks. Comparing the information presented by Lehne with the information from the Colour Index, a slightly different observation can be made. The dyes derived from the R-acid have a light fastness of ISO 3–4, while those derived from the G-acid have a light fastness of ISO 4–5, depending on the concentration used. It should be noted that Lehne did not perform a standardised test, but exposed the dyes to the elements in different months of the year. That could have affected the result.

### *Recipes*

The Ponceaus belong to the group of acid dyes. Since they are very similar, the dye recipes described can be used for all dyes in this class. Lehne wrote uniform dye recipes for wool (V3A and V3B) and silk (V8) using the Ponceaus in his 1893 book. The concentration of the dye varies: 1.5% is given for Ponceau 2G, Ponceau 2R and Orange G while for the other dyes 2% is used when dyeing wool. For silk, the concentration is generally higher, up to 3%, although, with the exception of Amaranth, no silk recipes are given for the dyes listed above.

## 15.12 Case Studies

### *Hofdamask from the Hofburg and Schloß Schönbrunn in Vienna, Austria*

In cooperation with Schloß Schönbrunn we did a project on Hofdamask,, damask especially made for the court. In 1842, Hofdamask silk was used for wall hangings for the decoration of the apartments of Archduke Franz Carl and the Archduchess Sophie, the parents of the later Emperor Franz Joseph. The bluish-red 'Ananasdamask' (damask with a pineapple pattern, see Fig. 15.16) was also used around 1855 for the interior of the rooms where persons would be presented to the Archduke and Archduchess in a formal setting. In 1885, it seems that decoration of the walls and furniture in the apartments of the Emperor and Empress in the Hofburg was carried out using bluish-red Hofdamask.

**Fig. 15.16** Hofdamask (HD2) dyed with Cochineal, Fuchsine and Orange ll. (© Schloß Schönbrunn Kultur- und Betriebsges. m.b.H./Sammlung Bundesmobilienverwaltung)

The silk dyeing firm Franz Bujatti in Vienna delivered a red Ananasdamask under the name Grosgrain to the court for the last time in 1899. Over the years there were changes in colour, in the working width (originally the fabric was woven by hand, later by machine) and the fibre material (originally silk, later viscose-rayon). Dye analyses on warps and wefts showed that the fabrics were dyed with the natural dye Cochineal and synthetic dyes (see Fig. 15.16) and later only with synthetics (see Figs. 15.17 and 15.18). Besides Ponceau G, Ponceau 2R and Cochineal red A, several components with similar spectra but different retention times were also detected. These remain unidentified and were labelled Ponceau equivalents. Most likely, a dye was used which contains many side products. In addition to the Ponceau dyes, other dyes were found in the wall hangings: Fuchsine, Safranine T, Methyl violet, Orange ll, Orange IV, Azo flavine 3R, Silk scarlet N, Rhodamine B and New Magenta. By means of the most recently discovered dye on each textile, it was possible to give a terminus post quem dating. An overview of the results is presented in Table 15.4.

**Fig. 15.17** Hofdamask (HD1) dyed with Fuchsine, Methyl violet, Orange I, Ponceau 2R and Azo flavine 3R. (© Sammlung Bundesmobilienverwaltung)

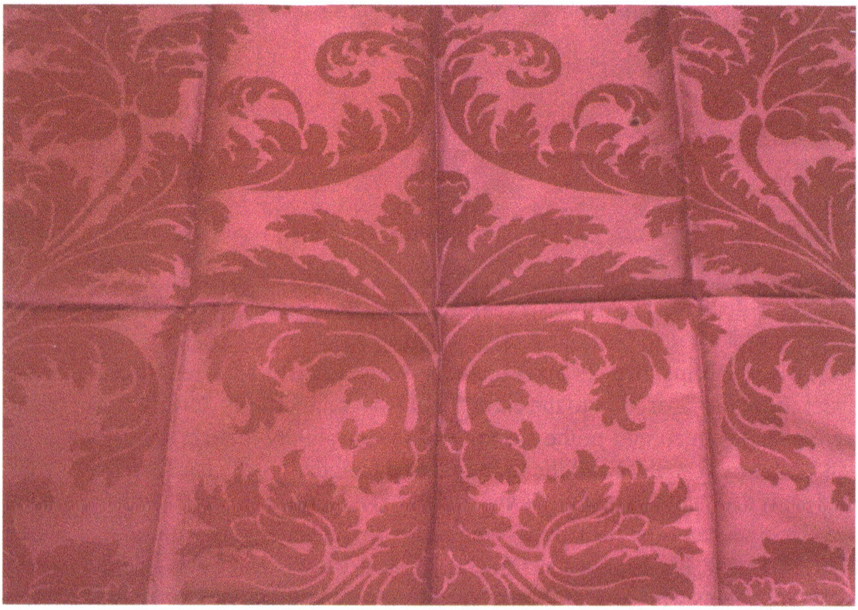

**Fig. 15.18** Hofdamask (HD3) dyed with Orange II, Cochineal red A, Silk scarlet N, Rhodamine B and New Magenta. (© Schloß Schönbrunn Kultur- und Betriebsges. m.b.H./Sammlung Bundesmobilienverwaltung)

**Table 15.4** Hofdamask in the Hofburg and Schloß Schönbrunn

| Hofdamask sample | HPLC results | Dating (estimated) | Dating (based on dye analysis) |
|---|---|---|---|
| Mobiliendepot<br>MD 177470 (HD 4) | Cochineal, Tannins<br>Safranine T (1859, produced by BASF since 1872) | Around 1880? | After 1872 |
| Mobiliendepot<br>MD 177523/4 (HD5) | Cochineal, Tannins<br>Fuchsine (1856)<br>Orange lV (1876/1877)<br>Ponceau G (1878)<br>Silk scarlet N (1884) | Around 1800 or 1880? | After 1884 |
| Schönbrunn<br>Fa. Backhausen<br>(HD2) | Cochineal<br>Fuchsine (1856)<br>Orange ll (1876) | Around 1900 or later | After 1876 |
| Mobiliendepot<br>MD 177270/000<br>Hofburg (HD1) | Fuchsine (1856)<br>Methyl violet (1861)<br>Orange ll (1876)<br>Ponceau 2R (1878)<br>Azoflavine 3R (1880) | Around 1900 or later | After 1880 |
| Schönbrunn<br>Fa. Backhausen<br>(HD3) | Orange II (1876)<br>Cochineal red A (1878)<br>Silk scarlet N (1884)<br>Rhodamine B (1887)<br>New Magenta (1889) | Around 1900 or later | After 1889 |

## *Paintings by Helene Maria Funke*

Helene Maria Funke (1869–1957) was born in 1869 in Chemnitz, Germany. In 1899 she moved to Munich to start her academic study as a painter. Her professor at the ladies' academy was Angelo Jank. After her studies, in 1905, she went to Paris where she came into contact with the art of Renoir, Seurat, Cézanne, Van Gogh, Gauguin and Matisse. In 1913 she went to Vienna. After her death, she fell into oblivion, but with the growth of the women's movement in the 1960s and 70s there was an increasing interest in her work from an art-historical point of view. She painted with inorganic pigments as well as natural and synthetic organic pigments. In the course of a thesis project, the painting technique of Helene Funke was examined for the first time [5]. In the painting *In the Loge*, owned by the Lentos Kunstmuseum Linz, she painted the hair of the central figure using a Madder lake. The synthetic version of Alizarin was identified in the paintings *Three Women* and *Girl with Doll*. In addition, in the painting *In the Loge* the artist used a pigment containing a pink dye belonging to the Ponceau group, known since 1878. In the book depicted in the painting *Girl with Doll* the hydroxy-phthalein dye Rhodamine B, discovered in 1887, was identified.

## *A Golden Bonnet with a Red Silk Bow from the Agnes Kinz Collection*

Agnes Kinz (1896–1939) was born Agnes Muther in Innsbruck, Austria, in 1896. When her father obtained a position at the finance office of the capital of the state of Vorarlberg the family moved to Bregenz. After her studies were completed, she became the secretary of Ferdinand Kinz, the mayor of Bregenz and in 1919 she married him; he was 24 years older. His sudden death in 1935 shattered the life of his family. A few days before Christmas 1939, Agnes died at 43 years old due to cardiovascular disease. Out of interest and from an awareness of tradition, she collected golden bonnets and costume hats. She was the leading person in Vorarlberg encouraging the revival and maintenance of authentic costumes. Her private collection contains 96 women's headdresses, such as bonnets and crowns, as parts of costumes originating mainly from Vorarlberg [6]. In 2007, the collection was donated to the Vorarlberger Landesmuseum in Bregenz, which is now the vorarlberg museum. The red bow of a Tyrolean bonnet, dated to the end of the eighteenth century, showed an especially interesting degradation process (see Fig. 15.19). Microscopic examination showed that the fibre material of the bow is silk [7]. Figure 15.20 shows the bonnet with the red bow after conservation.

Surprisingly a natural organic red dye was not detected on the silk using thin layer chromatography. On the warp and the weft, two synthetic dyes were identified with HPLC-PDA and fluorescence detection: Ponceau 2G and Rhodamine B, which was discovered in 1887. This means that the red bow was dyed after 1887. The fragile condition of the silk threads was investigated by SEM-EDX. The elements tin, phosphorus, silicon and aluminium were detected. They correspond with a textile manufacturing process called silk weighting according to the Neuhaus process, developed in 1893 [8]. The reason for adding metal compounds to silk fabric is to increase the weight of the fabric. Silk fabrics were sold in those days by the pound, so the extra weight increased the selling price of the fabric. The disadvantage of weighted silk is that it loses much of its strength and durability. Based on these analytical results, the production of the red bow must have been after 1893.

## *Other Case Studies*

The Van Gogh Museum is the owner of a Chinese lacquer box filled with coloured balls of wool by Vincent van Gogh. The balls of wool were mainly dyed with synthetic dyes, including Ponceau 2R discovered in 1878. This case study is discussed in more detail in Chap. 10. Several embroideries were designed by Émile Bernard and were part of the former collection of Andries Bonger; they belong today to the Van Gogh Museum, Amsterdam. In the four embroideries the Ponceau dyes Ponceau 2R and Cochineal red A, both discovered in 1878, were identified. The results are discussed in more detail in Chap. 13.

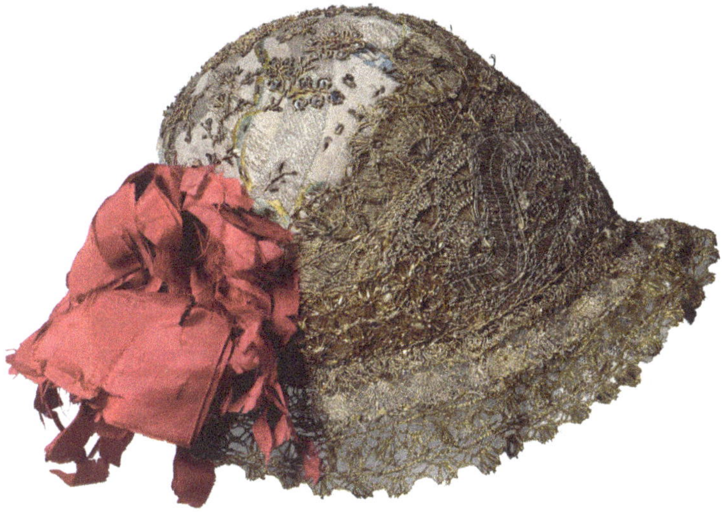

**Fig. 15.19** Golden bonnet with red silken bow before conservation vorarlberg museum, Bregenz, Austria (Inv 26). (© Institute of Conservation, University of Applied Arts Vienna. Photograph: Nina Zangerl and Stefan Oláh)

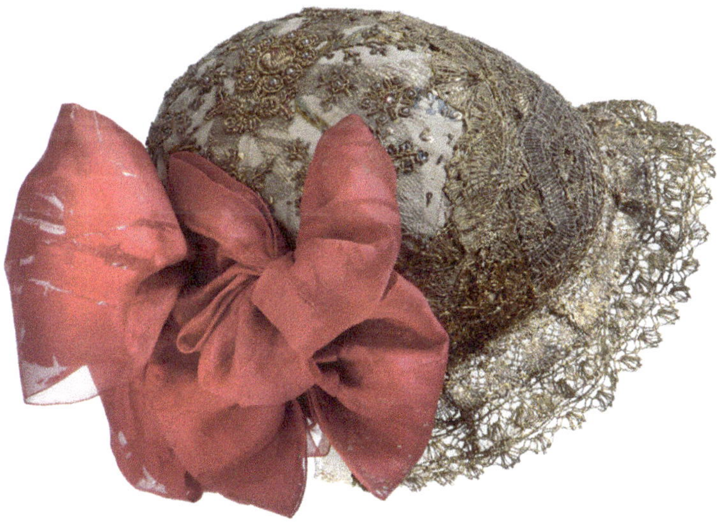

**Fig. 15.20** Golden bonnet with red silken bow after conservation. vorarlberg museum, Bregenz, Austria (Inv 26). (© Institute of Conservation, University of Applied Arts Vienna. Photograph: Nina Zangerl and Stefan Oláh)

# References


1. Farbwerke vorm. Meister Lucius & Brüning 1863–1913. Hergestellt in der Hausdruckerei der Farbwerke vorm. Meister Lucius & Brüning, Höchst am Main
2. I.G. Farbenindustrie Aktiengesellschaft. In: Farbwerke vorm. Meister Lucius & Brüning 1863–1938. p. 43. Gesamtherstellung in Buch- und Kupfertiefdruck von den Graphischen Kunstanhalten F. Bruckmann KG., München (1938)
3. Bäumler, E.: Ein Jahrhundert Chemie. Econ Verlag GmbH, Düsseldorf (1963)
4. Baum, H.: Ponceau R (R acid with commercially available xylidine)/(R acid with crude xylidine); Ponceau 2R (essentially with m-xylidine and as pure as possible R-salt/pure 2-naphthol-3,6-disulfonic acid); Ponceau G, GR, Brillantponceau G, Ponceau J (crude xylidine. G.P. 3.229 (1878)
5. Fischer, E., de Keijzer, M., Erlach, R.: Helene Funke – Studie zur Maltechnik. In: Helene Funke 1869–1957, Lentos Kunstmuseum Linz. pp. 211–219. Verlag für moderne Kunst Nürnberg, Nürnberg (2007)
6. Eisenhardt, B.: "...unter die Haube gekommen..." Die Haubensammlung von Agnes Kinz - Konservierung und Restaurierung von 94 Goldhauben und Trachtenhüten, Diploma thesis. University of Applied Arts Vienna, Institute of Conservation, (2008)
7. Zangerl, N.: Konservierung einer dreiteiligen Seidenmasche einer Goldhaube der Sammlung Agnes Hinz aus dem Vorarlberger Landesmuseum, Bregenz, Student project. University of Applied Arts Vienna, Institute of Conservation, (2008)
8. van Oosten, T.: De degradatie van fibroine onder invloed van de verzwaring. Central Research Laboratory for Objects of Art and Science, Amsterdam (1991)

# Chapter 16
# Bright Organic Red Colours: The Xanthenes

## 16.1 Introduction

This chapter focuses on the xanthenes, an important group used as dyes and pigments. This chemical group of synthetic dyes was discovered by Adolf Baeyer in 1871. He discovered Fluorescein, a yellow colourant which had limited use as a textile dye, but is still currently used as a biological stain. However, the invention of Fluorescein was very important because it is a building block for other components of this dye class which can be used as textile dyes, for inks and in some cases as organic pigments. First, the Eosins and their derivatives the Phloxines were developed; the dye-class was then extended some years later by the Erythrosines, with their derivatives the Rose Bengals. These were all acid dyes. At the end of the nineteenth century, the basic Rhodamines and their derivatives were invented. A list of relevant colourants is presented, demonstrating the swift development of this important dye class (see Table 16.1).

In 1860, Baeyer became a lecturer in organic chemistry and director of the new organic chemical laboratory at the Gewerbeinstitut, a trade college in Berlin. After his appointment as professor at the German University of Strasbourg in 1871 he began the experiments resulting in the discovery of a new dye-class. By heating phthalic anhydride with phenols and sulfuric acid, he prepared Gallein, Phenolphthalein and Coerulein, which are hydroxy-phthaleins, and Fluorescein, which is the first xanthene dye. Since their chemical structure resembles that of the natural dyes Brazilein (Brazilwood) and Haematein (Logwood, Campeche wood), Baeyer believed that these products could be used as dyes. However, at the beginning of the 1870s there was a lack of information about the exact chemical structures of the hydroxy-phthaleins and xanthenes. In September 1873, Baeyer met


M. R. van Bommel, M. de Keijzer, *Bright Colours from the Past*, Cultural Heritage Science, https://doi.org/10.1007/978-3-031-88664-5_16

**Table 16.1** Fluorescein, Eosins, Phloxines, Erythrosines, Rose Bengals, Rhodamines and their derivatives

| Name | C.I. name | C.I. number | Year of discovery | Discoverers |
|---|---|---|---|---|
| Fluorescein | Acid Yellow 73 | 45350 | 1871 | Adolf Baeyer |
| Eosin | Acid Red 87 | 45380 | 1873 | Heinrich Caro |
| Spirit Eosin | Solvent Red 44 | 45385 | 1874 | Heinrich Caro |
| Eosin S | Solvent Red 45 | 45386 | 1874 | Heinrich Caro |
| Eosin BN | Acid Red 91 | 45400 | 1875<br>1876 | Heinrich Caro<br>Adolf Baeyer |
| Phloxine | Acid Red 98 | 45405 | 1875 | Emilio Noelting |
| Erythrosine G | Acid Red 95 | 45425 | 1875 | Emilio Noelting |
| Rose Bengal | Acid Red 93 | 45435 | 1875 | Emilio Noelting |
| Erythrosine | Acid Red 51 | 45430 | 1876 | Kussmaul |
| Cyanosin spritl. | Solvent dye | 45415 | 1876 | Emilio Noelting |
| Chrysoline | Acid Yellow 74 | 45360 | 1877 | Frédéric Reverdin |
| Phloxine B | Acid Red 92 | 45410 | 1882 | Robert Gnehm |
| Cyanosin B | Acid dye | 45420 | 1882 | Robert Gnehm |
| Rose Bengal B | Acid Red 94 | 45440 | 1882 | Robert Gnehm |
| Rhodamine B | Basic Violet 10 | 45170 | 1887<br>1888 | Maurice Cérésole<br>Benno Homolka, Hermann Boedeker |
| Rhodamine S | Basic Red 11 | 45050 | 1888 | Kahn and Majert |
| Violamine B | Acid Violet 30 | 45186 | 1888 | Hermann Boedeker |
| Violamine R | Acid Violet 9 | 45190 | 1888 | Hermann Boedeker |
| Violamine G | Acid Blue 19 | 45205 | 1889 | Hermann Boedeker |
| Rhodamine 3B | Basic Violet 11 | 45175 | 1891 | Pierre Prosper Monnet |
| Rhodamine G | Basic Red 8 | 45150 | 1891 | Maurice Cérésole |
| Rhodamine 6G | Basic Red 1 | 45160 | 1892 | Heinrich August Bernthsen |
| Rhodamine 3G | Basic Red 3 | 45210 | 1895 | Carl Oscar Müller |

Heinrich Caro at the meeting of the Society of German Natural Scientists and Physicians (Gesellschaft Deutscher Naturforscher und Ärzte) in Wiesbaden. He handed over some Fluorescein and Caro continued working on the dye, which can be seen as the start for the development of the xanthenes as a significant dye class [1].

## 16.2 Fluorescein

Colour Index name: Acid Yellow 73; Colour Index number: 45350
IUPAC name: 3′,6′-dihydroxyspiro[2-benzofuran-3,9′-xanthene]-1-one
Discoverer: Adolf Baeyer, 1871
Contemporary terminology: Fluorescein (En), Fluorescéine (Fr), Fluorescein (Ge), Fluoresceine (NL)
Obsolete terminology: Uranine (Fr) and Uranin (Ge)
Chemical formula: $C_{20}H_{12}O_5$ (Fig. 16.1)

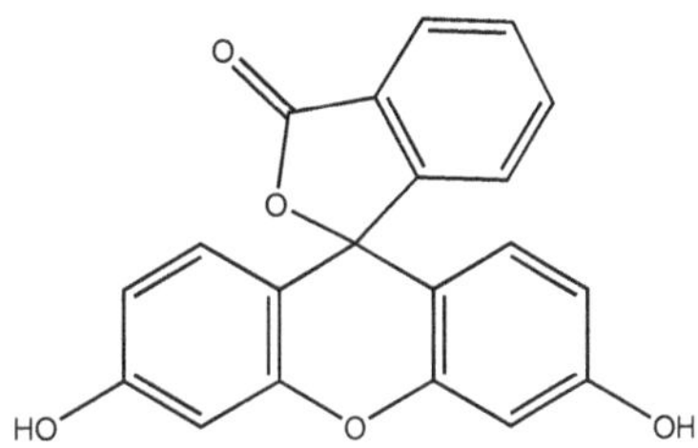


**Fig. 16.1** Structural formula of Fluorescein

### *History*

Fluorescein (or resophthalein) was discovered by the German chemist Adolf Baeyer (1835–1917) in 1871 in the course of his researches on the condensation of phthalic anhydride with various phenols. It was prepared by first heating phthalic anhydride, resorcinol (3-hydroxyphenol) and a small quantity of zinc chloride, which acts as a condensing agent, in an oil-bath at a temperature of 195–200 °C. The product is then made alkaline with a dilute sodium hydroxide solution [2–5]. Baeyer called his yellow product Fluorescein because of its strong yellowish-green fluorescence in an aqueous or dilute alkaline solution.

### *Trade Names*

German dye firms in particular, including BASF, Farbwerke ML&B, AGFA, Leonhardt, Weiler-ter Meer and Jäger, produced the sodium salt of Fluorescein and brought it to the market under the trade name Uranin [6, 7]. Sandoz and Schoellkopf gave it a similar name: Uranine. Cassella, Leonhardt, Durand-Huguenin and Soc. Anon. St. Denis marketed it as Fluorescein.

## Recipes

Although Fluorescein was hardly used for textile dyeing, Adolf Lehne wrote in his book that silk can be dyed following recipe V8, using 2% dyestuff after degumming the silk. The colour can be brightened using acetic acid or tartaric acid. Wool can be dyed with the addition of 2.5% acetic acid, followed by 25 min boiling.

## Identification Using UHPLC-PDA

The analysis of Fluorescein (RCE no. 3824) shows a clear peak; it must be noted that it absorbs much more strongly in the UV (Fig. 16.2).

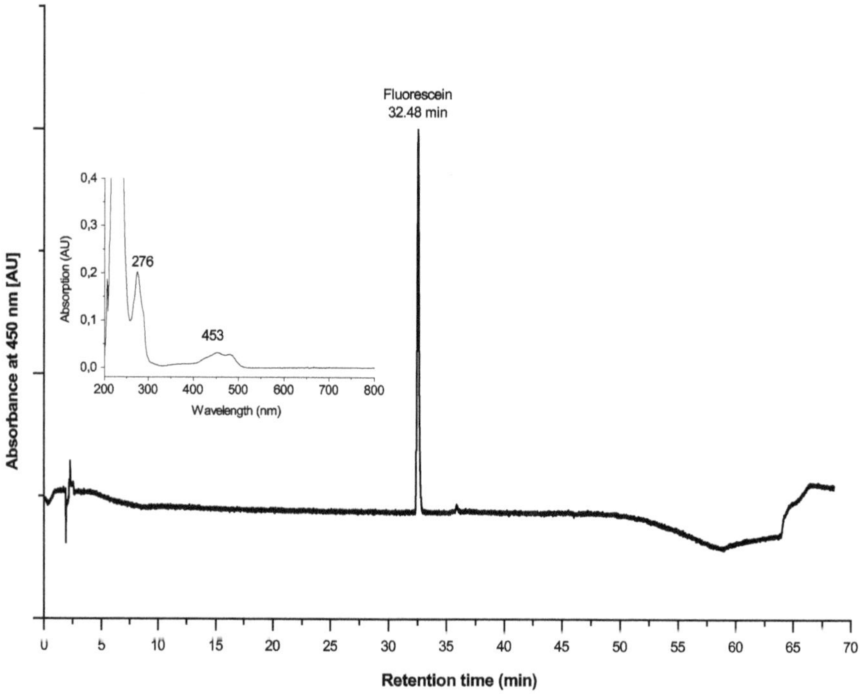


**Fig. 16.2** Chromatogram and spectrum of Fluorescein (RCE no. 3824)

## 16.3 Chrysoline

Colour Index name: Acid Yellow 74; Colour Index number: 45360
IUPAC name: 7′-benzyl-3′,6′-bis(sodiooxy)-3H-spiro[2-benzofuran-1,9′-xanthen]-3-one
Discoverer: Frédéric Reverdin, 1877
Chemical formula: $C_{27}H_{16}Na_2O_5$ (Fig. 16.3)

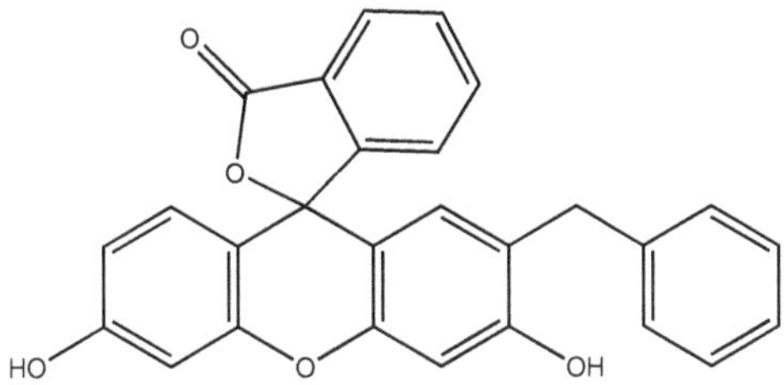


**Fig. 16.3** Structural formula of Chrysoline

### *History*

Another yellow dye of this group is Chrysoline, discovered by the Swiss chemist Frédéric Reverdin (1849–1931) and patented (F.P. 113.695) in 1877. Chrysoline is the sodium salt of a compound prepared by condensing resorcinol, phthalic anhydride and benzyl chloride in the presence of sulfuric acid. This yellow dye also shows a green fluorescence in solution.

### *Trade Names*

The dye was produced by P. Monnet, Soc. Anon. St. Denis and Sandoz, but was of limited use for dyeing.

### *Recipes*

Similarly to Fluorescein, Chrysoline was hardly used for textile dyeing. However, Lehne does give a dye recipe for silk, which can be dyed following recipe V8, using 2% dyestuff after degumming of the silk. The colour can be brightened using acetic acid or tartaric acid.

## *Identification Using UHPLC-PDA*

The analysis of Chrysoline (RCE no. 6789) reveals a complex mixture and in Fig. 16.4 the chromatogram and the spectra of the four main compounds are shown. These spectra look very similar; probably this mixture is formed during synthesis.

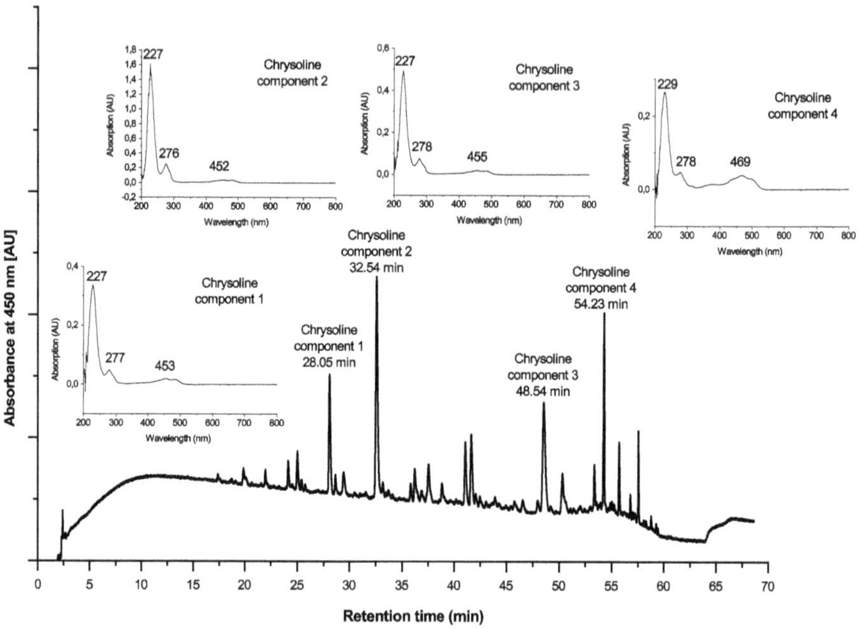


**Fig. 16.4** Chromatogram and spectra of Chrysoline (RCE no. 6789)

## 16.4 Eosin

Colour Index name: Acid Red 87; Colour Index number: 45380
IUPAC name: 2-(2,4,5,7-tetrabromo-6-oxido-3-oxo-3H-xanthen-9-yl)benzoate
Discoverer: Heinrich Caro, 1873
Contemporary terminology: Eosin (En), Éosin (Fr), Eosin (Ge), Eosine (NL)
Chemical formula: $C_{20}H_6Br_4Na_2O_5$ (Fig. 16.5)

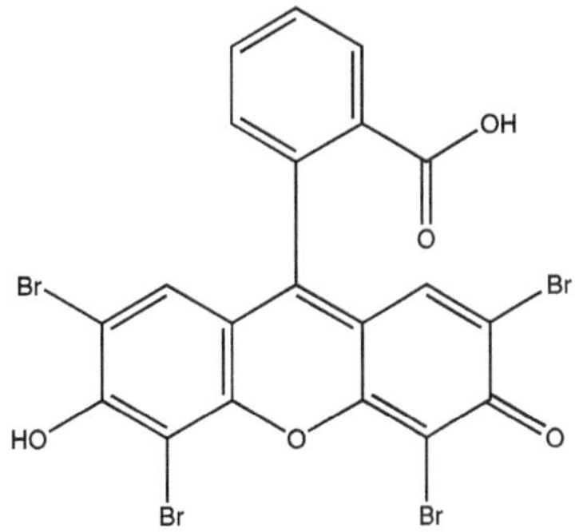


**Fig. 16.5** Structural formula of Eosin

## *History*

Soon Fluorescein attracted attention due to the beautiful fluorescence of its alkaline solutions as well as from its employment as a yellow dye. As previously mentioned, Baeyer met the German chemist and BASF research-director Heinrich Caro (1834–1910) in the autumn of 1873 at the meeting of the Society of German Natural Scientists and Physicians (Gesellschaft Deutscher Naturforscher und Ärzte) in Wiesbaden, Germany, during which Baeyer handed a sample of Fluorescein over to Caro for further examination. A few days later Caro reported to Baeyer that he had obtained a beautiful dye by the complete bromination of Fluorescein in an aqueous sodium hydroxide solution, which dyed silk and wool excellently, giving it a brilliant red hue [4, 5]. From the chemical point of view, Eosin is 2,4,5,7-tetra-bromofluorescein. An incorrect discovery year for Eosin, 1871, is mentioned in the Colour Index and in some secondary literature.

Caro named the red dye Eosin after the nickname 'rosenfingerige Eos' (rosy-fingered Dawn) given to Anna Peters, a girl he had admired in his youth [1]. Eos, the goddess of the dawn, was represented by the ruddy beams of the morning light.

After solving various problems, such as scaling up the production of the necessary intermediates (especially resorcinol, which was an expensive laboratory product) to a technical scale and discovering the method of purifying Eosin by means of its potassium salt, BASF was able to put the salt on the market in July 1874.

Sometime later, Caro visited Baeyer in Strasbourg and handed over a sample of Eosin. The red Eosin dye was also shown to Baeyer's research students, among them the German Hermann Emil Fischer (1852–1919). Baeyer suggested to the student Fischer that he should examine Fluorescein in more depth and as a result it became one of Fischer's first research investigations in organic chemistry, ending up as the theme of his doctoral thesis [8].

Eosin was not patented in Germany, in England and other countries. In those days, the patent law in Germany offered little protection. In 1869, some years before the discovery of Eosin, Caro had been repeatedly disappointed by earlier attempts to obtain German patents for the production of synthetic Alizarin (by the sulfonation process) and benzene, an important raw material for producing synthetic dyes, which had become scarce and therefore rather expensive. While in England it was possible to patent both production processes, in Germany the requests were declined by the Prussian Patent Office. Because of these bad experiences in Germany, Baeyer and Caro decided to keep the manufacturing process of Eosin secret. But it turned out that this was not an adequate protection: by December 1874, the Swiss chemist Robert Gnehm (1852–1926) already suspected that Eosin was related to Baeyer's xanthene dyes [9].

Unfortunately for both chemists, the German August Wilhelm Hofmann (1818–1892), professor and director of the laboratory of the Friedrich Wilhelms University in Berlin and a consultant for AGFA, analysed a sample of Eosin, determining and publishing the chemical formula and details of the Eosin synthesis in 1875 [10]. In this way the secret, which had been kept for almost one year, was revealed and the production process of Eosin became public property, with the result that other firms were able to produce this dye as well and could market it. Otto Mühlhäuser and the Basel dye firm Bindschedler & Busch developed processes for the technical preparation of Eosin on a manufacturing scale [11, 12]. In addition, in 1886, Usine P. Monnet et Cie. was acquired by Soc. Chim. Us. Rhône in Saint-Fons, a municipality in Lyon, and obtained Eosin by the anodic oxidation of a solution of Fluorescein and bromine in dilute sodium carbonate solution (G.P. 108.838, filed on the 12th of February in 1899) [13]. Eosin gives brilliant red hues and in spite of its high price at that time, 800 German Marks (nowadays approximately 5000 Euros) for 1 kg, the silk dyers used it immediately [14]. One of the more desirable qualities of Eosin was that 1 kg could dye 100–200 kg of silk.

## Trade Names

Depending on the production process, various eosin types were produced with a broad palette of nuances between yellowish and bluish-red, such as Eosin by Jäger; Eosin A and Eosin G by BASF; Eosin gelblich by AGFA and Kalle; Eosin G extra and 2G by Weiler-ter Meer; Eosin GGF by Cassella; Eosin extra wasserlöslich by Farbwerke ML&B and Weiler-ter Meer; Eosin B, Eosin A extra and Eosin DH by Durand-Huguenin; Eosin 3J, 4J extra and OO extra by Leonhardt; Eosin KS ord. by Sandoz; Eosin JJF by Manufacture Lyonnaise; Eosine by Soc. Anon. St. Denis; Eosine J extra and Eosine B extra by P. Monnet; Eosin I gelblich and Eosin S extra gelblich by Bayer.

## Trade and Commercial Availability of Eosin as Dye and Artist's Pigment

In the 1870s, Germany became successful as an industrial power and had at that time 50% of the global synthetic dye market; during the 1880s and 1890s it covered 75% of the world demand. In the 1888 and 1891 editions of *Tabellarische Übersicht der künstlichen organischen Farbstoffe*, by the German chemists Gustav Schultz and Paul Julius, and in the 1893 book by Adolf Lehne, the German dye factories

AGFA, BASF, Cassella, Farbwerke ML&B and Leonhardt are listed as suppliers of Eosin.

To produce Eosin for the German dye factories in France, it was necessary to establish a manufacturing subsidiary because of the 1844 patent legislation. Products patented in France had to be manufactured there, otherwise the relevant patent lost its validity. BASF therefore decided to enter into a partnership with the renowned firm of Thomas Frères from Avignon in 1872. In 1878, it withdrew from the contract and instead leased a branch factory from the Thomas firm in Neuville-sur-Saône near Lyon under the company name Succursale de la Badische Anilin-& Soda-Fabrik, which was bought in 1882. Farbwerke ML&B bought the firm Coupier near Creil, north of Paris, and named its branch factory Compagnie Parisienne de Couleurs d'Aniline in 1883. In the same year Bayer founded its subsidiary factory Produits Fréderic Bayer et Cie. in Flers near Roubaix to produce their patented dyes. They produced the patented dyes in small quantities in France while exporting large quantities from their German firms.

Eosin was not only produced in Germany, but also in Switzerland by the dye firms Sandoz, Bindschedler & Busch and Durand-Huguenin. All these dye firms were situated in Basel. Furthermore, Durand-Huguenin had subsidiary companies in Saint-Fons near Lyon and Huningue in France. At the beginning of the 1870s, the political situation was dominated by the Franco-Prussian War of 1870–1871. One of the consequences was that France lost the city of Strasbourg and the Germans gained possession of Alsace and the northern part of Lorraine.

In France, the French dye factory Renard, Villet & Bunand in Lyon started to produce Eosin shortly after 1874. Other French dye firms, such as P. Monnet and Manufacture Lyonnaise, both in Lyon, and Soc. Anon. St. Denis also prepared this red dye. The most important French dye factory for Eosin at that time was Patry & Cie in Puteaux near Paris, which produced it from 1878 until 1890. The firm was renamed as A. George in 1891 and was located in Campdeville, a village 60 kilometres north of Paris. Another important Eosin firm was J. Ruch & Fils in the city of Pantin, northwest of Paris (now the northern suburbs of Paris). This firm produced Eosin after 1892.

The paint manufacturers did not employ Eosin as an artists' pigment immediately; it took almost 20 years before it became available in this form. The red dye was put on the market by BASF in the summer of 1874. Heinrich Caro wrote[1] in 1892 that in spite of its high price, 800 German marks for 1 kg, the silk dyers used it immediately: that 1 kg could dye 100–200 kg of silk, one of its more desirable qualities. However, according to Caro, the price dropped soon after its introduction and the range of applications went up accordingly [14].

[1] 'Trotz seines enorm hohen Preises findet es sofort Eingang in den Seidenfärberei. Das zarte Rosa, schimmernd in gelblichem Schein, erwirbt sich schnell die Huld der modebestimmenden Frauenwelt. Bald fällt der anfängliche Preis und die Anwendungen mehren sich, neue Handelssorten von bläulicherem Farbenton folgen nach, man färbt und druckt auf Wolle und Seide bis zu den reichsten Scharlachnüancen.' Anfänglich 800 Mark das Kilo.

Studies dedicated to BASF trade reports show that the production of Eosin increased from 0.7 tons in 1875 to 35 tons in 1889 and that the average price between 1880 and 1889 decreased by 56.5% [15]. Thanks to this decrease in price, it became profitable to use it for artists' paint. Comparison of price lists issued by the German dye firm Bayer between 1875 and 1900 shows that Eosin was still more expensive than the other red synthetic organic dyes, such as synthetic Alizarin, Fast red types, Bordeaux types, Tuchrot types and Congo red. An 1877 price list of Bayer shows that the price of Eosin dropped enormously from 800 German marks in 1874 to 35 German marks in 1877 (Eosin ((Aurora)) chemisch rein, gelblich and bläulich). The lesser quality of Eosin (Eosin ((Aurora)) No. ll gelblich and bläulich) dropped even more, to 18 German marks [15]. In 1874 1 kg of Eosin cost 800 German marks. Taking into account a factor of 1.23,[2] 1 kg of Eosin corresponds to 984—nearly 1,000—French francs; so 1 g of Eosin would cost 1 French franc in the mid-1870s. For one 15 ml tube of paint, around 2 g of Eosin are needed. According to a price list of about 1888–1890, Fabricant de couleurs fines Tanguy in Paris charged 0.2–1.5 French francs for paint tubes, which means that the costs of Eosin in 1874 would have largely exceeded the regular price of a tube. One can conclude that in the beginning of its production Eosin was too expensive to be put into a paint tube.

## *Recipes*

Silk can be dyed with Eosin following recipe V8, using 0.5% dyestuff. For wool a recipe is given by Lehne stating that it dyes using 2% acetic acid and 4% alum, followed by 25 min boiling. Slightly more vivid colours can be achieved, if the wool is first treated with 5% alum, 5% tartaric acid and 5% acetic acid for 1½ h. Next, the bath is allowed to cool down to 40 °C and the coloured solution (no concentration given) is then added. The mixture is brought to the boil and boiled for a quarter of an hour. After that, the textile is rinsed. This shows that although Eosin can dye a textile directly, mordanting was done as well.

Cotton can be dyed in a lukewarm salt bath of 5% according to the following procedure: Add the colouring solution; dry after dyeing without rinsing. The bath must be very concentrated, but there are no concentrations given.

[2] After the Franco-Prussian War the Reichsmark was introduced in Germany, which was based on a strict gold standard. The gold standard is a system in which international currencies are tied to a specific amount of gold. The foundation of the gold standard is that a currency's value is supported by some weight in gold. A pure gold standard was used between 1879 and 1914 by many modern trading nations. Under the gold standard system, all participating currencies were convertible based on its gold value (one kilogram of gold corresponds to 2.790 German marks and to 3.444 French francs; factor 1.23). Between 1879 and the end of the 1880s these currencies did not fluctuate much, the relation stayed constant.

## Identification Using UHPLC-PDA

The analysis of a sample of Eosin (RCE no. 4328) shows a clear response; the compound has a very distinctive UV-Vis absorption spectrum (Fig. 16.6).

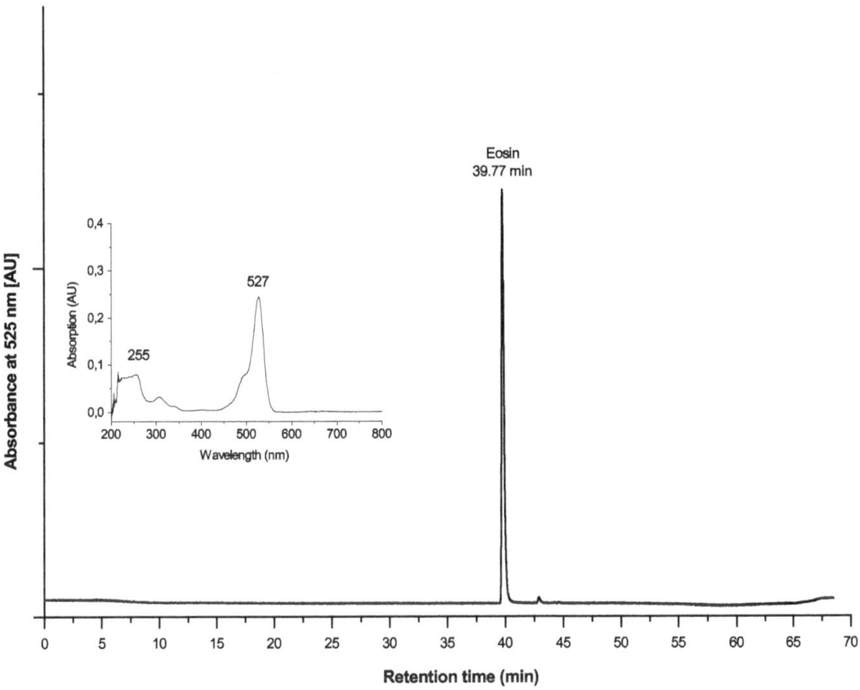


**Fig. 16.6** Chromatogram and spectrum of Eosin (RCE no. 4328)

## Eosin Derivatives

Other dyes related to Eosin with the same Fluorescein structure vary in the number of methyl, ethyl and nitro groups, or bromine and chlorine atoms attached. An example is Spirit Eosin (C.I. name: Solvent Red 44; C.I. number: 45385), developed by Caro in 1874. He discovered that by the methylation and ethylation of Eosin, bright bluish reds were obtained.

The methyl ester of Eosin is named Spirit Eosin (Spritlösliches Eosin) by BASF, which was also sold by other companies under slightly different names, such as Eosin spritlöslich by Weiler-ter Meer; Methyleosin by P. Monnet and Bindschedler & Busch and Methyleosin J extra by Farbwerke ML&B.

The ethyl ester of Eosin is called Eosin S (C.I. name: Solvent Red 45; C.I. number: 45386) by BASF; Primerose à l'alcool DH by Durand-Huguenin; Spriteosin by

Weiler-ter Meer; Aethyleosin, Eosin spritlöslich by Farbwerke ML&B; Eosin BB, Eosin à l'alcool and Rose JB, à l'alcool by Bindschedler & Busch.

In 1875 Caro, and one year later in 1876, Baeyer discovered Eosin BN (C.I. name: Acid Red 91; C.I. number: 45400) by the nitration of dibromofluorescein or by the bromination of dinitrofluorescein, followed by conversion to the sodium or ammonium salt; a bright bluish red dye was obtained, which was brought onto the market by BASF.

Other products derived from Eosin were also put on the market, including: Methyleosin B extra by AGFA; Safrosin by Bindschedler & Busch; Eosinscharlach B by Cassella; Eosinscharlach BB extra by Farbwerke ML&B; Eosin B by Leonhardt and Weiler-ter Meer; Ecarlate J, JJ and V by P. Monnet; Eosin scharlach by Weiler-ter Meer; Nopalin G by Weiler-ter Meer and Jäger; Eosin BW and DHV by Durand-Huguenin; Ecarlate d'Eosine B by Manufacture Lyonnaise; Eosin I bläulich and Eosin S extra bläulich by Bayer and Lutetienne by Soc. Anon. St. Denis.

## 16.5 Phloxines

Colour Index name: Acid Red 98; Colour Index number: 45405
IUPAC name: Dipotassium 2′,4′,5′,7′-tetrabromo-4,7-dichloro-3-oxospiro [2-benzofuran-1,9′-xanthene]-3′,6′-diolate
Discoverer: Emilio Noelting, 1875
Contemporary terminology: Phloxine (En), Phloxine (Fr), Phloxin (Ge), Phloxine (NL)
Chemical formula: $C_{20}H_4Br_4Cl_2K_2O_5$ (Fig. 16.7)

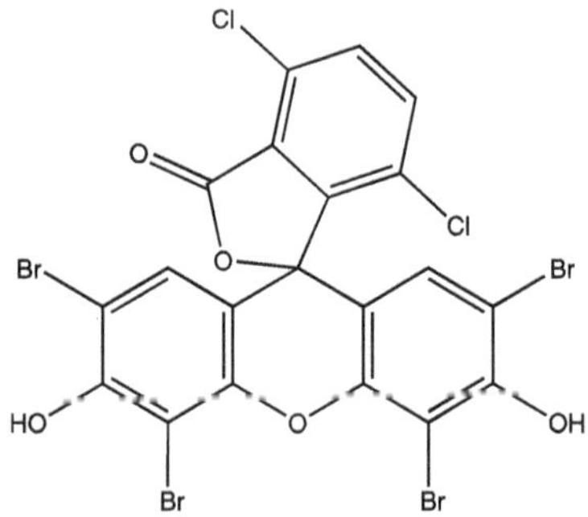


**Fig. 16.7** Structural formula of Phloxine

### *History*

In order to modify the shade of Eosin, in 1875 the French chemist Emilio Noelting (1851–1922) introduced the use of chloro-derivatives of phthalic acid which were condensed with resorcinol as an alternative to Fluorescein. This resulted in the Phloxines, a dye class strongly related to Eosin. Phloxine is prepared by the complete bromination of 4,7-dichlorofluorescein, obtained by the condensation of resorcinol with 3,6-dichlorophthalic anhydride and converted to the potassium salt. As a result, the potassium salt of tetra-bromo-di-chlorofluorescein is obtained, which has a pink colour. The sodium salt was introduced by the French dye firm P. Monnet in Dardagny-La Plaine near Geneva, also under the name Phloxine. Alexandre Le Royer gives a brief account of the substance [16].

Trade names are Phloxine by the dye companies Cassella, Farbwerke ML&B and Durand-Huguenin; Phloxine P by BASF and Erythrosine BB by Durand-Huguenin.

## *Recipes*

For Phloxine, Lehne refers to the recipe used for Eosin, however 1.5% dye needs to be used.

## *Identification Using UHPLC-PDA*

The analysis of Phloxine (RCE no. 4961) shows a clear main peak and several minor compounds which are probably side products of the synthesis (Fig. 16.8).

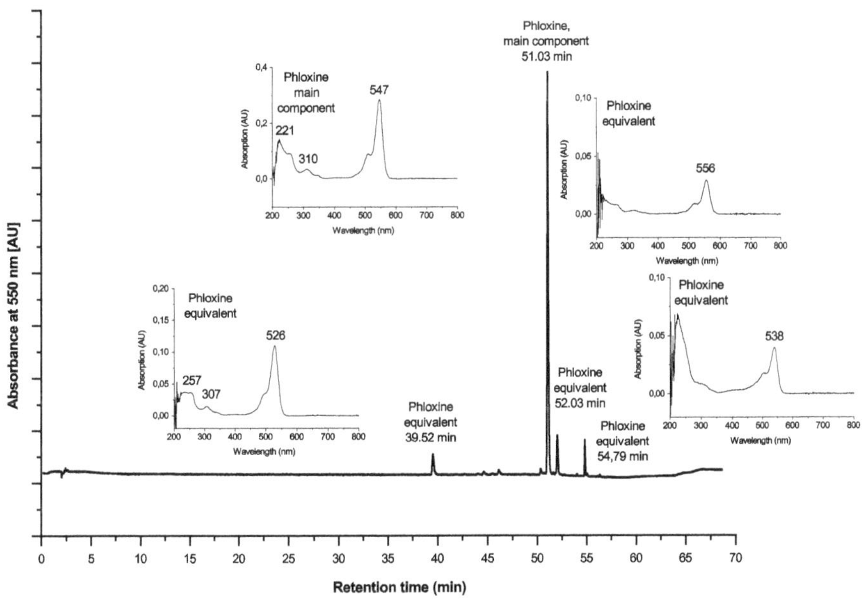


Fig. 16.8 Chromatogram and spectra of Phloxine (RCE no. 4961)

## 16.6 Phloxine B

Colour Index name: Acid Red 92; Colour Index number: 45410
IUPAC name: Disodium 2′,4′,5′,7′-tetrabromo-4,5,6,7-tetrachloro-3-oxo-3H-spiro[[2]benzofuran-1,9′-xanthene]-3′,6′-bis(olate)
Discoverer: Robert Gnehm, 1882
Contemporary terminology: Phloxine B (En), Phloxine B (Fr), Phloxin B (Ge), Phloxine B (NL)
Chemical formula: $C_{20}H_4Br_4Cl_4Na_2O_5$ (Fig. 16.9)

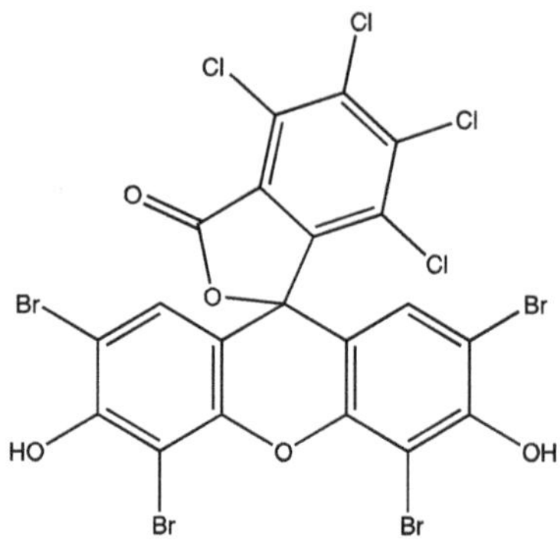


**Fig. 16.9** Structural formula of Phloxine B

### *History*

In 1876, Noelting discovered that by the methylation of Phloxine its methyl ester is formed (tetra-bromo-di-chloro-fluoresceinmethylester), which was produced under the trade name Cyanosin spritlöslich (C.I. name: Solvent dye; C.I. number: 45415) by Farbwerke ML&B, Kalle and Sandoz.

Another Phloxine derivative, Phloxine B (C.I. name: Acid Red 92; C.I. number: 45410), was prepared by the complete bromination of 4,5,6,7-tetra-chloro-fluorescein obtained by the condensation of resorcinol with tetrachlorophthalic anhydride and converted to the sodium salt. Tetra-bromo-tetra-chloro-fluorescein itself was discovered by Gnehm in 1882 [16]. His invention was assigned to the Basel firm CIBA by G.P. 32.564, patented on the 1st of February 1885, and by U.S.P. 322.368, filed on the 26th of February 1885. In 1882 Gnehm discovered by the ethylation of Phloxine B its ethyl ester; this was named Cyanosin B (C.I. name: Acid dye; C.I. number: 45420) by CIBA.

## *Trade Names*

These bluer shades were named Phloxine by Farbwerke ML&B; Phloxine TA by P. Monnet; Phloxin N and Phloxine BB by Jäger; Eosine 10B by Cassella and Manufacture Lyonnaise; Eosine bleu by Sandoz; Cyanosin by Durand-Huguenin and Erythrosine B by Bindschedler & Busch.

## *Recipes*

For Phloxine B, Lehne refers to the recipe used for Eosin; however, 1% dye needs to be used.

# 16.7 Erythrosine G

Colour Index name: Acid Red 95; Colour Index number: 45425
IUPAC name: Disodium 2-(4,5-diiodo-6-oxido-3-oxoxanthen-9-yl)benzoate
Discoverer: Emilio Noelting, 1875
Contemporary terminology: Erythrosine G (En), Érythrosine G (Fr), Erythrosin G (Ge), Erythrosine G (NL)
Chemical formula: $C_{20}H_8I_2Na_2O_5$ (Fig. 16.10)

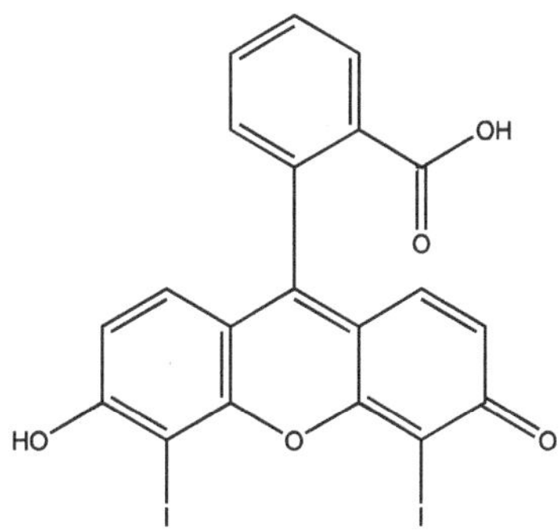


**Fig. 16.10** Structural formula of Erythrosine G

## *History*

Erythrosine G was discovered by Noelting in 1875. This yellowish-red dye is obtained by the iodination of Fluorescein in an aqueous or alkaline solution.

## *Trade Names*

This dye, 4,5-di-iodofluorescein, was sold under the name of Erythrosine extra yellowish by BASF. Other trade names were Erythrosine R by CIBA; Erythrosine G by Leonhardt; Erythrosin gelblich by Cassella; Erythrosin extra gelb N by Farbwerke ML&B; Jodeosin and Jaune d'orient. The sodium salt of di-iodofluorescein was sold by Usine P. Monnet et Cie. in Saint-Fons near Lyon under the name of Pyrosine J. The ammonium salt was known as Erythrosine G as well as Dianthine G by BASF.

## *Recipes*

Erythrosine G can be dyed following the recipe given for Eosin, using 1% dye.

## 16.8 Erythrosine

Colour Index name: Acid Red 51; Colour Index number: 45430
IUPAC name: 2-(6-hydroxy-2,4,5,7-tetraiodo-3-oxo-xanthen-9-yl)benzoic acid
Discoverer: Kussmaul, 1876
Contemporary terminology: Erythrosine (En), Érythrosine (Fr), Erythrosin (Ge), Erythrosine (NL)
Chemical formula: $C_{20}H_6I_4Na_2O_5$ (Fig. 16.11)

**Fig. 16.11** Structural formula of Erythrosine

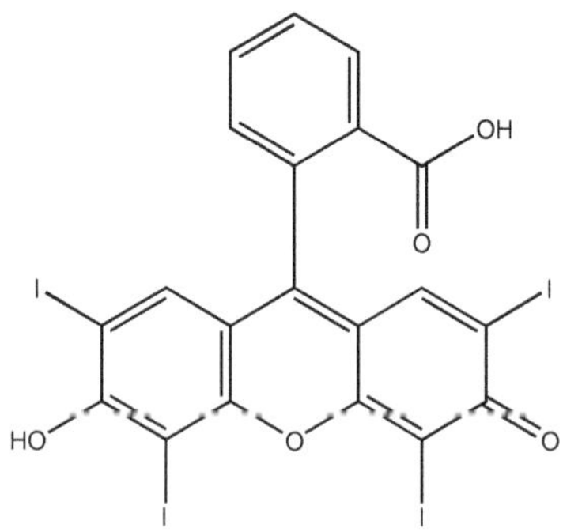

## History

In 1876, Kussmaul prepared a bluish-pink Erythrosine by the complete iodination of Fluorescein in an aqueous or alkaline solution, resulting in 2,4,5,7-tetra-iodofluorescein. The chemical structure of Erythrosine is similar to that of Eosin, with the exception that the bromine atoms of Eosin are replaced by iodine atoms. Dyes related to Erythrosine with the same basic structure vary in the number of iodine and chlorine atoms attached.

Eosine bleuâtre (soluble à l'eau) was introduced by the Basel firm of Bindschedler & Busch in 1876, the process being based on the experiments of Kussmaul. Otto Mühlhäuser states that better results were obtained when an organic acid is used to liberate the iodine. The electrolytic preparation of Erythrosine has been described by the Soc. Chim. Us. Rhône [11, 12, 17].

## Trade Names

Tetra-iodofluorescein was marketed in the form of the sodium salt under the trade names of Dianthine B and Eosin J by BASF; Erythrosine by BASF and Farbwerke ML&B; Erythrosine extra bläulich by BASF, Weiler-ter Meer, Jäger and Durand-Huguenin; Erythrosine bläulich by Farbwerke ML&B; Erythrosine B by AGFA, Cassella and CIBA; Erythrosine D by Cassella and Manufacture Lyonnaise; Rose B à I'eau by Bindschedler & Busch and CIBA; Erythrosine J extra, JNV, W extra and Pyrosine B by P. Monnet; Primerose soluble by Durand-Huguenin and Jodeosin B, Eosine bleuâtre by Soc. Anon. St. Denis.

## Recipes

Erythrosine can be dyed following the recipe given for Eosin, using 1% dye for wool and 1.5% for silk.

## Identification Using UHPLC-PDA

The analysis of Erythrosine (RCE no. 5303) shows a clear peak and spectrum (Fig. 16.12).

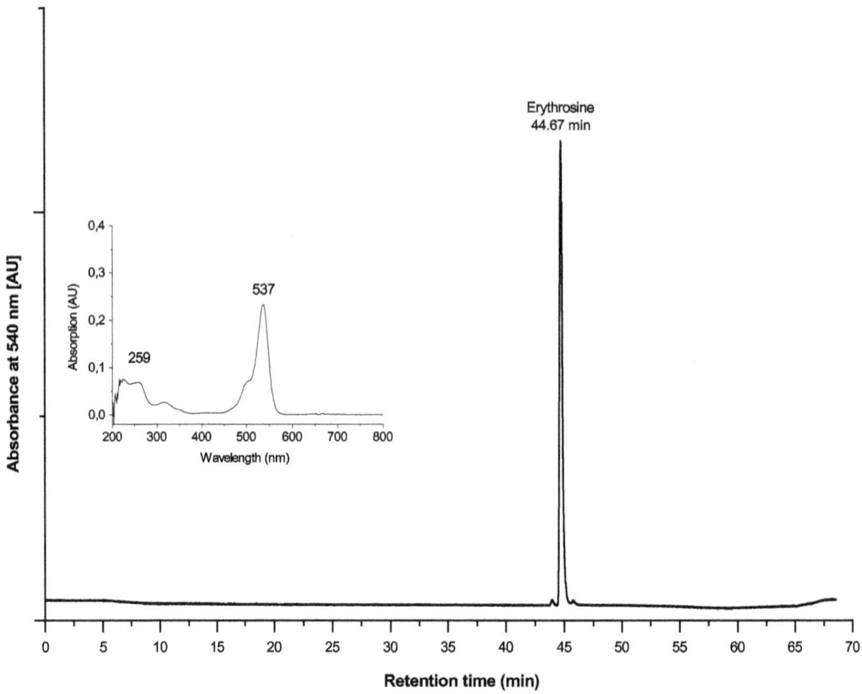


**Fig. 16.12** Chromatogram and spectrum of Erythrosine (RCE no. 5303)

## 16.9 Rose Bengal

Colour Index name: Acid Red 93; Colour Index number: 45435
IUPAC name: 4,5,6,7-tetrachloro-3′,6′-dihydroxy-2′,4′,5′,7′-tetraiodo-3H-spiro[[2] benzofuran-1,9′-xanthen]-3-one
Discoverer: Emilio Noelting, 1875
Contemporary terminology: Rose Bengale (En), Rouge Bengale (Fr), Rose Bengal (Ge), Bengaalsrood (NL)
Chemical formula: $C_{20}H_2Cl_2I_4Na_2O_5$ (Fig. 16.13)

**Fig. 16.13** Structural formula of Rose Bengal

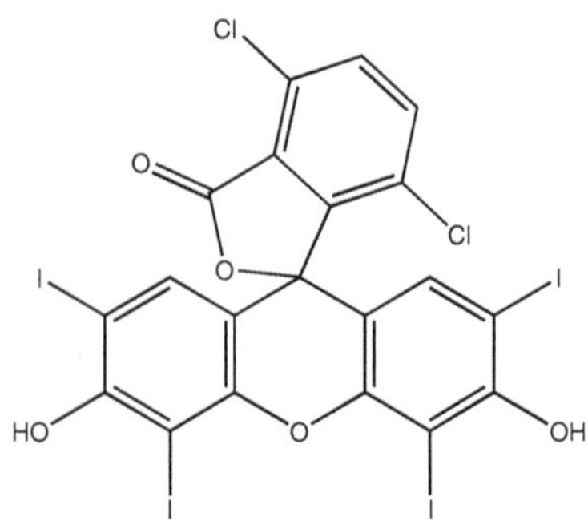

### History

Two dyes related to the Erythrosines are the Rose Bengals. Rose Bengal was discovered by Noelting in 1875 and is prepared by the complete iodination of 4,7-dichlorofluorescein in the presence of potassium or sodium chlorate and copper (I) chloride. The name derives from rose (flower) and Bengal (region). The chemical structure of these components is similar to that of the Phloxines, with the exception that the bromine atoms of the Phloxines are replaced by iodine atoms. Rose Bengal is the alkali salt of tetra-iodo-di-chlorofluorescein. This bluish-red dye was introduced as a sodium or potassium salt under the name of Rose Bengal. Le Royer prepared the substance by heating an alkaline solution of dichlorofluorescein on the water-bath and adding a solution of iodine in dilute caustic potash until the fluorescence produced by the dichlorofluorescein disappeared [18, 19].

### Trade Names

Rose Bengal was produced by BASF, AGFA, Weiler-ter Meer and Durand-Huguenin. Rose Bengal N was listed by Cassella and Manufacture Lyonnaise. The compound was marketed as Rose Bengal AT by BASF; Rose Bengal G by Farbwerke ML&B and Leonhardt; Bengalrosa by Durand-Huguenin and Bengalrosa B by Jäger.

### Recipes

No dye recipes are given by Lehne; however, samples are present indicating a dye concentration of 0.5–1%, both for wool. Presumably the same recipe as for Eosin has been used.

## 16.10 Rose Bengal B

Colour Index name: Acid Red 94; Colour Index number: 45440
IUPAC name: 4,5,6,7-tetrachloro-3′,6′-dihydroxy-2′,4′,5′,7′-tetraiodo-3H-spiro[[2]benzofuran-1,9′-xanthen]-3-one
Discoverer: Robert Gnehm, 1882
Contemporary terminology: Rose Bengale B (En), Rouge Bengale B (Fr), Rose Bengal B (Ge), Bengaalsrood B (NL)
Chemical formula: $C_{20}H_4Cl_4I_4O_5$ (Fig. 16.14)

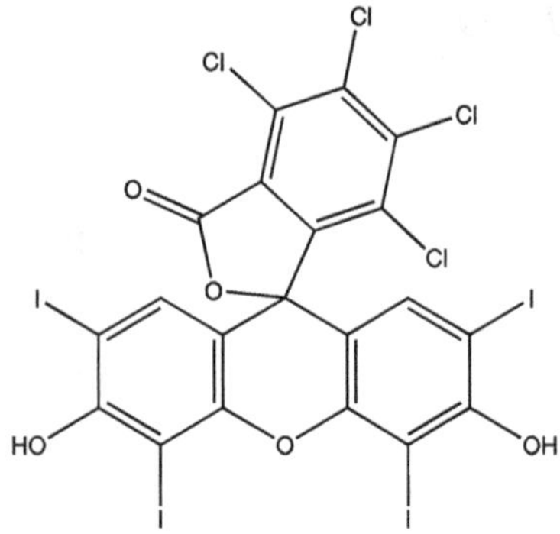


**Fig. 16.14** Structural formula of Rose Bengal B

## *History*

Another rose Bengal type obtained by the complete iodination of 4,5,6,7-tetrachlorofluorescein prepared by the condensation of resorcinol with tetrachlorophthalic anhydride and converted into the potassium salt. This Rose Bengal B is tetraiodo-tetrachlorofluorescein and was discovered by Gnehm in 1882 [12]. The same Phloxine patents are listed for the dye Rose Bengal B (G.P. 32.564 and U.S.P. 322.368, both filed in 1885). The dye was used for the dyeing of wool.

## *Trade Names*

The potassium salt of this bright bluish-pink compound was marketed as Rose Bengal by BASF and Cassella; Rose Bengal B by Leonhardt; Rose Bengal 2B and Bengalrosa 2B by Jäger; Rose Bengal 3B konz. by Farbwerke ML&B and Durand-Huguenin and Rose Bengal extra by Durand-Huguenin.

## *Recipes*

Rose Bengal B can be dyed following the recipe given for Eosin, using 1% dye for wool and 1.5% for silk.

## *Identification Using UHPLC-PDA*

The analysis of Rose Bengal B (RCE no. 7261) showed that this is composed of a mixture of closely related compounds (Fig. 16.15).

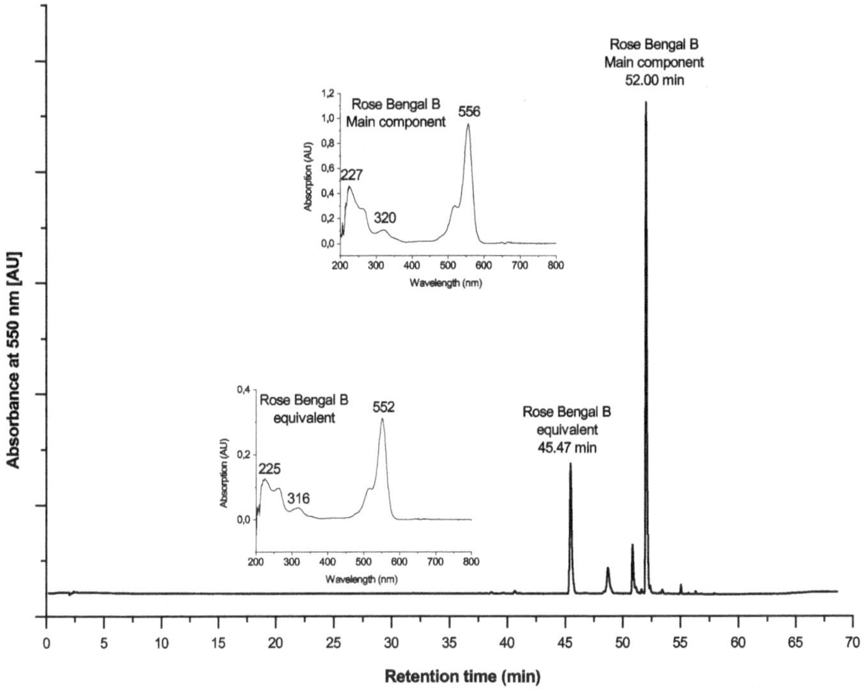


**Fig. 16.15** Chromatogram and spectra of Rose Bengal B (RCE no. 7261)

## 16.11 Rhodamine B

Colour Index name: Basic Violet 10; Colour Index number: 45170)
IUPAC name: 9-(2-carboxyphenyl)-6-(diethylamino)-N,N-diethyl-3H-xanthen-3-iminium chloride
Discoverer: Maurice Cérésole, 1887
Contemporary terminology: Rhodamine B (En), Rhodamine B (Fr), Rhodamin B (Ge), Rhodamine B (NL)
Chemical formula: $C_{28}H_{31}ClN_2O_3$ (Fig. 16.16)

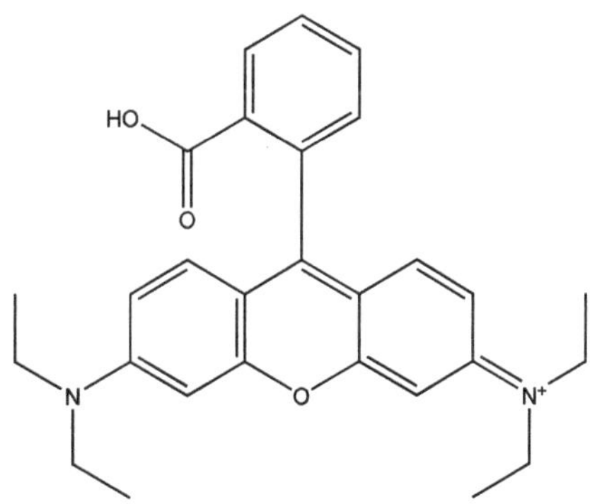


**Fig. 16.16** Structural formula of Rhodamine B

## *History*

The Rhodamines form a class of dyes very closely related to Fluorescein and its derivatives: whereas Fluorescein is 3,6-dihydroxyfluoran, Rhodamine B is the corresponding tetraethyl-3,6-diaminofluoran. The Rhodamines are the representatives of the basic type of the xanthene group.

The first Rhodamine, Rhodamine B, was discovered in 1887 by the Swiss chemist Maurice Cérésole (1860–1936) in the BASF laboratories. His invention was assigned to BASF and patented by G.P. 44.002, filed on the 13th of November 1887 and B.P. 15.374, filed in 1887, followed by patents in France and the United States of America the following year [20–22]. This bluish-red dye was prepared by condensing phthalic anhydride with 3-diethylaminophenol (*m*-diethylaminophenol) (diethyl-*m*-aminophenol) in the presence of zinc chloride. One year later, in 1888, the German chemists Benno Homolka (1860–1925) and Hermann Boedeker of Farbwerke ML&B patented an alternative method of preparation for Rhodamine B by reacting 3,6-dichlorofluoran with diethylamine under pressure (G.P. 48.367, patented on the 3rd of July 1888 and B.P. 9.600, filed in 1888). A number of additional patents were taken out by BASF and Farbwerke ML&B.

## *Trade Names*

The dye was marketed as Rhodamine B and B extra by BASF, CIBA, Bayer, Farbwerke ML&B, AGFA, Cassella and Kalle; Rhodamine O by Farbwerke ML&B, and Brillantrosa B by Leonhardt.

## *Recipes*

Lehne mentioned that Rhodamine is especially used to dye silk and wool; for silk recipe V7 is mentioned. Wool is dyed with the addition of 4% acetic acid (from a 52% solution) and 4% alum, which is boiled for about half an hour. Lehne mentions that copper dyeing kettles are unsuitable, probably because copper is released from the kettle during dyeing and might affect the colour. In combination with Ponceau R, Quinoline yellow and similar dyes, scarlet colours can be produced which, according to Lehne, are as beautiful as Cochineal and which cannot be achieved with other synthetic colourants.[3]

---

[3] Wolle wird unter Zusatz von 4% Essigsäure von 8 °Be und 4% Alaun ausgefärbt; etwa 1/2 Stunde kochen lassen. Kupferne Färbekessel sind ungeeignet. In Verbindung mit Ponceau R, Chinolingelb etc. lassen sich Scharlachnüancen herstellen, deren Schönheit den mit Cochenille erhältlichen mindestens gleichkommen und die mit anderen künstlichen Farbstoffen nicht zu erzielen sind.

## *Identification Using UHPLC-PDA*

In Fig. 16.17, the analysis of Rhodamine B (Sigma-Aldrich CAS 81-88-9) is shown, the UV-Vis spectra of the main compound and a side product are very similar.

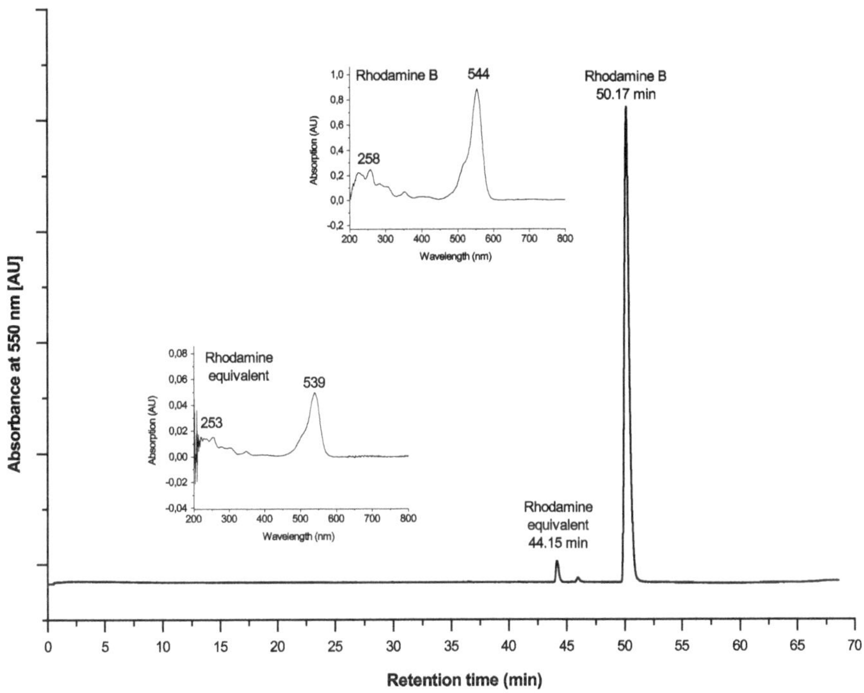


**Fig. 16.17** Chromatogram and spectra of Rhodamine B (Sigma-Aldrich CAS 81-88-9)

## *Rhodamine Derivatives*

In 1891 French chemist Pierre Prosper Monnet (1834–1914) of the French dye firm Gilliard, P. Monnet & Cartier in Lyon prepared and patented Anisoline (C.I. name: Basic Violet; C.I. number: 45175), the ethyl ester of Rhodamine B, by esterification (obtained by replacing the hydrogen of the carboxyl group by an ethyl group). Other trade names are Rhodamine 3B and 3B extra by BASF and CIBA [23]. An interesting dye belonging to the Rhodamines is prepared by condensing succinic anhydride with dimethyl-*m*-amidophenol and is known as Rhodamine S (C.I. name: Basic Red 11; C.I. number: 45050). It is a substantive dye, dyeing unmordanted cotton a pink colour. Rhodamine S was invented by Kahn and Wilhelm Majert in 1888. This dye was patented, between 1888 and 1892, by Bayer, BASF and CIBA. It was marketed as Rhodamine S and S extra by Bayer, BASF and CIBA [24].

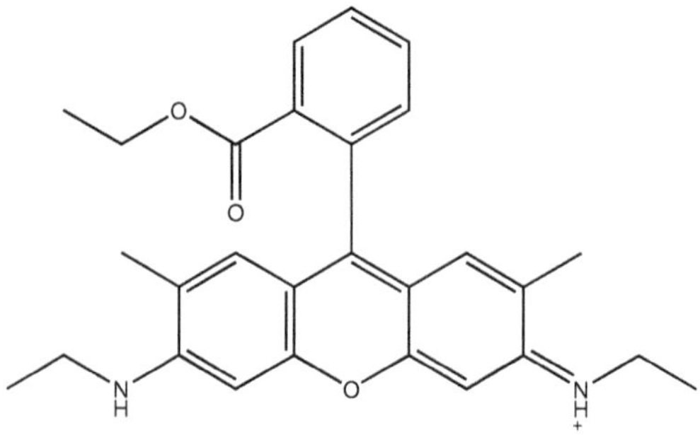


**Fig. 16.18** Structural formula of Rhodamine 6G

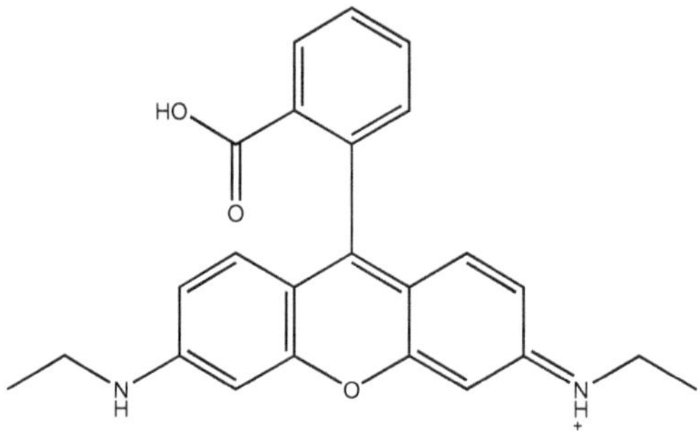


**Fig. 16.19** Structural formula of Rhodamine G

An important basic-type is Rhodamine 6G (C.I. name: Basic Red 1; C.I. number: 45160). It has a scarlet-red tone and was discovered by the German BASF chemist Heinrich August Bernthsen (1855–1931) in 1892. It is prepared by condensing 2-monoethylaminophenol (monoethyl-*m*-aminophenol) (*m*-ethylaminophenol) with phthalic anhydride, and then esterifying the product with ethanol and hydrochloric acid [25]. The resulting ethyl ester forms the dye Rhodamine 6G. This dye was protected by German, English, French and American patents taken out by BASF in the same year [26] (Figs. 16.18 and 16.19).

## *Trade Names*

Trade names are Rhodamine 6G and 6G extra by CIBA and Farbwerke ML&B and Brillantrosa 5G by Leonhardt. The compound was marketed as Rhodamine G and Rhodamine G extra by BASF, Farbwerke ML&B, Kalle, CIBA and Sandoz and Brillantrosa G by Leonhardt [27].

The corresponding ethyl ester is known as Rhodamine 3G (C.I. name: Basic Red 3; C.I. number: 45210) and was discovered by the German chemist Carl Oscar

Müller in 1895. His invention was assigned to Bindschedler & Busch in Basel [28]. The dye was marketed as Rhodamine 3G and Rhodamine 3G extra by BASF and Irisamin G and G extra by Cassella.

### *The Violamines: Rhodamine Derivatives as Acid Dyes*

In a short period of time, a series of Rhodamines followed. Generally, the Rhodamines are employed as basic dyes, but by sulfonation acid dyes can be prepared. In the period 1888–1889, Hermann Boedeker discovered and patented a series of sulfonic acids of Rhodamines converted into sodium salts and produced by Farbwerke ML& under their trade name Violamine. These included Violamine B (C.I. name: Acid Violet 30; C.I. number: 45186), also known as Fast acid violet B; Violamine R (C.I. name: Acid Violet 9; C.I. number: 45190), known as Fast acid violet A2R; Violamine G (C.I. name: Acid Blue 19; C.I. number: 45205), also sold under the trade name Acid Rosamine A, and Violamine 3B, also known as Fast acid blue R [29].

## 16.12 Application and Properties of the Xanthenes as Dyes and Pigments

The xanthenes as dyes give colours ranging from yellow, orange scarlet, bluish red to violet; the dyes can be converted into lake pigments as described below. The xanthenes can be used to dye textiles, such as wool and especially silk. The silk dyeings of Fluorescein and Chrysoline have a poor washing fastness compared to the other dyes of this class. Arranged in this shade order, they are Fluorescein, Eosins, Phloxines, Erythrosines, Rose Bengals and Rhodamines, and they yield colours of an extreme brilliance and bright, pure tone. However, both the dyes on textiles and the lake pigments have a poor light fastness (see Table 16.2). In his 1893 book, Adolf Lehne writes on the light fastness of Fluorescein, Eosins, Phloxines, Erythrosines and Rose Bengals[4]: 'After 5 days of light exposure, in June, paler or much paler and in 18 days nearly completely faded'.

Compared to the Eosins, Erythrosines and Rose Bengals, the light fastness of the Rhodamine types is better:[5] 'Rhodamine B: After 5 days of light exposure, in June, [the colour has] not changed so much. After 18 days strongly faded, brighter and more yellow tone'.

[4] Echtheit: ´Nach 5 tägiger Einwirkung der Witterung (Juni) viel heller, nach 18 Tagen vollständig verblasst´.

[5] Rhodamnin B: Echtheit: ´Nach 5 tägiger Einwirkung der Witterung (Juni) sehr wenig verändert, nach 18 Tagen ziemlich stark verblasst, heller und gelber´.

**Table 16.2** Overview of the light fastness according to the Colour Index

| Name | Light fastness | Method |
|---|---|---|
| Fluorescein | n.a. | – |
| Eosin | 2 | ISO |
| Spirit eosin | 3 | ISO |
| Eosin S | 5–6 | ISO |
| Eosin BN | 3 | ISO |
| Phloxine | Very low | – |
| Chrysoline | n.a. | – |
| Cyanosin spritl. | n.a. | – |
| Phloxine B | 1 | AATCC |
| Cyanosin B | n.a. | – |
| Erythrosine G | n.a. | – |
| Erythrosine | 1 | AATCC |
| Rose Bengal | 1 | ISO |
| Rose Bengal B | 1 | ISO |
| Rhodamine B | 2 | ISO |
| Rhodamine S | 1 | C* |
| Rhodamine 3B | 1 | B* |
| Rhodamine 6G | 2 | ISO |
| Rhodamine G | 2 | C* |
| Rhodamine 3G | 1 | B* |
| Violamine B | 3–4 | ISO |
| Violamine R | 4 | ISO |
| Violamine G | 3–4 | ISO |

The dyes and lake pigments prepared using Rhodamine 6G are quite similar in properties and in usefulness to Rhodamine B, but since the carboxyl group is esterified, the stability of the molecule is appreciably increased. The light fastness of Rhodamine 6G:[6] 'After 6 weeks of light exposure, in March and April, much duller and more yellow tone'.

## 16.13 The Synthetic Organic Dyes as Artists' Paints

The dyes discussed in this group can be used as artists' pigments by making lake pigments through precipitation on co-ordination metals. Precipitation is accomplished by adding mineral salts, such as alum and lead acetate, to an aqueous solution of the dye. The xanthenes give lakes ranging in colour from an orange scarlet to bluish red. Arranged in this shade order they are Eosins, Erythrosines, Phloxines, Rose Bengals and Rhodamines.

[6] Rhodamin 6G: Echtheit: ´Nach 6 wöchentlicher Einwirkung der Witterung (März, April) viel matter und gelber´.

In 1893, Lehne described the preparation of Eosin lake:[7] 'Eosin is widely used in the production of lakes, which are used in lithography, book and wallpaper printing and in the manufacture of coloured paper. A not very lightfast vermilion substitute is obtained when an Eosin solution is mixed with red lead. A solution of sugar of lead is added to precipitate the poorly soluble lead salt. The so-called geranium lakes are produced in a similar way by precipitating the lead lake onto hydrated alumina. For wallpapers and paints, hydrated alumina is wholly or partly replaced by a cheaper white, such as barytes. To precipitate 1 kg of Eosin requires about 1 kg sugar of lead'.

Before the nineteenth century, artists generally prepared their own paints, grinding the pigments with the desired paint medium as required. This situation changed after English, German and French artists' paint manufacturers, such as Winsor & Newton (est. 1832), Künstlerfarbenfabrik Dr. Fr. Schoenfeld (founded in 1862), Firma H. Schmincke & Co. (founded in 1881) and Lefranc Frères (founded in 1840) started to produce paint tubes, and it was no longer necessary for artists to do this. Paint factories were instructed to prepare so many different colour combinations that in a short period of time an extensive range of artists' paints became available. The paint industry was also responsible for many new and imaginative trade names. This also resulted in the use of the uncomplimentary terms 'coal tar' and/or 'aniline' colours.

Catalogues offered at least 600 different colour shades, including 10–20 names for the colour white, 130 yellow tones, about 90 brown colours and approximately 90 black tones. An interesting example in one catalogue gave a total of 8 cochineal carmines, 39 alizarin tones and 34 other red paints, all listed for the colour red [30–32].

However, a major problem was that most of the invented synthetic organic dyes and organic pigments during this period were insufficiently permanent for the purpose of art. Seduced by the brilliance of the new colours, artists used them but with disastrous results.

At the end of the nineteenth century new brilliant red lakes, such as Geranium lake, began appearing under various labels and fantasy names, each manufacturer inventing its own name. In the trade catalogues of the paint manufacturers, such as in the Lefranc & Cie. catalogue of 1890, the A.S. Aloe Company artist' instruction book and the Winsor & Newton trade catalogue of 1896, Geranium lake was mentioned as extremely fugitive or as not considered stable. Artists of that time were aware that certain reds ran the risk of fading [33]. In the early 1890s the French

---

[7] 'Eosin findet ausgedehnte Verwendung zur Darstellung von Lackfarben, welche im Stein-, Buch- und Tapetendruck sowie in der Buntpapierfabrikation gebraucht werden. Ein allerdings wenig lichtechter Zinnoberersatz wird erhalten, indem Eosinlösung mit Mennige angerührt und durch Zusatz von Bleizuckerlösung das schwerlösliche Bleisalz niedergeschlagen wird. Die sogenannte Geraniumlacke werden in analoger Weise dargestellt, indem der Bleilack auf Thonerdehydrat gefällt wird. Für Tapeten- und Anstrichfarben wird Thonerdehydrat theilweise oder ganz durch ein billigeres Weiss wie Schwerspath u.s.w. ersetzt.—Zum Fällen von 1 kg Eosin ist etwa 1 kg Bleizucker erforderlich.'

painter Jehan-Georges Vibert (1840–1902) advised against synthetic reds (and other synthetic colourants) that had recently been made available to artists [34]. The discolouration was rapid enough for some painters to live to see it for themselves. A story goes that when the French painter Pierre-Auguste Renoir (1841–1919) saw his 1876 painting *Ball at the Moulin de la Galette* hanging in the Louvre after World War l, he joked: 'This painting is not mine, I painted it in pink, this one is blue'.

In an attempt to remedy this situation and to create a system of paint quality, the skilled craftsman and researcher Adolf Wilhelm Keim (1851–1913), together with the German chemist Max von Pettenkofer (1818–1901) and the German painter Franz von Lenbach (1836–1904), tried to restore order to this chaotic field and founded the Deutsche Gesellschaft zur Beförderung rationeller Malverfahren (German Society for the Promotion of Rational Painting Techniques) on the 14th of April 1886 in Munich. The society comprised not only painters, but also a group of manufacturers who felt responsible for the products used by their artistic clients.

A list of 21 pigments deemed to be 'absolutely safe' or 'indispensable' was drawn up and published on the 18th of February 1887. It contained only one synthetic organic dye, Alizarin red, also known as dark violet [35].

This had an immediate and dramatic effect on the range of artists' colours being produced, and by the early twentieth century the number of colour combinations had been reduced from 600 to as few as 50. This was reflected in both the literature and pigment catalogues of that time. However, even in his book *The Permanent Palette*, dated 1930, the German-North American physician and artist Martin Fischer (1879–1962) warned artists: 'It is a good rule for the painter to eliminate everything from the manufacturer's catalogue which bears a name unfamiliar to him or the content of which is not wholly disclosed to him'. [35] In the chapter on the natural and synthetic lakes he condemns many pigments of the xanthene group such as Fluorescein, Eosin, Geranium lake, Rose Bengal and Rhodamine. Their use, Fischer says, should be forbidden in works of art that aspire to permanence [35]:

> 'The only usable pigments in this list are the red and rose madders and, since synthetic chemistry has produced substitutes (the alizarins) for these which are purer, the painter has best use them. The remaining pigments are unreliable whether obtained from natural or as coal tar substitution products. The following are chemical derivatives from different fractions of coal tar. The entire list, with the exception of pure alizarin red, should be learned only to be forgotten for, however tempting the first effects produced by these pigments, their life is not long. The lakes prepared from aniline fade in weeks. The azo dyes stand up better but are still far from satisfactory. The dye chemist offers the painter a new tone almost daily. While such recent discoveries are frequently guaranteed as permanent, it is obvious that not enough time has elapsed since their discovery to prove that such is the case. Thus far every adoption by the oil painter has proved disappointing. The fine art worker does well, therefore to hold fast only to such as time has proved good, leaving experiments with new pigments to those who are not writing their life story with them.'

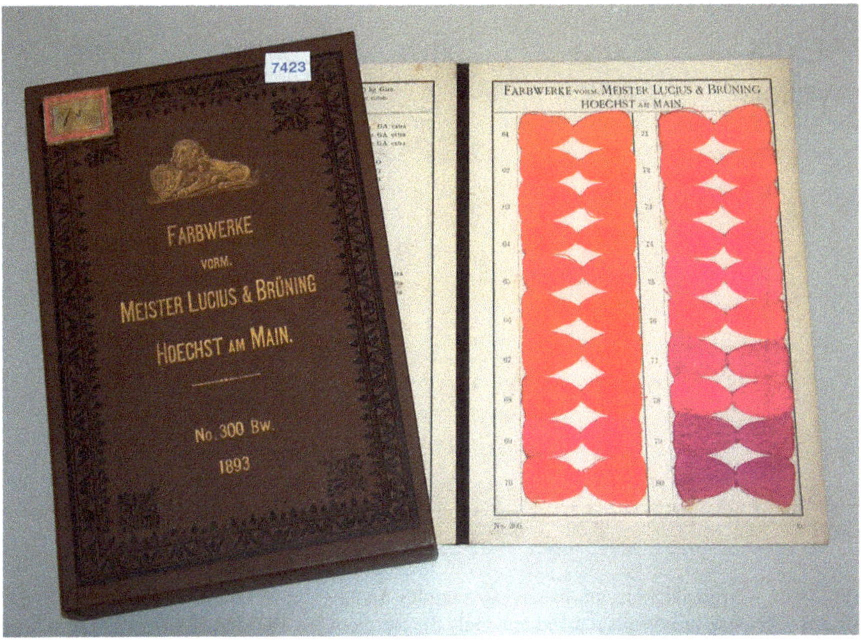


**Fig. 16.20** Sample book from Farbwerke vorm. Meister Lucius & Brüning, Hoechst am Main, 1893. RCE reference collection, no. 7423. Photograph: Maarten R. van Bommel

## 16.14 Case Studies

### *The Sample Book from Farbwerke vorm. Meister Lucius & Brüning*

A sample book from Farbwerke ML&B, dated 1893, which was used by salesmen was selected for investigation (see Fig. 16.20).

This book contains silk samples dyed with different synthetic dyes. Among them, five different eosins are mentioned, such as Eosin extra yellow, Eosin extra and Eosin extra BB. The colour differences between them are very small and just visible with the naked eye. To determine the differences of chemical constitutions between these eosins, thin silk threads of 1 cm were taken from each eosin sample for analysis. The analysis was performed with the HPLC system using a water, methanol and phosphoric acid gradient.

The HPLC-PDA chromatograms of the five eosins are presented in Fig. 16.21. Only the section between 20 and 28 min is of interest. Eosin is marked red in all graphs and was identified as the main component in each of the samples. However, there are significant differences in composition between them. In addition to Eosin, two minor products were found in all samples. Eosin extra is the purest; the concentration of Eosin is the highest. In Eosin extra yellow, a third minor component was

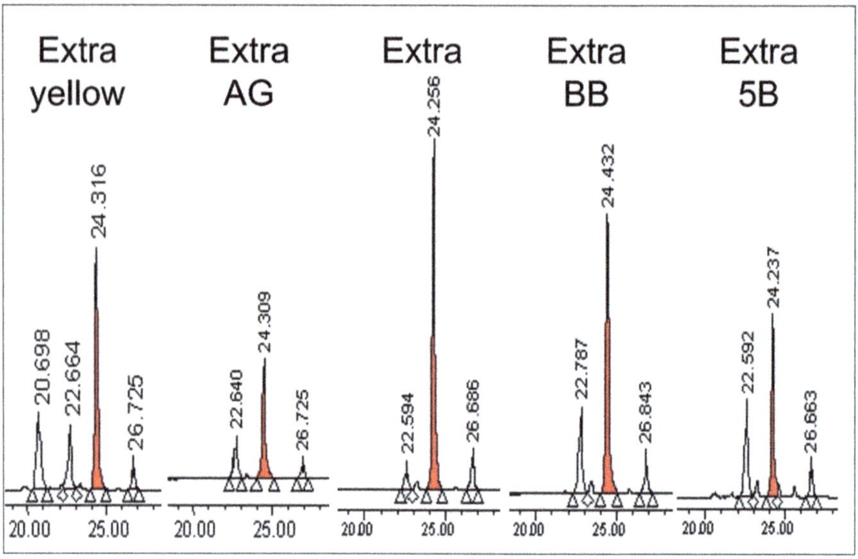


**Fig. 16.21** Chromatograms of five eosin samples taken from a sample book of Farbwerke ML&B. The peak of Eosin is marked red. Only the part between 19.0 and 28.0 min is shown

found, eluting at 20.6 min. The ratio of Eosin and the minor products differs in each sample, and these differences could be used to distinguish between the reference materials. However, it is not likely that determination of the ratios between Eosin and by-products in reference materials can be used for the exact identification of dyes on art objects. First, other mixtures of Eosin, with different constitutions, were available on the market at the time and it would be an impossible task to investigate them all. In addition, the ratio between the different components can be altered during ageing, which makes identification even more complicated. Nevertheless, as Eosin was the main component in these references, at least an Eosin dye can be identified in historical objects.

## *Paint Analysis on the Late Period Works of Vincent Van Gogh*

In the year 2000 an intensive research programme was initiated to examine paintings by Vincent van Gogh, from the period November 1885 to February 1888 when he lived in Antwerp and Paris. The main goal of this collaborative project involving the Van Gogh Museum, the Instituut Collectie Nederland (ICN) and the Shell Research and Technology Centre in Amsterdam, was to gain knowledge of the artist's working methods in this period, since this had not been investigated in depth previously.

A close examination of van Gogh's paintings provides some insight into his use of red lakes. In Antwerp and especially in Paris his palette evolved, under the influence of his new environment, into a much more modern and brighter palette. Within a month of his arrival in Antwerp he bought more durable and brilliant colours, including a red lake, which he described on the 28th of December 1885 as follows in his letter 550: 'Carmine is the red of wine, and it's warm, spirited as wine'. During his stay in Paris (from 1886 until 1888) he typically painted with a variety of natural organic pigments, such as Kopp's purpurin (a madder derivative), Cochineal and Redwood lakes. Mixtures of these natural organic pigments were found in this period. Examinations also showed that the natural red lakes were often mixed with different concentrations of the red pigment vermilion (mercury(ll) sulfide). The natural red lakes were also mixed with the blue pigment synthetic ultramarine, allowing the artist to vary the tone to a more red-violet or violet colour [36]. Severe discolouration has been observed in areas of pink, red or purple paint where he used certain red lake pigments, often mixed with white, especially in paintings at the end of his lifetime. The change in colour had an immense effect on the overall balance of colour in the paintings and the impression that was intended by van Gogh is no longer conveyed. The original colour is only revealed in areas of paintings protected from light by the frame.

After van Gogh moved to Arles and Saint-Rémy in the south of France, where he stayed from 1888 to 1890, he came in contact with a synthetic organic dye, namely *laque géranium*, a translucent paint made from Eosin, which is bright red in colour. While busily at work in Arles on his blossoming orchards (F0403, F0404, F0555) in April 1888, he wrote to his brother Theo to order a long list of paints (letter 593, dated 5th of June, Arles) from the colour merchant Tasset et L'Hôte in Paris. In a subsequent letter (number 889, dated 17th of June 1890, Auvers-sur-Oise) he explained why Theo had to order the paints of this artist's paint company: van Gogh preferred the products of Tasset et L'Hôte to those of Tanguy because in general the colours were not so dull [37].

In eight letters written in Arles, Saint-Rémy and Auvers-sur-Oise between April 1888 and June 1890, Vincent asked Theo to send *laque géranium* from Tasset et L'Hôte besides the various aluminium- and tin-containing carmine lake pigments. A total of 38 paint tubes of *laque géranium* of different sizes is mentioned in his letters [38–44].

As part of the current study, a tube of Tasset et L'Hôte tube *laque géranium* was analysed and found to be an Eosin lake. In many paintings by van Gogh in Arles, Saint-Rémy and Auvers-sur-Oise (April 1888–July 1890), such as *Small Pear Tree in Blossom* (F0405), mid-April 1888; *The Sower* (F422, Kröller-Müller Museum, Otterlo), second half of June 1888; *The Pink Peach Tree* (F404), May 1888; *The Bedroom* (F0482, Van Gogh Museum, Amsterdam), mid-October 1888; *The Sower*

(F0451, Van Gogh Museum, Amsterdam), late November 1888; *Undergrowth* (F0746), first half of July 1889; *Thresher (after Millet)* (F0692), mid-September 1889; *Pietà (after Delacroix)* (F0630), mid-September 1889; *The Bedroom* (F484, Art Institute of Chicago, Chicago), mid–late September 1889; *Evening (after Millet)* (F0647), late October–early November 1889; *Garden with Butterflies* (F0402), spring 1890; *The Raising of Lazarus (after Rembrandt)* (F0677), early May 1890; *Le Docteur Doctor Gachet* (F754, Musée d'Orsay, Paris), 1st of June 1890; *Daubigny's Garden* (F0765), mid-June 1890; *Mademoiselle Gachet dans son jardin à Auvers-sur-Oise* (F756, Musée d'Orsay, Paris), June 1890; *View of Auvers-sur-Oise* (F800, Rhode Island School of Design, Providence), June–July 1890; *Tree Roots* (F0816), July 1890; *Wheatfield under Thunderclouds* (F0778), first half of July 1890 and *Deux fillettes* (Two little girls) (F783, Musée d'Orsay, Paris), July 1890, Eosin precipitated onto an aluminium-based substrate was identified.[8] [45–47] Chlorine and iodine were not detected, meaning that no halogenated fluorescein products, such as the Phloxines, Erythrosines and the Rose Bengals, were found.

In the Eosin paint of *Pietà (after Delacroix)* (F0630), painted in September 1889 in Saint-Rémy, and *Wheat Field under Thunderclouds* (0F778), painted in July 1890 in Auvers-sur-Oise, some particles contain the element iodine, detected using SEM-EDX, suggesting that Eosin was mixed with some Erythrosine, although HPLC-PDA analysis was not able to confirm this.

Van Gogh was aware that the new colours were unstable and could fade. On the 11th of April 1888, he wrote to his brother from Arles (letter 595). He considered trying to overcome the problem of fading by applying excessively pronounced colours, but it is clear that this technique was not successful [48]. Today, the results of this phenomenon can be seen on several of his paintings.

## *The Banners of the Nativity of Saint John the Baptist in Hoogwoud*

One of the banners of the Nativity of Saint John the Baptist in Hoogwoud depicts the Immaculate Conception of Mary, surrounded by a rose cane, lily branches and passion flowers. The other banner shows John the Baptist surrounded by stems with leaves and ears of corn executed in raised silver appliqués. Colour bleeding in the banners may have occurred when the banners were temporarily stored elsewhere due to a move around 2005. In 2001, the colours had not yet run. The colour

[8] All paintings listed are in the collection of the Van Gogh Museum, Amsterdam (Vincent van Gogh Foundation), unless otherwise stated.

bleeding in the textiles occurs in the red lining and in various red embroidery threads. The red dye from the chenille contour thread in the Saint John's banner has run out over the surrounding surface. In a sample from this banner, Rhodamine B and a Ponceau red were found. The main component of the chenille yarn in the Saint John's banner is Rhodamine B, a basic dye with a poor wash fastness. Based on research, the two banners can be dated. The dye analysis of Rhodamine B provided a terminus post quem for the banner: this basic violet dye was discovered in 1887. These results support the dating of around 1900 as previously done based on form, style, material and technique [49].

## *Other Case Studies*

The Van Gogh Museum is the owner of a Chinese lacquer box filled with coloured woollen balls from Vincent van Gogh. Several dyes were identified, one of them being Eosin. This case study is described in Chap. 10. Rhodamine B was detected in the so-called Hofdamask from the Hofburg and Schloß Schönbrunn in Vienna, described in Chap. 15.

## References

1. Reinhardt, C., Travis, A.S.: Heinrich Caro and the Creation of Modern Chemical Industry. Kluwer Academic Publishers, Dordrecht/Boston/London (2000)
2. Baeyer, A.: Ueber eine neue Klasse von Farbstoffen. Berichte der deutschen chemischen Gesellschaft zu Berlin. Ferd. Dümmler's Verlagsbuchhandlung, Berlin. 4, 558 (1871)
3. Baeyer, A.: Ueber die Phenolfarbstoffe. Berichte der deutschen chemischen Gesellschaft zu Berlin. Ferd. Dümmler's Verlagsbuchhandlung, Berlin. 4, 662 (1871)
4. Baeyer, A.: Zur Geschichte des Eosins. Berichte der deutschen chemischen Gesellschaft zu Berlin. Ferd. Dümmler's Verlagsbuchhandlung, Berlin. 8, 146–148 (1875)
5. Baeyer, A.: Ueber die Verbindungen der Phtalsäure mit den Phenolen. Justus Liebig's Annalen der Chemie. C.F. Winter'sche Verlagshandlung, Leipzig und Heidelberg. 183, 38–42 (1876)
6. Lehne, A.: Tabellarische Uebersicht über die künstlichen organischen Farbstoffe und ihre Anwendung in Färberei und Zeugdruck. Verlag von Julius Springer, Berlin (1893)
7. Schultz, G., Julius, P.: Tabellarische Übersicht der künstlichen organischen Farbstoffe. R. Gaertner's Verlagsbuchhandlung, Berlin. 56–57 (1888)
8. Fischer, E.: Ueber Fluorescein und das Phthalein des Orcins. Berichte der deutschen chemischen Gesellschaft zu Berlin. Ferd. Dümmler's Verlagsbuchhandlung, Berlin. 7, 1211 (1874)
9. Hewitt, T.C.: Synthetic Colouring Matters, Dyestuffs derived from Pyridine, Quinoline, Acridine and Xanthene. Longmans, Green & Co., New York/Toronto/Bombay/Calcutta/Madras (1922)

10. Hofmann, A.W.: Ueber das Eosin. Berichte der deutschen chemischen Gesellschaft zu Berlin. Ferd. Dümmler's Verlagsbuchhandlung, Berlin. 8, 62–66 (1875)
11. Bindschedler, R., Busch, A.: Sur la fabrication de la résorcine, de l'éosine et d'autres dérivés de la résorcine. Le Moniteur scientifique. 3, 1169–1172 (1878)
12. Mühlhäuser, O.: Fabrikation der Eosinfarbstoffe. Dingler's Polytechnisches Journal, Verlag der J.G. COTTA'schen Buchhandlung, Stuttgart. 263, 100–106 (1887)
13. Société chimique des Usines du Rhône: G.P. 108.838 (1899)
14. Caro, H.: Referate, "Ueber die Entwicklung der Theerfarben-Industrie." Berichte der deutschen chemischen Gesellschaft zu Berlin, Commissionsverlag von R. Friedländer & Sohn, Berlin. 25, 955–1105 (1892)
15. Bayer Farben Revue, Centenary of Farbenfabriken Bayer AG, 'Notes on the History of Farbenfabriken Bayer AG and on Bayer Dyes for Textiles', Verkauf Farben, Farbenfabriken Bayer AG, Leverkusen, No. 5, 1963, 7
16. Le Royer, A.: Ueber ß-Dichlorphtalsäure. Justus Liebig's Annalen der Chemie. C.F. Winter'sche Verlagshandlung, Leipzig und Heidelberg, 238, 350–361 (1887)
17. Mühlhäuser, O.: Zur Fabrikation des Jodeosins. Dingler's Polytechnisches Journal. Verlag der J.G. COTTA'schen Buchhandlung, Stuttgart. 283, 258 (1892)
18. Gesellschaft für chemische Industrie: U.S.P. 322.368 (1885); G.P. 32.564 (1885).
19. Juvalta, N.: G.P. 50.177 (1889)
20. Meister, Lucius & Brüning: G.P. 48.367 (1888); B.P. 9.600 (1888) and F.P. 192.589
21. Weingärtner, E.: Rhodamin. Chemiker Zeitung, Central-Organ für Chemiker, Techniker, Fabrikanten, Apotheker, Ingenieure, Verlag der Chemiker Zeitung, Cöthen. 11, 1620 (1887)
22. Badische Co.: B.P. 15.374 (1887), G.P. 44.002 (1887), G.P. 54.684 (1890), U.S.P. 377.349 (1888), U.S.P. 377.350 (1888), F.P. 186.697, F.P. 198.173
23. Monnet: F.P. 216.407 (1891), F.P. 225.341 (1891), U.S.P. 499.927 (1891), B.P. 4.677 (1892), B.P. 7.298 (1892); BASF: G.P. 66.238 (1891), G.P. 71.490 (1892) and G.P. 73.451 (1892)
24. Bayer: G.P. 51.983 (1888) and F.P. 194.908; Gnehm: U.S.P. 402.436 (1888) and U.S.P. 425.504 (1888); CIBA: B.P. 2.635 (1889), B.P. 10.047 (1890), B.P. 7.298 (1892), F.P. 195.930 and G.P. 54.997 (1890); BASF: G.P. 66.238 (1891) and G.P. 71.490 (1892)
25. Bernthsen, A.: Chemische Gesellschaft zu Heidelberg. Chemiker Zeitung, Central-Organ für Chemiker, Techniker, Fabrikanten, Apotheker, Ingenieure. Verlag der Chemiker Zeitung, Cöthen. 16, 1956–1957 (1892)
26. Badische Co.: B.P. 9.633 (1892), F.P. 225.341, U.S.P. 516.584 (1894), G.P. 73.573 (1892), G.P. 73.880 (1892)
27. BASF: B.P. 14.723 (1891), G.P. 63.325 (1891), F.P. 215.700 (1891); Cérésole: U.S.P. 516.588 (1891), U.S.P. 516.589 (1891).
28. Chem. Fabr. Bindschedler: B.P. 4.985 (1895), B.P. 12.180 (1897), U.S.P. 584.119 (1897), U.S.P. 695.441 (1897), F.P. 245.593 (1895), G.P. 85.931 (1894), G.P. 87.068 (1894), G.P. 96.108 (1896), G.P. 132.066 (1901).
29. Boedeker, ML&B: Violamine B (Fast Acid Violet B), G.P. 46.807 (1888), G.P. 49.057 (1888), B.P. 9.600 (1888), F.P. 195.917 (1888), Violamine R (Fast Acid Violet A2R), G.P. 46.807 (1888), G.P. 49.057 (1888), B.P. 9.600 (1888); Violamine 3B (Fast Acid Blue R), G.P. 53.300 (1888), F.P. 201.660; Violamine G (Acid Rosamine A), G.P. 63.844 (1888), F.P. 201.660
30. Trillich, H.: Das deutsche Farbenbuch, II Teil, Die Künstler-Farb- und Malmittel. Verlag B. Heller, München (1925)
31. Sessa, C., Steuer, C., Quintero Balbas, D., Sciutto, G., Prati, S., Stege, H.: Analytical studies on commercial artists' colour charts from Das Deutsche Farbenbuch (1925)—identification of synthetic and natural organic colourants by Raman microscopy, surface-enhanced Raman spectroscopy and metal underlayer ATR-FTIR spectroscopy. Herit Sci. 10, (2022). https://doi.org/10.1186/s40494-022-00740-3
32. Doerner, M.: Malmaterial und seine Behandlung im Bilde. Weizinger & Co., München (1922)

33. Russell, W.J., de Abney, W.: Report to the Science and Art Department of the Committee of Council on Education on the Action of Light on Water Colours, London, Her Majesty's Stationery Office. (1888)
34. Vibert, J.G.: La science de la peinture. P. Ollendorf, Paris (1891)
35. Fischer, M.: The Permanent Palette. National Publishing Society, New York (1930)
36. van Bommel, M.R., Geldof, M., Hendriks, E.: An investigation of organic red pigments used in paintings by Vincent van Gogh (November 1885 to February 1888). Art Matters 3, 111–137
37. Letter 889 to Theo van Gogh, Auvers-sur-Oise, 17th of June 1890
38. Van Gogh, V.: Letter 863, 29th April 1890, 2 medium tubes
39. Van Gogh, V.: Letter 654, 3rd of August 1888, 4 small tubes
40. Van Gogh, V.: Letter 877, 3rd of June 1890, 2 medium tubes
41. Van Gogh, V.: Letter 710, 22nd of October 1888, 10 medium tubes
42. Van Gogh, V.: Letter 593, 5th of April 1888, 6 small tubes of laque géranium
43. Van Gogh, V.: Letter 758, between 14th and 17th of April 1889, 2 medium tubes
44. Van Gogh, V.: Letter 674 and 687, 4th and 25th of September 1888, 6 small tubes respectively 6 medium tubes
45. Cadorin, P.: Colour Fading in Van Gogh and Gauguin. In: Peres, C., Hoyle, M., and van Tilborgh, L. (eds.) A Closer Look, Technical and Art-Historical Studies on Works by Van Gogh and Gauguin. Waanders Publishers, Zwolle, Cahier Vincent 3. pp. 12–19 (1991)
46. Rioux, J.P.: Caractérisation de pigments décolorés dans des tableaux de Van Gogh peints à Auvers-sur-Oise. In: Bridgland, J. (ed.) ICOM Committee for Conservation. pp. 403–408. , London (1999)
47. Geldof, M., de Keijzer, M., van Bommel, M.R., Pilz, K., Salvant, J., van Keulen, H., L., M.: Van Gogh's Geranium Lake. In: Vellekoop, M., Geldof, M., Hendriks, E., Jansen, L., and de Tagle, A. (eds.) Van Gogh's Studio Practice. pp. 268–289. Mercatorfonds, Brussels (2013)
48. Van Gogh, V.: Letter 595 to Theo van Gogh, Arles, 11th of April 1888
49. de Bruijne, M.: BLOEDEND ROOD Uitlopende rode kleurstof in textiel rond 1900 (Bleeding Red. Red Dye Migration in Textiles around 1900), Master thesis, University of Amsterdam, (2015)

# Chapter 17
# The Other Yellow Synthetic Organic Dyes

## 17.1 Introduction

This chapter provides a concise overview of various synthetic organic yellow dyes which are not discussed elsewhere. The synthetic dyes belonging to the yellow colour spectrum can be categorised into three distinct organic groups. The first group encompasses the nitro dyes, namely Picric acid, Martius yellow and Naphthol yellow S. These dyes are described in Chap. 10: 'The nitro dyes'. The second group consists of azo dyes, including Aniline yellow. Aniline yellow was the first azo dye, hence it is described in Chap. 14: 'The early azo dyes: Aniline yellow, Bismarck brown, Chrysoidine, Orangés, Fast Red A'. Finally, the third group comprises miscellaneous dyes, such as Fluorescein, Chrysoline, Quinoline yellow and Auramine. Chap. 16, entitled 'Bright organic red colours: the history and analysis of Fluorescein, Eosins, Phloxines, Erythrosines, Rose Bengals, Rhodamines, and their derivatives', offers an exploration of Fluorescein and Chrysoline, highlighting their historical significance and analytical aspects in relation to vibrant organic red colours.

The present chapter centres on the remaining yellow dyes: Metanil yellow, Alizarine yellow GG, Brilliant yellow, Tartrazine yellow, Flavazine L, Auramine and derivatives and Quinoline yellow, providing an overview of their history, characteristics and properties.



M. R. van Bommel, M. de Keijzer, *Bright Colours from the Past*, Cultural Heritage Science, https://doi.org/10.1007/978-3-031-88664-5_17

## 17.2 The Yellow Azo Dyes

In the chapter 'The early azo dyes: Aniline yellow, Bismarck brown, Chrysoidine, Orangés, Fast red A', the history of azo dyes is presented, including the most important dyes developed before 1875. After 1875, several other azo dyes were discovered, and they soon became the most important dye class. In Table 17.1, an overview is presented of the most relevant yellow azo dyes, including their patents, which clearly show that these dyes were produced in different countries.

These five azo dyes will be further discussed, followed by three other important yellow dyes, Quinoline yellow, Auramine O and Auramine G (see Table 17.2).

**Table 17.1** Discoverers and patents of the yellow azo dyes

| Name | C.I. name, C.I. number | Year of discovery | Discoverers | Patents |
|---|---|---|---|---|
| Metanil yellow | Acid Yellow 36, 13065 | 1879<br>1882 | C. Rumpff<br>E. Hepp | B.P. 1.226 / 1879<br>B.P. 4.966 / 1880<br>Austrian Patent. v. 4 I. / 1882 |
| Tartrazine yellow | Acid Yellow 23, 19140 | 1884 | J.H. Ziegler<br>BASF<br>BASF<br>BASF<br>BASF<br>BASF<br>BASF<br>Bennert<br>Bennert | B.P. 9.858 / 1885<br>U.S.P. 324.630 / 1885<br>F.P. 169.964 / 1885<br>G.P. 34.294 / 1885<br>B.P. 5.693 / 1893<br>B.P. 765 / 1897<br>G.P ap. B200.84<br>B.P. 8.504 / 1887<br>G.P ap. B 8.43 |
| Brilliant yellow | Direct Yellow 4, 24890 | 1886 | F. Bender | B.P. 4.387 / 1886<br>U.S.P. 350.229 / 1886<br>U.S.P. 360.553 / 1887<br>F.P. 175.360 / 1886<br>G.P. 38.735 / 1886<br>G.P. 40.575 / 1886 |
| Alizarine yellow GG | Mordant Yellow 1, 14025 | 1887 | R. Nietzki | B.P. 17.583 / 1887<br>U.S.P. 424.019 / 1890<br>F.P. 187.821 / 1887<br>G.P. 44.170 / 1887<br>B.P. 13.920 / 1888<br>F.P. 193.190 |
| Flavazine L | Acid Yellow 11, 18820 | 1892 | C. Möllenhoff | - |

**Table 17.2** The other yellow dyes Quinoline yellow and the Auramines

| Name | C.I. name, C.I. number | Year of discovery | Discoverers | Patents |
|---|---|---|---|---|
| Quinoline yellow | Acid Yellow 3, 47005 | 1882 | E. Jacobsen | G.P. 23.188/1882 |
| | | | E. Jacobsen | G.P. 25.144/1883 |
| | | | – | B.P. 1.362/1883 |
| | | | E. Jacobsen | U.S.P. 290.585/1883 |
| | | | – | F.P. 154.512 |
| Auramine O | Basic Yellow 2, 41000 | 1883 | H. Caro and A. Kern (BASF) | G.P. 29.060/1884 |
| | | | Ewer & Pick | G.P. 31.936/1884 |
| | | | Farbwerke ML&B | G.P. 41.751/1887 |
| | | | Farbwerke ML&B | G.P. 44.077/1887 |
| | | | A. Feer | G.P. 53.614/1889 |
| | | | BASF | G.P. 58.277/1890 |
| | | | BASF | G.P. 70.908/1893 |
| | | | BASF | G.P. 71.320/1893 |
| | | | Bayer | G.P. 77.329/1892 |
| | | | – | B.P. 5.512/1884 |
| | | | H. Caro and A. Kern (BASF) | U.S.P. 301.802/1884 |
| | | | – | B.P. 5.741/1884 |
| | | | – | B.P. 9.632/1884 |
| | | | – | B.P. 12.022/1886 |
| | | | T. Sandmeyer | B.P. 12.549/1889 |
| | | | T. Sandmeyer | B.P. 16.666/1890 |
| | | | – | B.P. 6.249/1893 |
| | | | – | B.P. 274.187 |
| | | | – | F.P. 160.990/1884 |
| | | | – | F.P. 164.099 |
| | | | – | F.P. 175.376 |
| | | | – | F.P. 181.351 |
| | | | – | F.P. 200.613 |
| Auramine G | Basic Yellow 3, 41005 | 1892 | BASF | G.P 67.478/1892 |
| | | | – | B.P. 10.465/1892 |
| | | | R. Gnehm and | U.S.P. 488.430/1892 |
| | | | J. Schmid (CIBA) | F.P 222.275 |

## 17.3 Metanil Yellow

Colour Index name: Acid Yellow 36; Colour Index number: 13065
IUPAC name: Sodium;3-[(4-anilinophenyl)diazenyl]benzenesulfonate
Discoverers: Carl Heinrich Christian Ludwig Rumpff, 1879
Eduard Hepp, 1882
Contemporary terminology: Metanil yellow (En), Jaune Métanile (Fr), Metanilgelb (Ge), Metanilgeel (NL)
Chemical formula: $C_{18}H_{15}N_3O_3S$ (Fig. 17.1)

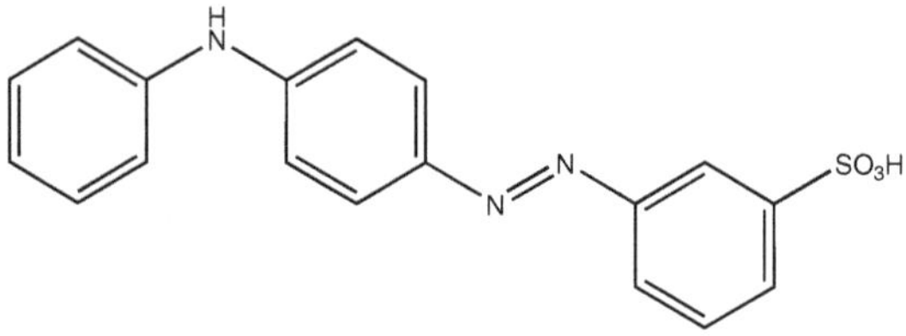


**Fig. 17.1** Structural formula of Metanil yellow

### *History*

Metanil yellow is an orange-yellow mono-azo dye which was independently discovered by the German chemist Carl Heinrich Christian Ludwig Rumpff (1839–1889), the first chairman of the dye firm Bayer, in 1879 (B.P. 1.276, filed in 1879) and by the German chemist Eduard Hepp (1851–1917) in 1882, when he was working for the K. Oehler Anilin- und Anilinfarbenfabrik in Offenbach am Main, Germany (B.P. 4.966, filed in 1880). It is prepared by the reaction of diazotised metanilic acid and diphenylamine. Several industrial processes for the manufacture of Metanil yellow are described by Ludwig Paul [1]. The splitting of azo dyes is known to be an excellent means of determining their constitution and has been used for Metanil yellow by the German chemist Richard Meyer (1846–1926) [2].

## Trade Names

The dye Metanil yellow was put on the market under the commonly used trade name Metanilgelb extra by AGFA, BASF, Cassella, Beyer & Kegel, Durand-Huguenin, Geigy, Kalle, Griesheim-Elektron, Farbwerk Amersfoort, Sévoz-Boasson, Sandoz, and Wülfing-Dahl. Other trade names were used, such as: Metanilgelb GR extra konz. by Weiler-ter Meer; Metanilgelb O by Leonhardt; Metanilgelb PL by BASF; Victoriagelb O doppelt konz. by Farbwerke ML&B; Orange MN, MNO by CIBA; Jaune M by Soc. Anon. St. Denis; Tropäolin G by Cassella; Jaune Métanile extra by Manufacture Lyonnaise; Jaune Métanile 230 by Wiescher, and Metanil Yellow by Schoellkopf.

## Applications

Metanil yellow is used to dye leather, wool, silk, nylon, jute and paper.

## Properties

The light fastness of Metanil yellow is good and it is also very fast to washing, according to Lehne:[1] 'After 40 days of light exposure, in July and August, the colour only slightly changed (darker)'. Unfortunately, the orange-yellow dye is acid-sensitive. The Colour Index indicates that a normal dyed textile shows a light fastness of ISO 3, the wash fastness is ISO 2.

## Recipes

Lehne reports that wool can be dyed following recipe V3, without specifying if this should be V3A or V3B. It is likely that both could provide a good colour. A concentration of 2% dye is mentioned. In addition, Lehne states that the addition of sulfuric acid should be limited as far as possible.

[1] Metanilgelb: Echtheit: ´Nach 40 tägiger Einwirkung der Witterung (Juli und August) nur wenig verändert (dunkler)´.

### *Identification Using UHPLC-PDA*

Analysis of a reference sample of Metanil yellow (Sigma-Aldrich CAS 587-98-4) using UHPLC-PDA results in a clear, single peak as presented in Fig. 17.2.

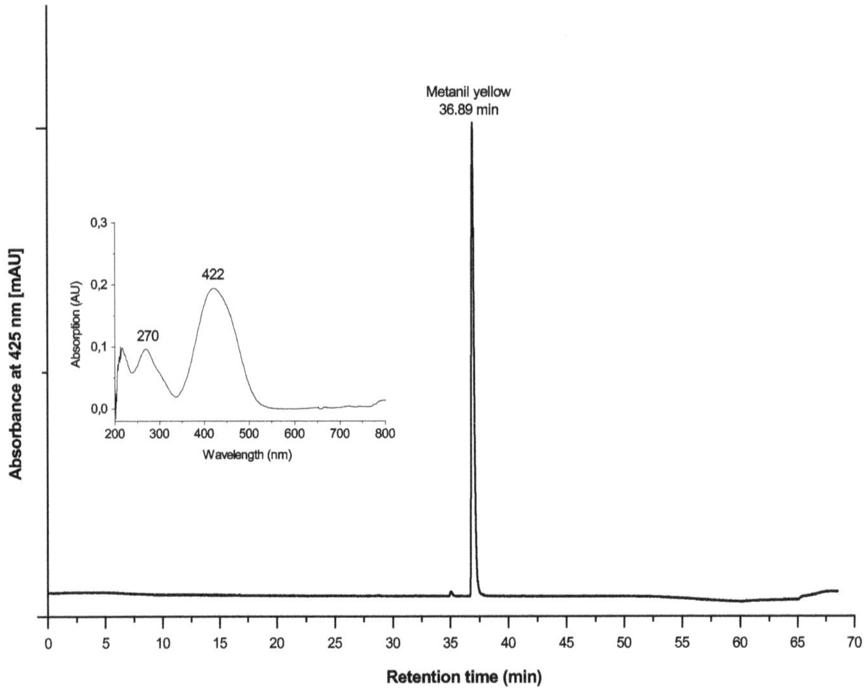


**Fig. 17.2** Chromatogram and spectrum of Metanil yellow (Sigma-Aldrich CAS 587-98-4)

## 17.4 Tartrazine Yellow

Colour Index name: Acid Yellow 23; Colour Index number: 19140
IUPAC name: Trisodium 5-hydroxy-1-(4-sulfonatophenyl)-4-[(E)-(4-sulfonatophenyl)diazenyl]-1H-pyrazole-3-carboxylate
Discoverer: Johann Heinrich Ziegler, 1884
Contemporary terminology: Tartrazine yellow (En), Jaune tartrazine (Fr), Tartrazin (Ge), Tartrazinegeel (NL)
Chemical formula: $C_{16}H_{12}N_4O_9S_2$ (Fig. 17.3)

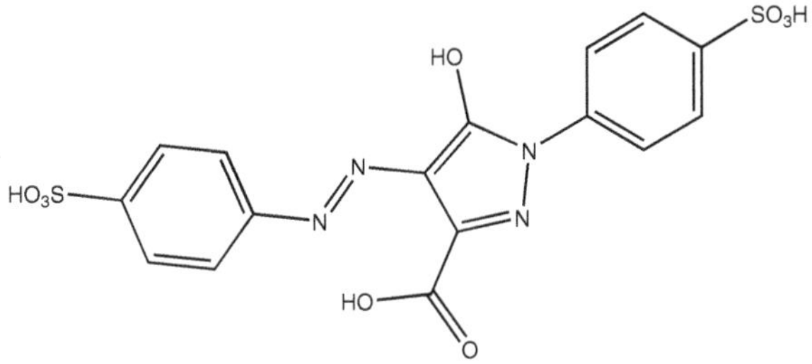


**Fig. 17.3** Chemical structure of Tartrazine yellow

## *History*

Tartrazine yellow and Flavazin L, which will be discussed later on, have nearly the same interesting coupling component, named pyrazolone, and these are the only important hydrazone dyes from a commercial point of view. In 1884, the Swiss chemist Johann Heinrich Ziegler (1857–1936) developed Tartrazine yellow in the laboratories of the Bindschedler & Busch factory in Basel. The yellow azo dye was patented and produced by BASF in Germany one year later, in June 1885 (G.P. 34.294, filed on the 18th of June 1885), first published in 1887 [3]. Tartrazine yellow was originally synthesised by reacting phenylhydrazine-4-sulfonic acid with tetrahydroxysuccinic acid under alkaline conditions. For the compound, Ziegler formulated the disodium salt of an osazone, a group of carbohydrates which are formed by the reduction of sugars using phenylhydrazine [4, 5].

Although the structure proposed was not confirmed, Ziegler was able to develop an alternative manufacturing process in 1893, based on the consideration that a hydrazone is the tautomeric form of an azo compound (described as azo-hydrazo tautomerism). He succeeded in synthesising Tartrazine yellow by azo coupling of diazotised sulfanilic acid and the diazonium salt is coupled to the pyrazolone compound. This is obtained by the condensation of phenylhydrazine-4-sulfonic acid with oxaloacetic acid ethyl ester in the presence of sodium acetate. The ester is then saponified with a sodium hydroxide solution to form Tartrazine yellow. This manufacturing process was patented in 1893 (B.P. 5.693, filed in 1893) [6].

In 1897, the German chemist Carl Johann Philipp Noé Richard Anschütz (1852–1937) was able to show that the structure published by Ziegler was not correct. He elucidated that the structure of tartrazine is a pyrazolone derivative [7, 8].

## *Trade Names*

Tartrazin was put on the market by BASF, Bayer, CIBA, Read Holliday, and Sandoz. Different trade names were used as well, such as: Tartrazin O by Farbwerke ML&B; Hydrazingelb O, L by Griesheim-Elektron; Echtwollgelb G by Kalle; Säuregelb AT by Cassella; Tartrabarin by Weiler-ter Meer; Buffalo Yellow by Schoellkopf, and Jaune tartrique by Soc. Anon. St. Denis.

## *Applications*

Tartrazine yellow was first used as a dye for wool and silk; later it was also used to dye nylon and as a food dye. It is still used as a food dye known as E102 (C.I. name: Food Yellow 4). The dye is also used for printing textiles. A pigment of Tartrazine yellow can be prepared by precipitation of the dye on an alumina hydrate base (C.I. name: Pigment Yellow 100).

## *Properties*

The light fastness of Tartrazine yellow is really good and it is also fast to washing according to Lehne who reported:[2] 'After 30 days exposed to the elements, in May and June, the colour did not change'. The Colour Index reports a light fastness of ISO 4 and the fastness to washing is ISO 2.

## *Recipes*

Tartrazine yellow can be used to dye wool (recipe V3) and silk (recipe V8). For wool a concentration of 1% should provide a good colour while for silk 2% is used. Both textiles can be also printed with this dye; in that case recipe V21 is used.

## *Identification Using UHPLC-PDA*

Analysis of a reference sample of Tartrazine yellow (RCE no. 5316) using UHPLC-PDA results in a single peak, but shows little retention in the system used (see Fig. 17.4). However, its spectrum has a typical profile which can be easily distinguished from other early eluting components.

[2] Tartrazin: Echtheit: ´Nach 30 tägiger Einwirkung der Witterung (Mai und Juni) nicht verändert. Sehr waschecht, halt auch leichtere Walke aus´.

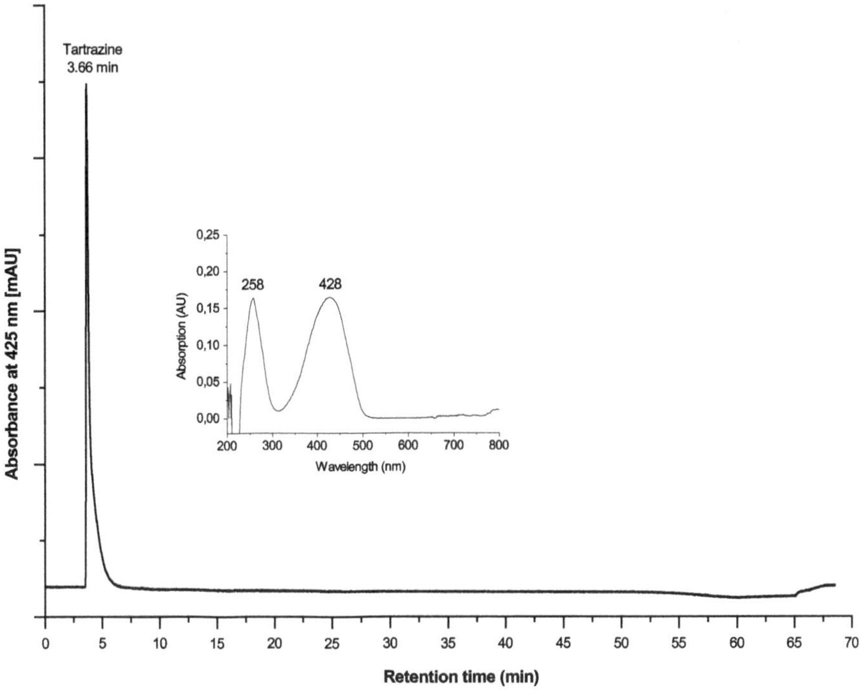


**Fig. 17.4** Chromatogram and spectrum of Tartrazine yellow (RCE no. 5316)

## 17.5 Case Study: Cabinet Designed By Piet Kramer

In the framework of our research into furniture designed by the Dutch architect and designer Pieter Lodewijk Kramer (1881–1961), we investigated the colours of a buffet, made between 1933 and 1936, shown in Fig. 17.5. The colour differences between the areas exposed to light and protected from light are striking, as can be seen in Fig. 17.6. Chemical analysis using UHPLC-PDA showed that the dye was composed of 53.7% Tartrazine yellow, 38.5% Orange G, 5.3% Cochineal red A, and 2.5% of a Ponceau-type dye. The chemical ratios were determined by calculating the peak areas of the compounds at 450 nm, between the optima of Tartrazine yellow and the other dyes. The calculation of an absolute amount of dye requires thorough calibration and was beyond the scope of the investigation. The presence of the now faded dyes in this twentieth-century object showed that these were applied long after their invention [9, 10].

Tartrazine yellow has been also identified in the embroidery *Plants and Flowers on an Orange Background* designed by Émile Bernard in 1927. This object is presented in more detail in Chap. 13.

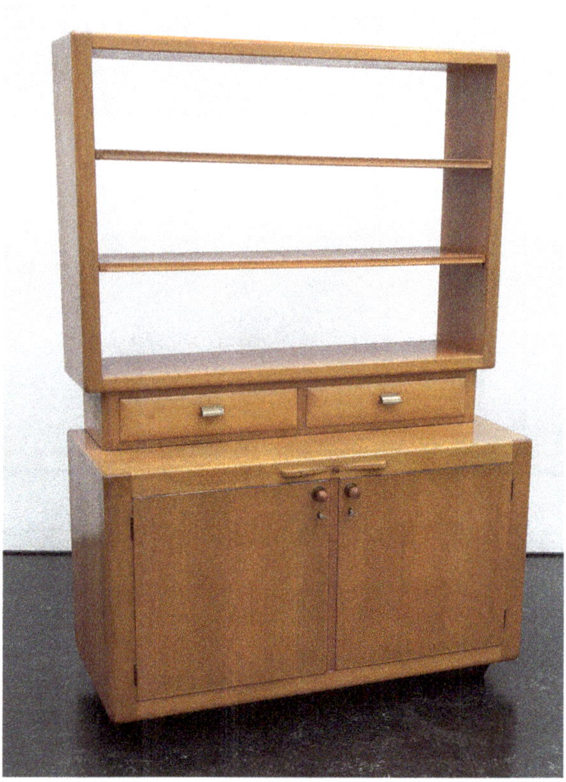

**Fig. 17.5** Piet Kramer, Buffet, 1933–1936, oak solid & veneer on plywood, solid spruce, 178 × 123 × 57 cm. Collection Cultural Heritage Agency of the Netherlands, inventory number AB 25457-B. Photograph: Cultural Heritage Agency of the Netherlands

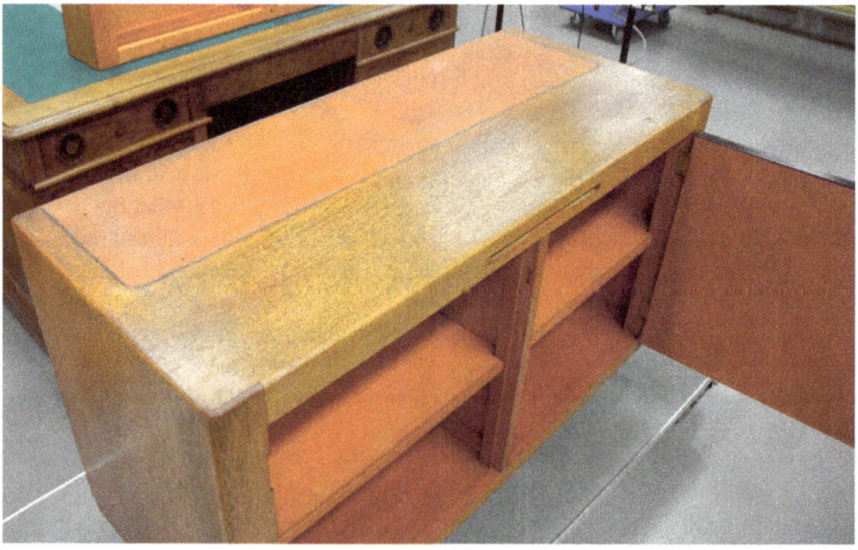

**Fig. 17.6** Detail of the buffet shown in Fig. 17.5. Photograph: Jasmijn Groeneveld

## 17.6 Brilliant Yellow

Colour Index name: Direct Yellow 4; Colour Index number: 24890
IUPAC name: Disodium;5-[(4-hydroxyphenyl)diazenyl]-2-[(E)-2-[4-[(4-hydroxyphenyl)diazenyl]-2-sulfonatophenyl]ethenyl]benzenesulfonate
Discoverer: Fritz Bender, 1886
Contemporary terminology: Brilliant yellow (En), Jaune brilliant (Fr), Brillantgelb (Ge), Briljantgeel (NL)
Chemical formula: $C_{26}H_{20}N_4O_8S_2$ (Fig. 17.7)

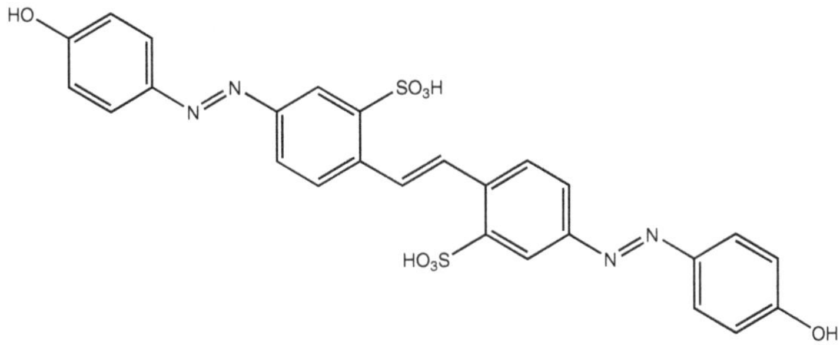


**Fig. 17.7** Chemical structure of Brilliant yellow

### *History*

In 1886, the German chemist Fritz Bender (1859–1908) working for the German dye factory Farbwerk Mühlheim vorm. A. Leonhardt & Co near Frankfurt am Main prepared and patented a yellow dye by first treating stilbene with sulfuric acid. He obtained diamino-stilbene-disulfonic acid. A bis-diazonium salt (containing two azo groups) is yielded on treatment with nitrous acid. By coupling two molecules of phenol to the bis-diazonium salt, Brilliant yellow is formed (G.P. 38.735, filed on the 29th of January 1886; B.P. 4.387, filed on the 29th of March 1886; F.P. 175.360, filed on the 19th of April 1886; U.S P. 350.229, filed on the fifth of October 1886; G.P. 40.575, filed on the 14th of December 1886 (additional patent for G.P. 38.735) and U.S.P. 360.553, filed on the fifth of April 1887).

### *Trade Names*

Brillantgelb was put on the market by AGFA, Bayer, Griesheim-Elektron, Leonhardt, Sandoz, Jäger, and CIBA. Different trade names were used as well, such as: Papiergelb 3G by BASF, and Renolbrillantgelb conc. by Weiler-ter Meer.

### *Applications*

Brilliant yellow is used to dye silk and wool, but can also dye cellulose and nylon according to the Colour Index.

### *Properties*

Lehne made the following observations on Brilliant yellow:[3] 'After 5 days weathering (in June), no changes, after 30 days (May and June) somewhat brownish and duller'. In the Colour Index, only the AATCC fastness properties are present, being 5 for the light fastness and 3 for the wash fastness.

### *Recipes*

Recipe V12 is given to dye cotton, using a concentration of 2% dyestuff. Since it is a direct dye, recipe V2 can probably be used to dye wool. For silk, no recipes are present for direct dyeing. However, Lehne reported the dyeing of 'Halbseide' (half silk), a fabric composed of silk and cotton. Using 15–20 g sodium sulfate in a litre of water at 60–80 °C dyeing takes between 30 and 45 min; acetic acid can be used to intensify the colour. Afterwards the textile should be rinsed with water acidified using acetic acid.[4]

### *Identification Using UHPLC-PDA*

Analysis of a reference sample of Brilliant yellow (RCE no. 5706) using UHPLC-PDA results in a single peak and a typical UV-Vis absorption spectrum (see Fig. 17.8).

[3] Brillantgelb: Echtheit: ´Nach 5 tägiger Einwirkung der Witterung (Juni) unverändert, nach 30 Tagen (Mai und Juni) etwas bräunlicher und matter´.

[4] Halbseide wird unter Zusatz von 15–20 g Glaubersalz auf das Liter Wasser bei 60–80 ° C. in ½–¾ Stunden ausgefärbt und stark avivirt mit Essigsäure. Spülen in Wasser, welches mit Essigsäure versetzt ist.

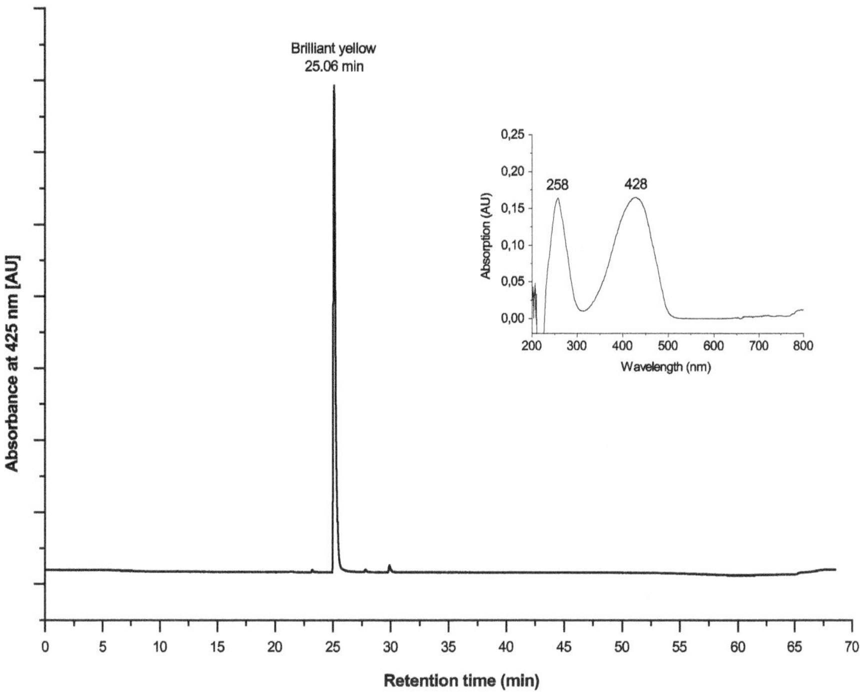


**Fig. 17.8** Chromatogram and spectrum of Brilliant yellow (RCE no. 5706)

## 17.7 Alizarin Yellow GG

Colour Index name: Mordant Yellow 1; Colour Index number: 14025
IUPAC name: Methyl 2-methoxy-5-[(3-nitrophenyl)diazenyl]benzoate
Discoverer: Rudolf Hugo Nietzki, 1887
Contemporary terminology: Alizarin yellow (En), Jaune d'alizarine (Fr), Alizaringelb (Ge), Alizarinegeel (NL)
Chemical formula: $C_{13}H_9N_3O_5$ (Fig. 17.9)

**Fig. 17.9** Chemical structure of Alizarin yellow GG

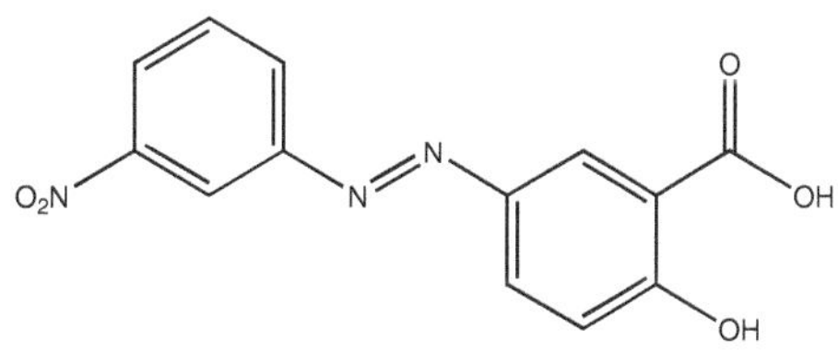

## History

The German chemist Rudolf Hugo Nietzki (1847–1917), who worked at Farbwerke ML&B, discovered the mordant dye Alizarin yellow GG in 1887 by diazotizing nitroaniline with salicylic acid (G.P. 44.170, filed on the 16th of November 1887). The invention relates to a new and useful improvement in the production of azo dyes using metal mordants, such as those based on chromium (U.S.P. 424.019, filed on the ninth of October 1888 and issued on the 25th of March 1890). By adjusting the coordination metal used as mordant, shades between yellow and brown are obtained. Other international patents on Alizarin yellow GG are F.P. 187.821, filed on the 12th of December 1887; B.P. 17.583, filed on the 21th of December 1887; B.P. 13.920, filed in 1888 and F.P. 193.190.

## Trade Names

Alizaringelb GG was put on the market by Farbwerke ML&B. Other trade names used are: Alizaringelb GGW by Farbwerke ML&B; Alizaringelb G by Levinstein; Alizaringelb 3G by Levinstein and Bayer; Beizengelb 2GT by BASF; Anthracengelb GG by Cassella; Jaune chrome R by Soc. Anon. St. Denis, and Alizaringelb G by Sandoz.

## Applications

The dye is used in particular as a substitute for Persian berries (unripe berries of *Rhamnus saxatilis* Jacq., a species of buckthorn) for cotton, but can also be used to dye wool and silk, using a mordant.

## Properties

The light fastness of Alizarin yellow GG is good. Adolf Lehne comments:[5] 'The colour on a chrome mordant was very little changed after 30 days, somewhat duller in appearance; the colour on an alum mordant had become much lighter and duller at the same time'. The Colour Index confirms the light fastness of Alizarin yellow GG on a chromium mordant: for both wool and silk ISO 4 to 5 is reported. The wash fastness is very good due to the mordant: for wool ISO 4 is given and for silk ISO 4 to 5. Unfortunately, no fastness properties are given for other mordants, such as aluminium.

[5] Alizaringelb GG: Echtheit: 'Die Färbung auf Chrombeize war nach 30 tägiger Einwirkung der Witterung (Mai und Juni) sehr wenig verändert, etwas matter in Schein; die Färbung auf Alaunbeize war in derselben Zeit ziemlich viel heller und matter geworden'.

## *Recipes*

Since this is a mordant dye, recipe V4 is used. Although an alum mordant recipe is not specified in Lehne's book, its use is mentioned for one of the textile samples present. Next to alum, chromium is also used for mordanting; it is assumed that mordant recipe V4 is used for both. Only 0.4% of dye was used. Lehne reports that the dye can also be used for dyeing cotton, using recipe V13A, and for printing cotton.

## *Identification Using UHPLC-PDA*

Analysis of a reference sample of Alizarin yellow GG (RCE no. 3147) using UHPLC-PDA results in a main peak with a typical UV-Vis absorption spectrum and a minor component which is not present in the blank analysis (see Fig. 17.10). This is likely to be a side-product from the synthesis process.

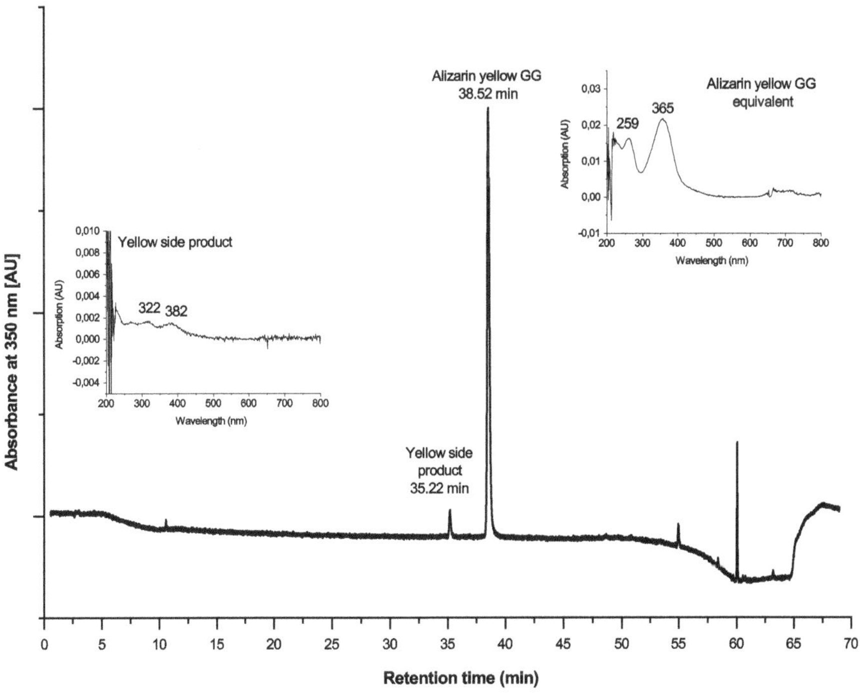


**Fig. 17.10** Chromatogram and spectra of Alizarin yellow GG (RCE no. 3147)

## 17.8 Flavazine L

Colour Index name: Acid Yellow 11; Colour Index number: 18820
IUPAC name: Sodium 4-{5-hydroxy-3-methyl-4-[(E)-phenyldiazenyl]-1H-pyrazol-1-yl}benzenesulfonate
Discoverer: Carl Möllenhoff, 1892
Contemporary terminology: Flavazine L (En), Flavazine L (Fr), Flavazin L (Ge), Flavazine L (NL)
Chemical formula: $C_{16}H_{14}N_4O_4S$ (Fig. 17.11)

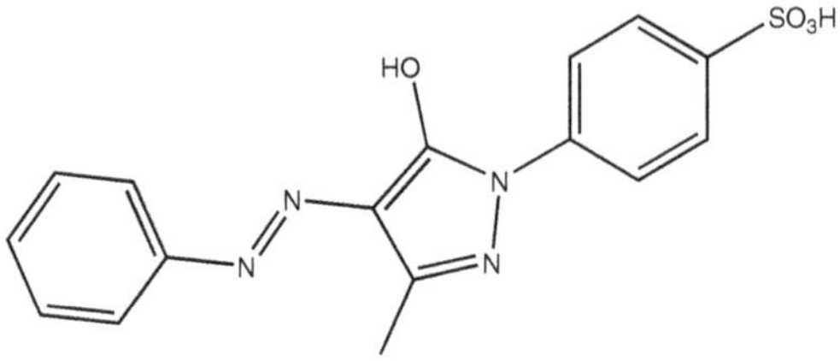


**Fig. 17.11** Chemical structure of Flavazine L

### *History*

Flavazine L was discovered by the German chemist Carl Möllenhoff, working for Farbwerke ML&B, in 1892 [11]. The yellow dye was prepared from 3-methyl-1-(*p*-sulfophenyl)-5-pyrazolone and aniline. It was also produced by Bayer under the tradename Echtlichtgelb G, 2G, 3G [12].

### *Trade Names*

Flavazine L was put on the market by Farbwerke ML&B. Another trade name, Echtlichtgelb G, 2G, 3G, was used by Bayer, and BASF. Other trade names are Echtwollgelb Gl, 2Gl, 3GL by Kalle, Echtgelb LG, LR by Jäger, Echtsäuregelb 3G by Cassella; Echtsäuregelb G, 3G by Österreichischer Verein für Chemische und Metallurgische Produktion, and Kitongelb S, JR by CIBA. The international trade name is Fast light yellow

## Applications

Flavazine L can be used to dye wool. However, it was not mentioned in the books by Lehne, hence it was probably not often used. As result, no recipes are known, although probably the same dyeing procedures as for Tartrazine yellow can be used. There are no properties presented in the Colour Index, which confirms the likelihood that this dye was little used, or not used at all.

## Identification Using UHPLC-PDA

Analysis of a reference sample of Flavazine L (RCE no. 5349) using UHPLC-PDA results in a single peak and a typical UV-Vis absorption spectrum (see Fig. 17.12).

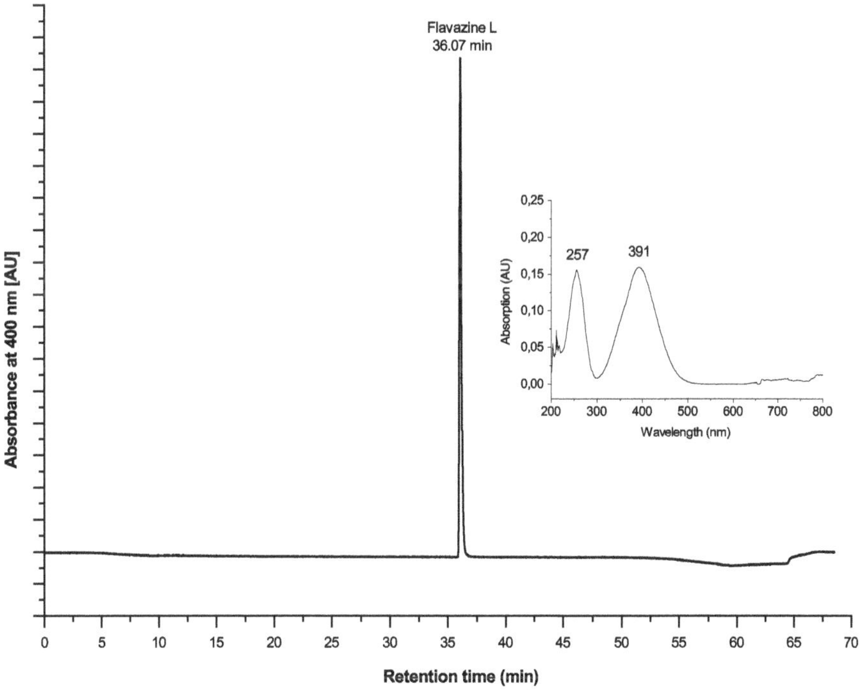


**Fig. 17.12** Chromatogram and spectrum of Flavazine L (RCE no. 5349)

## 17.9 Quinoline Yellow

Colour Index name: Acid Yellow 3; Colour Index number: 47005
IUPAC name: 2-Quinolin-2-ylindene-1,3-dione
Discoverer: Emil Jacobsen, 1882
Contemporary terminology: Quinoline yellow (En), Jaune de Quinoléine (Fr), Chinolingelb (Ge), Chinolinegeel (NL)
Chemical formula: $C_{18}H_{11}NO_2$ (Fig. 17.13)

**Fig. 17.13** Chemical structure of Quinoline yellow

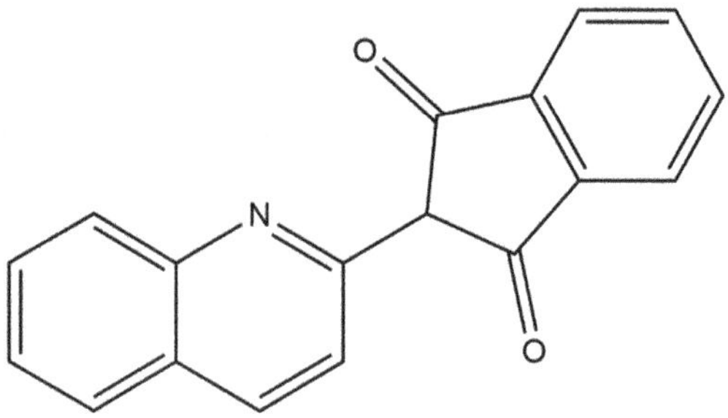


### *History*

The chromophore of this class is the conjugated system represented by quinophthalone. This forms when 2- and 4-methylquinolines are condensed with phthalic anhydride at 200 °C in the presence of zinc chloride. The German chemist Emil Jacobsen (1836–1911) discovered and patented the dye Quinoline yellow (G.P. 23.188, filed on the 4th of November 1882) in 1882 [13–24]. The dye was also patented in other countries such as England (B.P. 1.362, filed on the 14th of March 1883), in France (F.P. 154.512, filed on the 27th of March 1883), in the United States of America (U.S.P. 290.585, filed on the 9th of April 1883) and in Germany (G. P. 25.144, filed in 1883).

### *Trade Names*

Chinolingelb (wasserlöslich, O, extra) was the trade name used by AGFA, BASF, Bayer, Ruch, Sandoz, Farbwerke ML&B, and Wiescher. Other manufacturers sold the dye using trade names like: Jaune de Quinoléone by Wiescher, and Chinaldingelb by CIBA.

## *Applications*

When Quinoline yellow is sulfonated, it produces valuable acid dyes for wool and silk; later it was also used for dyeing nylon. Unfortunately, Quinoline yellow has only found limited use in wool dyeing because of its high price.

## *Properties*

Adolf Lehne writes about the light fastness of Quinoline yellow:[6] 'After five days of light exposure, in June, the colour did not change, after 3 weeks lighter and paler'. According to the Colour Index, Quinoline yellow has a light fastness of ISO 2, so rather poor, while the wash fastness is ISO 1.

## *Recipes*

Lehne reports dye recipes for wool (V3A, using 1.5% dye), silk (V8, using 3% dye) and jute (V18, no mention of the concentration). It can be also used to print wool using recipe V21 with 30 g/l of dye.

## *Identification Using UHPLC-PDA*

Analysis of a reference sample of Quinoline yellow (RCE no. 6531) using UHPLC-PDA results in two peaks with comparable UV-Vis absorption spectra. These components are likely to be isomers, formed during the synthesis process (Fig. 17.14).

---

[6] Chinolingelb: Echtheit: ´Nach 5 tägiger Einwirkung der Witterung (Juni) unverändert, nach 3 Wochen heller und fahler´.

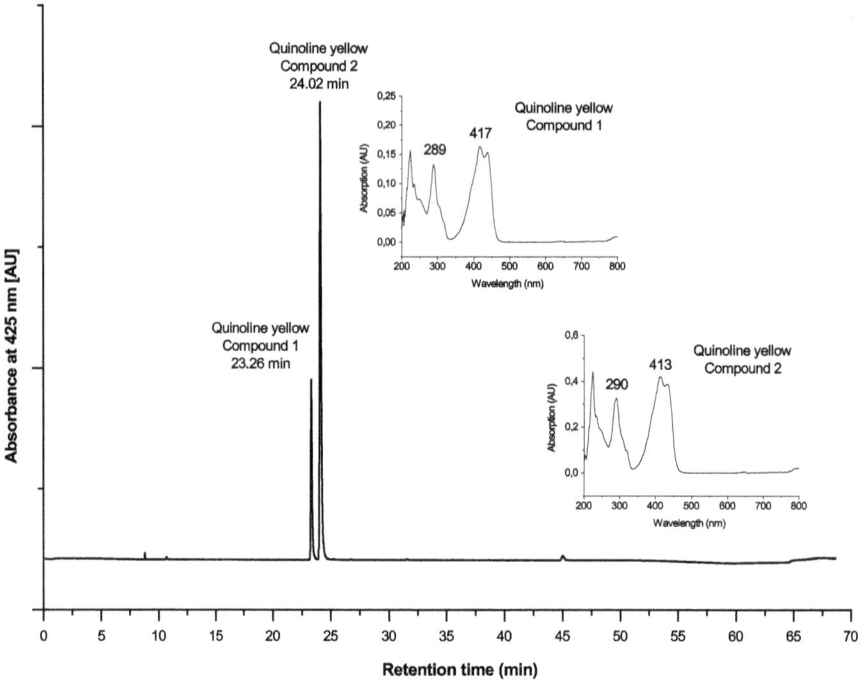


**Fig. 17.14** Chromatogram and spectra of Quinoline yellow (RCE no. 6531)

## 17.10 Case Studies

### *The Backhausen Archive at the Leopold Museum in Vienna*

The Backhausen firm started in 1810, when the weaver Franz Backhausen moved from Cologne to Vienna. His sons Karl and Johann founded the company 'Karl Backhausen & Co.' in 1849, renamed 'Johann Backhausen & Söhne in 1860'. In 1870 a factory was built and the company moved its production to Hoheneich in the Upper Waldviertel. From that time onwards, Backhausen took over the development and production of textiles for the interiors of numerous representative and official buildings in Vienna and across the world, such as the State Opera, the Parliament, the Burgtheater, the City Hall, the Musikverein, Imperial palaces, many theatres, concert halls, hotels and restaurants. In 1888, the family-run company was awarded the title 'K. & K. Hoflieferant': suppliers to the Imperial family. The firm was not only a key supplier to the Wiener Werkstätte (1903–1932), but also a member since its foundation in 1903. With more than 3500 original designs by over 300 contemporary artists, such as Josef Hoffmann (1870–1956), Koloman Moser (1868–1918) and Otto Wagner (1841–1918), the company holds the world's most comprehensive Backhausen archive of historical textile designs [25].

The Backhausen archive, bequeathed from Dr. Louise Kiesling, is on permanent loan at the Leopold Museum and is a listed heritage object since 2022. It includes

**Fig. 17.15** Fabric 2197 from the Backhausen pattern book 1890. © The Backhausen Archive, bequest from Dr. Louise Kiesling, on permanent loan at the Leopold Museum in Vienna. Photograph: Tanya Niedermüller

design books of wall and furniture decorations, from 1861 onwards. Studying the design books of the period 1861–1870 shows that yellow was a popular colour in which only the natural yellow dye Weld (*Reseda luteola* L.) was identified. The last quarter of the nineteenth century is characterised by dark colours, with hardly any yellow decorations; only one fabric from the pattern book 1890 had a yellow ground colour with a blue floral pattern (see Fig. 17.15). The yellow silk was dyed with Quinoline yellow and Azo flavine 3R, the blue silk with Methyl violet and a Patent blue equivalent. During the Jugendstil period (about 1895–1910), the colour yellow was rediscovered. In these decorations, four yellow dyes were found of which the origin remains unclear. They show a match with Azo flavine 3R and Flavazine, but the retention times are significantly different, so their identity is not confirmed.

### *Paulina Bisdom van Vliet's Dress*

A greenish evening dress belonged to Paulina Bisdom van Vliet (1840–1923), presented in Fig. 17.16.

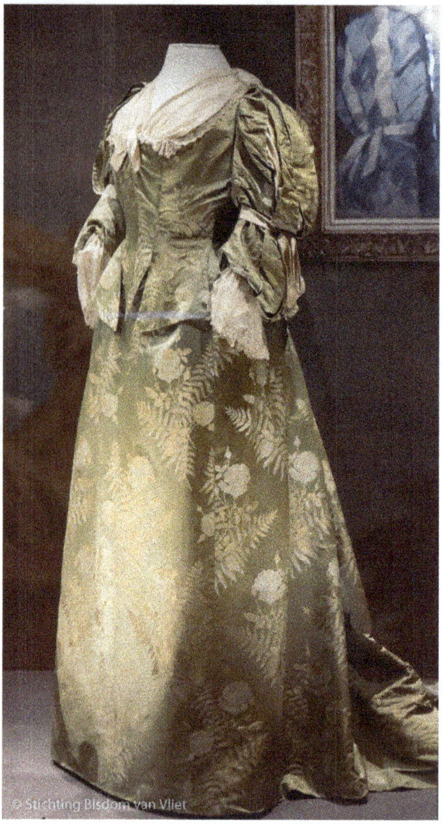


**Fig. 17.16** Gown, Silk, Maison Hirsch & Cie. (c. 1890). Inv. No. 9527, Collection Museum Paulina Bisdom van Vliet, donated by Paulina Maria le Fèvre de Montigny—Bisdom van Vliet

In 1892, the Dutch portrait painter Thérèse Schwartze (1851–1918) portrayed Paulina in a pastel. The weave of the garment is satin and *cannelé simpleté* at the selvedge. However, in the pastel she wears a blue dress. The Museum Paulina Bisdom van Vliet in Haastrecht, near the city of Gouda in the province of South Holland, was interested to understand whether the silk satin dress was blue in order to match the Thérèse Schwartze pastel of Paulina Bisdom van Vliet, or if the colour has changed over time.

Therefore, the colourants of the warp and weft were examined. The colourants found in the silk satin were a mixture of blue and yellow natural and synthetic dyes. The yarns of the weft are dyed with natural Indigo, the natural yellow dye Berberin and ellagic acid as a tannin. Ellagic acid was used for pre-mordanting the silk fibres or as a colour fixing agent. The yarns of the warp contain the blue semi-synthetic dye Indigo carmine, the synthetic organic dye Quinoline yellow and ellagic acid, which indicates the use of tannins.

The light fastness of the natural and synthetic dyes varies. Natural Indigo has a good light fastness. The light fastness of the natural yellow dye Berberin is moderate. In the warp, Indigo carmine was found, and this colourant has a poor light fastness. The synthetic dye Quinoline Yellow also has a poor light fastness. Under the microscope the yarns of the warp looked greenish and discoloured. So, the conclusion was that the dress has undergone a colour change. The sample analysed came from a part that was not exposed to direct daylight and so shows less discolouration. The fabric was intended to be green from the beginning: the dress was never blue and does not match the colour of the dress in the pastel of Paulina Bisdom van Vliet [26].

Quinoline yellow has been also identified in three embroideries designed by Émile Bernard. These embroideries are described in Chap. 13.

## 17.11 Auramine

Colour Index name: Basic Yellow 2; Colour Index number: 41000
IUPAC name: 4-[4-(Dimethylamino)benzenecarboximidoyl]-N,N-dimethylaniline
Discoverers: Heinrich Caro and Alfred Kern, 1883
Walter, 1887
Traugott Sandmeyer, 1889
Contemporary terminology: Auramine (En), Auramine (Fr), Auramin (Ge), Auramine (NL)
Chemical formula: $C_{17}H_{21}N_3$ (Fig. 17.17)

**Fig. 17.17** Chemical structure of Auramine

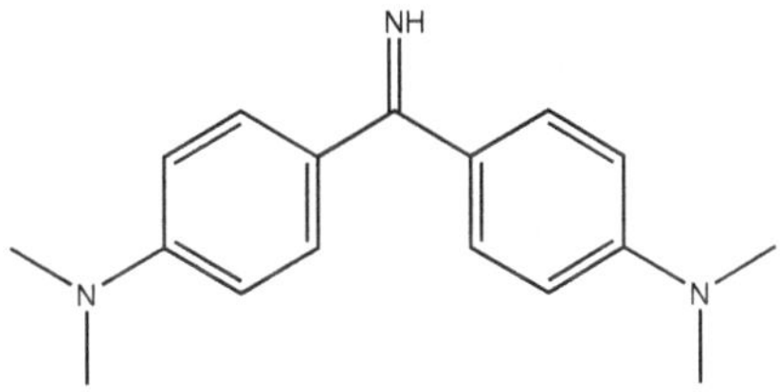


### *History*

The co-operation between the aniline factory Bindschedler & Busch in Basel and the BASF in Ludwigshafen am Rhein also opened the way to an entirely new dye class, the di-arylmethane dyes, of which the yellow Auramines were the most successful products. Auramine O, the hydrochloride of 4,4′-tetramethyldiaminobenzophenonimine, is the most important of these, as can be seen from the large number of patents listed in Table 17.2. This dye was discovered by the German chemist Heinrich Caro (1834–1910), the research-director of the BASF, and the Swiss chemist Alfred Kern (1850–1893), the plant manager of Bindschedler & Busch in 1883. Their invention was assigned to BASF and patented, G.P. 29.060 and B.P. 5.512, both filed on the 11th of March 1884, followed by patents in Germany, Britain, France and the United States of America the same year. Both chemists prepared the yellow dye by heating Michler's ketone (4,4′-tetramethyldiamidobenzophenonimine) with ammonium chloride and zinc chloride at 150–160 °C, and it was introduced by BASF under the name of Auramine in 1884 [27–32].

In the same year, 1884, the firm Ewer & Pick in Berlin-Grünau condensed dimethylaniline with urea in the presence of zinc chloride, which gives first the ketone and then the ketoimine (G.P. 31.936, filed on the 9th of May 1884). The yield was very small and in adapting this method to the preparation of 4,4′-diaminobenzophenonime according to directions in the patents, Auramine was not formed, but instead the zinc chloride salt of an unidentified compound.

Five years later, in 1889, another commercial process was prepared which consisted of heating 4,4′-tetramethyldiamidodiphenylmethane with sulfur, ammonium chloride and common salt in gaseous ammonia, patented by the German chemist Adolf Feer (1862–1913) (G.P. 53.614, filed on the 8th of August 1889). This process, although patented in the name of Feer, is credited to the Swiss chemist Traugott Sandmeyer (1854–1922). In the period 1889–1890, Sandmeyer patented the same method in B.P. 12.549, filed in 1889, and B.P. 16.666, filed in 1890.

Another dye of this dye class is Auramine G (C.I. name: Basic Yellow 3; C.I. number: 41005), discovered by the Swiss chemists Robert Gnehm (1852–1926) and Jakob Schmid (1862–1918) and patented by CIBA in Germany, Britain, France and the United States of America in 1892 (see Table 17.2). Auramine G was produced by BASF, Farbwerke ML&B, Geigy, and CIBA. A third dye from this dye class is named Calcozine Yellow (C.I. name: Basic Yellow 37; C.I. number: 41001).

### Trade Names

Auramin was the trade name used by BASF, CIBA, Geigy, Farbwerke ML&B, Sandoz, Bayer, Weiler-ter Meer, AGFA, Leonhardt, Read Holliday, and Cassella.

### Applications

The basic dye Auramine has been used for dyeing and printing of wool, silk, cotton, paper and leather. The Auramines posed a serious competitive threat to the yellow natural dyes, such as buckthorn, turmeric, and quercitron.

### Properties

Auramine's light fastness is moderate according to Lehne:[7] 'After eight days of light exposure, in June, the colour is much paler, the dyeings are however much faster than those of Curcuma (Turmeric)'. The Colour Index reports a light fastness and a wash fastness of ISO 1. Auramine is readily hydrolysed by boiling water, hence in its use, care must be taken not to allow the dye bath to exceed a temperature 60–70 °C.

---

[7] Auramin: Echtheit: ´Nach 8 tägiger Einwirkung der Witterung in Juli viel heller, die Färbungen sind aber erheblich echter wie solche mit Curcuma´.

## Recipes

Although a basic dye, Lehne presents recipe V12, which is a recipe for direct dyeing, to dye cotton and linen using a concentration of 1% dye. Although he mentions that wool can also be dyed, he does not elaborate on which recipe should be used. Silk can be dyed using recipe V7 at 70 °C and 0.75% of dye. The colourant can also be used to print wool, silk (both recipe V19) and cotton (V23A).

## Identification Using UHPLC-PDA

Analysis of a reference sample of Auramine acquired from Sigma-Aldrich (CAS 492-80-8) using UHPLC-PDA results in a clear peak with UV-Vis absorption spectrum as shown in Fig. 17.18. Analysis of historical reference materials, for example a sample taken from a textile patch present in the book written by Lehne, showed that there are often three or four components present in Auramine with comparable UV-Vis spectra. These components are likely to differ in the number of methyl groups present, formed during the synthesis process.

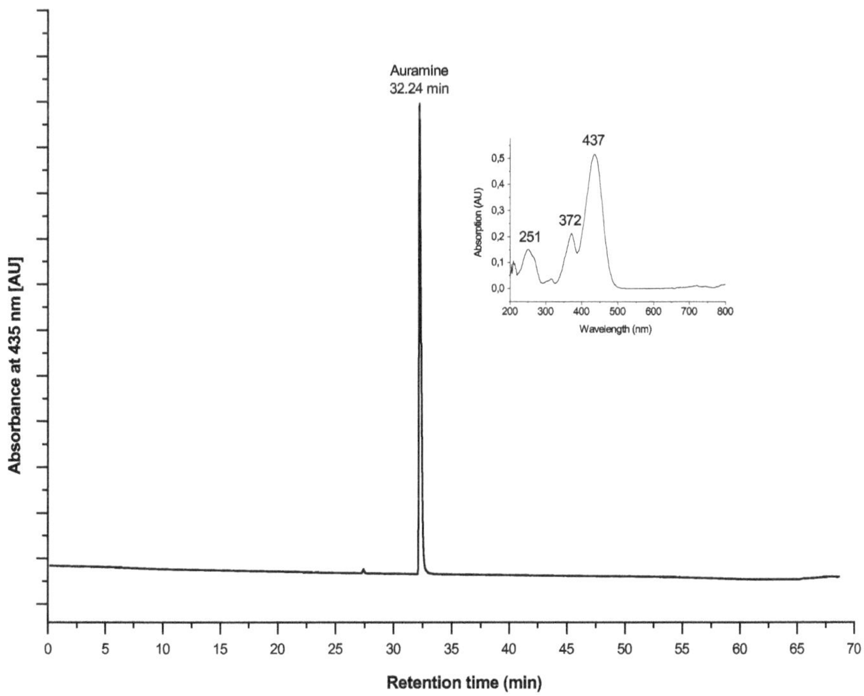


**Fig. 17.18** Chromatogram and spectrum of Auramine (Sigma-Aldrich, CAS 492-80-8)

## Other Case Studies

Auramine was identified in the coloured balls of wool from a Chinese lacquer box by Vincent van Gogh, dated 1887. More information is provided in Chap. 10. Quinoline yellow and Flavazine L were found in samples from billiard cloth which is described in Chap. 9. Auramine was also detected in two embroideries designed by Émile Bernard: *Breton Women, Fire Screen* (1892) and in *Plants and Flowers on an Orange Background* (1927). These objects are described in Chap. 13.

## References

1. Paul, L.: Erfahrungen aus der Farbenfabrikation ll. Zeitschrift für angewandte Chemie. Verlag von Julius Springer, Berlin. 9, 679–692 (1896)
2. Meyer, R.: Hydrierende Spaltung organischer Farbstoffe. Berichte der deutschen chemischen Gesellschaft zu Berlin. Ferd. Dümmler's Verlagsbuchhandlung, Berlin. 53, 1274 (1920)
3. Ziegler, J.H., Locher, M.: Ueber die Tartrazine, eine neue Klasse von Farbstoffen. Berichte der deutschen chemischen Gesellschaft zu Berlin. Commissionsverlag von R. Friedländer & Sohn, Berlin. 20, 834–844 (1887)
4. Tafel, J.: Ueber die Reaction der Diphenylhydrazindioxyweinsäure. Berichte der deutschen chemischen Gesellschaft zu Berlin. Commissionsverlag von R. Friedländer & Sohn, Berlin. 20, 244–249 (1887)
5. Knorr, L.: Ueber die Identität des Phenylmethylpyrazolonazobenzols mit dem Phenylhydrazinketophenylmethylpyrazolon und über die innere Anhydridbildung der Diphenylhydrazinacetylglyoxylsäure und Diphenylhydrazindioxyweinsäure. Berichte der deutschen chemischen Gesellschaft zu Berlin. Commissionsverlag von R. Friedländer & Sohn, Berlin. 21, 1201–1205 (1888)
6. Bernthsen, A.: Chemische Gesellschaft zu Heidelberg. Chemiker-Zeitung. Verlag der Chemiker-Zeitung, Cöthen. 22, 456–457 (1898)
7. Anschütz, R.: Ueber die Constitution des Tartrazins. Justus Liebigs Annalen der Chemie. C.F. Winter'sche Verlagshandlung, Leipzig und Heidelberg. 284, 219–243 (1897)
8. Anschütz, R.: Ueber die Constitution des Tartrazins. Justus Liebigs Annalen der Chemie. C.F. Winter'sche Verlagshandlung, Leipzig und Heidelberg. 300, 1–5 (1899)
9. van Adrichem, F., van Bommel, M.R.: Retouching without touching. Creating the illusion of recoloured furniture through light projection. In: Dias, M.V. (ed.) Material imitation and imitation materials in furniture and conservation, Stichting Ebenist. pp. 33–47 (2016)
10. van Bommel, M.R., van Adrichem, F.B., Boonstra, J.J.: Bringing Back Color: Retouching Faded Furniture With Colored Light. AIC Wooden Artifacts Group Postprints, Houston, Texas, 2018. 95–100 (2018)
11. Möllenhoff, C.: Ueber Sulfosäuren des Phenylmethylpyrazolons und seiner Derivate. Berichte der deutschen chemischen Gesellschaft zu Berlin. Commissionsverlag von R. Friedländer & Sohn, Berlin. 25, 1941–1951 (1892)
12. Meyer, R.: Le groupe du pyrazol et ses applications au point de vue des matières colorants et des produits pharmaceutiques. Chimie et Industrie. 2, 780–787 (1919)
13. Traub, M.C.: Ueber die Einwirkung von Phthalsäureanhydrid auf Chinolin. Berichte der deutschen chemischen Gesellschaft zu Berlin. Ferd. Dümmler's Verlagsbuchhandlung, Berlin. 16, 297–299 (1883)

14. Jacobsen, E., Reimer, C.L.: Ueber die Einwirkung von Phthalsäureanhydrid auf Chinolinbasen. Berichte der deutschen chemischen Gesellschaft zu Berlin. Ferd. Dümmler's Verlagsbuchhandlung, Berlin. 16, 513–514 (1883)
15. Jacobsen, E., Reimer, C.L.: Zur Kenntniss des Steinkohlentheerchinolins. Berichte der deutschen chemischen Gesellschaft zu Berlin. Ferd. Dümmler's Verlagsbuchhandlung, Berlin. 16, 1082–1087 (1883)
16. Jacobsen, E., Reimer, C.L.: Ueber Condensationsprodukte methylirter Chinoline und Pyridine. Berichte der deutschen chemischen Gesellschaft zu Berlin. Ferd. Dümmler's Verlagsbuchhandlung, Berlin. 16, 2602–2608 (1883)
17. Eibner, A., Lange, O.: Zur Constitution des Chinophtalons und der beiden isomeren Chinophthaline. Justus Liebigs Annalen der Chemie. C.F. Winter'sche Verlagshandlung, Leipzig. 315, 303–356 (1901)
18. Eibner, A., Simon, E.: Zur Constitution des p-Toluchinophtalons. Berichte der deutschen chemischen Gesellschaft zu Berlin. Commissionsverlag von R. Friedländer & Sohn, Berlin. 34, 2303–2311 (1901)
19. Eibner, A., Merkel, H.: Zur Constitution des Chinophtalons: Ueber die Alkaliverbindungen desselben und seines Isomeren. Berichte der deutschen chemischen Gesellschaft zu Berlin. Commissionsverlag von R. Friedländer & Sohn, Berlin. 37, 3006–3011 (1904)
20. Eibner, A., Hofmann, K.: Zur Kenntniss der Isochinophthalone. Berichte der deutschen chemischen Gesellschaft zu Berlin. Commissionsverlag von R. Friedländer & Sohn, Berlin. 37, 3011–3018 (1904)
21. Eibner, A., Hofmann, K.: Zur Frage des Existenz des Isopyrophtalons. Berichte der deutschen chemischen Gesellschaft zu Berlin. Commissionsverlag von R. Friedländer & Sohn, Berlin. 37, 3023–3026 (1904)
22. Eibner, A., Hofmann, K.: Weitere Beiträge zur Kenntniss der Chinophtalone, Isochinophthalone und Chinophtaline. Berichte der deutschen chemischen Gesellschaft zu Berlin. Commissionsverlag von R. Friedländer & Sohn, Berlin. 37, 3018–3023 (1904)
23. Eibner, A.: Zur Kenntnis des Chinophtalons. Chemiker-Zeitung. Verlag der Chemiker-Zeitung, Cöthen. 28, 1206–1209 (1904)
24. Eibner, A.: Mechanismus und Grenzen der Phtalonreaction. Berichte der deutschen chemischen Gesellschaft zu Berlin. Commissionsverlag von R. Friedländer & Sohn, Berlin. 37, 3605–3612 (1904)
25. Oswald-Graf, U., Steinwender, A.M., Wipplinger, H.-P. eds: Poesie des Ornaments. Das Backhausen Archiv / Poetry of the Ornament. The Backhausen Archives. Exhibition Catalogue, Leopold Museum, Vienna. Verlag der Buchhandlung Walther und Franz König, Köln (2024)
26. Cultural Heritage Agency of The Netherlands, Research Report, 19th Century Gown, Museum Paulina Bisdom van Vliet, Haastrecht, RCE Werknummer: 2017-032, Februari 2018
27. Fehrmann, W.: Zur Kenntniss der Auramine. Berichte der deutschen chemischen Gesellschaft zu Berlin. Commissionsverlag von R. Friedländer & Sohn, Berlin. 20, 2844–2862 (1887)
28. Stock, A.: Zur Kenntniss der Auramine. Journal für die praktische Chemie. 47, 401–413 (1893)
29. Stock, A.: Ueber die Constitution des Auramins. Berichte der deutschen chemischen Gesellschaft zu Berlin. Ferd. Dümmler's Verlagsbuchhandlung, Berlin. 33, 318–320 (1900)
30. Graebe, C.: Ueber Imine des Benzophenons und über die Constitution des Auramins. Berichte der deutschen chemischen Gesellschaft zu Berlin. Ferd. Dümmler's Verlagsbuchhandlung, Berlin. 32, 1678–1683 (1899)
31. Semper, L.: Die Konstitution des Auramins. C.F. Winter'sche Verlagshandlung, Leipzig und Heidelberg. 381, 234–264 (1911)
32. Reinhardt, C., Travis, A.S.: Heinrich Caro and the Creation of Modern Chemical Industry. Kluwer Academic Publishers, Dordrecht / Boston / London (2000)

# Chapter 18
# The First Synthetic Substantive Cotton Dyes: Congo Red, Benzopurpurine B, Benzopurpurine 4B and Some of Their Derivatives

## 18.1 Introduction

In 1885 the German dye industry was in a severe economic crisis. This was mainly due to fluctuations in the prices of the raw materials used for the production of synthetic dyes. In the early 1880s the prices of benzene and aniline had risen sharply. As a result, 100 kg of benzene cost 250–300 German marks on the German market between 1880 and 1882. Just 1 year later, in 1883, the price dropped to 90 and finally to 40 German marks in 1885. In 1884 the price for 100 kg of aniline was 330 German marks, decreasing 1 year later to 160 German marks. The dye factories which had bought raw materials at high prices saw their large stocks devalued and were forced to reduce the prices of the dyes produced. Cheap dyes were sold at a loss and hardly any new dyes were marketed at that time. Even with synthetic Alizarin, the dye factories could no longer earn money and, last but not least, the 1881 Alizarin Convention between the big German dye factories BASF, Farbwerke ML&B and Bayer collapsed in August 1885 [1]. As an example, the Bayer dye factory in the Ruhr Valley could not pay a dividend and for the first time since the firm was set up in 1863, the number of workers had to be reduced [2, 3]. By November 1885, matters had become so bad for Bayer that the Berlin financial newspaper *Allgemeine Börsen-Zeitung* forecast the company's imminent doom and advised investors to sell their shares of Bayer stock. The Berlin stock exchange paper wrote the following:

> 'Those who wish to protect themselves from major losses should sell their [Bayer *sic*] shares, since neither the state of the alizarin and aniline markets, nor the ability of this company's management, point to an improvement in the situation. It seems more than doubtful that any dividend can be paid over the next few years – assuming the company can last that long' [2, 3].



M. R. van Bommel, M. de Keijzer, *Bright Colours from the Past*, Cultural Heritage Science, https://doi.org/10.1007/978-3-031-88664-5_18

Surprisingly, one German dye firm was exceptionally successful in this period. AGFA in Berlin was still making profits by exporting their synthetic organic dye Congo red to the textile manufacturers, first in Europe and later to India and the Far East. Congo red was such a major success that the other German dye companies soon found themselves unable to compete and some competitors had to close their doors.

## 18.2 Direct or Substantive Dyes

The success of AGFA was due to the introduction of new, direct dyes, also called substantive dyes. This is a class of coloured, water-soluble compounds that have an affinity for all fibres and are taken up directly by the textiles. They are easily applied and can yield bright colours. Before the discovery of the synthetic direct dyes only a few natural direct dyes were available, including Henna (C.I. name: Natural Orange 6; C.I. number: 75486), Orchil (C.I. name: Natural Red 28; C.I. number: 758600), Annatto (C.I. name: Natural Orange 4; C.I. number: 75120), Turmeric (C.I. name: Natural Yellow 3; C.I. number: 75300) and Saffron (C.I. name: Natural Yellow 6; C.I. number: 75100). They all have a low fastness to light and washing [4].

Most synthetic dyes developed earlier, before the 1880s, possessed a great affinity for wool and silk, but unfortunately they had virtually no attachment to cotton and linen. It was always necessary to use a mordant to bind the dye permanently to the cotton or linen fibre, a requirement that added an extra step to the dyeing process. The same applies to the mordant dyes: both natural and, later on, synthetic colourants require a coordination metal to bind to the textile fibre. Dyeing a durable red with the mordant dye Madder on cotton and linen is more difficult than on wool and silk. As an example, one particular method, the Turkey red process, was developed whereby cotton and linen were treated by a complicated and time-consuming process. A recipe from the end of the eighteenth century shows that about 20 steps were needed for dyeing cotton and linen yarns red. However, the red-dyed cotton possesses a brilliant colour and fastness against light and washing and as a result was very popular [4–6].

But that changed in 1883. Congo red allowed the dyers to dye cotton and linen without the use of a mordant for the first time. The dyeing process is much simpler and required less skilled dyers; as a consequence, it was less laborious and thus cheaper. Congo red was the first member of a new class of dyes known as direct or substantive dyes and it caused a revolution in the cotton-dyeing industry. Its ability to fix itself directly onto the cotton fibre is due to the presence of diphenyl in its molecule. To prepare direct cotton dyes, the base from which they are derived must be a diamine, since monoamino-derivatives of diphenyl do not yield diazo compounds substantive to cotton. The two amino-groups must also be in the para-position to the coupling bond.

## 18.3 Congo Red

Colour Index name: Direct Red 28; Colour Index number: 22120
IUPAC name: Disodium 4-amino-3-[4-[4-(1-amino-4-sulfonato-naphthalen-2-yl) diazenylphenyl]phenyl]diazenyl-naphthalene-1-sulfonate
Discoverer: Paul Böttiger, 1884
Contemporary terminology: Congo red (En), Rouge Congo (Fr), Congorot (Ge), Kongorood (NL)
Chemical formula: $C_{32}H_{24}N_6O_6S_2$ (Fig. 18.1)

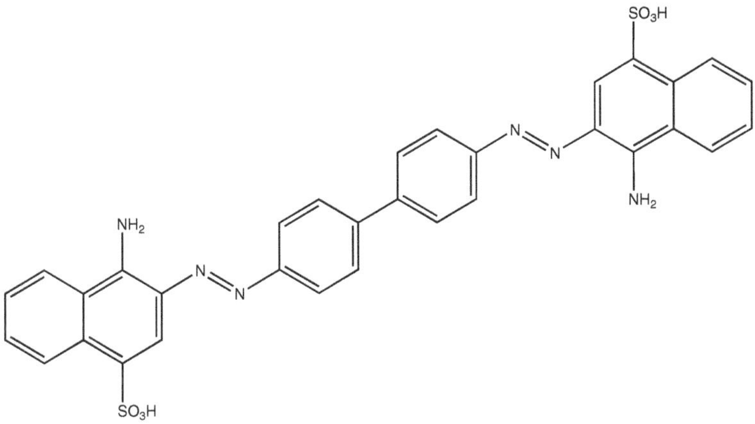


**Fig. 18.1** Structural formula of Congo red

### *History*

The first incentive for the interest in benzidine, which is an important starting material for the synthesis of direct dyes, occurred in the mid-1860s when the German chemists Heinrich Caro (1834–1910) and Johann Peter Griess (1829–1888) observed the compound while preparing dyes during their collaboration in Manchester, England. It was included in one of Caro's patent applications of 1866. Benzidine (4,4′-diaminobiphenyl) was made from nitrobenzene, and contains two free amino groups. In 1869 Caro, who was at the time research director of BASF, invented Palatine Orange (C.I. name: Acid dye; C.I. number: 10311), based on the reaction of nitric acid with benzidine [1, 7–9].

Caro visited Griess in 1882 at the Samuel Allsopp & Sons Brewery laboratory in Burton-on-Trent, Straffordshire, England, where he was employed. Griess advised Caro that a benzidine-derived azo dye had the property to dye cotton without the use of a mordant; however, Caro was not interested. One year later, in 1883, Henry

Theodore Böttinger (1848–1920), of the dye company Bayer in Elberfeld, visited his birthplace Burton-on-Trent and met his friend Griess. Böttinger was Bayer's son-in-law and a senior figure in the Bayer company. He was the son of Heinrich Wilhelm Böttinger (1820–1874), the chief chemist at Samuel Allsopp & Sons Brewery when Griess started working there in 1862. When the Böttinger family returned to Germany in 1866, Griess became chief chemist. Griess showed Henry Theodore Böttinger his collection of dyed cotton yarns and explained to him how he had coloured them without prior mordant treatment. Böttinger may have not been an expert in dyes, but a blue dye, based on benzidine-sulfone-disulfonic acid, that Griess showed him looked so interesting that he took a few samples back to the Bayer firm in Elberfeld.

After his return to Germany, Böttinger asked Griess in a letter how he had dyed the cotton yarns. Griess wrote on the 19th of September 1883,[1] 'The dye is dissolved in hot water with some sodium carbonate and in this solution cotton and wool are dyed homogeneously without any mordant'. Unfortunately, the results achieved at the Bayer firm were unsatisfactory, so the German chemist Paul Seidler was sent to Griess to obtain more information regarding the process. Seidler obtained the rights to Griess's invention on behalf of Bayer for 250 lb sterling in November 1883. However, it soon proved that the commercial manufacture of the dye was very problematic. The production of the intermediate benzidine-sulfone-disulfonic acid caused unexpected difficulties: it could be produced in the laboratory, but not on a large scale. The yield remained too low and in the end the production was discontinued [10].

Nevertheless, Bayer and Griess registered patents for compounds of benzidine disulfonic acid in 1883 (G.P. 27.954, filed on the 5th of December 1883),[2] and in 1884 (B.P. 1.075, filed on the 9th of January 1884).[3] These were the first patents for a new class of azo dyes: the direct and substantive dyes, without the use of mordants [1, 3, 8–11].

During Seidler's visit in November 1883, Griess told him that he had solved the constitution of various products of benzidine and that he had developed methods for preparing dyes in different colours, including red, to dye cotton directly. Once back in Germany, Seidler did not report this information to his executives. At that same time, his friend the German industrial chemist Paul Böttiger (not to be confused with Böttinger) was working at Bayer. Not much is known about Böttiger. He was born in Łódź and worked at Bayer from July 1881 till the end of December 1883.

---

[1] 'Sie fragen wie ich die blauen Farbstoffe auf die Baumwolle gebracht habe. Alle die betreffenden Farbstoffe sind in heissem Wasser, das etwas Soda enthält, etwas löslich, und es färben die Lösungen sowohl Baumwolle als Wolle ohne irgendwelche Beizmittel, und zwar sehr gleichmässig'.

[2] Farbenfabriken vorm. Friedr. Bayer & Co., G.P. 27.954, 5th of December 1883, Verfahren zur Darstellung von Sulfosäuren des Benzidins und Benzidinsulfons, sowie zur Darstellung neuer Azofarbstoffe aus den Tetrazoverbindungen derselben und Aminen, Phenolen oder deren Sulfosäuren.

[3] Johann Peter Griess, B.P. 1.075, 9th of January 1884, Improvements in the Manufacture or Production of Colouring Matters from the Tetrazo Compound of Benzidine Disulphonic Acid.

Between Seidler and Böttiger there were no secrets. Böttiger dealt secretly with Griess's invention, without saying one word to his employers. Having learned from Seidler about Griess's red dye, he further developed the method. He worked on a production process and invented the first synthetic substantive red dye by coupling 1-naphthionic acid (4-aminonaphthalene-1-sulfonic acid) with doubly diazotised benzidine (4,4′-diaminobiphenyl) [1, 3, 8, 9, 11, 12]. After the discovery of Congo red, Böttiger presented his findings to his supervisors. Unfortunately for Bayer, Böttiger's supervisors showed little interest and rejected the new red dye as its fastness to acid was poor. Bayer's chemists probably failed to recognise the value of the direct dye because they were much more interested in finding a synthetic replacement for natural Indigo. Böttiger kept the details to himself and left Bayer at the end of December 1883, ostensibly to work for his father, who owned a small dye factory in the industrial city of Łódź (nowadays Poland). A few months later Seidler also left Bayer and joined Böttiger in Łódź.

At the end of February and in the beginning of March 1884, Böttiger filed German, French and British patents in Łódź for his new red dye in his own name (G.P. 28.753, filed on the 27th of February 1884; B.P. 4.415, filed on the fifth of March 1884 and F.P. 166.722, filed on the fourth of March 1884, but never published) [1, 10, 12]. After the patents had been issued, he attempted to sell rights, first to Bayer, perhaps to avoid legal action, but they declined, as did the other large dye companies BASF and Farbwerke ML&B, because of the dye's high sensitivity to acids [1, 3, 8–13]. Some time later, in the years 1884 and 1885, Böttiger filed two patents in the United States of America, U.S.P. 365.666, filed on the 22nd of November 1884, patented on the 28th of June 1887, and U.S.P. 365.667, filed on the 24th of April 1885 and issued on the 28th of June 1887, which were granted in 1887. The two U.S. patents taken by Böttiger were licensed to AGFA (see Table 18.1).

Böttiger's patent (G.P. 28.753) was a success, but Griess's patent (B.P. 1.075) of 1884 was not. The difference between the patents for the dyes invented by Griess and Böttiger is minimal: a small difference in the chemical structure of the two dyes as, unlike Griess, Böttiger used the cheaper, unsubstituted benzidine in his preparation. The absence of sulfone groups on the benzidine molecule results in a better adherence to cotton [14]. He subsequently sold the patent for an undisclosed amount of money to AGFA and began selling the dye under the trade name Congo red in 1885. Initially, AGFA was not interested in Congo red and was disposed to reject the patent, but a master dyer in Berlin urged the firm to buy it, believing that the brilliance of the dye would make the dye attractive to dyers despite its sensitivity to acids [1–3, 10].

## *The Berlin West Africa Conference*

Although it is not entirely clear how Congo red received its name, it was introduced within the time frame of the scramble for Africa: the invasion, occupation, colonisation and annexation of African territory by European powers during the

**Table 18.1** International patents of Congo red

| Discoverer, company | Patent | Title |
|---|---|---|
| Paul Böttiger in Łódź | G.P. 28.753 filed 27th of February 1884, issued 20th of August 1884 | Verfahren zur Darstellung von Azofarbstoffen durch Combination von Tetrazodiphenylsalzen mit α und β Naphtylamin oder deren Mono- und Disulfonsäuren |
| Paul Böttiger | F.P. 160.722 filed 4th of March 1884, never published | |
| Paul Böttiger | B.P. 4.415 filed 5th of March 1884, issued 20th of November 1884 | The Coupling of Tetrazotised Benzidine with Naphthylamine and its Derivatives |
| Paul Böttiger in Łódź, Russia, assignor to AGFA | U.S.P. 365.666 filed 22th of November 1884, issued 28th of June 1887 | Manufacture of New Red Dye-Stuffs or Coloring-Matters |
| Paul Böttiger in Gössnitz, Germany, assignor to AGFA | U.S.P. 365.667 filed 24th of April 1885, issued 28th of June 1887 | Combination of the salts of Tetrazo-Diphenyl and the Naphthylamines |
| BASF | B.P. 6.697 filed in 1895 | |
| BASF | B.P. 17.260 filed in 1895 | |
| BASF | F.P. 248.210 | |
| BASF | G.P. 84.893 filed 27th of March 1895, issued 30th of December 1895 | Verfahren zur Darstellung von Azofarbstoffen der Benzidinreihe aus Monoazofarbstoffen |
| BASF | G.P. 88.597 filed 17th of July 1895, issued 22th of September 1896 | Verfahren zur Darstellung von Azofarbstoffen der Benzidinreihe aus Monoazofarbstoffen |

New Imperialism Period (1870–1914). The name 'Congo red' has a strong link to a geopolitical event later known as the Berlin West Africa Conference. From the 15th of November 1884 until the 26th of February 1885, a diplomatic conference took place, one of the most significant political events of the decade. Germany organised the conference apparently to help resolve a trade dispute between a number of European colonial powers about the Congo River basin in Central Africa. High-ranking representatives of the major European powers and the United States of America were invited to Berlin by the German 'Iron Chancellor', Otto von Bismarck (1815–1898). Germany had only just established its first African colony in 1884, so could not be described as a significant colonial power; Bismarck may have called the conference to validate his country's new imperialist leanings. The official report of the conference called for free trade in the Congo basin, but this did not happen. The actual result may have been a decision about how the colonial powers would divide the continent of Africa among themselves.

The central topic of the Berlin conference was the Congo: such an exotic location for Europeans in the 1880s; a name on the tip of every tongue. Perhaps it is not surprising, therefore, that AGFA, a Berlin dye company, gave the name Congo to their synthetic substantive red dye. The name turned out to be an effective marketing tool: AGFA filed the G.P. 35.615 for a modification of the Congo red dye on the 17th of March 1885, 3 weeks after the conference ended, an application that mentions that Congo red was already 'well-known'.

From its exotic name, one might think that the production of Congo red required raw materials imported from Africa, or perhaps that the dye was named to celebrate colourful native African textiles, but there is no evidence for this. The chemicals used in making Congo red were derived from coal-tar products from Germany's Ruhr Valley and traditional textiles from the Congo River basin are typically black, very rarely red. The name of Congo red was soon to be joined by other synthetic dyes with African place names, such as Somalia yellow, Sudan black, Sudan brown, Sudan red, Sudan grey, Nile blue, Sambesi blue and Sambesi brown [15]. By 1910 there were many textile dyes bearing the word Congo: Congo rubine, Congo Corinth, Brilliant Congo, Congo orange, Congo brown, Congo violet and Congo blue.

## *Preparation of Congo Red*

Explicit details of the manufacturing process of Congo red, which involves coupling of diazotised benzidine (4,4′-biphenyl diamine) with the sodium salt of naphthionic acid (4-aminonaphthalene-1-sulfonic acid), were published in the patents of Böttiger (G.P. 28.753, U.S.P. 365.666 and U.S.P. 365.667). A solution of 25.7 kilos of benzidine hydrochloride in 500 L of water and 30 kg of hydrochloric acid (20 °Bé, 32%) is diazotised by adding a solution of 13.8 kg of sodium nitrate (100%) in 150 L of water, and the solution of tetrazodiphenyl chloride is slowly run into a solution of 55 kilos of sodium naphthionate and 12 kilos of sodium acetate in 1,000 L of water, stirring well. After 24 h, the dye is filtered off, converted into the sodium or ammonium salt and dried.

## *Trade Names*

In the literature the main patents related to Congo red are dated in the mid 1880s (see Table 18.1). The red dye was named Congo red by AGFA. Other trade names were used, such as: Congo extra by Sandoz; Congorot by Beyer & Kegel, Bayer, Griesheim-Elektron, Leonhard, Österreichischer Verein, Weiler-ter Meer, CIBA and Sandoz; Congorot B by Kalle; Dianilrot R by Farbwerke ML&B; Baumwollrot B by Kalle; Baumwollrot 4B by Griesheim-Elektron; Baumwollrot A by Cassella; Baumwollrot C by CIBA. More trade names for Congo red are Baumwollrot 4 BC; Kosmosrot extra; Azidinrot C; Cottonrot conc.; Directrot C; Congo red; Congo red L; Congo red R; Congo red extra; Congo red SD; Congo red W; Amanil Congo YY extra; Erie Congo 4B; Chokusetsu red; Rouge Congo; Rouge Congo direct; Rouge Congo P, P extra; Rouge direct and Rosso Congo [16–20].

## *Applications*

Congo red was used for dyeing cotton, linen and wool. The dye can be used in one bath without the aid of a mordant.

## *Properties*

Congo red appears in commerce in the form of a rich brown powder, which dissolves very easily in water providing a red colour. It is sparingly soluble in alcohol, but insoluble in ether and benzene. Cotton, immersed into a boiling solution of Congo red, is rapidly dyed a brilliant scarlet resembling a yellow shade of Turkey red. The colour produced seems to be actually dyed on the fibre: it will not come out with washing and is only slightly affected by boiling soap solution.

The advantages of Congo red are the direct and easy application to cotton goods without using mordants and the low price. Congo red has, however, several disadvantages. In the first place the colour has a poor light fastness, a poor acid fastness, becomes dark on exposure on air and thus loses much of its brilliancy, changing from a bright scarlet to a dull red. Lehne reported:[4] 'After 6 days of light exposure (September) the colour of Congo red changed strongly brown'. On wool, Congo red produces a good colour in a neutral bath, which, however, cannot compete with the brilliant azo-scarlets. By adding 5% of sulfuric acid to the dye bath, the wool is dyed brown. The aqueous solution is so sensitive to acids that a single drop of dilute sulfuric acid converts the liquid to a blue, a fact which makes this colouring matter useful as an indicator in volumetric analysis. The colour change is from blue to red at pH 3.0–5.2 [21, 22]. As a result, textiles dyed with Congo red turn blue when washed with an acidic solution, but this can be reversed simply by washing the fabric in mildly alkaline soaps. The Colour Index reports a light fastness of ISO 1 and a wash fastness of ISO 2.

## *Recipes*

Lehne wrote that for wool dyeing, recipe V2 is used with 5% dyestuff, while for cotton and linen, recipe V12 can be used, although no specific dye concentration is given.

Holsboer presented recipes for cotton, wool and silk [23]. He writes:[5] 'The general prescription for dyeing cotton with direct dyes is as follows: for light colours

---

[4] Congo: Echtheit: 'Nach 6 tägiger Einwirkung der Witterung (September) stark verändert braun'.

[5] Het algemene voorschrift voor directe kleurstoffen is als volgt: Verfmethode voor katoen: Voor lichte kleuren 0,01–1% kleurstof, 1% soda, 1–2% zeep, 3–5% $Na_2SO_4$ of $Na_2HPO_4$. Voor middelkleuren: 1–3% kleurstof, 2% soda, 10% $Na_2SO_4$. Voor donkere kleuren: 3–7% kleurstof, 2% soda, 20–30% $Na_2SO_4$ of NaCl. Het verfbad wordt aan de kook gebracht en daarnaan wordt de

0.01–1% dye, 1% soda, 1–2% soap and 3–5% sodium sulfate ($Na_2SO_4$) or disodium hydrogen phosphate ($Na_2HPO_4$); for medium colours: 1–3% dye, 2% soda and 10% $Na_2SO4$ and for dark colours: 3–7% dye, 2% soda, 20–30% $Na_2SO_4$ or salt (NaCl). The dye bath is brought to boiling and then the soda or soap is added. The dye is dissolved in a little water, this solution is added to the dye bath and, after stirring, the material is introduced into the dye-bath. The dyeing is carried out at boiling point for 1 h and then left to soak for another hour without heating. Sodium sulfate ($Na_2SO_4$) or sodium chloride (NaCl) is only added after half an hour of dyeing. Lighter shades are obtained by dyeing lukewarm or cold'.

The general prescription for dyeing wool with direct dyes is as follows:[6] 'After adding the dye solution (5% of dye), sodium sulfate is added (10% for light colours and 20% for darker colours) and 5% ammonium acetate to the bath at 60 °C, the goods are placed in the dye-bath and heated slowly to boiling. After ¾ of an hour acetic acid (2% of 30% acetic acid for light colours, 5% of 30% acetic acid for darker colours) is added and then boiled for another half hour; and finally rinsed'.

The general prescription for dyeing silk with direct dyes is as follows:[7]

> 'Silk is dyed in a bast soap bath with dilute acetic acid (1/3 bark soap, 2/3 water) or only with acetic acid. Very light colours are advantageously dyed in a fat soap bath. One starts at 40 °C, then slowly heated to boiling and dyeing for three quarters of an hour. If necessary, acetic acid can be added.'

Soxlett wrote:[8] 'Dissolve the Congo red in boiling water in a ratio of 5:1000. The dyeing is done on wooden skids or vats with 2.5% soap, 5% soda and 3% Congo red at a temperature of 40 °C and raise the temperature to 100 °C and kept so for 1 h. Then take the goods out, cool and place them in a cold bath, to which is added a

---

soda of zeep toegevoegd. In een klein gedeelte van dit water lost men de kleurstof op, doet deze oplossing in het verfbad en nadat men heeft doorgeroerd, brengt men het material in het vlot. Men verft 1 uur kokend en laat daarna nog ½ uur natrekken zonder verwarmen. Eerst nadat men ½ uur heeft geverfd, voegt men het $Na_2SO_4$ of NaCl toe. Lichtere nuancen verkrijgt men door lauw of koud uit te verven.

[6] Verfmethode voor wol: Nadat aan het 60 °C warme bad de kleurstofoplossing, het $Na_2SO_4$ (10% voor lichte, 20% voor donkerder kleuren) en 5% ammoniumacetaat zijn toegevoegd, brengt men het material in het vlot en verwarmt langzaam tot koken. Na ¾ uur voegt men azijnzuur (2% van 30% azijnzuur voor lichte, 5% van 30% azijnzuur voor donkerder kleuren) toe en kookt dan nog een half uur na; tenslotte spoelen.

[7] Verfmethode voor zijde: Deze wordt in een met verdund azijn zuur zwak gebroken bastzeepbad (1/3 bastzeep, 2/3 water) of alleen met azijnzuur geverfd. Zeer lichte kleuren verft men met voordeel in een vet zeepbad. Men begint bij 40 °C, verwarmt langzaam tot koken en verft ¾ uur uit. Zo nodig kan azijnzuur toegevoegd worden.

[8] Man löst das Congorot in kochendem Wasser im Verhältnis von 5: 1000. Das Färben geschieht auf hölzernen Kufen oder Bottichen mit 2,5% Seife, 5% Kristalsoda und 3% Congorot; man geht lauwarm mit den Stücken ein und lässt bei fortwährendem Behandeln eine Stunde kochen, hierauf herausnehmen, verkühlen und in ein kaltes Bad bringen, welchem man eine Lösung von 10% Kristallsoda in 150% Wasser und 50% Türkischrotöl zusetzt; man behandelt die Stücke 10 Minuten darin, lässt auf der Centrifugalmachine (ohne vorheriges Spülen) entwässern und dann trocknen. Die Türkischrotölpassage kann man auch weglassen, das Rot wird dann indessen nicht so lebhaft und so widerstandsfähig gegen die Wäsche.

solution of 10% soda in 150% water and 50% Turkey red oil; the pieces are treated in it for 10 min, drained in the centrifugal machine (without prior rinsing) and dried. The Turkey red oil treatment can also be disregarded, but then the red will not be as lively and not as resistant to washing. Congo red is a substantive dye, so mordants are not necessary to use' [24].

## *Identification Using UHPLC-PDA*

Analysis of Congo red (RCE no. 5528) using liquid chromatography showed a clear, single peak as shown in Fig. 18.2. The spectrum obtained is rather typical.

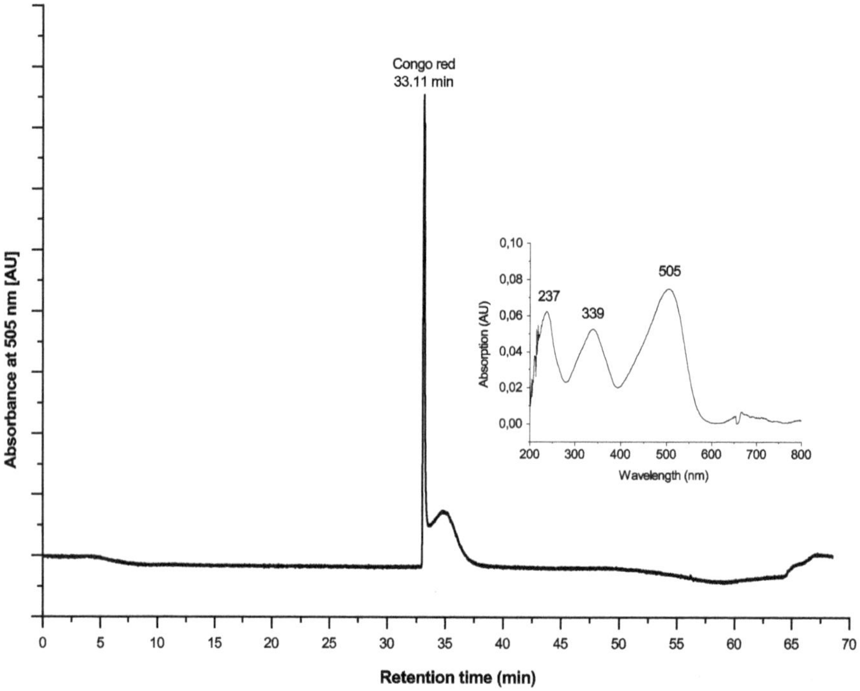


**Fig. 18.2** Chromatogram and spectrum of Congo red (RCE no. 5528)

## 18.4 Benzopurpurine B

Colour Index name: Direct Red 21; Colour Index number: 23560
IUPAC name: Sodium;6-amino-5-[[4-[4-[(2-amino-6-sulfonaphthalen-1-yl)diazenyl]-3-methylphenyl]-2-methylphenyl]diazenyl]naphthalene-2-sulfonic acid
Discoverer: Carl Duisberg, 1885
Contemporary terminology: Benzopurpurine B (En), Benzo purpurine (Fr), Benzopurpurin B (Ge), Benzopurpurine B (NL)
Chemical formula: $C_{34}H_{28}N_6O_6S_2$ (Fig. 18.3)

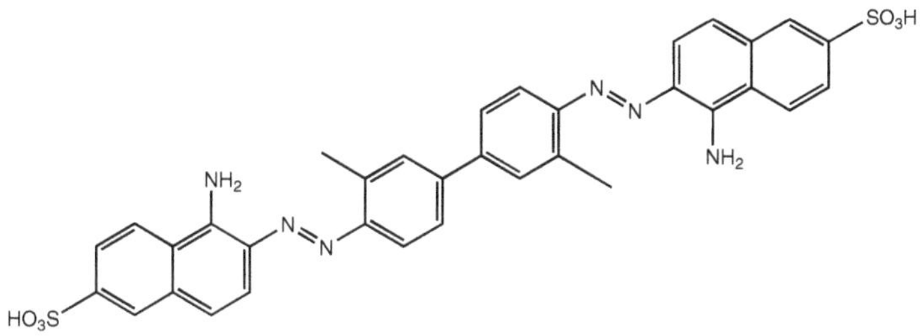


**Fig. 18.3** Structural formula of Benzopurpurine B

### *History*

The interest in benzidine-based azo dyes at the dye factory Bayer was revived by the success of Congo red. In 1884 the German chemist Eugen Frank (1854–1914), who worked for Bayer, discovered the direct yellow dye Chrysamine G (C.I. name: Direct Yellow 1; C.I. number: 22250) by coupling salicylic acid with diazotised benzidine (4,4′-diaminobiphenyl) (G.P. 31.658, patented on the 14th of June 1884, U.S.P. 329.638, filed on July 15, 1884 and B.P. 9.162 and 9.606, both filed in 1884). This dye could be used for dyeing cotton and linen [1, 3, 8–10, 25]. At that time the German chemist Carl Duisberg (1861–1935) joined Bayer and investigated similar coupling reactions to those used by Frank when preparing Chrysamine G, but without much success. Duisberg then decided to try the coupling component 2-naphthionic acid, also known as Brönner's acid (6-amino-2-naphthalenesulfonic acid). This was coupled with doubly diazotised *o*-tolidine (3,3′-dimethylbenzidine) resulting in Benzopurpurine B [1, 3, 8–10, 12]. Brönner's acid was discovered by

Julius Brönner (1823–1888) who founded a chemical factory for letterpress inks in Frankfurt am Main, Germany, in 1847. He manufactured tar distillates and in the beginning of the 1870s he produced aniline colours. In 1888, the factory was liquidated due to the Alizarin crisis and its patents, including Brönner's acid, were purchased by AGFA. However, Duisberg had chosen *o*-tolidine (3,3′-dimethylbenzidine) instead of benzidine, a substance not mentioned in Böttiger's patents. Duisberg also discovered that Bayer had found a different way to obtain 2-naphthionic acid before Brönner had patented his process. Duisberg saw his chance and registered a patent on behalf of Bayer for tolidine dyes on the 17th of March 1885. The G.P. 35.615 became an addition to AGFA's Congo red patent no. 28.753 and was transferred to AGFA (see Table 18.2).

With much publicity, the dye with the name Benzopurpurine B was on sale from the autumn of 1885 [3, 8, 9]. However, it was not received with the same enthusiasm as Congo red because it was less useful. Although faster to light and to acid than Congo red, Benzopurpurine B had a dull, brick-red tone and lacked Congo red's colouring power [1, 8, 12].

**Table 18.2** International patents of Benzopurpurine B

| Discoverer, company | Patent | Title |
|---|---|---|
| AFGA | G.P. 35.615 filed 17th of March 1885, issued 13th of May 1886 | Verfahren zur Darstellung von Azofarbstoffen durch Combination von Tetrazoditolyol mit α und β Naphtylamin oder deren Mono- und Disulfonsäuren |
| Bayer & Co. | B.P. 3.803 filed in 1885 | |
| Carl Duisberg | U.S.P. 329.633 filed 29th of July 1885, issued 3rd of November 1885 | Coloring-Matter obtained from Tetrazo-Ditolyl |
| Bayer & Co. | F.P. 167.876 | |
| BASF | B.P. 6.697 filed in 1895 | |
| BASF | B.P. 17.260 filed in1895 | |
| BASF | G.P. 84.893 filed 27th of March 1895, issued 30th of December 1895 | Verfahren zur Darstellung von Azofarbstoffen der Benzidinreihe aus Monoazofarbstoffen |
| BASF | G.P. 88.597 filed 17th of July 1895, issued 22th of September 1896 | Verfahren zur Darstellung von Azofarbstoffen der Benzidinreihe aus Monoazofarbstoffen |
| BASF | F.P. 248.210 | |

## *Trade Names*

Benzopurpurin B was put on the market under this name by AGFA, Bayer, Griesheim-Elektron, Leonhardt, Weiler-ter Meer, Read Holliday and Levinstein. Other trade names were used, such as Baumwollrot B by CIBA and Amidine Fast Scarlet B [16, 26–29].

## *Applications*

Benzopurpurine B is used for dyeing cotton, linen and silk. The shades produced are not very intense, which is why the dye is rarely used in dyeing.

## *Properties*

Benzopurpurine B is faster to light and acid than Congo red, but has a dull, brick-red colour. It dissolves readily in water with a yellowish-red colour. It is sparingly soluble in alcohol, but insoluble in ether and benzene. Although better than Congo red, Benzopurpurine B still has a poor light and acid fastness and an inadequate fastness to washing. Lehne reported:[9] 'After 6 days of light exposure (September) the colour of Benzopurpurine B changed strongly light brown'. This corresponds with the information provided by the Colour Index: the light fastness is ISO 1 while the wash fastness is ISO 1–2.

## *Recipes*

The same dyeing recipes can be used for Benzopurpurine B as given for Congo red.

## *Identification Using UHPLC-PDA*

Analysis of a sample of Benzopurpurine B (RCE no. 7186) showed the presence of three compounds, with similar spectra (Fig. 18.4).

[9] Benzopurpurin B: Echtheit: 'Nach 6 tägiger Einwirkung der Witterung (September) stark verändert, hellbraun'.

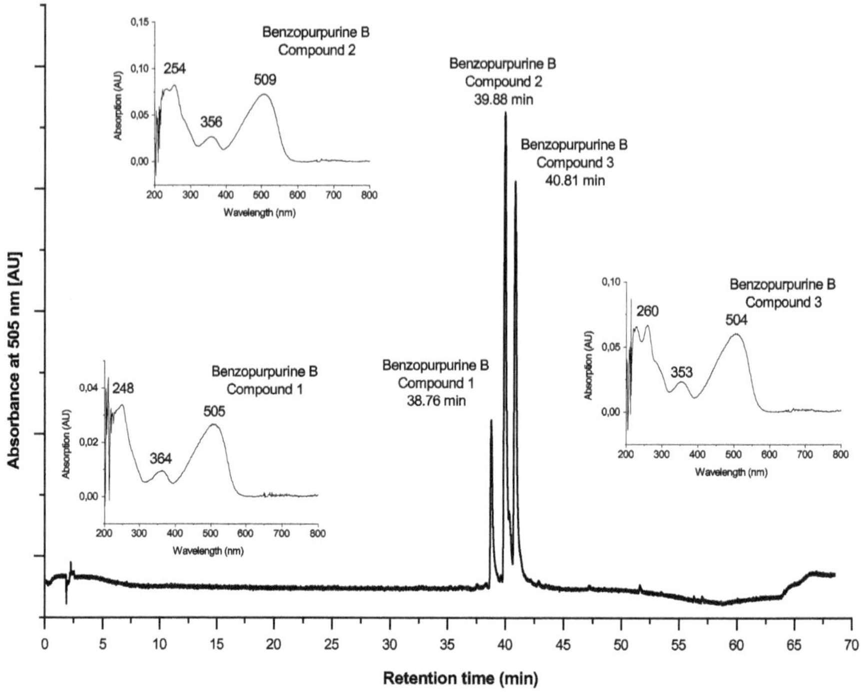


**Fig. 18.4** Chromatogram and spectra of Benzopurpurine B (RCE no. 7186)

## 18.5 Benzopurpurine 4B

Colour Index name: Direct Red 2; Colour Index number: 23500
IUPAC name: Disodium;4-amino-3-[[4-[4-[(1-amino-4-sulfonatonaphthalen-2-yl)diazenyl]-3-methylphenyl]-2-methylphenyl]diazenyl]naphthalene-1-sulfonate
Discoverer: Carl Duisberg, 1885
Contemporary terminology: Benzopurpurine 4B (En), Benzo purpurine 4B (Fr), Benzopurpurin 4B (Ge), Benzopurpurine 4B (NL)
Obsolete terminology: Eclipseroth, Echtscharlach, Azamin 4B and Victoriaroth (Ge)
Chemical formula: $C_{34}H_{28}N_6O_6S_2$ (Fig. 18.5)

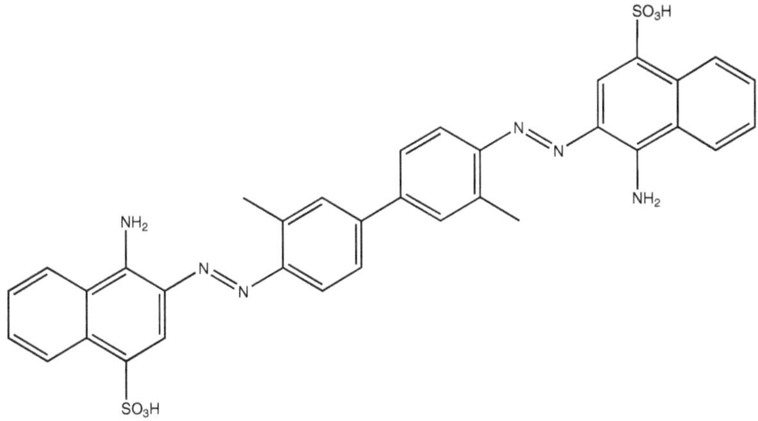


**Fig. 18.5** Structural formula of Benzopurpurine 4B

## *History*

As mentioned above, Benzopurpurine B was less useful than Congo red due to its dull colour, so the Bayer chemists experimented with alternative coupling agents. A little later, Duisberg achieved a great success. One product turned out to be an even better dye than Congo red: it had a more beautiful shade than Congo red and also had better fastness properties. It was made by coupling 1-naphthionic acid (4-aminonaphthalene-1-sulfonic acid) with doubly diazotised *o*-tolidine (3,3′-dimethylbenzidine) and was sold as Benzopurpurine 4B (G.P. 35.615). This patent became an addition to AGFA's Congo red G.P. 28.753 and was transferred to AGFA (see Table 18.3). The improved property as a dye arose from the fact that it was less sensitive to acid due to the two methyl groups in the *o*-tolidine molecule [1, 3, 8–10, 12].

Although 1 kg of Benzopurpurine 4B cost 12.50 German marks, Bayer could not meet the tremendous demand. For many years Benzopurpurine 4B remained the most well-known and profitable of all Bayer products. This invention was the last-minute breakthrough that saved the Bayer firm: not only was it crucial for the future of the company, but also for Duisberg himself. His contract was extended for a further 6 years and his salary was raised from 2,700 to 6,000 German marks per year. In addition, he was promised a 2.5% share of the next income from his inventions. His first share equalled 9,000 German marks. With that money, he bought shares in the Bayer company and because of the financial crisis, he could obtain them for 80% of their nominal value. In his memoirs Duisberg wrote, 'I felt like a partner—even like an owner—and acted accordingly' [1, 3, 12]. At the age of 26, he was already in charge of nine chemists. With his commitment as a 'joint owner', he made two decisions that would play a decisive role in the company's future. One was to concentrate scientific research in a single central laboratory. The other suggestion was to subject patents, including those from competitors, to systematic

**Table 18.3** International patents of Benzopurpurine 4B

| Discoverer, company | Patent | Title |
|---|---|---|
| AGFA | G.P. 35.615 filed 17th of March 1885, issued 13th of May 1886 | Verfahren zur Darstellung von Azofarbstoffen durch Combination von Tetrazoditolyol mit α und β Naphtylamin oder deren Mono- und Disulfonsäuren |
| Bayer & Co. | B.P. 3.803 filed in 1885 | |
| Carl Duisberg | U.S.P. 329.632 filed 29th of July 1885, issued 3rd of November 1885 | Coloring-Matter obtained from Tetrazo-Ditolyl |
| Bayer & Co. | F.P. 167.876 | |
| Dawson & Hirsch | B.P. 14.432 filed in 1890 | |
| BASF | B.P. 6.697 filed in 1895 | |
| BASF | B.P. 17.260 filed in 1895 | |
| BASF | G.P. 84.893 filed 27th of March 1895, issued 30th of December 1895 | Verfahren zur Darstellung von Azofarbstoffen der Benzidinreihe aus Monoazofarbstoffen |
| BASF | G.P. 88.597 filed 17th of July 1895, issued 22th of September 1896 | Verfahren zur Darstellung von Azofarbstoffen der Benzidinreihe aus Monoazofarbstoffen |
| BASF | F.P. 248.210 | |

**Table 18.4** Dividend of the Farbenfabriken Friedr. Bayer & Co., period 1885–1891

| Year | Dividend (%) |
|---|---|
| 1885 | 0 |
| 1886 | 4 |
| 1887 | 7 |
| 1888 | 12 |
| 1889 | 15 |
| 1890 | 17 |
| 1891 | 18 |

evaluation. In 1883 alone, the Imperial Patent Office in Berlin had granted 3,062 patents in the field of chemical technology and of these patents, 1,791 were interesting for Bayer [3]. Readers of the Berlin financial newspaper *Allgemeine Börsen-Zeitung* who took the newspaper's advice and sold their Bayer shares would be sorry: the dividend rose from zero in 1885 to 18% in 1891 (see Table 18.4) [3].

## *Trade Names*

Benzopurpurine 4B was put on the market under this trade name by AGFA, Bayer, Griesheim-Elektron, Leonhardt, Weiler-ter Meer, Levinstein, Beyer & Kegel, Sandoz, Farbwerk Amersfoort, Österreichischer Verein für Chemische und Metallurgische Produktion, Steiner, and Wiescher. Other trade names were used, such as Baumwollrot 4b by BASF; CIBA and Kalle; Dianilrot 4B by Farbwerke ML&B; Diaminrot 4B by Cassella; Sultan 4B by Read Holliday; Direct Red 4B by Schoellkopf; and Rouge Cotton BP by Soc. Anon. St. Denis. Further names are Rouge Diamine 4B; Azidinrot 4B; Alkalipurpurrot 4B; Diphenylrot 4B; Benzopurpurine 4B conc. For printing; Benzopurpurine 4B new; Amanil Purpurine 4B; Benzanil Purpurine 4B; Diazine Red 4B; Sultan 4B; Cotton Red 4B; Purpurine 4B, 4B conc; Erie Red 4B; Chokusetsu Red 4B; Benzopurpurine P4B; Rouge direct 4B; and Purpurine diazole N4B [16, 26–28, 30].

## *Applications*

Benzopurpurine 4B has mainly been used to dye cotton and linen in a single bath without the aid of a mordant.

## *Properties*

Benzopurpurine 4B forms a dark-red powder which dissolves easily in water, with a red-orange colour. The advantages of Benzopurpurine 4B are its direct and easy application to cotton goods without using mordants and the low price. The disadvantages are a poor light and acid fastness and an inadequate fastness to washing. In his 1893 book, Lehne wrote:[10] 'After 6 days of light exposure (September) the colour of Benzopurpurine 4B changed strongly brownish'. The Colour Index report a light fastness of ISO 1 and a wash fastness of ISO 2.

## *Recipes*

For dyeing wool, the same dyeing recipe is used as for Congo red.

Soxlett wrote:[11] 'All the preconditions, such as dissolving the dye, being careful in the case of calcareous water, and removing soap that may form, remain the same as Congo red. To obtain the red, 2.5% Marseille soap, 5% water glass (sodium silicate) and 3% Benzopurpurine 4B are boiled for 1 h. The red is very sensitive to acids and has also a poor light fastness. To improve the acid-fastness the goods are given a slightly alkaline finish after their completion' [24].

[10] Benzopurpurin 4B: Echtheit: 'Nach 6 tägiger Einwirkung der Witterung (September) stark verblasst (bräunlich)'.

[11] Alle Vorbedingungen, wie Lösen des Farbstoffes, Vorsicht wegen des kalkhaltigen Wassers, und Abnahme der sich vielleicht bildenden Kalkseife bleiben dieselben als bei Congorot. Zu dem hier gegebenen Rot wurden: 2,5% Marseiller Seife, 5% Wasserglas und 3% Benzopurpurin 4B; eine Stunde kochend angefärbt. Das Rot ist sehr säureempfindlich und auch sehr schwach gegen Lichteinflüsse, die Säurewiderstandfähigkeit kann verbessert werden, wenn man den Waren nach ihrer Fertigstellung eine etwas alkalische Appretur gibt.

### *Identification Using UHPLC-PDA*

Analysis of a sample of Benzopurpurine 4B (RCE no. 7170) showed the presence of a clear, single peak (Fig. 18.6).

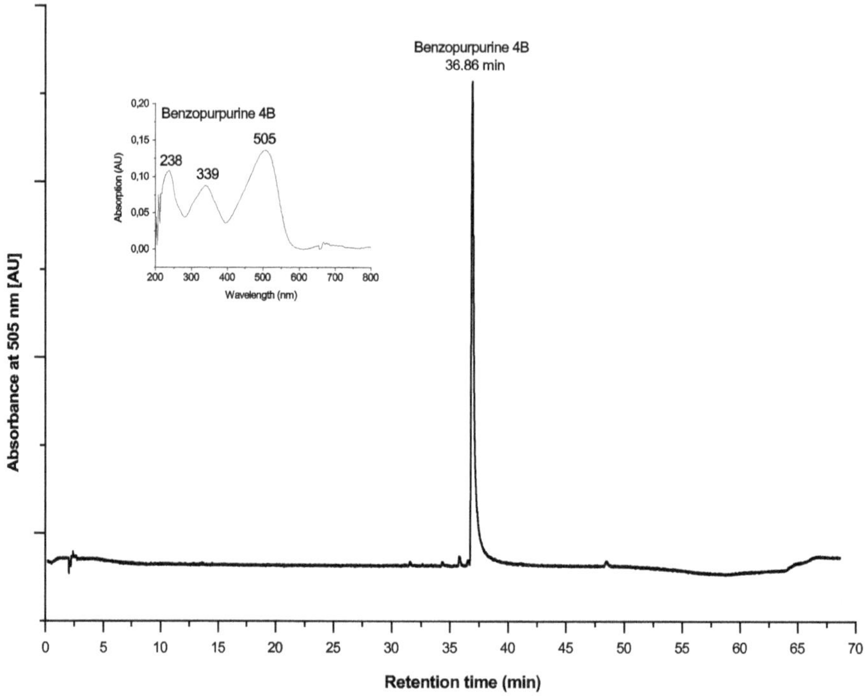


**Fig. 18.6** Chromatogram and spectrum of Benzopurpurine 4B (RCE no. 7170)

## 18.6 Patent Struggles

Benzopurpurine 4B was not only a dangerous competitor for Congo red and AGFA: it also soon surpassed Congo red because it was brighter in colour and less sensitive to acid. AGFA therefore filed opposition against the Benzopurpurine patent because the two dyes were too similar in their chemistry. The only difference, AGFA claimed, lay in the diazo-compound, benzidine in the case of Congo red and *o*-tolidine for Benzopurpurine 4B. The German court ruled that Benzopurpurine was not patentable and decided in favour of AGFA. Bayer's management, especially Duisberg, advised strongly against appealing against the judgment in favour of AGFA. Instead, executives of Bayer and AGFA met in the offices of the Deutsche Bank in Berlin and decided to cooperate. After prolonged secret negotiations, the result was that Bayer's German Benzopurpurine B and Benzopurpurine 4B patent no. 35.615 became an addition to AGFA's Congo red patent, G.P. 28.753. In this way the Benzopurpurine dye patents were transferred to AGFA and a costly patent lawsuit was avoided. Both firms agreed on a cross-licensing arrangement, forming a cartel known as the Benzopurpurine Convention, to protect their mutual interests for the manufacture and sale from each other's dyes, including Congo red and Benzopurpurines. In this way both companies secured an almost impregnable position in the field of substantive cotton dyes, leaving other manufacturers hardly any chance in this area.

Some years later, in 1889, a greater threat for both companies came from the small firm Ewer & Pick in Berlin-Grünau. This company brought Benzopurpurine 4B and Congo red on the market produced through slightly different processes. The Benzopurpurine Convention claimed infringement by Ewer & Pick, and the first judgment ruled in favour of the convention. Ewer & Pick appealed, claiming that the original Congo red patent was invalid. The appeal court rejected their claim and Ewer & Pick filed a second appeal. In 1889, the dispute reached the highest German court, the Imperial Court of Justice in Leipzig. The final day in court took place on the 20th of March and, from the very beginning of the day, there was a tense atmosphere in the courtroom. In the end, the judges decided that the Congo red patent was valid because it represented a new technical development. Accordingly, Ewer & Pick was denied access to Congo red and the Benzopurpurine dyes, as were all other competitors [1, 9, 11, 12].

## 18.7 Other Direct or Substantive Cotton Dyes in Different Colours

Further research by both Bayer and AGFA followed. In 1885 the first direct blue cotton dye Azo blue was invented and patented by Carl Duisberg. Azo blue (C.I. name: Direct Violet 39; C.I. number: 23680) was formed by the coupling of Nevile-Winther acid (1-naphthol-4-sulfonic acid, NW-acid) with diazotised *o*-tolidine (G.P. 35.341, patented on the first of August 1885; U.S.P. 366.078, filed on the 22nd of August 1885; B.P. 9.510, patented in 1885, and F.P. 171.133, patented in 1885) [1, 8–10, 12, 16, 31–34]. (Fig. 18.7)

Soon Azo blue was followed by Benzoazurine G (C.I. name: Direct Blue 8; C.I. number: 24140), made by coupling Nevile-Winther-acid with diazotised dianisidine, which was faster than Azo blue and had a very clear and attractive blue shade. (Fig. 18.8) This dye was also discovered by Duisberg and patented on the 19th of November 1885 (G.P. 38.802). Other international Benzoazurine G patents are the U.S.P. 357.273, filed on the 14th of December 1885; U.S.P. 481.934, filed on the 16th of July 1891; B.P. 14.424, filed on the 24th of November 1885; F.P. 173.042, filed on the 19th of December 1885; Austria-Hungary patents 2.432 and 2.587, both filed on the 7th of May 1886, and the Italian patent no. 19.967, filed on the 26th of May 1886. In the summer of 1886 Benzoazurine G was commercialised and introduced to the market. After it was put on the market, however, buyers found it would fade rapidly under the influence of sunlight. Unfortunately,

**Fig. 18.7** Structural formula of Azo blue

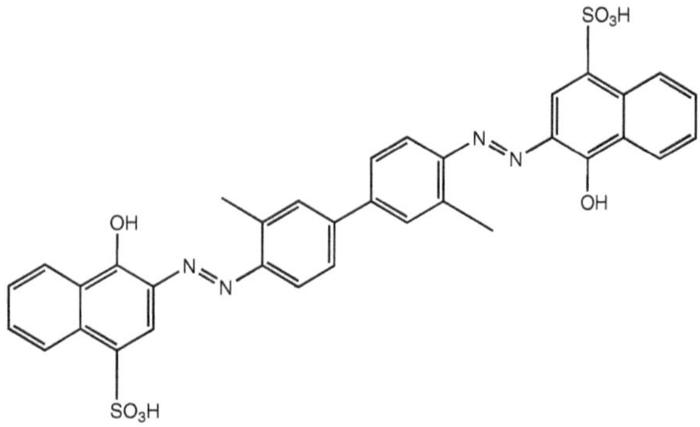


**Fig. 18.8** Structural formula of Benzoazurine G

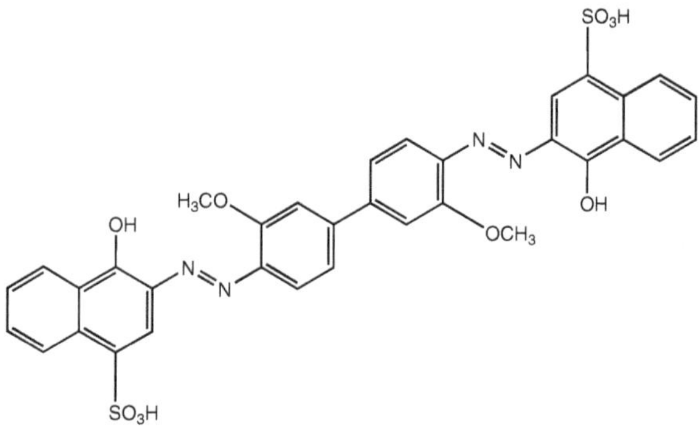


the fastness to light was not quite satisfactory, so Benzoazurine G could not compete with natural Indigo. A patent dispute with AGFA on Benzoazurine G very nearly took place, but negotiations with the active participation of Duisberg solved this issue [1, 8–10, 12, 34–37].

Lehne writes on Azo blue:[12] 'After 6 days of weather exposure, redder and slightly dull, after 14 days strongly faded'. On Benzoazurine G he wrote:[13] 'After 14 days of weather exposure (September) much redder'. According to the Colour Index, the light fastness of Azo blue is ISO 1–2 and for Benzoazurine G it is ISO 1. This resembles the information presented by Lehne.

Around 1890, substantive dyes in different colours were marketed using the name Congo, including Congo rubine, Congo Corinth, Brilliant Congo, Congo orange, Congo brown, Congo blue and Congo violet, as mentioned above (see Table 18.5). Other dye factories also produced direct dyes under their trade names: Cassella named them Diamine; BASF used the word Oxamine, and Farbwerke ML&B named them Dianil.

[12] Azoblau: Echtheit: 'Nach 6 tägiger Einwirkung der Witterung röther und etwas matter, nach 14 Tagen stark verblasst'.

[13] Benzoazurin G: Echtheit: 'Nach 14 tägiger Einwirkung der Witterung (September) viel röther'.

**Table 18.5** Some important substantive dyes produced by AGFA

| Name | C.I. name | C.I. number | Discoverers | Year of discovery |
|---|---|---|---|---|
| Congo red | Direct Red 28 | 22120 | P. Böttiger | 1883 |
| Benzopurpurine B | Direct Red 21 | 23560 | C. Duisberg | 1885 |
| Benzopurpurine 4B | Direct Red 2 | 23500 | C. Duisberg | 1885 |
| Benzopurpurine 10B | Direct Red 7 | 24100 | C. Duisberg | 1885 |
| Benzopurpurine 6B | Direct Red | 23530 | S. Pfaff, C. Duisberg, AGFA | 1885 |
| Chrysamine G | Direct Yellow 1 | 22250 | E. Frank | 1884 |
| Chrysamine R | Direct Yellow 2 | 23640 | E. Frank | 1884 |
| Congo Rubine | Direct Red 17 | 22150 | G. Schultz | 1891 |
| Congo Corinth G | Direct Red 10 | 22145 | S. Pfaff | 1885 |
| Congo orange R | Direct Orange 13 | 23605 | O. Borgmann | 1889 |
| Congo orange G | Direct Orange 101 | 22190 | O. Borgmann | 1889 |
| Congo brown (G) | Direct Brown 6 | 30140 | J. Strasburger | 1888 |
| Congo brown (R) | Direct Brown 21 | 30155 | J. Strasburger | 1888 |
| Congoechtblau B | Direct Blue 30 | 31955 | G. Schultz | 1890 |
| Congoechtblau R | Direct Blue 63 | 31910 | G. Schultz | 1890 |
| Azo blue | Direct Violet 39 | 23680 | C. Duisberg | 1885 |
| Benzoazurine G | Direct Blue 8 | 24140 | C. Duisberg | 1885 |
| Congo GR | Direct dye | 22000 | S. Pfaff, C. Martius | 1885 |
| Brilliant Congo (R) Brilliant Congo (G) | Direct Red 34 | 23570 | R. Krügener | 1886 |
| Congo 4R | Direct Red 68 | 23515 | S. Pfaff | 1885 |
| Congo Korinth B | Direct Violet 21 | 23520 | S. Pfaff | 1885 |

# References


1. Werkgeschichte Farbenfabrikn VORM. Friedr. Bayer & Co. 1863–1938 Der Gefolgschaft der Werke Leverkusen, Elberfeld und Dormagen zur Erinnerung an die 75. Wiederkehr des Gründungstages der Farbenfabriken vorm. Friedr. Bayer & Co. F. Bruckmann KG., München
2. Verg, E., Plumpe, G., Schuttheis, H.: Benzopurpurine 4B: Bayer's life-saver. In: Verg, E. (ed.) Milestones – The Bayer Story 1863–1988, Bayer AG Leverkusen. pp. 74–79. Informedia-Verlags-GmbH (1988)
3. Verg, E., Plumpe, G., Schuttheis, H.: Benzopurpurin 4B: Die Rettung in der Not. In: Verg, E. (ed.) Meileinsteine. 125 Jahre Bayer 1863–1988. pp. 74–79. Informedia-Verlags-GmbH (1988)
4. Hofenk de Graaf, J.H.: The Colourful Past, Origins, Chemistry and Identification of Natural Dyestuffs. Abegg Stiftung/Archetype Publications Ltd., Riggisberg / London (2004)
5. Cardon, D.: Natural Dyes: Sources, Tradition, Technology and Science. Archetype Publications, London (2007)

6. Chenciner, R.: MADDER RED: A history of luxury and trade: Plant dyes and pigments in world commerce and art. Taylor and Francis (2003)
7. Schmidt, H., Schultz, G.: Ueber Benzidin (alpha-Diamidodiphenyl). Justus Liebig's Annalen der Chemie, C.F. Winter'sche Verlagshandlung, Leipzig und Heidelberg. 207, 320–345 (1881)
8. Homburg, E., Travis, A.S., Schröter, H.G.: The Chemical Industry in Europe, 1850–1914, Industrial Growth, Pollution, and Professionalization. Kluwer Academic Publishers, Dordrecht/Boston/London (1998)
9. Reinhardt, C., Travis, A.S.: Heinrich Caro and the Creation of Modern Chemical Industry. Kluwer Academic Publishers, Dordrecht/Boston/London (2000)
10. Geschichte und Entwicklung der Farbenfabriken vorm. Friedr. Bayer & Co. Elberfeld in den ersten 50 Jahren. Meisenbach Riffarth & Co., München (1918)
11. Murmann, J.P.: Knowledge and Competitive Advantage, The Coevolution of Firms, Technology, and National Institutions. Cambridge University Press, New York (2003)
12. Sonnemann, R., Etzold, H.: Patent und Monopol, Eine Studie zur Herausbildung von Monopolen unter dem Einfluß des Patentwesens. Jahrbuch für Wirtschaftsgeschichte/Economic History Yearbook. 6, 147–203 (1965)
13. Abelshauser, W.: Die BASF, Eine Unternehmensgeschichte, Zweite Auflage. Verlag C.H. Beck oHG, München (2003)
14. Cooksey, C.J., Dronsfield, A.: Peter Griess: a critical evaluation of his contribution to azo dye chemistry. The Diversity of Dyes in History and Archaeology. 354–357 (2017)
15. Steensma, D.P.: "Congo" Red, Out of Africa? Arch Pathol Lab Med. 125, 250–252 (2001)
16. Schultz, G., Julius, P.: Tabellarische Übersicht der künstlichen organischen Farbstoffe. R. Gaertner's Verlagsbuchhandlung, Berlin (1888)
17. Schultz, G., Julius, P.: Tabellarische Übersicht der künstlichen organischen Farbstoffe, Zweite verbesserte und vermehrte Auflage. R. Gaertner's Verlagsbuchhandlung, Berlin (1891)
18. Schultz, G., Julius, P.: Tabellarische Übersicht der im Handel befindlichen künstlichen organischen Farbstoffe, Vierte verbesserte und stark vermehrte Auflage. R. Gaertner's Verlagsbuchhandlung, Berlin (1902)
19. Schultz, G.: Farbstofftabellen, Fünfte Auflage. Weidmannische Buchhandlung, Berlin (1914)
20. Schultz, G.: Farbstofftabellen, Siebente Auflage. Akademische Verlagsgesellschaft M.B.H., Leipzig (1931)
21. Communication from Edmund Knecht PH.D., 'Congo Red', The Journal of the Society of Dyers and Colourists. 146–147 (1885)
22. New Colouring Matters, Communication from Edmund Knecht PH.D., 'Benzopurpurine'. The Journal of the Society of Dyers and Colourists. 267 (1885)
23. Holsboer, H.B.: De synthetische kleurstoffen, hare toepassing in de ververij en hare bereidingswijzen. P. Noordhoff, Groningen (1918)
24. Soxhlet, V.H.: Die Färberei der Baumwolle mit direkt färbenden Farbstoffen. Verlag der J.G. Cotta'schen Buchhandlung, Stuttgart. 47–50 (1891)
25. Bayer & Co., B.P. U.S.P. 401.024 (1889), F.P. 162.880
26. Schultz, G., Julius, P.: Tabellarische Übersicht der künstlichen organischen Farbstoffe, Zweite verbesserte und vermehrte Auflage. R. Gaertner's Verlagsbuchhandlung, Berlin (1891)
27. Schultz, G., Julius, P.: Tabellarische Übersicht der im Handel befindlichen künstlichen organischen Farbstoffe, Vierte verbesserte und stark vermehrte Auflage. R. Gaertner's Verlagsbuchhandlung, Berlin (1902)
28. Schultz, G.: Farbstofftabellen, Fünfte Auflage. Weidmannische Buchhandlung, Berlin (1914)
29. Schultz, G.: Farbstofftabellen, Siebente Auflage. Akademische Verlagsgesellschaft M.B.H., Leipzig (1931)
30. Schultz, G.: Farbstofftabellen, Siebente Auflage. Akademische Verlagsgesellschaft M.B.H., Leipzig (1931)

31. Schultz, G., Julius, P.: Tabellarische Übersicht der künstlichen organischen Farbstoffe, Zweite verbesserte und vermehrte Auflage. R. Gaertner's Verlagsbuchhandlung, Berlin (1891)
32. Schultz, G.: Farbstofftabellen, Fünfte Auflage. Weidmannische Buchhandlung, Berlin (1914)
33. Schultz, G.: Farbstofftabellen, Siebente Auflage. Akademische Verlagsgesellschaft M.B.H., Leipzig (1931)
34. Verg, E., Plumpe, G., Schuttheis, H.: Echter als Indigo, aber nur im Winter. In: Meilensteine - 125 Jahre Bayer 1863-1988, Bayer AG Leverkusen. pp. 80–83. Informedia-Verlags-GmbH (1988)
35. Schultz, G., Julius, P.: Tabellarische Übersicht der künstlichen organischen Farbstoffe, Zweite verbesserte und vermehrte Auflage. R. Gaertner's Verlagsbuchhandlung, Berlin (1891)
36. Schultz, G.: Farbstofftabellen, Fünfte Auflage. Weidmannische Buchhandlung, Berlin (1914)
37. Schultz, G.: Farbstofftabellen, Siebente Auflage. Akademische Verlagsgesellschaft M.B.H., Leipzig (1931)

# Chapter 19
# Sulfur Dyes in Nearly Every Colour and Shade

## 19.1 Introduction

The sulfur dyes represent a huge dye-class and were the last great invention in the field of the early synthetic dyes before the First World War. In two editions of the book *Farbstofftabellen* by the German chemist Gustav Schultz (1851–1928), published in 1914 (and again in 1931), respectively 52 and 76 different sulfur dyes are listed, with hundreds of trade names and protection by almost 1000 international patents filed by many dye factories [1–3]. This clearly shows their importance. The sulfur dyes were first produced by heating organic materials, such as sawdust, bran, straw and other lignin-containing matter, with sulfur and sodium sulfide. Later on, aromatic or heterocyclic products, from the coal tar industry, for example, were used as the source and heated either with sulfur or with sulfur-containing compounds. The sulfur dyes usually consist of complex mixtures; as a result, their chemical composition is not well-defined and is poorly understood [4]. Yet these dyes were widely used for dyeing cotton and played an important role, particularly in black dyeing. They are produced in all colours, but the black, brown, blue and green dyes are of the greatest importance, hence this chapter will mainly focus on these colours.

## 19.2 History of Black Dyes

Before the sulfur dyes entered the market, other synthetic black colourants were already being used for dyeing and printing, the first being Aniline black (C.I. name: Pigment Black 1; C.I. number: 50440), discovered by the English calico printer



M. R. van Bommel, M. de Keijzer, *Bright Colours from the Past*, Cultural Heritage Science, https://doi.org/10.1007/978-3-031-88664-5_19

John Lightfoot (1832–1872) in 1859. He patented this process in 1863 (B.P. 151). In the following decades, the process underwent many innovations and improvements in order to achieve stable dyes, which cause minimum damage to the fibre when applied. In the 1860s Aniline black was already favoured over the other black colourants then in use, such as natural dyes extracted from tannins and Logwoods, which require long dyeing procedures and had a negative effect on the textile fibre. In 1867, the French chemist Jean Théodore Coupier (1820-1908) discovered Nigrosine (C.I. name: Acid Black 2; C.I. number: 504200), an azine dye with a good light fastness of ISO 5. This dye was patented in the same year in both Britain and France (B.P. 3657, filed in 1867 and F.P. 77854) [5]. Another way of producing black was by overdyeing a textile using the natural dyes Madder, Indigo and/or Weld, which was obviously a laborious process as well. This process was used for centuries, and although laborious did not harm the textile fibre as is the case with tannins for example [6]. It must be noted that it was also possible to create a black colour by overdyeing with synthetic dyes, such as Malachite green (C.I. name: Basic Green 4; C.I. number: 42000) and Fuchsine (C.I. name: Basic Violet 14; C.I. number: 42510). At the end of the nineteenth century, the use of Aniline black reached enormous proportions for textile printing, but it then suffered a considerable decrease in popularity, due to the introduction of the sulfur blacks, which were cheaper and easier to use [7].

In the second half of the 1880s two basic black dyes were discovered by the German chemist Fritz Bender (1859–1908), working for Farbwerk Mühlheim vorm. A. Leonhardt & Co. near Frankfurt am Main. The first was Diazine black (C.I. name: Basic Black 2; C.I. number: 11825), developed in 1886, and in 1889, Fast black (C.I. name: Basic Black 7; C.I. number: 51215) was discovered (G.P. 50.612, B.P. 8.264, both filed in 1889, and F.P. 198.511). Regrettably, both dyes gave poor results. In general, these dyes were only used for products that were exposed to low lighting conditions, such as lining fabrics.

However, acid black dyes were used more often in the last quarter of the nineteenth century. In 1870, the German industrialists Fritz Gans (1833–1920), his brother and chemist Leo Gans (1843–1935) and their brother-in-law Bernhard Weinberg (1815–1877), together with the German chemist August Leonhardt (1827–1899), founded the Frankfurter Anilinfarbenfabrik von Gans & Leonhardt. In 1883, the German chemist Arthur Weinberg (1860–1943) joined this dye factory, run by his uncle Leo Gans. This factory supplied dyes to Cassella in Frankfurt am Main and in 1894 the factory was integrated into the Leopold Cassella company.

In 1885, Arthur Weinberg, together with the German chemist Meinhard Hoffmann (1853–1936), succeeded in inventing the first black acid diazo dye Naphthol black B (C.I. name: Acid Black 5; C.I. number: 27260), which was patented and marketed by Cassella [8]. Three years later, in 1888, Weinberg had much greater success with the synthesis of Naphthylamine black D [9]. However, this dye had one disadvantage: due to its purple hue it had to be nuanced by yellow or green synthetic acid dyes. Hoffmann developed Amido black in 1891, which is also a diazo dye.

The following three acid dye groups discovered in the early 1890s are also noteworthy. Firstly, the Naphthol blue-black types of Cassella, discovered and patented by Meinhard Hoffmann in 1891 [10]. Similar dyes include Coomassie blue-black from the dye factory Levinstein Limited in Blackley near Manchester, and the Sulfone Cyanines by Bayer in Leverkusen in 1892 (G.P. 118.655 and G.P. 122.065, both filed in 1892) [11, 12].

At the same time as the first black azo dyes were discovered, the first black substantive (also called direct) cotton dyes were developed. They have the advantage of dyeing cotton directly in one bath without the aid of a mordant. An example is the diazo dye Diamine black BO (C.I. name: Direct dye; C.I. number: 24040), discovered by Gans in 1889. In the following years numerous similar products were discovered, but no fundamental advances were made.

Progress was made 10 years later with Direct deep black E (C.I. name: Direct Black 11; C.I. number: 30249). This dye was discovered for the first time by the German chemist Myrtil Kahn (1863–1939) and produced by Bayer in 1898. However, the company overlooked making an application for a patent and in 1901 it was rediscovered by the German chemist Oscar Mueller, who filed and assigned the patent to the American dye company Schoellkopf, which named the dye Erie Direct black GX (G.P. 153.557 and U.S.P. 688.478, both filed in 1901). Other dye firms attempted to produce equivalent black dyes by circumventing patents, which led to patent difficulties. In 1900, CIBA in Basel brought Carbidschwarz E onto the market, and Cassella produced Oxydiaminschwarz JW. In 1903, AGFA in Berlin followed with Columbiaschwarz EAW. But actually, these are all different trade names for Direct Black 11 [13, 14]. The dispute only came to an end with the expiry of the main patent in 1916. The black dye was cheap, and even though it had only moderate fastness, it was used in vast quantities. In 1945, the production of Direct deep black E amounted to approximately five million kilograms [15].

At the end of the nineteenth century, in 1896, Farbwerke ML&B in Frankfurt-Hoechst took on the production and marketed their first substantive dye under the trade name Dianil black R (C.I. name: Direct dye; C.I. number: 30180). This dye was discovered by the German chemists Friedrich Schmidt and Otto Ernst (1870–1936) in 1894 (G.P. 89.285, filed in 1894, and U.S.P. 582.958, filed in 1897). One year later, in 1895, the same chemists expanded the Dianil group with Dianil black N (C.I. name: Direct dye; C.I. number: 30170) and Dianil black G (C.I. name: Direct dye; C.I. number: 30175) (G.P. 91.036, filed in 1895) [16].

## 19.3 Cachou de Laval

Colour Index name: Sulphur Brown 1; Colour Index number: 53000
IUPAC name: not available
Discoverers: Louis Marie François Bretonnière and Edouard Croissant, 1873
Contemporary terminology: Cachou de Laval (En), Cachou de Laval (Fr), Cachou de Laval (Ge), Cachou de Laval (NL)

Chemical formula: The chemical formula and structure of Cachou de Laval are unknown. It contains a mixture of products which have in common that one or more sulfide bonds are integrated into the chromophore.

## History

The black sulfur dyes attained even more importance than the substantive blacks. The first sulfur dye, Cachou de Laval, was prepared in France by the French chemist Louis Marie François Bretonnière (1842-1902) and his business partner Edouard Croissant. According to their patents (F.P. 98.915, filed on April 12, 1873, U.S.P. 142.892, filed on May 15, 1873 and B.P. 1.489, filed in 1873) they heated various organic materials—such as sawdust, starch, straw, bran, horn, rubber, wool and other organic waste products—with sodium sulfide and sulfur. The component responsible for the colour is lignin, which reacts with the sulfides. The products obtained were porous black masses with an odious sulfurous smell of mercaptan.

The only product that gained importance was the dye Cachou de Laval, made by heating sawdust with sodium sulfide between 100–350 °C; the higher the temperature, the darker the colour. Cachou de Laval was patented by several firms and manufactured for many years, but was replaced by stronger and purer products of analogous types [17–19] (see Table 19.1).

During the First World War, the Cachou de Laval dyes regained their importance. In 1916, the import of German synthetic dyes to France ceased, so sawdust, bark, hay, starch, herbs and similar starting materials were again used for the manufacturing of sulfur dyes. The best product was a brown dye obtained by using peat, sodium sulfide and sulfur at 120 °C, and a black-brown by fusing pea starch in the presence of copper sulfate [20].

## Trade Names

**Table 19.1** Cachou de Laval dyes and their manufacturers

| Cachou de Laval types | International dye companies |
|---|---|
| Cachou de Laval | Société Anonyme des matières colorantes et produits chimiques de St. Denis |
| Cattu Italiano | Lepetit, Dollfuß & Gansser |
| Kryogenbraun D | Badische Anilin- & Soda-Fabrik |
| Katigenschwarzbraun | Farbenfabriken vorm. Friedr. Bayer & Co. |
| Sulfanilinbraun B | Chemische Fabrik Kalle & Co. |
| Pyrogenbraun D | Gesellschaft für chemische Industrie |
| Thionalbraun B | Chemische Fabrik vorm. Sandoz |
| Sulphur Brown BF | Kern & Sandoz |
| Thiogene Drab G | Read Holliday & Co. |
| Atlantic Brown K | Atlantic Dyestuff |
| Cross Dye Deep Brown R | British Alizarine Company, Ltd. |

## Applications

Cachou de Laval dyes cotton, linen, hemp and jute with greenish-yellow shades, which change to brown on exposure to the air.

## Properties

The colour is fast to washing, but only moderately fast to light. The light fastness is improved by after-treatment with iron, copper or chromium mordanting salts. Adolf Lehne wrote in his 1893 book:[1] 'After 8 days of exposure to the weather (July) unchanged. – Wash fast'. This is likely to be the dye which was treated with a chromium salt afterward as presented in the recipe below. According to the Colour Index, the light fastness is ISO 3 to 5, while the wash fastness is ISO 4 to 5.

## Recipes

Lehne mentions that dyeing was often done in combination with basic or direct cotton dyes, or with Alizarin. However, the recipe given does not elaborate on the exact procedure and focuses on dyeing with Cachou de Laval alone:[2] 'For dyeing cotton or linen (jute, hemp), mostly in combination with basic or substantive cotton dyes, or with Alizarin according to Erban & Specht's dyeing process. Dye with the addition of about 5% sodium chloride in 20–25 min; rinse; treat in a warm potassium chromate bath (approximately 3%), rinse and dry´.

---

[1]Cachou de Laval: Echtheit: ´Nach 8 tägiger Einwirkung der Witterung (Juli) unverändert.—Waschecht´.

[2]Zum Färben von Baumwolle oder Leinen (Jute, Hanf), meistens in Combination mit basischen oder substantiven Baumwollfarbstoffen, mit Alizarin nach dem Erban & Specht'schen Verfarhen. Man färbt unter Zusatz von etwa 5% Kochsalz in 20–25 Minuten aus, spült, behandelt in warmen Chromkalibad (ca. 3%), spült und trocknet.

## 19.4 Vidal Black

Colour Index name: Sulphur Black 3; Colour Index number: 53180
IUPAC name: not available
Discoverer: Henri Raymond Vidal, 1893
Contemporary terminology: Vidal black (En), Noir Vidal (Fr), Vidalschwarz (Ge), Vidalzwart (NL)
Chemical formula: Unknown. The dye consists of a mixture of dyes; the chemistry is poorly understood.

### *History*

In the 1890s, the invention of three black dyes led to a boost for the sulfur dyes. The first sulfur dye produced from a known intermediate was Vidal black. In October 1893, the French chemist Henri Raymond Vidal (1862–1938) obtained a number of new dyes by fusing different derivatives of benzene and naphthalene, such as phenols and quinones, with alkali sulfides at a temperature of 160–210 °C. Oxidation by air produced a blackish-green which, with an after-treatment using oxidising substances, such as potassium dichromate or iron chloride, resulted in a blue-black colour [21].

Two months later, in December 1893, the organic derivatives *p*-phenylenediamine and *p*-aminophenol were introduced. Vidal found that *p*-aminophenol was an especially good substance for producing a black sulfur dye when treated with sodium sulfide and sulfur at a temperature of 180–210 °C. The product had a dark greenish-black colour and, following an after treatment with oxidising substances such as potassium dichromate, iron sulfate or copper sulfate, a solid black dye was obtained. Vidal black was marketed by Société Anonyme St. Denis in France using the trade name Noir Vidal (G.P. 85.330, filed in 1893, and U.S.P. 618.152, filed in 1899).

Vidal black was the breakthrough for the sulfur dyes and during the next 10 years in the laboratories of the dye factories, especially in Germany and Switzerland, almost every organic substance was subjected to the sulfide condensation. As an example, at the same time that Vidal black appeared, the French-German-Swiss chemist René Bohn (1862–1922), who was working at BASF in Ludwigshafen am Rhein, discovered Fast black B. This was obtained by heating 1,8-dinitronaphthalene with an aqueous solution of sodium sulfide [22].

### *Trade Names*

The dye was put on the market under the trade name Noir Vidal by Soc. Anon. St. Denis.

## *Applications*

Vidal black is especially used to dye cotton.

## *Properties*

Lehne writes in his 1899 book about the fastness:[3] 'Not changed after 6 weeks of exposure to the weather (May, June).' According to the Colour Index, Vidal black has a good light fastness (ISO 6) and wash fastness (ISO 5).

## *Recipes*

In the 1899 book of Lehne, the following recipe can be found:[4] 'Vidal black cotton dyeing. Dyeing with the addition of 10% sodium chloride and 10% sodium carbonate, with 5% potassium chromate and 5% sulfuric acid. According to F.P. 276612 of Hoelken & Co., the cotton should be taken out of the boiling bath (per kilo) and quickly wrung out several times. After the developing bath, the yarn is exposed to humid air at 40–65 °C for 6–12 hours and then rinsed thoroughly. It should be possible to obtain uniform colouring in this way.' Note that the concentration of Vidal black is rather high, 50% of the weight of the textile.

A printing recipe is given as well:[5] 'Cotton print. Printing ink: 200 g Vidal black S in paste, 40 g chromium acetate, 1 litre Traganth thickener (100:1000). After printing, steam for 1 hour without overpressure; wash with soap at 50 °C for 15 minutes.'

[3] Vidal-Schwarz: Echtheit: ´Nach 6 wöchentlicher Einwirkung der Witterung (Mai, Juni) nicht verändert´.

[4] Vidal-Schwarz in der Baumwollfärberei. Färben unter Zusatz von 10% Kochsalz und 10% Soda, mit 5% Chromkali und 5% Schwefelsäure entwickeln. Nach dem Franz. patent No. 276612 der Firma Hoelken & Co. soll die Baumwolle kilogrammweise aus dem kochenden Bad herausgenommen und schnell mehrmals ausgerungen werden. Nach dem Entwicklungsbad setz man das Garn in der Oxydationhänge feuchter 40–65 °C. heisser Luft während 6–12 stunden aus und spült sodann gründlich. Er sollen auf diese Weise gleichmässige Färbungen zu erhalten sein.

[5] Baumwolldruck. Druckfarbe: 200 g VidalSchwarz S in Teig, 40 g essigsäures chrom, 1 liter Traganth-Verdickung (100:1000) Nach dem Aufdruck 1 Stunde ohne Ueberdruck dämpfen, waschen und 15 Minuten bei 50 °C. seifen.

## 19.5 Sulphur Black T

Colour Index name: Sulphur Black 1; Colour Index number: 53185
IUPAC name: 2,10-Dinitro-12H-[1,4]benzothiazino[3,2-b]phenothiazin-3-one
Discoverer: Henri Raymond Vidal, 1896
Contemporary terminology: Sulphur black T (En), Noir soufre T (Fr), Schwefelschwarz T (Ge), Zwavelzwart T (NL)
Chemical formula: $C_{18}H_8N_4O_5S_2$. Note that in many sources sulfurised dinitrophenol is mentioned, but this is not the final product. It is likely that this consists of a mixture of sulfurised compounds (Fig. 19.1).

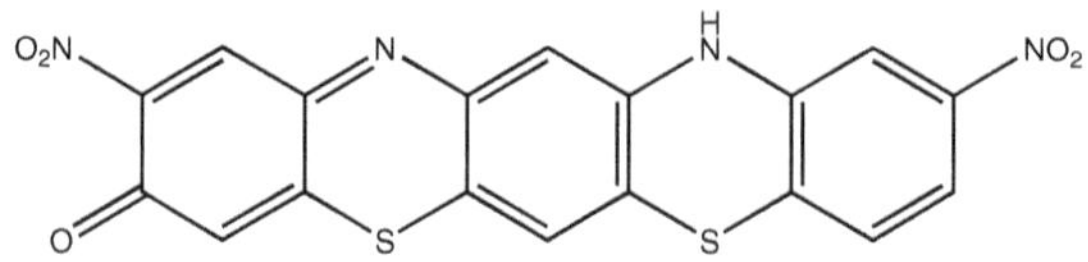


**Fig. 19.1** Structural formula of the main compound in Sulphur black T

### *History*

In 1896, Vidal obtained another black dye by heating diaminophenol with sodium sulfide and sulfur. In the patent, he mentions that diaminophenol can be replaced by 2,4-dinitrophenol. Together with sodium sulfide, 2,4-dinitrophenol is heated at 140 °C and after the reduction sulfur is added [23]. Vidal could not exploit this black dye because the chemical compound 2,4-dinitro-chlorobenzene (1-chloro-2,4-dinitrobenzene) necessary for the preparation of 2,4-dinitrophenol was not produced in France at that time. It seems that sometime later, around 1898, Soc. Anon. St. Denis produced it on an industrial scale.

Three years after Vidal's invention, in 1899, the German chemists Bernhard Priebs and Oskar Kaltwasser obtained the same black dye by boiling 2,4-dinitrophenol, sodium sulfide, sulfur and water for about 15 hours in a vessel fitted with a reflux condenser. The Berlin dye firm AGFA marketed it as Sulphur black T [24]. This is the most important member of the black sulfur dye class; many equivalent products were patented and manufactured by a number of dye firms, and it became one of the most widely used dyes worldwide (see Table 19.2) [25–30].

## Trade Names

**Table 19.2** Trade names of Sulphur black T and their manufacturers

| Sulphur black T types | International dye companies |
|---|---|
| Schwefelschwarz T | Actien-Gesellschaft für Anilin-Fabrikation |
| Thiogenschwarz | Farbwerke vorm. Meister, Lucius & Brüning |
| Immedialschwarz NN | Leopold Cassella & Co. |
| Kryogenschwarz TGO | Badische Anilin- & Soda-Fabrik |
| Katigenschwarz 2B | Farbenfabriken vorm. Friedr. Bayer & Co. |
| Auronalschwarz | Chemische Fabriken Weiler-ter Meer |
| Thionschwarz 2B | Chemische Fabrik Kalle & Co. |
| Pyrogenschwarz C | Gesellschaft für chemische Industrie |
| Thionalschwarz T | Chemische Fabrik vorm. Sandoz |
| Atlantic Black B | Atlantic Dyestuff |
| Thionone Black FA | Read Holliday & Co. |
| Thional Black BLG | British Dyestuff Corp. |
| Thionol Black O | Levinstein Limited Blackley |
| Thiophenol Black T | Clayton Aniline Co. |

## Applications

Sulphur black T can be used to dye cotton and linen. It can be also used for printing.

## Properties

In 1906 Adolf Lehne wrote [31]):[6] 'Unchanged after 14 days of exposure to the weather (July and August).' According to the Colour Index, the light fastness of Sulphur black B is between ISO 6 and 7, wash fastness is ISO 5.

## Recipes

In the book of Lehne from 1906, the following recipe is given:[7] 'To dye cotton and linen (recipe V12) in wooden vats or iron kettles (copper must be avoided); for 100 kg of material, use 2,500 litres of water, 5 kg of sodium carbonate, 40 kg of

---

[6] Immedialschwarz: Echtheit: ´Nach 14 tägiger Einwirkung der Witterung (Juli und August) unverändert´.

[7] Zum Färben von Baumwolle und Leinen (V12) in Holzkufen oder eisernen Kesseln (Kupfer muß vermieden werden); für 100 kg Material sind erforderlich: 2500 Liter Wasser, 5 kg Solvaysoda, 40 kg Farbstoff, 10 Minuten kochen, 100 kg Kochsalz, Dampf abstellen, mit der gut ausgekochten Baumwolle eingehen, umziehen, dann zum Kochen treiben, 1 Stunde ausfärhen. Satteres Schwarz

dye, boil for 10 minutes. Next, add 100 kg of sodium chloride, and turn off the heat; add the well (pre-)boiled cotton; move, then bring to the boil and dye for 1 hour. A richer black is achieved by post mordanting with a fresh bath of 5 kg bichromate and 5 kg sulfuric acid for 20 minutes at 60 °C – For printing on cloth, use 90 g Direct Sulphoblack 2 B, 108 cc sodium hydroxide of 42%, 72 g glycerine, 100 cc water and 72 g dextrin. After printing, steam, treat with chromate and soap.'

## 19.6 Immedial Black FF

Colour Index name: Sulphur Black 9; Colour Index number: 53230
IUPAC name: not available
Discoverer: Georg Kalischer, 1897
Contemporary terminology: Immedial black FF (En), Noir Immedial FF (Fr), Immedialschwarz FF (Ge), Immediaalzwart (NL)
Chemical formula: In a similar way to the other black sulfur dyes, the chemical composition is unknown. It probably consists of a mixture.

### *History*

In 1897, another and better black sulfur dye was discovered and patented by the German chemist Georg Kalischer (1873–1938). This was made by heating 4-hydroxy-2,4-dinitro-diphenylamine (obtained by condensing *p*-aminophenol with 1-chloro-2,4-dinitrobenzene) with sodium sulfide and sulfur at about 140 °C. Although this dye did not require any after-treatment, potassium dichromate improved the lightfastness and a darker colour was obtained [32]. Cassella marketed it as Immedial black FF, but also produced this dye under different trade names such as Immedial black G extra, NR and V extra. Blue dyes are produced by treating Immedial black with a mixture of air and steam at a temperature exceeding 100 °C, in the presence of an alkali (G.P. 118.087, filed in 1899). These were sold under the trade names Immedial blue C, CB, CBL and CR.

### *Trade Names*

The dye was put on the market by Cassella under the commonly used trade name Immedial black FF extra, but also under different trade names such as Immedial black G extra, NR and V extra. Other trade names were used, such as Immedial blue C, CB, CBL and CR by Cassella, and Sulfinschwarz by Griesheim G.m.b.H.

---

wird durch Nachchromieren auf frischem Bad mit 5 kg Bichromat und 5 kg Schwefelsäure—20 Minuten bei 60 °C.—erzielt.—Im Zeugdruck Druckfarbe: 90 g Direkt-Sulfoschwarz 2 B, 108 cc Aetznatron 45 ° Bé, 72 g Glyzerin, 100 cc Wasser, 72 g Dextrin. Nach dem Drucken dampfen, chromieren und seifen.

## Applications

Immedial black FF is mainly used for dyeing cotton and linen.

## Properties

Lehne (1906) wrote for Immedial black FF extra that it remains unaltered after exposure for 14 days to weathering conditions (in September).[8] According to the Colour Index, the light fastness of Immedial black FF is between ISO 3 and 4, so less compared to the other sulfur dyes. The wash fastness is ISO 5.

## Recipes

Lehne provide the following recipe in his 1906 manuscript:[9] 'For dyeing cotton, linen (Recipe V12). Dye in a short bath and use 10–12% Immedial black, 8–10% sodium sulfide, 2% sodium carbonate and 6% sodium chloride. Dye for about 1 hour near to the boiling point, [take it out] squeeze and rinse immediately. The coloured textiles can be treated with 3% potassium chromate or a mixture of 1.5% potassium chromate and 1.5% chromium (III) potassium sulfate (chrome alum) with the addition of 3% acetic acid, which makes the shade more vivid or greenish. Finally, rinse well, adding 2–3 g sodium acetate per litre of water to the last rinsing bath. No apparatus containing copper may be used for dyeing, and the fabric must be squeezed well immediately after removal so that the sodium sulfide used as a solvent does not oxidise and the excess dye then deposits dye on the fabric. Immedial black FF extra provides a deep black with a blue sheen, while Immedial black G extra gives a greenish black.'

---

[8] Marke FF extra: Echtheit: ´Nach 14 tägiger Einwirkung der Witterung (September) unverändert´.

[9] Zum Färben von Baumwolle, Leinen (V12). Man färbt in kurzer Flotte und benötigt im stehenden Bad 10–12% Immedialschwarz, 8–10% krist. Schwefelnatrium, 2% Soda, 6% Kochsalz, färbt etwa 1 Stunde nahe der Kochhitze, quetscht ab und spült sofort. Die Färbungen können mit 3% Chromkali oder einem Gemisch von 1,5% Chromkali und 1,5% Chromalaun unter Zusatz von 3% Essigsäure nachbehandelt werden, wodurch die Nuance lebhafter bezw. Grünlicher wird. Schließlich wird gut gespült, unter Zusatz von 2–3 g Natriumacetat im Liter Wasser zum Ietzten Spülbade. Beim Färben dürfen keine kupferhaltigen Apparate gebraucht werden, und die Ware ist unmittelbar beim Herausnehmen gut abzuquetschen, damit das als Lösungsmittel verwendete Schwefelnatrium sich nicht oxydiert und die überschüssige Färbeflotte dann Farbstoff auf der Ware abscheidet. Immedialschwarz FF extra liefert ein tiefes Schwarz mit blauem Schein, die Marke G extra grünliches Schwarz.

## 19.7 Different Colours of the Sulfur Dyes

The discovery of the sulfur blacks mentioned above led to a rapid development of this dye class between 1898 and 1904. The German dye factories Cassella, Kalle and Bayer and the Swiss firms CIBA, Geigy and Sandoz in Basel, in particular, performed further research. As a result, many sulfur dyes in an almost complete range of colours were discovered, the most important being the black, brown, blue and green dyes. The yellow and orange sulfur dyes were of limited importance. Reds were represented only by dull maroons and Bordeaux shades; bright reds were lacking (see Table 19.3). The sulfur dyes in different colours were marketed in the first decade of the twentieth century.

At the beginning of the twentieth century, Arthur Weinberg, who worked at Cassella, invented several sulfur dyes, such as Immedial maroon B (C.I. name: Sulphur Red 3; C.I. number: 53710) in 1900, followed by Immedial yellow D

**Table 19.3** List of the most important sulfur dyes

| Name | C.I. name | C.I. number | Discoverers | Year of discovery |
|---|---|---|---|---|
| Cachou de Laval | Sulphur Brown 1 | 53000 | L.M.F. Bretonnière and E. Croissant | 1873 |
| Immedialbraun B | Sulphur Brown 5 | 53245 | M. Hoffmann and G. Kalischer | 1899 |
| Vidal black | Sulphur Black 3 | 53180 | H.R. Vidal | 1893 |
| Immedial black FF | Sulphur Black 9 | 53230 | G. Kalischer | 1897 |
| Sulphur black T | Sulphur Black 1 | 53185 | H.R. Vidal<br>B. Priebs, and O. Kaltwasser | 1896<br>1899 |
| Immedialreinblau | Sulphur Blue 9 | 53430 | A. Weinberg and Herz | 1900 |
| Immedialindon RF | Sulphur Blue 7 | 53440 | Herz | 1902 |
| Hydronblau R | Vat Blue 43 | 53630 | Haas and Herz | 1908/1909 |
| Hydronblau G | Vat Blue 42 | 53640 | Herz | 1909 |
| Kryogengelb G | Sulphur Yellow 3 | 53125 | BASF | 1903 |
| Immedialorange N | Sulphur Yellow 6 | 53105 | A. Weinberg and O. Lange | 1903 |
| Immedialgelb GG | Sulphur Yellow 4 | 53160 | Schmidt | 1906 |
| Immedialorange C | Sulphur Orange 1 | 53050 | A. Weinberg and O. Lange | 1902 |
| Immedialorange N | Sulphur Yellow 6 | 53105 | A. Weinberg and O. Lange | 1903 |
| Immedialbordo G | Sulphur Red 1 | 53721 | Schmidt | 1905 |
| Immedialmarron B | Sulphur Red 3 | 53710 | A. Weinberg and Schmidt | 1900/1905 |
| Immedialoliv GN | Sulphur Green 11 | 53165 | Lepetit, Dollfuß & Gansser | 1895 |
| Immedialgrün GG | Sulphur Green 3 | 53570 | Böniger | 1904 |

(C.I. name: Sulphur Brown 26; C.I. number: 53090) and Immedial orange C (C.I. name: Sulphur Orange 1; C.I. number: 53050) in 1902. The most famous dye of this class, next to the blacks, was Immedial pure blue (C.I. name: Sulphur Blue 9; C.I. number: 53430), a clear bright blue, prepared from 4-dimethylamino-4′-hydroxydiphenylamine (obtained by heating dimethyl-*p*-phenyleneamine hydrochloride with *p*-aminophenol). This dye was discovered by Weinberg and the German industrial chemist Richard Herz (1867–1936) in 1900 [33–36]. The dye was brought onto the market by Cassella under the commonly used trade name Immedialreinblau.

The number of manufactured dyes is indicative of the importance of a dye class. For example, in the first decade of the twentieth century alone, Bayer produced more than 80 different sulfur dyes [37]. The dye firms Kalle and CIBA each produced around 50 sulfur dyes in this period, using the process known as sulfurisation or thionation. Dye-producing factories adopted the word Thio as an abbreviation for their trade sulfur products, such as Thion, Thiogene, Thional, Thionol, Thiophor and Thioxine. The best-known manufactured brands and producers of the sulfur dyes are listed in Table 19.4.

The significance of the sulfur dyes is clearly shown by the evidence of export statistics, such as the large volumes exported to Japan. In the Meiji era (1868–1912), Japan was entirely dependent on imports from Germany with respect to synthetic

**Table 19.4** Best known manufactured brands and producers of the sulfur dyes

| Manufactured brands | International dye companies |
|---|---|
| Auronal | Chemische Fabriken Weiler-ter Meer |
| Clayton | Clayton Anilin Company |
| Cross Dye | Read Holliday & Sons Limited |
| Eclipse | Aktiengesellschaft J.R. Geigy |
| Immedial | Leopold Cassella & Co. |
| Katigen | Farbenfabriken vorm. Friedr. Bayer & Co. |
| Kryogene | Badische Anilin- & Soda-Fabrik |
| Pyrol | Farbwerke vorm. A. Leonhardt |
| Sulphur | Actien Gesellschaft für Anilin-Fabrikation |
| Pyrogene | Gesellschaft für chemische Industrie |
| Thiophenol | Gesellschaft für chemische Industrie |
| Thiogene | Read Holliday & Co. |
| Thion | Chemische Fabrik Kalle & Co. |
| Thional | Chemische Fabrik vorm. Sandoz |
| Thionol | Levinstein Limited Blackley |
| Thiophor | Carl Jäger G.m.b.H. Anilinfarbenfabrik |
| Thioxine | Chemische Fabrik Griesheim-Elektron |
| Vidal | Société Anonyme des matières colorantes et produits chimiques de St. Denis |

dyes. The term 'aniline dye' appeared in Japanese import statistics from 1883, and the colours black and blue were of special interest because fabrics for everyday kimonos were usually blue, dark blue and black. The import of synthetic Indigo into Japan started in 1902 and grew steadily. Natural and synthetic Indigo (C.I. name: Vat Blue 1; C.I. number: 73000) were durable, but expensive: it is said that the dyeing cost of Indigo was twice that of the sulfur dyes. For this reason, the blue and black sulfur dyes became inexpensive alternatives to synthetic Indigo and to the aniline dyes in general: in fact, 90% of the sulfur dyes imported into Japan were blue and black [38]. For instance, Bayer exported nine tons of the sulfur dye Katigen blue B on a monthly basis in 1907 [39]. About 6,000 tons of synthetics were imported annually before 1914, of which synthetic Indigo (2,500 tons) and the sulfur dyes (2,000 tons) formed the largest imports.[10] [38]

### *Blue Sulfur Dyes Versus Synthetic Indigo*

The blue sulfur dyes were the strongest competitors for synthetic Indigo, which had been produced since 1897 by BASF. Three developments led to a rapid fall in the price of synthetic Indigo: the Pfleger indigo process exploited by Farbwerke ML&B; the newly developed synthetic blue sulfur dyes; and the discovery of Indanthrene blue (C.I. name: Pigment Blue 60; C.I. number: 69800).

For everyday clothing, such as blouses, aprons and shirts, lower fastness standards were sufficient, and the blue sulfur dyes fulfilled this demand. In particular, the inexpensive and easy to use blue sulfur dyes from the German dye companies Cassella, Kalle and Bayer were strong competitors for synthetic Indigo. Immedial pure blue (C.I. name: Sulphur Blue 9; C.I. number: 53430), Hydron Blue R (C.I. name: Vat Blue 43; C.I. number: 53630), Hydron Blue G (C.I. name: Vat Blue 42; C.I. number: 53640) and Kalle's Thion Blue, were especially suitable for the production of navy and dark shades on cotton [40].

## 19.8 Patents

The rapid development of the sulfur dyes is also illustrated by the numerous patents that were filed. In the book *Schwefelfarbstoffe* by the German chemist Otto Lange (1875–1936), published in 1925, a total of 747 international, mainly German, patents are listed. Between January 1900 and July 1902 an average of two patents per week were published. Due to the large number of sulfur dyes, patent protection was

[10] The details are as follows: Indigo, 2500 tons; aniline dyestuffs, 3500 tons (direct dye, 800; basic dye, 250; acid dye, 250; sulfur dye, 2000; and aniline oil, 200).

problematic. The general and often vague patent information made it difficult to determine whether the starting materials and procedures were already known, or were new developments. A claim should be made based on an innovation or a substantial improvement, but at that time the value of individual patents was very difficult to judge, especially when international patents were involved. Of the 500 sulfur dyes patented, only 100 were truly innovative or a significant improvement of previous dyes [20]. The struggles between dye firms over patents reached the courts, with cases of innovation and improvement being based on:

- using different chemical products to produce the same sulfur dye;
- changing the temperature;
- changing the melting time, and
- adding different chemicals to the melt.

A good example is a contested patent for the dye Immedial black: although dinitro-oxydiphenylamine and its reduction product diamino-dioxydiphenylamine are two completely different substances, they were both used as starting materials. A patent dispute regarding this erupted between two (unknown) dye factories in 1901 and 1903. Another example was the effect of changing the temperature in the preparation of Immedial pure blue. CIBA used a temperature of 180 °C, which only achieved an insignificant bluish dye (G.P. 132.212, filed in 1898), whereas at a lower temperature, 140 °C or preferably 115–120 °C, the bright Immedial pure blue is formed instead. BASF added naphthols as a starting material to the melt in the production of Immedial pure blue, instead of glycerine, to preserve the homogeneity of the melt (G.P. 150.546, filed in 1902). AGFA obtained the same dye by adding glycerine to the melt (G.P. 141.752, filed in 1901). Since glycerine and naphthols are chemically and physically unrelated, it was interpreted as a patentable improvement to the melt of the blue sulfur dyes.

Bypassing patents was frequently done very elegantly, meaning that assigning the patent to an opponent could not be prevented. For the production of the sulfur dyes using azo dyes, patents with minor changes, for example the use of a different azo dye, were granted without any hesitation. In the heyday of the sulfur dyes, the first results of innovation and invention were immediately patented, with improvements and simplifications achieved at a later date. For instance, Soc. Anon. St. Denis initially acquired the German patent (G.P. 113.893, filed in 1899) for the melt, followed by three additional patents (G.P. 120.467, filed in 1900; G.P. 131.468 and G.P. 131.567, both filed in 1901) which show improvements to G.P. 113.893. To produce its yellow sulfur dyes, AGFA used a homologue or a derivative of the original starting material 4′-methylacetanilide (G.P. 166.865, filed in1904). Similar or better results were obtained using the derivatives 2′-methylacetanilide (additional G.P. 166.981, filed in 1905) and 3′-methylacetanilide (additional G.P. 171.118, filed in 1905). Patents dealt with the same subject until a final method was eventually achieved. A classic example of the versatility of some inventors is B.P. 11.370 from 1896 in which nearly all the starting materials of the aromatic series are mentioned [20].

## 19.9 Chemical Constitution of the Sulfur Dyes

The sulfur dyes consist of a mixture of products because sulfurisation involves various reactions, such as substitution, ring formation, reduction and oxidation, resulting in insoluble, high molecular weight species. The sulfur dyes are complex substances of an indeterminate chemical constitution, which can hardly be decomposed. Regarding their constitution, an exact chemical structure remains unknown and it is not possible to assign precise chromophores. There is very little knowledge about the constitution of the blacks, but it is likely they form molecules of long chains similar to the synthetic dye aniline black.

In 1893, Henri Raymond Vidal first presented some theoretical considerations on the origin and probable structure of these dyes, although these were not based on any detailed investigation (G.P. 84.632, filed in 1893). He followed these up with more theoretical information in other patents (G.P. 99.039, filed in 1896; G.P. 103.301, filed in 1897; G.P. 111.385, filed in 1897, and B.P. 13.093, filed in 1896) and publications in *Moniteur Scientifique* and *Chemisches Central-Blatt* in the years 1897 and 1903 [41–44].

The manufacturing process for sulfur dyes introduces sulfur into the organic, aromatic ring structures of the intermediates. A significant part of the total sulfur is attached to the aromatic ring in the form of disulfide or disulfoxide bridges. Research carried out by several chemists has shown that these dyes contain chromophores with ring structures of the thiazole, thiazine or thianthrene types. Generally, the constitution of the blue and green dyes consists of thiazine derivatives, while the yellow and brown are derived from thiazole and the red from azines. The Swiss chemist Robert Gnehm (1862–1926) and the Austrian chemist Felix Kaufler (1878–1957) found that Immedial pure blue was converted into tetrabromo-methylene violet (tetrabromo-dimethylaminothiazine) when heated with potassium bromate and hydrogen bromide. This was identical to the product obtained from Methylene blue (C.I. name: Basic Blue 9; C.I. number: 52015) using a similar treatment, thereby demonstrating the presence of the thiazine ring in the dye [45, 46].

The main contribution to the chemistry of the sulfur dyes was made by the Swiss chemist Hans Eduard Fierz-David (1882–1953) and his collaborators, who collected valuable data regarding the type of linkages and ring systems occurring in the sulfur dyes. Fierz-David formulated a scheme for the purification of these dyes. Together with Eduard Bernasconi working for CIBA, he removed the impurities by exhaustive extraction with dilute hydrochloric acid, dilute ammonia, water, alcohol and ether, making a reliable elemental analysis possible. This method allowed the formulation of some of the green, blue and purple sulfur dyes to be deduced.

## 19.10 Dyeing of the Sulfur Dyes

The dyeing process used for the sulfur dyes is similar to that for the vat dyes. They are water-insoluble compounds, but are soluble in alkaline liquids. In the presence of a reducing agent such as dilute aqueous sodium sulfide, and at a temperature of around 80 °C, the dye particles convert into water-soluble *leuco* compounds and can be absorbed by the fabric: sodium chloride facilitates the absorption. After the fabric is removed from the dye solution, the *leuco* compounds are oxidised by the oxygen in the air or by using an oxidation agent and are converted back into their original water-insoluble form. The dyes achieved demonstrate good fastness properties and many can be improved by an after-treatment with metal salts.

## 19.11 Properties of the Sulfur Dyes

The sulfur dyes are especially suitable for dyeing cotton, but not wool and silk due to the alkaline condition of the dye bath. They possess a good fastness to light and perspiration, and are easy to apply and are inexpensive. Being water-insoluble they give excellent fastness to washing: they will not bleed or stain other clothes when washed in water. Unfortunately, however, the dyes may have a poor fastness to rubbing and to chlorine bleaches. Bronzing, the tendency to show a bronze appearance, can be observed for Sulphur black T.

The first sulfur dyes were hygroscopic due to the content of hygroscopic salts, therefore they had to be used immediately. In addition, there was also the risk of self-oxidation, which could lead to spontaneous ignition. Another disadvantage was tendering—the loss of strength or degradation of cellulosic materials upon storage. Tendering is caused by the acid formation from the free sulfur present in the dyed materials by the action of moisture and air. The acid produced reacts with cellulose and degrades it, resulting in loss of strength. For items that are frequently washed this is of no consequence but rarely washed textiles, such as sewing threads and linings, gradually become brittle over time, so these dyes should not be used [15].

## References


1. Schultz, G.: Farbstofftabellen, Fünfte Auflage. Weidmannische Buchhandlung, Berlin (1914)
2. Schultz, G.: Farbstofftabellen, Sechste Auflage. Weidmannische Buchhandlung, Berlin (1923)
3. Schultz, G.: Farbstofftabellen, Siebente Auflage. Akademische Verlagsgesellschaft M.B.H., Leipzig (1931)

4. Hunger, K. (ed.): Industrial Dyes, Chemistry, Properties, Applications. WILEY-VCH Verlag GmbH & Co. KGaA, Weinheim (2003)
5. Society of Dyers and Colorists and American Association of Textile Chemists and Colourists: The Colour Index. Society of Dyers and Colourists, Bradford (1971)
6. Cardon, D.: Chapter 9. Into darkness: tannin plants. In: Natural Dyes: Sources, Tradition, Technology and Science. pp. 409–484. Archetype Publications Ltd., London (2007)
7. Travis, A.S.: From Manchester to Massachusetts via Mulhouse: The Transatlantic Voyage of Aniline Black. Technol Cult. 35, 70–99 (1994)
8. Leopold Cassella & Co.: G.P. 39,029 (1885), B.P. 9,214 (1885), F.P. 170.342, U.S.P. 345.901 (1886)
9. Leopold Cassella & Co., Arthur Weinberg: G.P. 50.907 (1888), B.P. 18.425 (1888), F.P. 170.342, U.S.P. 412.440 (1889)
10. Leopold Cassella & Co., Meinhard Hoffmann: G.P. 65.651 (1891), B.P. 1.742 (1891), B.P. 6.970 (1891), F.P. 201.770 (1891), U.S.P. 480.326 (1892). Leopold Cassella & Co.: Naphthol Blue-Black S, SB, S2B, S3B, 6B.
11. Levinstein Limited Blackley: Coomassie Blue-Black G, NN and XX
12. Farbenfabriken vorm. Friedr. Bayer & Co.: Sulfone Cyanine G, GR extra, 3R, 5R extra
13. Geschichte und Entwicklung der Farbenfabriken vorm. Friedr. Bayer & Co. Elberfeld in den ersten 50 Jahren, Wissenschaftliche Entwicklung der Azofarben. Meisenbach Riffarth & Co., München (1918)
14. Actien-Gesellschaft für Anilin-Fabrikation: B.P. 12.305 (1902), F.P. 321.626 (1902), U.S.P. 717.550 (1903)
15. Wizinger, R.: Gezielte Forschung. Rundschau, CIBA-GEIGY AG., Division Farbstoffe und Chemikalien. 2, 20–21 (1973)
16. Farbwerke vorm. Meister Lucius & Brüning 1863-1913, Hausdruckerei der Farbwerke vorm. Meister Lucius & Brüning, Höchst am Main
17. Société Anonyme des matières colorantes et produits chimiques de St. Denis: G.P. 88.392 (1895), B.P. 3.612 (1895), F.P. 244.585 (1895), U.S.P. 549.036 (1895), U.S.P. 549.082 (1895)
18. Farbenfabriken vorm. Friedr. Bayer & Co.: G.P. 101.541 (1895), G.P. 102.897 (1895), B.P. 22.417 (1895), F.P. 253.213 (1896), U.S.P. 603.755 (1898)
19. A. Redlich and G. Deutsch: G.P. 240.522 (1910), B.P. 28.862 (1910).
20. Lange, O.: Die Schwefelfarbstoffe ihre Herstellung und Verwendung, Zweite Auflage. Verlag von Otto Spamer, Leipzig (1925)
21. Henri Raymond Vidal: G.P. 84.632 (1893), B.P. 19.880 (1893), B.P. 23.578 (1893), F.P. 206.405, F.P. 236.405, F.P. 236.406, F.P. 244.585, U.S.P. 532.484 (1895), U.S.P. 532.503 (1895)
22. Badische Anilin- & Soda-Fabrik, René Bohn: G.P. 84.989 (1893), B.P. 10.996 (1893), F.P. 237.610 (1894), F.P. 319.790 (1902), U.S.P. 545.336 (1895), U.S.P. 545.337 (1895)
23. Société Anonyme des matières colorantes et produits chimiques de St. Denis, Henri Raymond Vidal: G.P. 98.437 (1896), G.P. 116.354 (1899), B.P. 16.449 (1896), B.P. 141.759, F.P. 231.118, F.P. 288.475, F.P. 288.476 and F.P. 288.477 (1897)
24. Actien-Gesellschaft für Anilin-Fabrikation, Bernhard Priebs and Oskar Kaltwasser: G.P. 127.312 (1900), G.P. 127.835 (1899), B.P. 1.151 (1900), U.S.P. 655.659 (1900), F.P. 299.721
25. Farbenfabriken vorm. Friedr. Bayer & Co.: B.P. 15.625 (1909), F.P. 404.719, U.S.P. 935.009 (1909)
26. Gesellschaft für chemische Industrie: B.P. 13.035 (1903), F.P. 333.096
27. Aktiengesellschaft Kalle & Co.: G.P. 186.860 (1902), B.P. 26.379 (1903), F.P. 337.278
28. Société Anonyme des matières colorantes et produits chimiques: G.P. 218.517 (1906)
29. Weiler-ter Meer: G.P. 208.377 (1905)
30. Leopold Cassella & Co.: B.P. 19.831 (1896), F.P. 259.509, F.P. 267.343
31. Lehne, A.: Tabellarische Übersicht über die künstlichen organischen Farbstoffe und ihre Anwendung in Färberei und Zeugdruck, Zweiter Ergänzungsband. Verlag von Julius Springer, Berlin (1906)

32. Leopold Cassella & Co., Georg Kalischer: G.P. 103.861 (1897), G.P. 118.087 (1899), B.P. 25.234 (1897), B.P. 4.069 (1899), F.P. 271.909, F.P. 286.287, Austrian P 3.224 (1897), U.S.P. 610.541 (1898), U.S.P. 678.884 (1901)
33. Leopold Cassella & Co., Arthur Weinberg and Richard Herz: G.P. 129.540 (1901), G.P. 134.947 (1900), G.P. 135.952 (1901), G.P. 136.188 (1900), G.P. 141.752 (1901), B.P. 16.247 (1900), B.P. 20.741 (1901), B.P. 21.310 (1900)
34. Leopold Cassella & Co., Arthur Weinberg and Richard Herz: F.P. 303.524, F.P. 305.494, F.P. 313.306, U.S.P. 693.632 (1902), U.S.P. 693.633 (1902)
35. Actien-Gesellschaft für Anilin-Fabrikation: B.P. 17.564 (1901), F.P. 313.947
36. Gesellschaft für chemische Industrie: F.P. 308.699.
37. Geschichte und Entwicklung der Farbenfabriken vorm. Friedr. Bayer & Co. Elberfeld in den ersten 50 Jahren, Geschichte der Katigenfabrik. Meisenbach Riffarth & Co., München (1918)
38. Hashino, T.: The Rise of the Japanese Synthetic Dye Industry during the First World War. Kobe University Economic Review. 53, 35–55 (2007)
39. Geschichte und Entwicklung der Farbenfabriken vorm. Friedr. Bayer & Co. Elberfeld in den ersten 50 Jahren, Die Entwicklung der Export-Abteilung Levante, Asien und Australien. Meisenbach Riffarth & Co., München (1918)
40. Abelshauser, W.: Die BASF, Eine Unternehmensgeschichte, Zweite Auflage. Verlag C.H. Beck, München (2003)
41. Vidal, R.: Neue substantive schwefelhaltige Farbstoffe. Chemisches Central-Blatt, Commissions-verlag von R. Friedländer & Sohn, Berlin. 68, 747–749 (1897)
42. Vidal, R.: Über den Einfluss der Substitution bei den Diphenylaminen auf die Erzeugung direkt färbender Schwefelfarbstoffe. Chemisches Central-Blatt, Commissionsverlag von R. Friedländer & Sohn, Berlin. 74, 266–267 (1903)
43. Vidal, R.: De l'influence des substitutions dans les diphénylamines génératrices de couleurs sulfurées directes. Moniteur Scientifique, Paris. 17, 427–430 (1903)
44. Vidal, H.R.: Nouveaux Colourants Substantivs Sulfurés. Moniteur Scientifique, Paris. 11, 655–657 (1897)
45. Dürig, R.: Zur Kenntnis der schwarzen Schwefelfarbstoffe, Promotionsarbeit, (1944)
46. Keller, E.: Zur Kenntnis der blauen und grünen Schwefelfarbstoffe, Promotionsarbeit, (1933)

# Chapter 20
# Copying Nature: Synthetic Alizarin

## 20.1 Introduction

Since prehistoric and ancient times, plants of the Rubiaceae family have been used worldwide for textile dyeing. Their rhizomes and roots contain various anthraquinone glycosides and red anthraquinone dyes, for example alizarin and purpurin, pseudopurpurin, munjistin and rubiadin. Dyer's madder (*Rubia tinctorum* L.) is the most prominent plant of this family, used throughout the world, primarily for dyeing textiles. While dyer's madder, which contains all five of the anthraquinones mentioned above, generally provides alizarin-rich dyeings, purpurin-rich dyeings are obtained with wild madder (*Rubia peregrina* L.), Indian madder (*Rubia cordifolia* L.) and many of the *Galium* (bedstraw) species. However, many late antique textiles from Egypt and the Mediterranean region, as well as European textiles, have purpurin-rich dyeings, possibly originating from dyer's madder. Current research is therefore focusing on special treatments and dyeing methods that could change the dye composition of this material observed on textiles [1–4].

The native range of dyer's madder is from Southeast Europe to Western Himalaya, including the eastern parts of the Mediterranean region, the Middle East, Central Asia and India. Through cultivation, the plant was introduced to other regions of Europe and Africa [4].

The oldest preserved recipes for dyeing with madder were found on a sixth century BCE Late Babylonian cuneiform clay tablet probably from Sippar, a city on the east bank of the river Euphrates in modern Iraq [4, 5].

Many ancient written sources mention this plant, its cultivation and quality. In the *Materia Medica* (Dsc. 3.160), the Greek physician, pharmacologist and botanist Pendanius Dioscorides (c. 40–c. 90 CE) refers to madder as *erythrodanon* and mentions its cultivation in Thabana (Galilee), Ravenna (Italy) and Caria (Cilicia, Asia Minor). In *Naturalis historia* (77 CE), the Roman politician and scholar Pliny the Elder (23/24–79 CE) states that *rubia* was widespread in the Roman provinces



M. R. van Bommel, M. de Keijzer, *Bright Colours from the Past*, Cultural Heritage Science, https://doi.org/10.1007/978-3-031-88664-5_20

and describes the excellent quality of madder grown near Rome (NH 19.47; 24.94). The *Papyrus Graecus Holmiensis*, also known as the Stockholm papyrus and dated to the third and fourth centuries CE, contains several recipes for mordanting and dyeing red, purple and pink shades with madder and a method for testing the quality of madder [4, 6–8].

In the seventh century madder was cultivated at Saint-Denis near Paris and later, in the fourteenth century, the best quality madder came from Holland. Well-known areas of cultivation in Europe were Zeeland in the Netherlands, the surroundings of the Alsace and around Avignon in the south of France [9, 10].

Madder was frequently used in Europe, the Middle East and India. In Egypt, the earliest finds of linen dyed with madder, but without an alum mordant, date from the 18th dynasty (1550–1292 BCE), but madder dyeing using alum as a mordant is not known before the 21st dynasty (1069–945 BCE) [11]. Madder or madder types have been identified in archaeological textiles from different countries, such as Egypt (from 1550 BCE onwards), Austria (Bronze Age Hallstatt, 1500–1200 BCE), Israel (from eleventh century BCE onwards) and Syria (from first century CE onwards) [4, 12, 13].

In the eighteenth century, madder was used to obtain a bright red known as Turkey red following a recipe originating from the Middle East. The method was very complicated, using a combination of mordants and multiple steps that took several weeks. Turkey red was highly valued for its brilliant colour and its fastness to light and washing [10].

The discussion above demonstrates the importance of madder as a textile dye, and explains why nineteenth-century scholars were interested in understanding the chemistry of the colourants found in madder and in developing synthetised analogues. The dyeing properties of madder had been investigated by several scientists early in the nineteenth century, but apart from the French chemist Charles Frédéric Kuhlmann (1803–1881), nobody tried to purify or isolate the main colourant from the madder root [14]. According to him the root contained a red and a brown colouring compound. The Paris pharmacists Pierre-Jean Robiquet (1780–1840) and Jean-Jacques Colin (1784–1865) were the first to isolate the main dye from madder in 1826 [15]. This was done by heating powdered root with sulfuric acid whereby it was converted into a black mass, the so-called *charbon de garance* which, on being carefully heated, yielded a sublimate of beautiful orange-coloured crystals [9, 16]. They called this compound alizarin, which is derived from *alizari*, the term formerly used in Mediterranean regions for roots of *Rubia tinctorum* imported from the Levant. The name *alizari* is derived from the Arabic *al-ʿaṣārah*, meaning 'the squeezed juice, extract'. During the research, Robiquet and Colin developed a new process, using a different extraction procedure, and published the results of their thorough chemical study [15]. By heating ground madder in a test tube with various acids and potash they obtained a yellowish vapour which crystallised into bright-red needles. Changes in colour observed during the different purification stages led Robiquet and Colin to suspect the presence of a second colouring matter in the roots of madder. Eventually, they isolated this compound, which they called purpurin because of its red-purple colour. The findings of their chemical study, like the elemental conditions and the chemical properties of alizarin and purpurin, were published in 1827 [17].

On the 26th of March, 1828, Robiquet and Colin, together with the business man A. Lagier from Avignon, acquired a 10-year French patent for the preparation of what they called garancine. By using sulfuric acid, an extract from madder roots was obtained with a concentration of alizarin three to four times higher than in the original madder powder [18]. Garancine was first introduced on the market in 1830, but the production of garancine only started to be carried out on a large scale in 1839, in Rouen and the Alsace. Garancine derivatives lit up the market. The French chemist Léonard Schwartz (1802–1885) from Mulhouse obtained a French patent to produce a derivative similar to garancine from by-products resulting from the treatment of madder with sulfuric acid on the 17th of April, 1843. He called this dyeing agent 'garanceux' [16]. Garancine was particularly popular in Lancashire, in northern England. Calico prints using madder and garancine were exhibited at the Great Exhibition in London in 1851. However, the preparation of pure alizarin from madder was too complex and time-consuming to be profitable.

A crucial moment in the history of alizarin was the epoch-making discovery in 1868 by Carl Graebe and Carl Theodore Liebermann that alizarin was a derivative of anthracene. They isolated the substance anthracene, determined its chemical structure and developed a way to synthesise Alizarin from coal tar by bromination.[1] The synthesis of Alizarin, which was still not suitable for industrial production, was published in 1869 in the German academic journal *Berichte der deutschen chemischen Gesellschaft zu Berlin*, so firms started looking for a suitable production process [19]. In May 1869, BASF (Heinrich Caro, Graebe and Liebermann) and William Henry Perkin independently discovered the same synthesis, the sulfonation of anthraquinone, but the BASF group filed their patent one day earlier (25 June 1869) than Perkin in London. After the discovery of synthetic Alizarin, the use of madder and garancine declined. In the 1870s the price of Alizarin dropped dramatically and the result was that the garancine factories gradually closed their doors. Within little more than a decade, madder plantations were forced to give up operations and many disappeared.

Alizarin was also used as a pigment, but has now been largely replaced by the more lightfast synthetic organic quinacridone pigments introduced by the American multinational chemical company DuPont de Nemours, Inc., in 1958 under the trade name Monastral. However, synthetic Alizarin is still used as a red pigment, Alizarin Crimson (PR 83), by artists in oil and water colour painting.

[1] We will use alizarin without capitals when we indicate the dye extacted from madder root, while Alzarin with a capital is used for the synthetic form.

## 20.2 Synthetic Alizarin

Colour Index name: Mordant Red 11; Colour Index number: 58000
IUPAC name: 1,2-Dihydroxyanthracene-9,10-dione
Discoverers: Jean-Jacques Colin and Pierre-Jean Robiquet (extracted from madder), 1826; Carl Graebe and Carl Theodore Liebermann, 1868; Heinrich Caro, Carl Graebe and Carl Theodore Liebermann, 1869; William Henry Perkin, 1869
Contemporary terminology: Synthetic alizarin (En), Rouge d´alizarine (Fr), Alizarin synthetisch (Ge), Synthetische Alizarine (NL)
As pigment: Permanent violet; Permanent crimson; Alizarin crimson; Crimson madder (En)
Chemical formula: $C_{14}H_8O_4$ (Fig. 20.1).

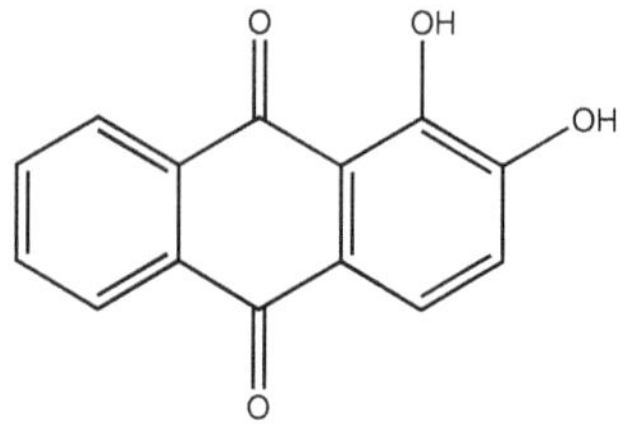


**Fig. 20.1** Structural formula of synthetic Alizarin

## 20.3 History

Although at the end of the 1860s a great number of synthetic dyes was on the market with new dyeing and fastness properties, there was still a great interest in natural dyes, for example by the German chemist Adolf Baeyer (1835–1917) who focused on alizarin and indigo, extracted from madder and indigo (*Indigofera* spp.) respectively. Between 1857 and 1864, the German chemist Friedrich August Kekulé (1829–1896) developed his carbon theories and, as such, contributed to the development of the understanding of the chemistry of dyes. Many scholars studied alizarin, first by investigating the composition of the dye in madder root, but later diverse synthesis routes for Alizarin were developed (see Table 20.1).

Around 1850, it was considered that alizarin was related to coal tar compounds. When the British chemist Henry Edward Schunck (1820–1903) purified madder by sublimation and crystallisation, he suggested that madder was a derivative of naphthalene. The search for a synthetic route for Alizarin led first to scientific studies into the chemical constitution of colourants and in 1866 a partial structure for alizarin could be drawn. The German chemist Adolph Friedrich Ewald Strecker (also known as Adolf Ludwig Strecker) (1822–1871) succeeded in a partial elucidation of the chemical constitution of alizarin and purpurin in 1867. The breakthrough occurred at the Gewerbe Institute, a trade college in Berlin, in early 1868, where the

**Table 20.1** Discoverers, year of discovery and patents of various synthetic Alizarin syntheses

| Discoverers | Year of discovery | Patent title | Patents |
|---|---|---|---|
| Jean-Jacques Colin and Pierre-Jean Robiquet | 1826 | – | – |
| Lagier, Jean-Jacques Colin and Pierre-Jean Robiquet | 1828 | Brevet d'invention sur la Garance | 26 March 1828 |
| Léonard Schwartz | 1843 | Brevet d'invention Production of garancine | 17 April 1843 |
| Carl Graebe and Carl Theodore Liebermann | 1868 | Improved process of preparing Alizarine | B.P. 3.850 |
| Heinrich Caro, Charles Graebe, Charles Liebermann | 1869 | | B.P. 1.936 |
| William Henry Perkin | 1869 | | B.P. 1.948 |
| William Henry Perkin | 1869 | | B.P. 3.318 |
| J. Brönner and H. Gutzkow | 1869 | | Bavarian patents of 29 September 1869 |
| Charles Graebe and Charles Liebermann | 5 October 1869 | Improved process of preparing Alizarine | U.S.P. 95.465 |
| J. Brönner and H. Gutzkow | 1870 | | Bavarian patent of 26 January 1870 |
| Heinrich Caro, Charles Graebe and Charles Liebermann | Filed 26 January 1870, dated 28 July 1874 | Improvement in the preparation of coloring matters from anthracene | U.S.P. 153.536 |
| Charles Graebe and Charles Liebermann | 4 April 1871 | Improvement in dyes in coloring matter from anthracene | U.S.P. 4.321 |
| Actiengesellschaft Farbwerke, vorm. Meister, Lucius and Brüning | 1881 | Neuerungen bei Herstellung von künstlichem Alizarin | G.P. 17.627 |
| Actiengesellschaft Farbwerke, vorm. Meister, Lucius and Brüning | 1894 | | B.P. 5.444 |

chemists Carl Graebe (1841–1927) and Carl Theodore Liebermann (1842–1914) were research assistants to Baeyer. Baeyer had already shown that complex carbon molecules could be broken into small fragments by heating them with zinc powder. Graebe had to try this with alizarin, since Baeyer wanted to know what the molecular structure of alizarin was. Once this was known, they could investigate the synthesis. Graebe established that the chemical structure of alizarin was probably a quinone and, together with Liebermann, showed that alizarin distilled with zinc powder is converted into anthracene rather than naphthalene [16, 20–24]. In the same year, 1868, they demonstrated that alizarin is a dihydroxyanthraquinone. In January 1869 they announced that they were successful in transforming anthracene into Alizarin [19]. The complete structure of alizarin was published in 1874 by Baeyer and Caro [25].

Their first synthesis, a three-step synthesis of Alizarin from anthracene was achieved at the end of 1868. This involved the oxidation of anthracene to anthraquinone using nitric acid. The anthraquinone thus obtained is treated with bromine, which replaces two equivalents of hydrogen and gives rise to dibromoanthraquinone. Next the two bromine atoms are substituted by hydroxyl groups (OH groups) in a boiling potassium hydroxide solution, followed by precipitation using a strong acid [23, 26, 27]. The reaction scheme is presented in Fig. 20.2.

Graebe and Liebermann succeeded in producing Alizarin using bromine (giving 1,2-dibromo-anthraquinone) in the laboratory. However, the bromination process was far too expensive for commercial use and gave a low yield of Alizarin. Still, the two inventors applied for patent protection for the production of Alizarin from anthracene. Graebe and Liebermann patented their discovery on the 18 December 1868 (B.P. 3.850), and then also in France, Prussia (Prussia on the 23 March 1869), most German states, Austria, Russia and the United States of America (U.S.P. 95.465, filed in 1869 and U.S.P. 4.321, filed in 1871) [28]. The Prussian patent was granted for five years, with the proviso, however, that it should be put into practical operation within 12 months. But on the 7 July, 1870, the patent was withdrawn because proof had not been produced that the patented methods had been carried out. The British patent for the bromine methods expired on the 18 December 1871, since the fee for continuation was not paid.

Graebe and Liebermann decided to sell the rights to their invention after abandoning plans to set up their own manufacturing facilities. The discoverers

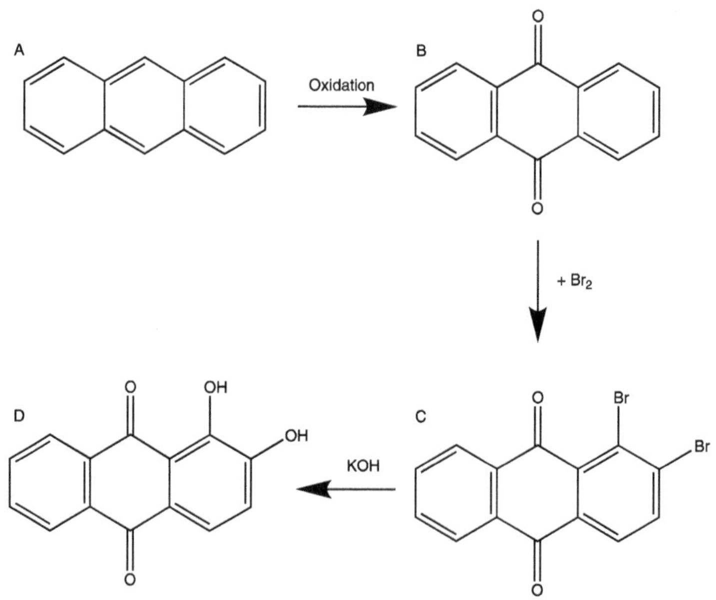


**Fig. 20.2** Conversion of anthracene (**A**) to Alizarin (**D**) via oxidation to anthraquinone (**B**) followed by bromination into 1,2-dibromo-anthraquinone (**C**)

approached several German and English companies, but they found that it was necessary to approach a large dye factory. They rejected an offer from Farbwerke ML&B (where Graebe had worked before joining Baeyer) to pay a percentage of the profits [23]. Graebe and Liebermann therefore entered into an alliance with BASF, the largest works of the kind in Germany, in fact on the continent of Europe. Both chemists consulted the German chemist Heinrich Caro (1834–1910) of BASF. On 29 May 1869, in exchange for a fourteen-year contract that would give the inventors three percent of the sales of Alizarin, Graebe and Liebermann agreed to transfer the patent rights to BASF and the two inventors agreed to assist the company with the technical development of Alizarin. In order to do this, Graebe was to work with Caro at BASF in Ludwigshafen and Mannheim during the summer of 1869.

Graebe and Liebermann had disclosed the synthesis of Alizarin in the German academic journal *Berichte der deutschen chemischen Gesellschaft zu Berlin*, so firms started looking for a suitable production process. Although the synthesis of Alizarin was now possible, there were still many issues to solve. One of the issues was that anthracene, the starting material for the synthesis of Alizarin, was scarce and it was unclear if this substance would ever be available in such large quantities that synthetic Alizarin could compete with the alizarin obtained from madder root. But in 1869 and 1870, the German chemists J. Brönner and H. Gutzkow applied for two Bavarian patents for the extraction of anthracene from coal tar by means of steam distillation, which was of great importance for the further development of the synthesis of Alizarin. In the second Bavarian patent in particular, they described an improved anthracene process to produce Alizarin [29].

Furthermore, the extremely complicated process used bromine in the form of a prohibitively expensive volatile liquid in small sealed tubes. Graebe and Liebermann were only able to produce Alizarin in small quantities. This synthetic pathway had been created in the laboratory, but infortunately this preparation process was not suitable for industrial purposes. The synthetic route had a low yield of Alizarin and, as large quantities of bromine were required, the bromination process was too expensive for commercial use [30].

Instead of bromination, the two chemists attempted to produce Alizarin by converting anthraquinone into anthraquinone disulfonic acid using sulfuric acid. This reaction would overcome the problems of the high cost of bromine and the low yield of the bromine route. Unfortunately, however, their attempt failed because the temperatures they employed were not high enough.

It was only through an accidentical discovery by Caro that Graebe and Liebermann were able to improve the production of Alizarin at BASF. During his investigations for a compound analogous to rosolic acid, Caro left an experiment unattended for a short time. The ingredients—anthraquinone, sulfuric acid and oxalic acid—were overheated in the process. The result was sulfuric acid smoke in the laboratory and a faint pink-coloured mass at the top of the dish, which suggested the formation of a sulfonic acid. It turned out that the sulfonation of anthraquinone only happens at a temperature over 200 °C. In this way Caro replaced the bromination step by a much cheaper sulfonation process using concentrated sulfuric acid. The final process of sulfonation of anthraquinone to anthraquinone 1,2-disulfonic acid was performed at 260 °C; this product is melted with potassium hydroxide and the two sulfuric acid residues are exchanged for hydroxyl groups. Caro's improvement became the basis for new patent applications under the names of Caro, Graebe and Liebermann. These were filed in London and Berlin. Graebe and Liebermann simplified the process again in 1871 by producing Alizarin from anthraquinone monosulfonic acid through oxidative fusion with potassium hydroxide.

Around the same time, several chemists independently discovered the same synthesis route, that is, the sulfonation of anthraquinone at an elevated temperature and the subsequent conversion to Alizarin using potassium hydroxide. As an example, William Henry Perkin (1838–1907) and his elder brother Thomas Dix Perkin (1831–1891) had developed two processes for the production of Alizarin, both without bromine, one of which was the method to heat anthraquinone with sulfuric acid followed by treatment with potassium hydroxide [20]. Almost simultaneously, at the end of April 1869 the German chemist Ferdinand Riese of Farbwerke ML&B also found that anthraquinone mixed with fuming sulfuric acid in an alkali melt produced Alizarin. Farbwerke ML&B did not file a patent, but registered its invention on the 19th of May, 1869 at the Hoechst magistrate's office in the form of a *pli cacheté*[2] [31, 32].

Several patents were granted for the manufacture of artificial Alizarin, and it seemed that the right moment for submitting the patent application was crucial. The sulfonation process was first patented by Caro, Graebe and Liebermann in London on the 25th of June, 1869 (B.P. 1.936, revised patent process), just one day before William Henry Perkin's application for the same production process arrived at the Patent Office (B.P. 1.948, Patented on June 26, 1869). A few years later, Perkin complained that the British patent office was slow and sloppy. BASF filed the full specification on the 13th of January, 1870; an application for a patent in the United States of America for producing artificial Alizarin by the sulfonation processes was filed by them on the 26 January 1870 and granted as U.S.P. 153.536. The full specification of the English patent and the U.S. patent are identical. The specifications state that the invention relates to improvements on the invention described in the English patent granted to Graebe and Liebermann on the 18th of December, 1868, and in U.S.P. 95.465, 'in which the preparation of artificial Alizarine is based upon the action of caustic alkalies upon dibromanthrachinone or dichloranthrachinone'. Then, during the interval of the delay of four and a half years in the issuing of U.S.P. 153.536, U.S.P. 95.465 was withdrawn and reissued in two parts on the 4th of April, 1871, one for the process and the other for the product. The claim in the latter, U.S.P. 4.321, was so worded as to cover 'artificial Alizarine, produced from anthracine or its derivatives, by either of the methods herein described, or by any other method which will produce a like result' (see Table 20.1) [33].

[2] *Pli cacheté*: the establishment of the priority of an idea or invention by depositing the details in a sealed envelope at the magistrate's office.

Although BASF filed the production process of Alizarin in Britain, the United States of America and in France, it was refused in Prussia by the Berlin patent office. The interpretation of the Prussian patent officers was that the description was too close to the original Graebe and Liebermann patent to suggest originality. The consequence was that there was no monopoly for the manufacture of Alizarin by the sulfonation process in Prussia, the principal German state. As a result, other German dye firms, like Farbwerke ML&B and Bayer, could also produce artificial Alizarin. Attempts were made to protect the processes by means of various patents, but in Germany alone there were 29 different patent laws, making it almost impossible to obtain protection. Besides, Ferdinand Riese at Farbwerke ML&B had also found a process for Alizarin production, 2 months before the British patent filings, and had already started production in 1869. BASF started the production of synthetic Alizarin in 1870. After BASF was refused the sulfonation Alizarin process patent in Prussia, Bayer in Elberfeld also entered into the market in 1872. Numerous new Alizarin factories were founded, of which there were already six in Elberfeld alone in 1873.

The situation between Perkin and BASF was resolved in a professional way. Although BASF's patent was achieved one day earlier than Perkin's one, in the end, both patents were granted. Furthermore, the BASF founder Friedrich Engelhorn (1821–1902) wanted to avoid legal disputes and an agreement on production was reached. With the help of the German-British chemist Hugo Müller (1833–1915) the two companies were brought together and this resulted in a cartel between Perkin and BASF for the division of the international market: Perkin would have a monopoly on the British market and BASF would control continental Europe and Northern America. They also agreed to collaborate, which resulted in cross licensing the June 1869 patents on the 13th of March, 1870 [34]. This provided BASF with valuable information on the successful purification of anthracene and on the production of anthraquinone from this substance. Still, the years 1869 and 1870 were not successful for the Alizarin synthesis at BASF. Graebe and Liebermann spent much time trying to develop the brominated-based Alizarin synthesis, so the sulfonation of anthraquinone was not the subject of serious research prior to the spring of 1870. As a result, Perkin in England and Farbwerke ML&B and Gebrüder Gessert (Gessert Frères) from Elberfeld (which was acquired around 1873 by Bayer) on the continent were in the lead in producing good quality Alizarin and were able to supply substantial quantities by late 1869. The first supply of synthetic Alizarin produced by the dye company Perkin & Sons left Harrow on the 4th of October, 1869. With competition from the aforementioned manufacturers, it became even more important for BASF to optimise the industrial production of Alizarin. Engelhorn decided to recruit other chemists, including the German chemists Carl Glaser (1841–1935) and Heinrich von Brunck (1847–1911), both of whom had worked with Kekulé in Ghent. Glaser and Caro started working on the Alizarin synthesis in the spring of 1870 and, in the following years, Glaser solved the problem of increasing the yield of Alizarin by improving production methods. The first difficulty to be solved was the purification of the raw material anthracene. The following step was the oxidation of anthracene to anthraquinone with potassium

**Table 20.2** Industrial production of synthetic Alizarin of BASF and Perkin & Sons between 1869 and 1873

| Year | Graebe and Liebermann (in tons) | Perkin & Sons (in tons[a]) |
|---|---|---|
| 1869 | – | 1 |
| 1870 | – | 40 |
| 1871 | 125–150 | 220 |
| 1872 | 400–500 | 300 |
| 1873 | 900–1,000 | 435 |

One ton = 1,016.05 kg, so about 16 kg more then the metric ton

dichromate, which was then sulfonated. A more efficient process for making Alizarin from anthracene was finally developed by Caro, Graebe and Liebermann in 1870 [35]. However, it was only towards the end of 1871 that BASF was capable of producing synthetic Alizarin at levels similar to the company's German competitors and Perkin & Sons and it was not until the end of 1875 that all the difficulties were finally overcome [23, 34].

In 1868, the amount of madder cultivated in all the countries of the world where madder was grown was estimated to be around 70,000 tons per year. BASF's efforts to produce synthetic Alizarin industrially were unsuccessful in the years 1869 and 1870. Until the end of 1870 the dye firm Perkin & Sons was practically the only producer of synthetic Alizarin. On the 4th of October, 1869 Perkin & Sons brought synthetic Alizarin onto the market and produced that year about one ton. BASF stated that they had only started producing 50 Ibs (22.7 kg) of Alizarin, 10% quality, per day in November 1870 and that no one else in that country supplied artificial Alizarin.[3] In 1870, Perkin & Sons already produced 40 tons; in 1871, 220 tons; in 1872, 300 tons; and in 1873, 435 tons. According to Graebe and Liebermann, synthetic Alizarin was not produced on the continent until 1871. In 1871, BASF manufactured 125–150 tons of synthetic Alizarin, these weights do not apply to dry colour, but to paste. In the following years, 1872 and 1873, the amount increased to 400–500 tons and 900–1,000 tons respectively (see Table 20.2). In November 1871, Gessert Frères announced to the Industrial Society of Mulhouse that they had produced 30,792 kg of Alizarin in paste (about 30 tons) [36].

At the end of 1873, Perkin retired from business and his firm was sold to Brooke, Simpson and Spiller in 1874. Two years later, in 1876, Burt, Boulton and Haywood took over Perkin's Greenford Green factory, and the rights to his processes. They closed Perkin's Greenford works and moved production to a new factory in Silvertown in the East End of London. In 1883, the new British Alizarine Company took over the Silvertown Alizarin factory.

In Perkin's detailed article describing the period from 1869 to the end of 1873, published in 1879, and supplemented by information in his Hofmann Memorial

[3] It is currently not clear what is meant with the term 10% quality of the paste. It can be interpreted that 10% of the weight of the paste consists of Alizarin but likely this varies between suppliers and batches. As a consequence, the exact amount of Alizarin can not be established.

Lecture, published in 1896, he gives an overview of the development of his Alizarin manufacture and the appropriate processing equipment used to obtain the starting products, not least anthracene, in sufficient quantities and the purification of the anthraquinone [36, 37]. Another detailed description of the production of anthraquinone and Alizarin is given by the Austrian chemist Rudolf Benedikt (1852–1896) [38, 39].

## 20.4 Trade Names

Synthetic Alizarin was mainly produced by the larger dye companies. In Great Britain, Perkin & Sons followed by Burt, Bolton & Haywood prepared Alizarin and the British Alizarine Company Ltd. (who took over the Burt, Bolton and Haywood factory in 1883) manufactured Alizarin under the trade name Alizarin P. In Germany the largest manufacturers were Gessert Frères of Elberfeld, Bayer, Farbwerke ML&B and BASF, under their trade names Alizarin V1, Alizarin No. 1 and Alizarin 1e. In Switzerland Sandoz produced Alizarin under their trade name Alizarinrot R.

## 20.5 The Alizarin Conventions of 1881 and 1900

The synthesis process for Alizarin had been improved over the years and as a result the price of Alizarin fell. At the beginning of the 1880s the Alizarin producers did not earn any money with synthetic Alizarin. The price per kilo of synthetic Alizarin, which had been 270 German marks in 1869, dropped to 23 Marks in 1877 (see Table 20.3). To control the Alizarin market, the surviving Alizarin firms formed the First Alizarin Convention in September 1881. This cartel agreed to share the market, to regulate the production quantities and to fix prices. The cartel included the British manufacturer Burt, Boulton & Haywood and nine German dye factories, including BASF, Farbwerke ML&B and Bayer. In 1883, the British Alizarine Company took over Burt, Boulton & Haywood and joined the cartel instead.

**Table 20.3** Synthetic Alizarin production between 1869 and 1924

| Year | Quantity[a] in 1000 kg | Price (marks per kg) |
|---|---|---|
| 1869 | – | 270. – |
| 1871 | 15 | 140.– |
| 1873 | 100 | 120. – |
| 1877 | 750 | 23. – |
| 1884 | 1,350 | 12.50 |
| 1900 | – | 6.30 |
| 1913 | 2,000 | 1.78 |
| 1924 | 1,120 | 1.15 |

Calculated based on 100% pure dyestuff. Most of the dye was applied as a paste, with different dye concentrations

In 1885 the German dye industry was in a severe economic crisis. This was mainly due to fluctuations in the prices of the raw materials, such as benzene and aniline, used for the production of synthetic dyes. The First Alizarin Convention did not show much stability and consequently this led to its collapse in August 1885 and a fierce competition ensued [24]. In 1900, the Alizarin market was again controlled by the Second Alizarin Convention, formed by the companies BASF, Farbwerke ML&B, Bayer and the manufacturer British Alizarine Company. The Second Alizarin Convention lasted until 1914 [24, 34].

## 20.6 Synthetic Alizarin versus Natural Alizarin

In 1869, chemists were able to produce a natural dye synthetically for the first time. Synthetic Alizarin soon began to replace the natural product. Shortly afterwards, the use of madder began to decline and by 1876 its cultivation had almost ceased. By 1877, 750 tons of synthetic Alizarin had already been produced; in the past, 50,000 tons of madder roots were needed to acquire the same amount of dye. Once the various side-products formed during synthesis had been isolated and characterised, greater control over the reaction became possible and the quality of Alizarin improved. The advantage was that the colour was consistent and easier to apply than the vegetable product. Synthetic Alizarin could be produced at less than half the cost of the natural product, and the market for madder collapsed.

As a result, within a few years, European madder farmers stopped production because it was no longer profitable. Although France tried to protect the cultivation of madder with subsidies, between 1870 and 1886 the European, especially the Alsatian, madder farmers had to switch their farms to other products. Thousands of farmers who had planted madder were robbed of their livelihoods. In 1873, 100,000 kg of artificial Alizarin were produced and at the beginning of the twentieth century two million kilograms, 80% of which were produced in Germany. With the development of the industrial synthesis, the extraction of alizarin from madder became irrelevant and by 1876 madder cultivation came to an almost complete stop [40, 41].

## 20.7 Applications

Synthetic Alizarin is used alone or in mixtures with other mordant dyes in very significant quantities in cotton printing, in the dyeing of cotton (Turkey red) and wool. It is also used to dye silk, in wool and silk printing and to produce madder varnishes applied on violins, very genuine pigment colours that are highly valued in oil painting and lithography [42]. The red dye is used for colouring toys and surface treatments of violins and other stringed instruments. Nowadays, synthetic Alizarin is mostly produced as a red pigment for artists in oil and water colour painting.

## 20.8 Properties

Synthetic Alizarin is an orange-red crystalline substance and is almost insoluble in water, but soluble in spirit with a yellow colour. Lehne wrote[4]: 'Fastness: unchanged after 30 days of exposure to the weather (June and July)'. According to the Colour Index, the light fastness of Alizarin on alum-mordanted wool is ISO 7 and on alum-mordanted silk ISO 6, showing that it has a strong light fastness.

## 20.9 Recipes

Lehne provided several recipes to dye with synthetic Alizarin, on some occasions referring to the general recipes, but he also provides specific recipes. Regarding the general recipes used, the following is written[5]:

> 'Used alone or mixed with No. 203 [Flavopurpurin] and 204 [Anthrapurpurin] and the other mordant dyes in very significant quantities in cotton printing and in the dyeing of cotton (Turkey red) and wool (V4, V5, V6A). Also used for fast dyeing of silk (V9 A and V9B, V10A and V 10B), in wool and silk printing and for the preparation of madder lakes, very stable pigments, highly valued in oil painting and lithography.'

In addition, he refers to recipe V13A and V13B for dyeing wool, using a chrome mordant. The most relevant recipes are presented in Chap. 4.

In addition, he described a recipe for dyeing cotton and linen. A recipe using alum as mordant reads as follows[6]:

> 'Dyeing of cotton (linen). a.) Turkish red and pink. Alumina mordant. V14 A. The recipe according to Erban & Specht (D.R.P. No. 54057) is chiefly used for producing uniform

[4]Alizarin: Echtheit: 'Nach 30 tägiger Einwirkung der Witterung (Juni und Juli) unverändert'.

[5]Wird allein oder in Mischung mit No. 203 und 204 und den übrigen Beizenfarbstoffen in sehr bedeutenden Mengen in der Baumwolldruckerei und in der Färberei der Baumwolle (Türkischroth) und Wolle (V4, 5, 6A) verwendet. Dient auch zum Echtfärben der Seide (V9 A und B, V10 A und B), in Woll- und Seidedruck und zur Darstellung von Krapplacken, sehr echten in der Oelmalerei und in der Lithographie hochgeschätzten Pigmentfarben.

[6]Farberei der Baumwolle (Leinen). a.) Türkischroth und Rosa. Thonerdebeize. V14 A. Das Verfahren nach Erban & Specht (D.R.P. No. 54057 vgl. Einleitung S. XXIII) dient vornehmlich zur Erzeugung von gleichmässigen Rosatönen sowie von Türkischroth in den Copsfärbemaschinen oder auf festgezwirnten Garnen (Leinenzwirn), welche nach dem älteren Verfahren nicht genügend im Innern gefärbt werden. Die rosarothe Färbung auf Baumwollstoff ist mit 0.27% Farbstoff hergestellt wie folgt: Waare in Bad I foulardiren, auf der Hotflue trocknen, in Bad II foulardiren, abermals trocknen, 2 Stunden bei 1½ Atm. dämpfen, zweimal je ½ Stunde bei 80 °C seifen mit 1 Kg Seife auf 1000 L Wasser. Bad I: 64 L kalkfreies Wasser, 2 L Türkischrothöl 80%, 6 L Alizarinlösung. Bad II: 71 ½ L Wasser, 100 cc Essigsäure von 8 °Bé, 648 cc essigsaure Thonerde von 12 °Bé, 114 cc essigsaurer Kalk von 18 °Bé.—Alizarinlösung: 250 g Alizarin No. 1 b 20% [M], 3 ¼ L kalkfreies Wasser, gut mischen, dann zusetzen ½ L Ammoniak 24%.

> shades of pink and Turkey red in the Cops dyeing machines[7] or on tightly twisted yarns (linen yarn), which are not sufficiently dyed [into the core of the yarn] by the older process. The pink dyeing on cotton fabric is produced with 0.27% dye as follows: Foulard[8] the fabric in bath I, dry on the Hotflue, foulard in bath II, dry again, steam for 2 h at 1½ atm., [rinse with] soap twice per ½ h at 80 °C with 1 kg of soap per 1000 L of water. Bath I: 64 L of lime-free water, 2 L of 80% Turkey red oil, 6 L of Alizarin solution. Bath II: 71½ L of water, 100 cc of 52% acetic acid, 648 cc of aluminium acetate 10 %, 114 cc of calcium acetate of 22%—Alizarin solution: 250 g Alizarin No. 1b 20% [M], 3¼ L of chalk-free water; mix well, then add ½ L of ammonia 24%.'

When using iron as a mordant, the following recipe can be used[9]:

> 'Polonowsky and Nitzberg recommend the following procedure for producing a violet background [using] vat blue (cf. *Färber-Zeitung*, 1893/94, p. 139): mordant 100 kilos of boiled cotton yarn with 6 kg of sumac extract, 1600 L of water at 50 °C, then treat for ½ min on the mordanting machine with wood acetic iron of 1.4 °Bé[10] (pour 30 L as a batch into the trough of the machine); add ½ L for each strand. Wring out strand by strand and leave for 15 h. Then [add] chalk, rinse in pure water and dye in a dye solution containing 7% Alizarin, prepare cold and heat to 70–75 °C in 2 h. Rinse and wring out. When pre-dyed yarn is dyed in the vat in this way, only about 1/3 of the amount of indigo is needed to produce a deep blue compared to using indigo alone.'

## 20.10 Identification Using UHPLC-PDA

The analysis of synthetic Alizarin (RCE no. 3034), shows in addition to the main peak of Alizarin also a few minor products which are probably side products of the synthesis. It must be noted that it is difficult to determine if synthetic Alizarin has

---

[7] 'The process of dyeing yarn upon the cop, in order to save the expense of reeling it into skeins and rewinding it upon bobbins. Many machines have been devised for this purpose but none has been entirely satisfactory.' Cited from https://www.wordnik.com/words/cop-dyeing. Visited April 19, 2024.

[8] 'The Foulard's process is the operation of impregnating a textile in a chemical solution containing a certain bath (dye, sizing or functional agent), and then drained by pressure cylinders. If it is a dye solution, the dye is trapped between the substrate, giving color to the fabric and if it is a product to give a specific finish, is impregnated, leaking excess liquid on the bucket.' Cited from https://medcraveonline.com/JTEFT/JTEFT-02-00082#:~:text=Foulard's%20process%3A%20The%20Foulard's%20process,then%20drained%20by%20pressure%20cylinders. Access April 19, 2024.

[9] Eisen Beize. Polonowsky und Nitzberg empfehlen zur Erzeugung eines violetten Untergrunds für Küpenblau folgende Vorschrift (vgl. Färber-Zeitung. 1893/94 S.139): 100 kg ausgekochte Baumwollgarn mit 6 kg Sumachextrakt, 1600 L Wasser von 50 °C beizen. auswinden, sodann auf der Beizmaschine ½ min lang durch holzessigsaures Eisen von 1,4 °Bé nehmen (30 L als Ansatz in den Trog der Maschine geben), für jeden Strang ½ L zusetzen. Strang für Strang auswinden und 15 Stunden liegen lassen. Hierauf kreiden, in reinem Wasser auswaschen und in eine Farbflotte, welche 7% Alizarin enthält, ausfärben, kalt ansetzen und in 2 Stunden auf 70–75 °C erwärmen. Spülen, auswinden. Wenn in dieser Weise vorgefärbtes Garn in der Küpe ausgefärbt wird, ist nur etwa 1/3 der Menge an Indigo für tiefes Blau erforderlich, wie bei Anwendung von Indigo allein.

[10] This is likely to be pyrolignite of iron, which is essentially iron dissolved in crude acetic acid. The concentration could not be converted but is probably about 1%.

been applied in a historical object, or just a pure form of madder (*Rubia tinctorum* L.), although in the last case usually purpurin is present as well. However, when this is not detected, it could due to a specific dyeing recipe and / or purpurin can be just below the detection limit. Therefore, care should be taken when conclusions are drawn based on the presence of Alizarin only (Fig. 20.3).

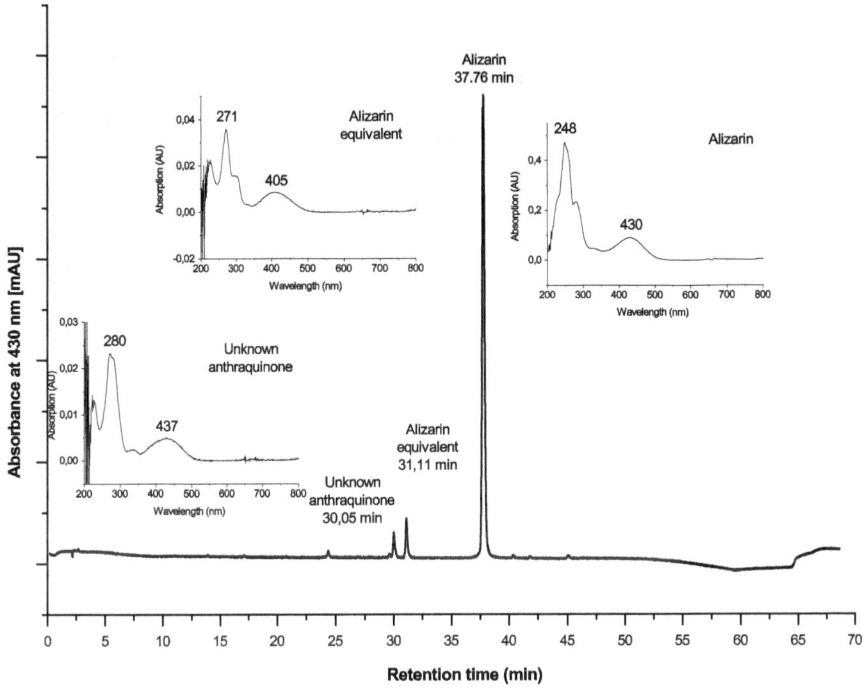


Fig. 20.3 Chromatogram and spectra of synthetic Alizarin (RCE no.3034)

## 20.11 Case Study: A Costume of a Dancing Tresterer

According to a winter tradition in Pinzgau since the mid-nineteenth century, the so-called Tresterer go from farm to farm on days between December and the 6 January. They dance, wish good luck and blessings and announce the approach of spring. At the end of the nineteenth century, Tresterer costumes came to the Volkskundemuseum Wien (see Fig. 20.4). The multipart Tresterer costume (ÖMV/5.306-5.311), which was purchased by Wilhelm Hein in 1896, is from Oberkrimml. Krimml is a municipality in the Zell am See District, in the Pinzgau region, in the federal state of Salzburg. Wilhelm Hein (1861–1903) was an Austrian linguist, folklorist, orientalist and ethnographer. Together with the Austrian folklorist and indologist Michael Haberlandt (1860–1940), with whom he worked in the anthropological-ethnographic

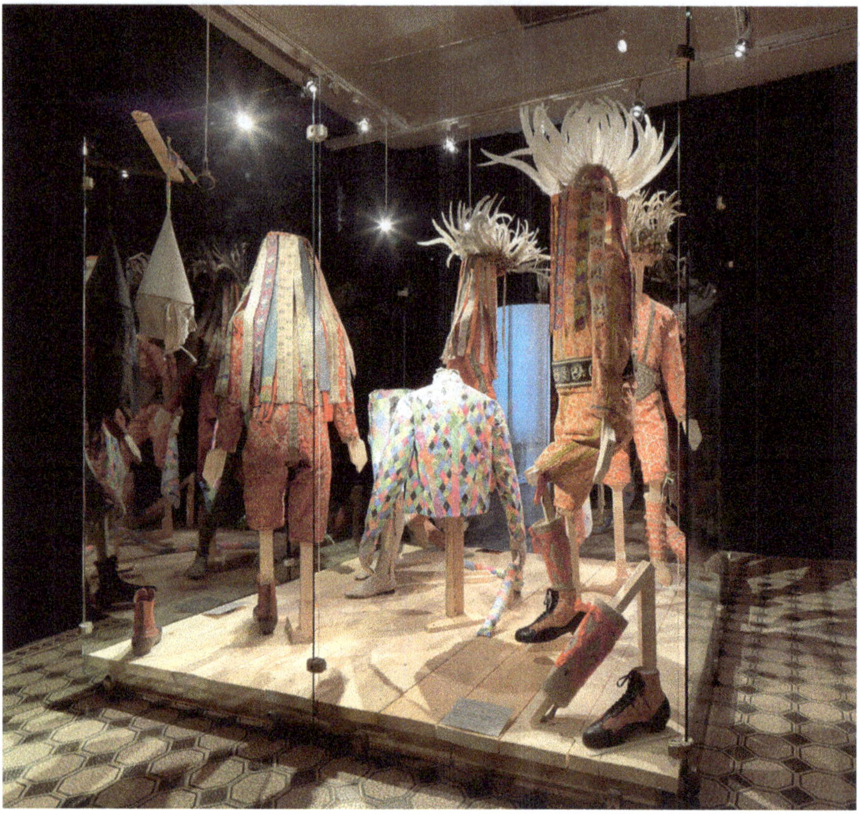

**Fig. 20.4** Tresterer costume (ÖMV/5.306-5.311) on the front left. © Volkskundemuseum Wien. Photograph: Birgit Kainz and Peter Kainz, faksimile digital

department of the Natural History Museum Vienna, he founded the Verein für Volkskunde in 1894, the *Zeitschrift für österreichische Volkskunde* in 1895 and the Museum für Volkskunde in Vienna in the same year [43]. The jacket (ÖMV/5.307) is made of light-coloured cotton, printed with a design of fruit baskets and tendril motifs in red (see Fig. 20.5). The short jacket has long, tight sleeves and a stand-up collar. The back, sleeves and front are decorated with silver threads and braids. The colours appear to have been applied using a direct printing process (see Fig. 20.6). Two samples were taken from a tear next to the buttonhole: the red sample for analysing the dyes; the beige sample to determine whether the fabric had been treated with dyes or tanning agents before printing. The aim of the research was to identify the dyes that were used to print the jacket and to determine whether they could provide information about the date of the object. If the colours were produced synthetically, this would be an indication that the multi-piece costume is from the second half of the nineteenth century.

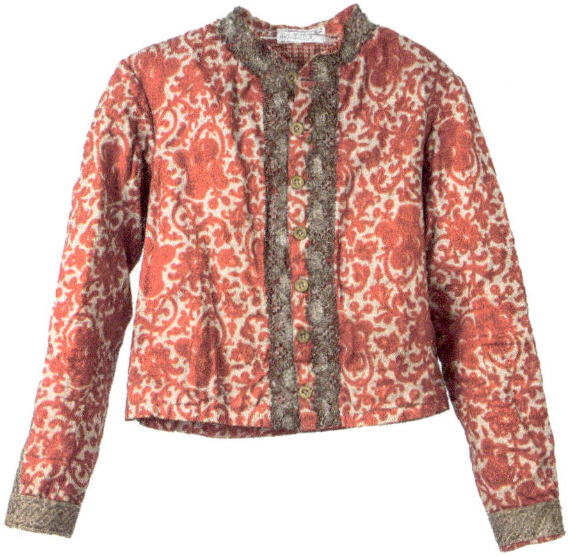

**Fig. 20.5** Jacket of a Tresterer costume (ÖMV/5.307). © Volkskundemuseum Wien. Photograph: Christa Knott

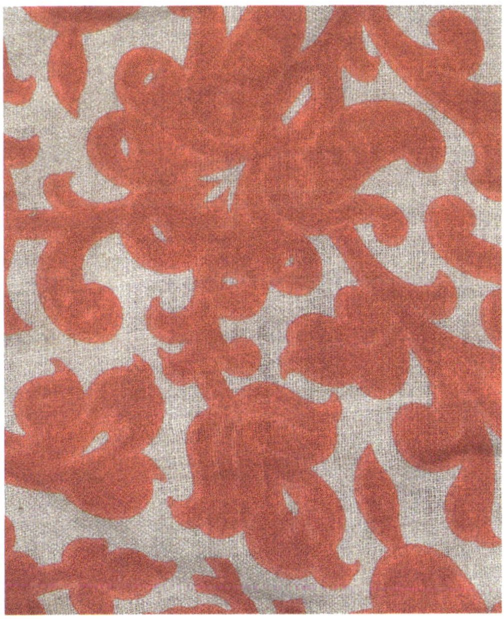

**Fig. 20.6** Detail of the printed fabric. © Volkskundemuseum Wien. Photograph: Monika Maislinger

The dyes were analysed using ultra-high performance liquid chromatography coupled with photo diode array detection (UHPLC-PDA). In the beige sample, analysis revealed that the cotton fabric was neither pre-treated with a dye nor a tanning agent before printing, although some red components were detected which are likely to be cross-contamination. In the fibre from the area printed in red, Alizarin and an Alizarin equivalent were detected as the main compounds, as well as two

**Table 20.4** The UHPLC analysis of the red printed treasterer costume

| Sample | Result | Conclusion | Comment |
|---|---|---|---|
| Tresterer-1 (red) | Alizarin<br>Alizarin equivalent<br>Indigotin (trace)<br>Two unknown anthraquinones | Dyed with Alizarin, probably a synthetic variety | The concentration of Alizarin and the equivalent are about the same, these are the main compounds. The Alizarin equivalent is likely to be anthrapurpurin |
| Tresterer-2 (beige) | Alizarin<br>Alizarin equivalent<br>Two unknown anthraquinones | Not dyed | Red dyes present at very low concentration, probably a cross-contamination |

unknown anthraquinones (see Table 20.4). If this colouring had been made using madder root (*Rubia tinctorum* L.), purpurin would have to be present as a secondary component. Since no purpurin could be detected, it can be assumed that the Alizarin is of synthetic origin. More importantly, the retention time and the spectrum of the Alizarin equivalent resembles that of Anthrapurpurin (Isopurpurin), one of the dyes mentioned by Lehne to be used in combination with synthetic Alizarin. The results of the dye analysis indicate that the cotton fabric of the multi-piece costume was probably printed with synthetic Alizarin [44].

## References

1. Groeneveld, I.: One size fits all (dyes). The development of a method for the simultaneous detection of natural and synthetic dyes by UHPLC-PDA, and the quantification of natural (over)dyeing process, master thesis, University of Amsterdam, (2018)
2. Groeneveld, I., van Bommel, M.R.: Quantification of dye uptake and loss by wool during (over)dyeing of weld, madder, redwood and indigo. In: Abstract Book of the 37th annual meeting of Dyes in History and Archaeology, Lisbon, Portugal. pp. 21–22 (2018)
3. Rösel-Mautendorfer, H., Bogensperger, I., Gaibor, A.N.P., Hofmann-de Keijzer, R., van Bommel, M.R.: Dyeing Experiments with Dyer's Madder (Rubia tinctorum L.). Experimentelle Archäologie in Europa. 21, (2022)
4. Hofmann-de Keijzer, R., van Bommel, M.R., Joosten, I., Bogensperger, I.: Late Antique Textiles from the Papyrus Collection of the Austrian National Library – Scientific investigation of fibres, dyes and dyeing techniques. MPER XXXIV, Band 2. de Gruyter, Boston / Berlin (2024)
5. Cardon, D.: Natural Dyes. Sources, Tradition, Technology and Science. Archetype Publications Ltd., London (2007)
6. "De Materia Medica" by Dioscorides. Library of Congress, Washington, D.C. 20540 USA.
7. Pliny the Elder: The Natural History Book VII, https://www.perseus.tufts.edu/hopper/text?doc=Plin.+Nat.+toc
8. Papyrus Graecus Holmiensis., https://www.loc.gov/item/2021668051/
9. Kirby, J., Spring, M., Higgitt, C.: The Technology of Eighteenth- and Nineteenth-Century Red Lake Pigments. National Gallery Technical Bulletin. 28, 69–95 (2007)

10. Hofenk de Graaff, J.H.: The Colourful Past : Origins, Chemistry and Identification of Natural Dyestuffs. Abegg-Stiftung / Archetype Publications Ltd., Riggisberg / London. 94–97 (2004)
11. Germer, Renate.: Die Textilfärberei und die Verwendung gefärbter Textilien im Alten Ägypten. 68–90 (1992)
12. Verhecken, A.: Relation Between Age and Dyes of 1st Millennium ad Textiles Found in Egypt. In: de Moor, A. and Fluck, C. (eds.) Methods of Dating Ancient Textiles of the 1st Millennium AD from Egypt and Neighbouring Countries. pp. 206–213. Lannoo, Antwerp (2007)
13. Hofmann-de Keijzer, R., van Bommel, M.R., Joosten, I., Hartl, A., Gaibor, A.N.P., Heiss, A.G., Kralofsky, R., Erlach, R., de Groot, S.: The colours and dyeing techniques of prehistoric textiles from the salt mines of Hallstatt. In: Grömer, K., Kern, A., Reschreiter, H., and Rösel-Mautendorfer, H. (eds.) Textiles from Hallstatt. Woven culture from Bronze and Iron Age salt mines. pp. 147–150. Archaeolingua, Budapest (2013)
14. Kuhlmann, C.F.: Analyse Chimique de la Racine de Garance (Rubia tinctorum). Annales de Chimie et de Physique. 24, 225–234 (1823)
15. Robiquet, P.J., Colin, J.J.: Sur un nouveau principe immédiat des vègétaux (l'alizarin) obtenu de la garance. Journal de pharmacie et des sciences accessoires. 12, 407–412 (1826)
16. Grace-Calvert, F., Stenhouse, John., Groves, C.E.: Dyeing and Calico Printing. Palmer & Howe, Manchester (1878)
17. Robiquet, P.J., Colin, J.J.: Nouvelles Recherches sur la Matière Colorante de la Garance. Annales de Chimie et de Physique. 34, 225–253 (1827)
18. Lagier, Robiquet, Colin, Brevet d'Invention sur la Garance, pris le 26 Mars 1828 pour 10 Ans, par M. Lagier, Négociant, Robiquet et Colin, Professeurs de Chimie à Paris, rue de l'Arbalère No. 13.
19. Graebe, C., Liebermann, C.: Ueber künstliche Bildung von Alizarin. Berichte der deutschen chemischen Gesellschaft zu Berlin. Ferd. Dümmler´s Verlagsbuchhandlung, Berlin. 2, 14 (1869)
20. Garfield, S.: Chapter 7: The terrible glare. In: Mauve: How one Man invented a Colour that changed the World. pp. 78, 83–84. Faber and Faber Ltd., London (2000)
21. Graebe, C., Liebermann, C.: Ueber Farbstoffe aus der Anthracengruppe,. Berichte der deutschen chemischen Gesellschaft zu Berlin. Ferd. Dümmler´s Verlagsbuchhandlung, Berlin. 1, 104 (1868)
22. Graebe, C., Liebermann, C.: Ueber Alizarin und Anthracen. Berichte der deutschen chemischen Gesellschaft zu Berlin. Ferd. Dümmler´s Verlagsbuchhandlung, Berlin. 1, 49–51 (1868)
23. Reinhardt, C., Travis, A.S.: Heinrich Caro and the Creation of Modern Chemical Industry. Kluwer Academic Publishers, Dordrecht / Boston / London (2000)
24. Morris, P., Travis, A.: A History of the International Dyestuff Industry. American Dyestuff Reporter. 81, (1992)
25. Baeyer, A., Caro, H.: Synthese von Anthrachinonabkömmlingen aus Benzolderivaten und Phtalsäure. Berichte der deutschen chemischen Gesellschaft zu Berlin. Ferd. Dümmler´s Verlagsbuchhandlung, Berlin. 7, 968 (1874)
26. Graebe, C., Liebermann, C.: Ueber künstliches Alizarin. Berichte der deutschen chemischen Gesellschaft zu Berlin, Ferd. Dümmler's Verlagsbuchhandlung, Berlin. 2, 332–334 (1869)
27. Graebe, C., Liebermann, C.: Ueber Anthracen und Alizarin. Annalen der Chemie und Pharmacie. C.F. Winter'sche Verlagshandlung, Leipzig und Heidelberg. 7, 257–322 (1870)
28. Carl Graebe and Carl Theodore Liebermann, B. P. 3.850, Improved process of preparing alizarine, 18th of December 1868.
29. Brönner, J., Gutzkow, H.: Verfahren zur Darstellung von Anthracen aus dem Pech von Steinkohlentheer, und zur Darstellung von Farbstoffen aus Anthracen. Dinglers Polytechnisches Journal. 201, 545–546 (1871)
30. Bäumler, E.: Die Rotfabriker, Familiengeschichte eines Weltunternehmens. R. Piper GmbH & Co. KG., München (1988)
31. 1863 ML&B 1938, I.G. Farbenindustrie Aktiengesellschaft, Werke: Farbwerke vorm. Meister Lucius & Brüning. F. Bruckmann KG., München (1938)

32. Bäumler, E.: Ein Jahrhundert Chemie. Econ Verlag GmbH, Düsseldorf (1963)
33. Cochrane v. Badische Anilin & Soda Fabrik, Syllabus U.S. Supreme Court, 111 U.S. 293 (1884)
34. Abelshauser, W.: Die BASF, eine Unternehmensgeschichte, Zweite Auflage. Verlag C.H. Beck, München (2003)
35. Caro, H., Graebe, C., Liebermann, C.: Ueber Fabrikation von künstlichem Alizarin. Berichte der deutschen chemischen Gesellschaft zu Berlin. Ferd. Dümmler's Verlagsbuchhandlung, Berlin. 3, 359–360 (1870)
36. Perkin, W.H.: The History of Alizarin and Allied Colouring Matters at their Production from Coal Tar. Journal of the Society of Arts. 27, 572–602 (1879)
37. Perkin, W.H.: Hofmann Memorial Lecture: The Origin of the Coal-Tar Colour Industry, and the Contributions of Hofmann and his Pupils. J Chem Soc. 69, 596–637 (1896)
38. Benedikt, R.: Die künstlichen Farbstoffe (Theerfarben), Ihre Darstellung, Eigenschaften, Prüfung, Erkennung und Anwendung, Band II. Verlag von Theodor Fischer, Kassel (1883)
39. Benedikt, R.: The Chemistry of the Coal-tar Colours. George Bell and Sons, London (1886)
40. Hofmann, R.: Färbepflanzen, PhD thesis, 5 volumes, Band 1, (1989)
41. Hofmann, R.: Färbepflanzen und Färbedrogen. Restauratorenblätter, Österreichische Sektion des IIC, Verlag Mayer & Comp., Klosterneuburg und Wien. 13, 41 (1992)
42. Madder Lake, genuine Pigments | Kremer Pigmente Online Shop, https://www.kremer-pigmente.com/en/shop/pigments/37202-madder-lake-genuine.html
43. Lecture: Tänze und Volksschauspiele in Tirol und Salzburg. Mitteilungen der Anthropologischen Gesellschaft Wien. 24, 45–48 (1894)
44. Hofmann-de Keijzer, R.: Naturwissenschaftliche Untersuchungen der Farbstoffe von einem bedruckten Gewebe aus einem Trestererkostüm. Endbericht. (2014)

# Chapter 21
# The Long Industrial Road to Synthetic Indigo

Versuchen Sie nicht nur, leichte Arbeiten zu machem,
wenden Sie sich vor allem schwierigen Fragen zu.
Adolf Baeyer

Don't just try to do easy work, but above all focus on difficult questions.
Adolf Baeyer

## 21.1 Introduction

A brief history of natural indigo[1] will show its extraordinary importance and explain why chemists dedicated themselves to the structural elucidation and synthesis of this pigment. Since prehistoric and ancient times, people around the world have discovered that a blue pigment can be obtained from some plants, belonging to different plant families. This pigment was suitable for both vat dyeing of textiles and for paintings [1]. The most important plants include *Indigofera* species (family Fabaceae), native to Africa, Asia, subtropical to tropical America and the Southwest Pacific, many of which are known as cultivated plants. Woad (*Isatis tinctoria* L., family Brassicaceae) is native to Southeastern and Eastern Europe and was cultivated in Europe, Asia, Africa, South and North America. Gara or Yoruba indigo (*Philenoptera cyanescens* (Schumach. & Thonn.) Roberty, syn. *Lonchocarpus cyanescens* (Schumach. & Thonn.) Benth., family Fabaceae) is native to the west of tropical Africa. Dyer's knotweed (*Persicaria tinctoria* (Aiton) Spach, syn. *Polygonum tinctorium* Aiton, family Polygonaceae) is the source of Japanese indigo. Indigo plants of Southeast Asia include Java indigo (*Marsdenia tinctoria* R.Br., syn. *Asclepias tinctoria* (R.Br.) Roxb., family Apocynaceae), Assam indigo (*Strobilanthes cusia* (Nees) Kuntze, family Acantheceae), and dyer's oleander (*Wrightia tinctoria* (Roxb.) R.Br., family Apocynaceae).

[1] In this chapter we will use indigo without capticals when indicating a natural origin while for synthetic Indigo we use a capital.



M. R. van Bommel, M. de Keijzer, *Bright Colours from the Past*, Cultural Heritage Science, https://doi.org/10.1007/978-3-031-88664-5_21

According to the recipes on a sixth-century BCE Late Babylonian cuneiform clay tablet probably from Sippar, a city on the east bank of the river Euphrates in modern Iraq, blue colours were dyed by a number of immersions in vats and a purple by double dyeing blue and madder red [1–3]. The occurrence of *indikon melan* as trade goods in the *Periplus Maris Erythraei* (Periplus of the Red Sea) indicates the import of indigo from India to Egypt in the mid-first century CE [3]. Indigo from India was known to ancient writers as a pigment, but not as a vat dye [4]. The Greek physician, pharmacologist and botanist Pedanius Dioscorides (c. 40–c. 90 CE) mentions indigo under the Greek term *indikon* (Dsc. 5.107) and the Roman politician and scholar Pliny the Elder (23/24–79 CE) by the Latin term *indicum* (NH 33.57.163). Dioscorides and Pliny the Elder describe two different origins for indigo, from plants native to India and from scum floating on top of copper kettles ('flower' from vats), and the latter was collected by artists (Plin. NH 35.46). Woad was the indigo plant of the Mediterranean and European region from prehistoric times while later indigo from the Tropics was also used. After the discovery of the sea route to India, indigo increasingly reached Europe, replacing woad and its cultivation [5].

In the nineteenth century, Britain dominated the indigo market, much to the displeasure of the German dye industry, as indigo was mainly derived from indigo plants grown on large plantations on the Indian subcontinent. Around 1900 the indigo market was huge. As an example, the indigo plant (*Indigofera tinctoria* L.) was cultivated on more than 700,000 hectares of land in Bengal. At that time, the annual production of natural indigo was around 8,000 tons, worth 100 million German marks in terms of pure dye [6].

Although a large number of synthetic dyes were already available on the market in the 1870s, no synthetic substitute for indigo existed that could compete with the natural product. In the period 1857–1864, Friedrich August Kekulé developed his carbon theories, a stimulus for the synthesis of alizarin in 1869, which replaced madder root as a dye source. Similarly, university laboratories and the chemical industry competed to find the structure of the indigotin molecule, aiming to develop a synthetic alternative for natural indigo. Already in 1865, Adolf Baeyer started investigating the molecular structure of indigo in order to synthesise it, but it took him 18 years to reveal the indigo formula. In his work, Baeyer was strongly supported by Heinrich Caro, then research director of BASF.

Between 1878 and 1882, Baeyer succeeded in establishing three routes for Indigo synthesis, using different starting materials such as isatin, cinnamic acid and *o*-nitrobenzaldehyde. However, these synthetic routes were too expensive to be commercially viable in competition with natural indigo. The two Indigo players,

BASF and Farbwerke ML&B, had to search for alternative, cheap starting materials. For several years there was little activity in the field of Indigo synthesis. In 1890, Karl Heumann discovered two new synthesis processes. In the first Heumann synthesis, *N*-phenylglycine, made from aniline, was fused with caustic potash. Unfortunately, the high temperature required for cyclisation led to a low yield of Indigo. The second Heumann synthesis, based on *N*-phenylglycine-*o*-carboxylic acid, made from anthranilic acid, was the economically attractive route. Having recognised Heumann's breakthrough, Farbwerke ML&B and BASF both acquired the rights to his invention and to adapt these chemicals for their production processes.

Between 1890 and 1897, many technical processes were developed to obtain the basic materials for the synthesis of Indigo. The *N*-phenylglycine-o-carboxylic acid process was applied by BASF in 1897 to produce the first synthetic Indigo on an industrial scale. Around 1900, and after 100 patents, the German dye factories BASF and Farbwerke ML&B were able to produce Indigo synthetically in an economically affordable way.

Three developments led to a rapid fall in the price of synthetic Indigo: the Pfleger process exploited by Farbwerke ML&B; the newly developed synthetic blue sulfur dyes of the German companies Cassella, Kalle and Bayer, which were strong competitors to synthetic Indigo, and the discovery of Indanthrene blue by René Bohn. In 1901, Johannes Pfleger, working for Degussa, discovered and patented a more efficient process that made it possible to synthesise Indigo for large-scale commercial production, which was taken over by Farbwerke ML&B. This process used the condensing agent sodium amide for the ring closure to form indoxyl at a lower temperature, resulting in a 90% yield of Indigo. A fierce competition began and in 1904, after the price had dropped to 7 German marks per kg, both dye factories formed the Second Indigo Convention in order to control the Indigo market.

In 1925, BASF developed an improved Indigo process based on phenylglycinonitrile, which was produced from aniline, formaldehyde and hydrogen cyanide [7]. Over time, the demand for synthetic Indigo decreased and many companies ceased production. By the 1960s, there was little interest in synthetic Indigo and even BASF wanted to stop the production. In the late 1960s, however, the worldwide demand increased significantly as it was used to dye fabrics for blue jeans. Even today, after more than 125 years, synthetic Indigo is still popular. This chapter will focus on synthetic Indigo routes, their advantages, problems and limitations seen from a historical and economical perspective (Tables 21.1 and 21.2).

**Table 21.1** Discoverers, year of discovery and patents of the various synthetic Indigo syntheses

| Discoverers | Year of discovery | Synthesis | Patents |
|---|---|---|---|
| Adolf Baeyer and Adolph Emmerling | 1869 | Baeyer-Emmerling indole synthesis | – |
| Adolf Baeyer | 1878 | Indigo synthesis of isatin from phenylacetic acid | – |
| Adolf Baeyer | 1880 | Indigo synthesis from *o*-nitrocinnamic acid (cinnamic acid, *o*-nitro-phenyl-propiolic acid) | G.P. 11.857/1880<br>G.P. 11.858/1880<br>G.P. 12.601/1880<br>G.P. 14.997/1881<br>G.P. 15.516/1881<br>G.P. 17.656/1881<br>G.P. 19.266/1881 |
| BASF | 1882 | Baeyer-Drewson process: Indigo synthesis by *o*-nitrobenzaldehyde | G.P. 19768/1882<br>B.P. 1.266/1882<br>U.S.P. 257.814/1882<br>U.S.P. 257.815/1882 |
| Karl Heumann | 1890 | 1st Heumann synthesis: Indigo synthesis from *N*-phenylglycine | G.P. 54.626/1890<br>G.P. 58.276/1890<br>G.P. 61.711/1890<br>G.P. 61.712/1890<br>G.P. 63.309/1891<br>G.P. 63.310/1891<br>G.P. 68.372/1891 |
| Karl Heumann | 1890 | 2nd Heumann synthesis: Indigo synthesis from *N*-phenylglycine-*o*-carboxylic acid | G.P. 56.273/1890<br>G.P. 73.687/1893 |
| Deutsche Gold- und Silber-Scheide-Anstalt vorm. Roessler | 1901 | Pfleger Indigo process: Indigo synthesis from *N*-phenylglycine fused with sodium amide | G.P. 117.623/1901<br>G.P. 137.955/1901<br>G.P. 141.749/1901<br>USP. 680.395/1901<br>G.P. 142.700/1902<br>G.P. 180.394/1904 |

**Table 21.2** Discoverers and patents of the various synthetic Indigo syntheses

| Discoverer, company | Patent | Title, reference | Starting materials | | | |
|---|---|---|---|---|---|---|
| | | | o-nitrocinnamic acid | o-nitrobenzaldehyde | *N*-phenylglycine | N-phenylglycine-*o*-carboxylic acid |
| Adolf Baeyer | G.P. 11.857<br>19 March 1880 | Darstellung von Derivaten der Orthonitrozimmtsäure, den Homologen und Substitutionsprodukten dieser Derivate und Umwandlung derselben in Indigblau und verwandte Farbstoffe | + | | | |
| Adolf Baeyer | G.P. 11.858<br>21 March 1880 | Neuerungen im Verfahren zur Darstellung der Materialien für die Fabrikation des künstlichen Indigos und Verfahren zur Erzeugung dieses Farbstoffes direkt auf der Faser | + | | | |
| Adolf Baeyer | G.P. 12.601<br>18 June 1880 | Darstellung von künstlichem Indigblau und verwandter Farbstoffe aus Orthonitrozimmtsäure, deren Homologen und Substitutionsprodukten | + | | | |
| BASF | G.P. 14.997<br>6 February 1881 | Neuerungen in dem Verfahren zur Darstellung des künstlichen Indigos und zur Erzeugung dieses Farbstoffs direkt auf der Faser | + | | | |
| BASF | G.P. 15.516<br>6 February 1881 | Neuerungen in dem Verfahren zur Darstellung des künstlichen Indigos und zur Erzeugung dieses Farbstoffs direkt auf der Faser | + | | | |
| BASF | G.P. 17.656<br>8 July 1881 | Neuerungen in dem Verfahren zur Darstellung des künstlichen Indigo's | + | | | |

(continued)

**Table 21.2** (continued)

| Discoverer, company | Patent | Title, reference | Starting materials | | | |
|---|---|---|---|---|---|---|
| | | | o-nitrocinnamic acid | o-nitrobenzaldehyde | *N*-phenylglycine | N-phenylglycine-*o*-carboxylic acid |
| BASF | G.P. 19.266 23 December 1881 | Neuerungen in dem Verfahren zur Darstellung des künstlichen Indigo's | + | | | |
| BASF | G.P. 19.768 24 February 1882 | Neuerungen in dem Verfahren zur Darstellung des künstlichen Indigos aus dem Orthonitrobenzaldehyd | | + | | |
| BASF | G.P. 54.626 6 May 1890 | Verfahren zur Darstellung künstlichen Indigos | | | + | |
| BASF | G.P. 58.276 15 August 1890 | Verfahren zur Darstellung von Farbstoffen aus der Indigoreihe | | | + | |
| BASF | G.P. 61.711 28 October 1890 | Verfahren zur Darstellung von Farbstoffen aus der Indigoreihe | | | + | |
| BASF | G.P. 61.712 30 December 1890 | Verfahren zur Darstellung von Farbstoffen der Indigoreihe | | | + | |
| BASF | G.P. 63.309 10 March 1891 | Verfahren zur Darstellung von Farbstoffen der Indigoreihe | | | + | |
| BASF | G.P. 63.310 10 March 1891 | Verfahren zur Darstellung von Farbstoffen der Indigoreihe | | | + | |
| BASF | G.P. 68.372 18 January 1891 | Verfahren zur Darstellung von Farbstoffen der Indigoreihe | | | + | |
| BASF | G.P. 56.273 11 July 1890 | Verfahren zur Darstellung von künstlichem Indigo | | | | + |
| BASF | G.P. 73.687 11 July 1893 | Verfahren zur Darstellung von carboxylierten Indigofarbstoffen | | | | + |

| | | | | | | |
|---|---|---|---|---|---|---|
| Deutsche Gold- und Silber-Scheideanstalt vorm. Roessler | G.P. 117.623<br>11 April 1900 | Verfahren zur Darstellung Von Alkali-Amid | | | + | |
| Deutsche Gold- und Silber-Scheide-Anstalt vorm. Roessler | G.P. 11.245<br>G.P. 137.955<br>18 January 1901 | Verfahren zur Darstellung von Indoxylderivaten aus aromatischen Glycinen | | | + | |
| Deutsche Gold- und Silber-Scheideanstalt vorm. Roessler | Austrian patent No. 8.097<br>17 August 1901 | Verfahren zur Condensation aromatischer Glycine zu Indoxylderivaten | | | + | |
| Deutsche Gold- und Silber-Scheideanstalt vorm. Roessler | G.P. 141.749<br>20 August 1901 | Verfahren zur Darstellung von Indoxylderivaten | | | + | |
| Johannes Pfleger, assignor to Deutsche Gold & Silber Scheide Anstalt vorm. Roessler | U.S.P. 680.395<br>13 August 1901 | Process of making indoxyl derivatives | | | + | |

(continued)

**Table 21.2** (continued)

| Discoverer, company | Patent | Title, reference | Starting materials | | | |
|---|---|---|---|---|---|---|
| | | | o-nitrocinnamic acid | o-nitrobenzaldehyde | *N*-phenylglycine | N-phenylglycine-*o*-carboxylic acid |
| Deutsche Gold- und Silber-Scheideanstalt vorm. Roessler | Austrian patent No. 10.691 26 August 1901 | Verfahren zur Darstellung von Indoxyl und dessen Homologen | | | + | |
| Deutsche Gold- und Silber-Scheideanstalt vorm. Rössler | G.P. 142.700 26 January 1902 | Verfahren zur Darstellung von Indoxyl bzw. dessen Derivaten und Homologen | | | + | |
| Deutsche Gold- und Silber-Scheide-Anstalt vorm. Roessler | G.P. 180.394 14 August 1904 | Verfahren zur Darstellung von Indoxylderivaten aus aromatischen Glycinen und ähnlich sich verhaltenden Verbindungen | | | + | |

## 21.2 Synthetic Indigo

Colour Index name: Vat blue 1; Pigment blue 66; Colour Index number: 73000
Natural indigo: Colour Index name: Natural Blue 1; Colour Index number: 75780
IUPAC name: 2-(3-Hydroxy-1H-indol-2-yl)indol-3-one
Discoverer: Adolf Baeyer, 1870
Contemporary terminology: Synthetic indigo (En), Indigo synthétique (Fr), Synthetischer Indigo (Ge), Synthetische indigo (NL)
Chemical formula: $C_{16}H_{10}N_2O_2$ (Fig. 21.1)

**Fig. 21.1** Structural formula of synthetic Indigo

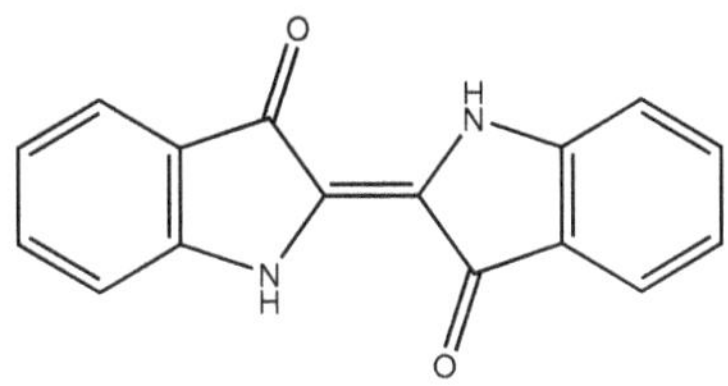


## 21.3 History

In the early days of the synthetic dyes, chemistry was largely empirical, with vague ideas about how atoms in a molecule were linked. The carbon theories developed by the German chemist Friedrich August Kekulé (1829–1896) provided chemists with a useful model for determining the structure of dyes. A number of constitutional formulae were proposed for indigotin by the chemists Baeyer (1868), A. Strecher (1868), A. Emmerling and C.O.V. Engler (1870), E. von Sommaruga (1878), E. Baumann and F. Tiemann (1879), N. Ljubawin (1879) and, again, Baeyer (1882) [8]. In 1865, the German chemist Adolf Baeyer (1835–1917), professor of chemistry at the Gewerbeschule (Technical College) in Berlin and later at the University of Munich, started investigating the determination of the structure and the synthesis of Indigo. After 18 years, in 1883, he finally elucidated a chemical formula for indigotin and provided this in a letter to the German chemist and BASF research director Heinrich Caro (1834–1910), but he proposed the *cis*-structure for the molecule, which was for a long time the generally accepted formula [9]. The correct structure of indigotin, known as the *trans* form, was determined in 1928 by X-ray diffraction and was confirmed by crystallographic studies in 1986 (see Fig. 21.1) [10].

## *The First Baeyer Indigo Synthesis (1878)*

Baeyer achieved his first success as early as 1866, when he identified the decomposition products indole and isatin from natural indigo by zinc dust distillation and three years later, in 1869, he proposed a formula for indole [11]. In 1869, Kelulé proposed a structural formula for isatic acid and also for isatin. He considered isatic acid to be *o*-aminophenylglyoxilic acid (*o*-aminobenzoylformic acid) and isatin as 2,3-dioxoindoline [12]. In the same year, in 1869, Baeyer and the German chemist Adolph Emmerling (1842–1906) synthesised indole by melting crude *o*-nitrocinnamic acid with potash and iron filings, hence this procedure was called the Baeyer-Emmerling synthesis.

In 1870, Baeyer prepared Indigo by the reduction of isatin (Fig. 21.2). Isatin had already been discovered by the French chemist Auguste Laurent (1807–1853) in France in 1840 as a product from the oxidation of indigo by nitric acid [13–16]. Unfortunately, isatin was rare and extremely expensive and as it was obtained by the oxidation of natural indigo, it was not an alternative to natural indigo. Therefore, it was necessary to find a feasible isatin synthesis [13]. In the same year, Emmerling and the German chemist Carl Engler (1842–1925) succeeded in preparing traces of Indigo by distilling nitroacetophenone with zinc dust and soda lime. However, this process was not reproducible and was not an effective method. Contrary to what is often reported, they did not develop the Indigo synthesis.

Eight years later, in 1878, Baeyer was successful in producing isatin from phenylacetic acid. Starting from phenylacetic acid, he obtained oxindole by way of several steps, which he further oxidised to isatin. By the chlorination of isatin with phosphorus pentachloride to isatin chloride and subsequent reaction with zinc in acetic acid, indigotin was produced [17–19]. The first real full synthesis of Indigo took place on the 6 June 1878 and in the same year Baeyer published the synthesis of oxindole and of isatin [19, 20]. Regrettably, Baeyer's synthetic route from phenylacetic acid was too expensive, but he had proved that the synthesis of indigotin was possible. Another isatin synthesis route, the Claisen-Shadwell synthesis, was *published* in 1879 [21, 22].

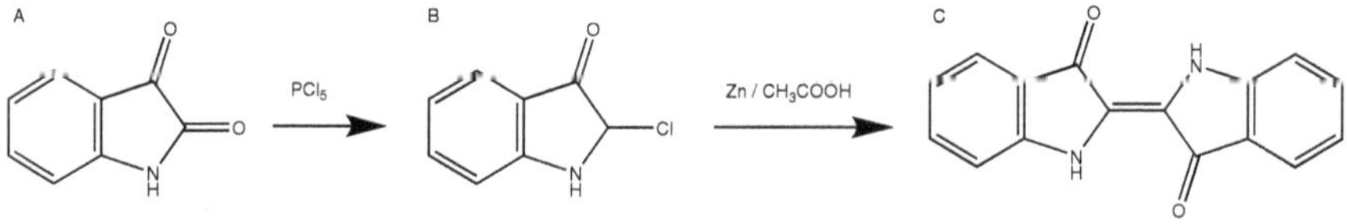


**Fig. 21.2** Chemical process of synthetic Indigo using isatin as staring material. (**A**) Isatin; (**B**) chlorinated isatin; (**C**) indigotin

## *The Second Baeyer Indigo Synthesis (1880)*

In 1880, Baeyer discovered another synthesis route based on cinnamic acid and soon it became clear that cinnamic acid was a better starting material than isatin and indole, previously used for the production of Indigo (Fig. 21.3). On the 19 March 1880, the Indigo synthesis starting from *o*-nitrocinnamic acid was patented (G.P. 11.857), including six additional patents [23–25]. Cinnamic acid is prepared by the reaction of sodium acetate with benzyl chloride, a product obtained from chlorine and toluene. Next, cinnamic acid treated with nitric acid forms *o*-nitrocinnamic acid, which is converted by bromination into the intermediate *o*-nitrophenylpropiolic acid. Finally, *o*-nitrophenylpropiolic acid is converted into Indigo by suitable reducing agents, such as grape sugar and some sulfur compounds under alkaline conditions. Baeyer and Caro were hopeful that this process could be industrialised.

This process was of importance for dye printing. In the summer of 1880, Caro of BASF succeeded in making Indigo on an industrial scale with the insoluble intermediate *o*-nitrophenylpropiolic acid, sodium xanthate as a mild reducing agent and borax as an alkali directly on the fibre under mild alkaline conditions, enabling direct printing on the fibre of cotton goods, producing dark Indigo prints. In April 1881, BASF started a small production enterprise and for several years marketed its *o*-nitrophenylpropiolic acid as 'propiolic acid' or 'little indigo' as a paste with 20% content for propiolic acid printing. For BASF, the introduction of 'little indigo' was more a matter of prestige than of dividends.

The synthetic preparation of Indigo from *o*-nitrocinnamic acid is protected by the additional G.P. 11.857, G.P. 11.858. G.P. 12.601, G.P. 14.997, G.P. 17.656 and G.P. 19.266, all patented between 1880–1881, and the method of printing using *o*-nitrophenylpropiolic acid (BASF), together with sodium xanthate and borax was also patented (G.P. 15.516 in 1881), and at one time employed. Unfortunately, there were some problems with this process. *O*-nitrocinnamic acid, the starting material for *o*-nitrophenylpropiolic acid, posed a great difficulty. First of all, only 40% of *o*-nitrophenylpropiolic acid is formed during the nitration of cinnamic acid, while the major part was converted into the *p*-nitro derivative, which is not suitable for the Indigo production. By using the cinnamic acid ester instead of the free acid, the percentage of *o*-nitrophenylpropiolic acid could be increased to 70%, but still, the yield of Indigo was not satisfactory. Furthermore, the propiolic acid process was unfavourable because of the extraordinarily unpleasant smell evolved during production and it was limited to cotton printing, used by a few printers for calico printing. The process failed to become profitable because the starting material was too expensive, the Indigo yield was too low, and only cotton goods could be printed: these factors led to the fact that the process could not compete with natural indigo. The production of 'little indigo' was discontinued three years later in 1884 [26].

In March 1880, Baeyer patented the synthetic route and sold the patent rights to BASF and Farbwerke ML&B for 100,000 German marks. In June 1880, both companies also signed a contract with Baeyer for cooperation on Indigo until the 30 June 1897 [27].

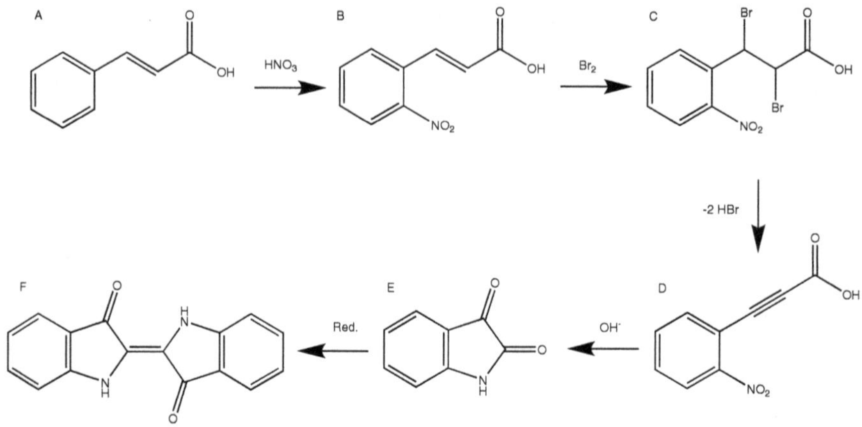


**Fig. 21.3** Baeyer's second Indigo synthesis based on cinnamic acid. (**A**) cinnamic acid; (**B**) *o*-nitrocinnamic acid; (**C**) brominated *o*-nitrocinnamic acid; (**D**) *o*-nitrophenylpropiolic acid; (**E**) isatin; (**F**) indigotin

## *The Baeyer-Drewson Indigo Process (1882)*

A third Indigo process was discovered by Baeyer and the Danish chemist Viggo Beutner Drewson (1858–1930) in 1882 [28] (Fig. 21.4). Drewson studied at the University of Munich under the supervision of Baeyer in 1879 and he received his doctorate in 1882. They developed the well-known Baeyer-Drewson process, in which Indigo is prepared by condensing *o*-nitrobenzaldehyde (derived from toluene) with acetone in a sodium hydroxide solution. This process was patented by Baeyer as the Baeyer-Drewson Indigo synthesis and the patent rights were acquired by BASF and Farbwerke ML&B (G.P. 19.768 in 1882; B.P. 1.266 in 1882; U.S.P.s 257.814 and 257.815 in 1882) [29, 30]. The Baeyer-Drewson process gave a good yield of indigotin, but turned out to be impractical for industrial-scale synthesis at that time because the starting material *o*-nitrobenzaldehyde was very difficult to produce; there was no cheap way to make it and it was therefore expensive. The routes examined, starting with toluene and applying various chlorination and nitration methods, were unsatisfactory. The nitration step was expensive and produced undesirable *m* and *p* derivatives and only small quantities of *o*-nitrobenzaldehyde (the formation of about 40% *p*-nitrotoluene was unavoidable). This strongly influenced the indigotin production and made the Baeyer-Drewson process impractical to apply on an industrial scale [31].

In the end, these three methods did, in fact, yield artificial Indigo, but at such a high price that it could not compete with natural indigo. For Baeyer, this marked the end of his work with Indigo. In 1883, probably influenced by the reorganisation in

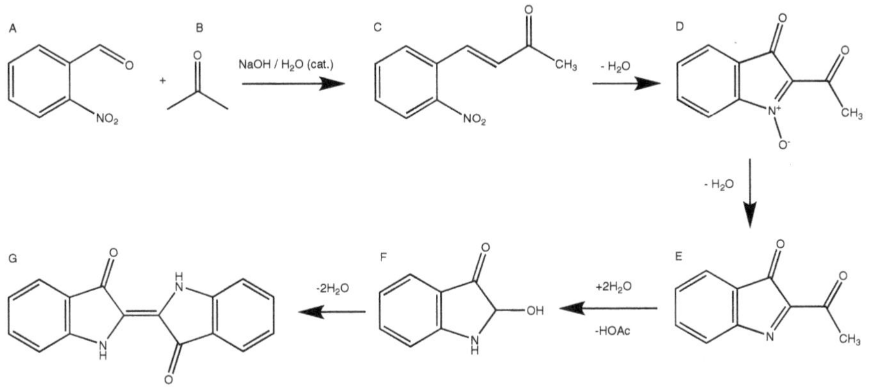


**Fig. 21.4** The Baeyer-Drewson Indigo synthesis based on *o*-nitrobenzaldehyde. (**A**) *o*-nitrobenzaldehyde, (**B**) Acetone; (**C**) 4-Hydroxy-4-(2-nitrophenyl)butan-2-one; (**D**) 2-acetyl-3-oxo-3H-indole 1-oxide; (**E**) 2-acetyl-3H-indole 3 one; (**F**) 2-hydroxyindolin-3-one; (**G**) Indigotin

the management of BASF, the laboratory cooperation of BASF with Baeyer and his assistants was discontinued.

## *The First Heumann Indigo Synthesis (1890)*

For several years there was little activity in the field of Indigo synthesis. The two industrial Indigo players BASF and Farbwerke ML&B had to look for alternative, cheap starting materials available in sufficient quantities. A new impulse was given in 1890 when the German chemist Karl Heumann (1850–1894), a professor working at the Zurich Federal Polytechnic (Eidgenossische Technische Hochschule Zürich), discovered two new synthesis routes. Heumann's first synthesis used cheap aniline as a starting material which reacts with monochloroacetic acid to form *N*-phenylglycine (Fig. 21.5). In an alkali melt at 300 °C, the cyclisation into indoxyl takes place and Indigo is produced by oxidation. This process, the first Heumann synthesis, was patented by BASF on 6 May 1890, G.P. 54.626, followed by the additional G.P. 58.276, G.P. 61.711, G.P. 61.712, G.P. 63.309, G.P. 63.310 and G.P. 68.372, all patented by BASF between 1890–1891 [32, 33]. However, the process led to a dead end. The main drawback of this process was the high temperature of 300 °C required for the ring closure. *N*-phenylglycine partially decomposes under these conditions, which had a negative effect on the formation of indoxyl and resulted in a low yield of a maximum of 10% indigotin, making the process uneconomical. The series of experiments with *N*-phenylglycine was discontinued by BASF in 1893.

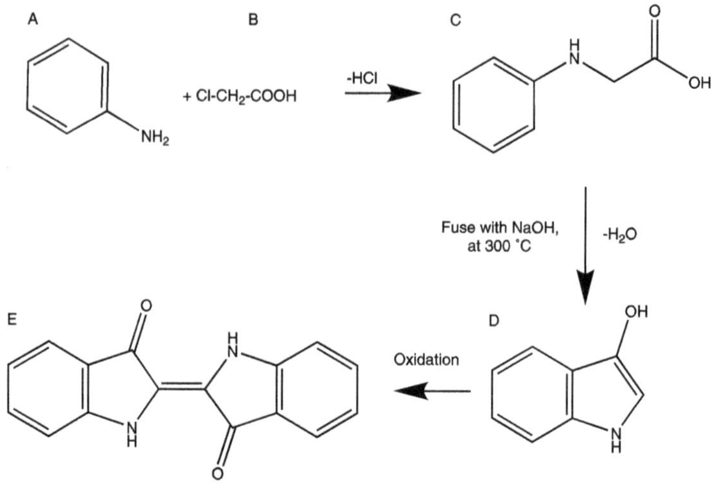


**Fig. 21.5** The First Heumann Indigo synthesis based on *N*-phenylglycine. (**A**) Aniline; (**B**) monochloroacetic acid; (**C**) *N*-phenylglycine; (**D**) indoxyl; (**E**) Indigotin

## *The Second Heumann Indigo Synthesis (1890)*

In the same year, the process was improved using the second Heumann synthesis, based on *N*-phenylglycine-*o*-carboxylic acid (Fig. 21.6). This component is prepared by the reaction of anthranilic acid with monochloroacetic acid; the reaction product *N*-phenylglycine-*o*-carboxylic acid is then converted to indigotin by heating up to 200 °C under alkaline conditions using caustic soda. Anthranilic acid was made from naphthalene, which was widely available in sufficient quantities from coal tar, by converting it to phthalic acid, next to phthalic anhydride and then to phthalimide. The advantage was that this process is almost quantitative and much higher Indigo yields of up to 90% were achieved, making this synthesis the economically viable route [34]. However, the first step in the second Heumann synthesis, the oxidation of naphthalene to phthalic acid, was at that time, in 1890, technically problematic and could only be carried out with expensive nitric acid, making the process too expensive to compete with natural indigo.

Heumann sold his inventions and the patents were acquired by BASF and Farbwerke ML&B (G.P. 56.273 in 1890 and the additional G.P. 73.687 in 1893) [32]. He did not live to see the commercial breakthrough of his process since he died in 1894. The *N*-phenylglycine-*o*-carboxylic acid process was applied by BASF in 1897 to produce the first synthetic Indigo on an industrial scale, which was sold under the trade name 'Indigo Pure BASF'. The production was carried out in Ludwigshafen am Rhein until 1924.

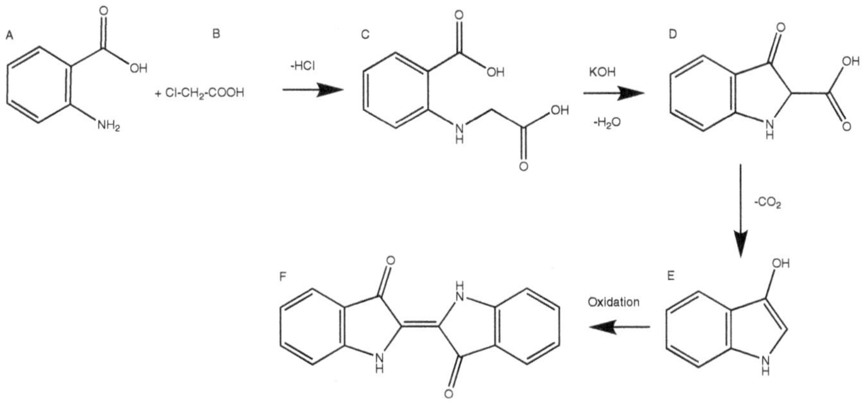


**Fig. 21.6** The Second Heumann Indigo synthesis based on *N*-phenylglycine-*o*-carboxylic acid. (**A**) Anthranilic acid, (**B**) monochloroacetic acid, (**C**) *N*-phenylglycine-*o*-carboxylic acid, (**D**) 2-indoxyl carboxylic acid, (**E**) indoxyl, and (**F**) Indigotin

However, to make the second Heumann synthesis economically efficient, it was essential to find an alternative to nitric acid to oxidise naphthalene to phthalic acid. To solve this problem BASF and Farbwerke ML&B adopted different processes. When the 1880 contract between Farbwerke ML&B and BASF expired on the 30 June 1897, the firms went their separate ways and the former partners became competitors. Farbwerke ML&B's first attempt was the oxidation of naphthalene with chromic acids. Although this was feasible, it was not economically viable since it was not possible to convert chromium hydroxide back into chromic acid at a low cost. This process required large amounts of electricity, which was expensive at the time. The electricity company Lahmeyer in Gersthofen, near Augsburg, had started to use hydroelectric power generated via the river Lech and was able to supply the required electricity, but needed expansion. In April 1900, Farbwerke ML&B signed a contract with the electricity company and the decision was made to build a new factory in Gersthofen to produce Indigo and its precursors. The cost of the project was estimated at 3.5 million German marks [6, 35].

BASF was working on a different synthetic Indigo process, namely the oxidation of naphthalene using fuming sulfuric acid, initially with less successful results: the yield was only 15%. This problem was coincidentally solved in 1891, when during an experiment a thermometer accidentically broke and mercury flowed into the reaction flask [6]. As a result, the German chemist Eugen Sapper (1858–1912), working for BASF, discovered that the catalytic oxidation of naphthalene with oleum (fuming sulfuric acid) and the addition of mercury sulfate formed phthalic acid and he realised that mercury was the ideal catalyst for low-cost oxidation of naphthalene to phthalic acid. The phthalic acid obtained in this way was the starting material for the synthetic Indigo production via anthranilic acid. However, at the end of the 1880s, fuming sulfuric acid cost about 25 German marks per 100 kilos, making it a relatively expensive product.

A little later, in 1893 the German chemist Rudolf Knietsch (1854–1906), continued the work of the German chemist Clemens Winkler (1838–1904), and developed the industrial sulfuric acid contact process. As a result, it was possible to produce 40% fuming sulfuric acid for about 5 German marks. This gave BASF a considerable advantage over Farbwerke ML&B in the synthesis of Indigo [26, 36].

During the years of work on synthetic Indigo, BASF realised that a successful Indigo process required a number of innovations and measures. Between 1891 and 1897, many technical processes were developed and plants were built to ensure the supply of the starting materials and precursors sulfuric acid, chlorine, anthranilic acid and chloroacetic acid [35]. The following activities were undertaken:

- The construction of an industrial-scale plant for synthesising phthalic acid by the oxidation of naphthalene with fuming sulfuric acid and a mercury catalyst;
- The expansion of the sulfuric acid factory to regenerate the sulfur dioxide produced to a capacity of 80,000 tons/annum;
- The erection of a new phthalimide and anthranilic acid plant. BASF purchased a patent from the Amsterdam quinine factory in order to produce anthranilic acid;
- The construction of a new chlorine-alkali electrolytic plant to produce the necessary quantities of chlorine and caustic soda;
- The expansion of the chlorine liquefaction plant to allow preparation of chlorine in the required purity for the production of anthranilic and chloroacetic acids, and
- The installation of a chloroacetic acid plant [7].

## 21.4 BASF's Indigo Manufacture from 1897 to 1904

In 1897, the new BASF Indigo production complex was complete and large-scale production started in February. On 10th July, the first Indigo Pure BASF was launched on the market as a paste containing 20% of Indigo suspended in water, with an annual capacity of 2,000 tons. It caused a sensation. It was admired, but it was also rejected and sometimes defamed by its competitors. The Prussian and French military authorities initially refused to use it, claiming that it was not synthetically produced and, above all, that it was not equal to the natural product. BASF countered with an intensive public relations campaign, providing information to customers throughout the world and setting up an Indigo dyehouse at the BASF works to advise customers on how Indigo should be applied. When synthetic Indigo entered the market in July 1897, it was slightly less expensive than natural indigo: Indigo Pure BASF came onto the market at a price of 17 German marks per kilo. By comparison, one kilogram of natural indigo cost about 20 German marks. In the end, the quality and application advantages spoke for themselves: it had a more brilliant hue; it was easier to handle and its quality was more consistent. From the first patent in 1880 to the market introduction of synthetic Indigo in July 1897, a period of 17 years passed, during which BASF invested 18 million German gold marks (1,000,000 English pounds), on research (for working out the process and in preparing for the commercial introduction of the product), more than the entire stock capital of the company at that time.

## 21.5 Development of Farbwerke ML&B's Indigo Production

For Farbwerke ML&B, synthetic Indigo had not yet been conquered; on the contrary, the road became more and more circuitous. They worked on various Indigo routes and brought Indigo synthesised from toluene onto the market. Since the chromic acid regeneration with naphthalene was too expensive, Farbwerke ML&B went back to the Baeyer-Drewson process of 1882, in which *o*-nitrobenzaldehyde was used. The disadvantage of *o*-nitrobenzaldehyde was that it was derived from the starting material toluene, which was only available in limited quantities. Toluene was very scarce, only present in small quantities in coal tar. Only half a kilogram of toluene is produced by distilling 100 kg of coal tar, which is not much compared to the other major hydrocarbons. Annually, only 5,000–6,000 tons of toluene were available and 4 tons of toluene were needed to produce 1 ton of Indigo [7]. Nevertheless, in 1897, Farbwerke ML&B introduced INDIGO F, derived from *o*-nitrobenzaldehyde, obtained from toluene, as the first viable Indigo process [37]. This process was simplified when Farbwerke ML&B acquired patents from the French firm Soc. Chim. Us. Rhône in Saint-Fons for the so-called Monnet process in 1899. This process allowed *o*-nitrobenzaldehyde to be obtained by direct oxidation of *o*-nitrotoluene, which was subsequently converted into indigotin. This product was launched as INDIGO M [27, 38]. However, the scarcity of toluene affected both processes, which together yielded 356 tons of Indigo. Some years later, in 1904, the board of Farbwerke ML&B decided to abandon INDIGO M because it had not produced the profits they had hoped for and expected [27].

Therefore, in 1900, Farbwerke ML&B decided on a third way of producing Indigo, based on naphthalene, since the electrolytic regeneration of chromic acid opened the way to the production of cheap phthalic acid and *N*-phenylglycine-*o*-carboxylic acid, and finally indigotin. This was mainly produced under the trade name INDIGO A at Frankfurt-Hoechst. Apparently, Farbwerke ML&B had become a little nervous and wanted to catch up with BASF. So far, Farbwerke ML&B had developed three Indigo processes. Although each process showed progress, none of them was perfect. While the plant in Gersthofen was under construction, an unexpected development occurred: the discovery of a new synthetic Indigo process, which changed Hoechst's situation completely [39].

## 21.6 The Pfleger Indigo Process (1901)

By 1900, four different commercial routes to produce synthetic Indigo were in use. The marketing of synthetic Indigo had just started when in 1901 the German chemist Johannes Pfleger (1867–1957), working as a scientist for Degussa in Frankfurt am Main, discovered and patented a more efficient variant of the first Heumann synthesis, based on the starting material aniline, that made it possible to synthesise Indigo for large-scale commercial production (Fig. 21.7). The manufacturing process employed the effective condensing agent sodium amide ($NaNH_2$) for the ring closure of *N*-phenylglycine to form indoxyl. Sodium amide was produced at Degussa as a by-product in the extraction of sodium cyanide, which is needed for the extraction of precious metals [6]. The great advantage of this synthetic route was that the temperature of the ring closure reaction, using sodium amide under alkaline conditions, could be reduced from 300 °C to 180–200 °C, resulting in a 90% yield of Indigo; this became, therefore, the more efficient process for producing Indigo. This means that the indoxyl formed by sodium amide prevents the harmful effects of water, such as thermal decomposition and saponification of *N*-phenylglycine. Another advantage was that the precursors used in the Pfleger method were cheaper than those used in the second Heumann synthesis. The Pfleger process was patented by Degussa on the 11 of April 1900, G.P. 117.623, followed by the G.P 137.955, G.P. 141.749, G.P. 142.700, G.P. 180.394 and U.S.P. 680.395, all patented between 1901–1904.

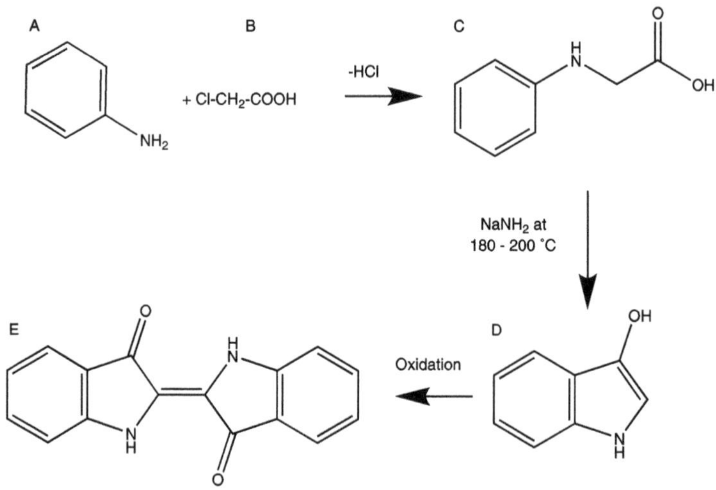


**Fig. 21.7** The Pfleger indigo synthesis. (**A**) Aniline, (**B**) monochloroacetic acid, (**C**) *N*-phenylglycine, (**D**) indoxyl, and (**E**) Indigotin

## 21.7 The Plant in Gersthofen

Farbwerke ML&B acquired the corresponding patent and together with Degussa set up the 'Indigo Company' to exploit the Pfleger process. In 1900, a pilot plant was started and soon there was no doubt that the location at Hoechst near Frankfurt am Main offered the best prospects for production. The product was sold as INDIGO G. As a result, in 1902 the decision was made that the completion of the Indigo plant in Gersthofen, which was almost ready for operation, should be discontinued. Between 1902 and 1904 the existing plant in Hoechst was extended at a cost of 11 million German marks to accommodate the Pfleger process, replacing all other Indigo processes at Hoechst. After the conversion, in 1902, the Gersthofen plant switched to the production of the necessary intermediates and auxiliaries: chromic acid, phthalic acid and sodium for the Hoechst Indigo factory (and hydrogen for Count Zeppelin's airships). By 1902, the extended plant at Hoechst produced 308 tons of INDIGO G, and one year later, in 1903, 765 tons were already brought onto the market. The Pfleger Indigo factory was not completed until 1910, but after that time operated almost unchanged until 1940 [6, 39, 40].

INDIGO G came onto the market at 16 German marks per kilogram and Farbwerke ML&B started an aggressive sales campaign to capture a larger market share. The rivalry between Farbwerke ML&B and BASF led to a rapid fall of the price of synthetic Indigo to 6.5–7 German marks per kilogram in 1904 [35, 40]. In 1904, BASF realised that Hoechst possessed the better Indigo production process. The two firms decided to share the Pfleger process: they divided the market and formed the Second Indigo Convention in order to control the Indigo market. As a result, the price could be stabilised at 8 German marks per kilogram, keeping it below that of the natural product and this price was held until 1913. The Pfleger patent proved highly lucrative with a large income by way of the licence, enabling Degussa to pay dividends of 30% up to 50% between 1906 and 1911.

## 21.8 International Patents

In 1902, the German chemist Herbert Eugen Albert von Meister (1866–1919), son of one of the founders of Farbwerke ML&B wrote a document which provided insight in the large number of Indigo patents available. He categorised 98 patents and 37 patent applications into 20 processes. These patents and applications are spread across 12 companies and 13 individual inventors. At that time, BASF led with 32 patents and 8 applications, followed by Farbwerke ML&B with 19 patents and 3 applications. These are followed by the dye companies Geigy in Basel with 13 patents, Kalle with 16 patents and 2 applications and Bayer with 9 patents and 5 applications. In 1902, Degussa is represented by only one patent and one application for a patent. This large number of patents showed the importance of synthetic Indigo for many dye factories [41].

## 21.9 BASF's Indigo Manufacture in the First Decades of the Twentieth Century

As mentioned above, by 1904, BASF realised that Farbwerke ML&B possessed the better economical production process. One year later, in 1905, BASF developed a variant of the Pfleger synthesis in which the expensive sodium amide was replaced by the cheaper dehydrating agent, finely ground calcium oxide, giving an Indigo yield of 75–80%, almost as high as the original Pfleger proces. The production using this process began in 1906 and continued until 1925, when a new process eliminated the use of finely ground calcium oxide, with its considerable dust problems, as dehydrating agent [7].

Yet, at the turn of the century, synthetic Indigo was not the key to success of which BASF had dreamed. BASF had only a few years to monopolise its expensive invention using the 1897 synthesis route. Three developments led to a rapid fall in the price of synthetic Indigo: the Pfleger process exploited by Farbwerke ML&B, the newly developed synthetic blue sulfur dyes, and the discovery of Indanthrene blue. For everyday clothing, such as blouses, aprons and shirts, lower fastness standards were sufficient and the blue sulfur dyes fulfilled this demand. In particular, the inexpensive and easy to use blue sulfur dyes produced by the German dye companies Cassella, Kalle and Bayer, especially Immedial Pure Blue (C.I. name: Sulphur Blue 9; C.I. number: 53430), Hydron Blue R (C.I. name: Vat Blue 43; C.I. number: 53630), Hydron Blue G (C.I. name: Vat Blue 42; C.I. number: 53640) and Kalle's Thion Blue, were strong competitors, drawing away potential Indigo customers for the production of navy and dark shades on cotton [42].

In addition, BASF created its own unexpected competitor with the discovery of the vat dye Indanthrene blue. In January 1901, in the BASF laboratories, the French-German-Swiss chemist René Bohn (1862–1922) developed the indanthrene dyes. He obtained the first indanthrene dye, which was named Indanthrene Blue RS (C.I. name: Pigment Blue 60, C.I. number: 69800) by the fusion of 2-aminoanthraquinone, derived from crude anthracene, with caustic potash at about 250 °C. Indanthrene was not sensitive to light and was resistant to washing and rubbing. Due to these excellent properties, Indanthrene blue was far superior to synthetic Indigo, at least for cotton and linen. Bohn named the product Indanthrene, a combination of Indigo and anthracene. In October 1901, BASF registered the trade name Indanthren and introduced the dye in 1903 as Indanthrenblau RS. Indanthrene blue was difficult to commercialise and costly to produce. As a consequence, the blue dye brought little profit. Dyers resisted the use of Indanthrene blue because they felt that dyeing with this dye was more difficult than with Indigo and required different methods. In order to boost sales of Indanthrene blue, Bohn and his colleagues wrote a brochure with precise instructions for chemists and dyers [42]. Bohn writes about it: 'This new blue can be dyed like indigo and has the same light fastness. But the dye sits on the fibre as if burnt in.'[2] In 1922, the dye companies Bayer, Farbwerke ML&B and BASF reached an agreement based on a BASF proposal. They decided to put all their vat dyes for cotton fibres on the market under a joint name. They chose the German 'Indanthren' in honor of the pioneer product [42, 43].

## 21.10 Synthetic Indigo in Britain

In 1908, the Swiss dye company CIBA joined BASF and Farbwerke ML&B as the world's third major supplier of synthetic Indigo. In 1907, the British national government changed its policy towards the supply of natural and synthetic Indigo. With the decline of Indian indigo exports and the growing dominance of German synthetic Indigo, the British government passed the *Patent and Designs Act* of 1907, which required foreign holders of British chemical patents to manufacture their process in Britain. It forced Farbwerke ML&B and BASF to open factories in Britain.

In 1907, BASF, Bayer and AGFA known as Dreibund and Little IG, built the BASF's Mersey Chemical Works at New Ferry, Birkenhead, and one year later, in 1908, the largest factory was established by Farbwerke ML&B at Ellesmere Port near Manchester. The costs of the factory at Ellesmere Port, including site and equipment was about 50,000 British pounds. Farbwerke ML&B's synthetic Indigo production rose from 9 tons in 1908 to 293 tons in 1913. It turned out that the companies were not producing the complete Indigo production process on site, but were playing the system by using intermediates imported from Frankfurt-Hoechst for the final stages of production.

At the outbreak of the First World War, Britain had no domestic production of synthetic Indigo. At Ellesmere Port all personel left the site by the summer of 1914. The supply of Indigo was quickly running out and the Germans had taken all knowledge and descriptions of the process with them. The important intermediate *N*-phenylglycine had to be imported and there was no manufacturer in Britain. The British government asked the industry for help and the Ellesmere Port works were sold, under the provisions of the *Trading with the Enemy Acts*, to the German-born British chemist Ivan Levinstein (1845–1916) for 70,000 British pounds, in 1916. The chemists of Levinstein Limited developed processes for *N*-phenylglycine and manufactured it on a scale sufficient to supply the Ellesmere Port factory. The first batches of synthetic Indigo were produced by November 1916, but it was not until March 1917 that Levinstein announced that the British demand for Indigo had been met [44].

---

[2] 'Dieses neue Blau lässt sich wie Indigo ausfärben und hat die gleiche Lichtechtheit. Aber der Farbstoff sitzt auf der Faser wie eingebrannt'.

## 21.11 Trade Names

In the first decade of the twentieth century the two German dye companies BASF and Farbwerke ML&B were the main producers of synthetic Indigo under their trade names: BASF as Indigo rein BASF Pulver L, Indigo rein BASF Teig 20%, Indigo BASF S Teig; Farbwerke ML&B as INDIGO F, INDIGO M, INDIGO A, Indigo MLB Pulver, Indigo MLB/Oe and Indigo MLB Teig 20%. Indigo was also traded as Indigo by CIBA in Basel. In 1910, the Rathjen company in Hamburg and the chemical factory von Heyden in Radbeul near Dresden started the synthetic Indigo manufacture, producing synthetic Indigo by the Sandmeyer process and in the same year by the *N*-phenylglycine process.

## 21.12 Synthetic Indigo Production Up to the Second World War

In 1903, Farbwerke ML&B produced 750 tons of synthetic Indigo, in 1905 already 1,500 tons and in 1907 3500 tons. Even before World War I, synthetic Indigo had practically displaced natural indigo and by 1914, natural indigo held only 4% of the world market. BASF and Farbwerke ML&B together exported nearly 10,000 tons of synthetic Indigo worth more than 50 million German marks. By 1913, BASF sold 4,900 tons of indigo manufactured by three simultaneously operated processes (starting with anthranilic acid, aniline and 2-anilinoethanol respectively) and in the same year Farbwerke ML&B reached its peak with 4,500 tons by the sodium amide process. By 1913, natural indigo had almost entirely been replaced by synthetic Indigo. The production of synthetic Indigo fell during the First World War and dropped to zero in 1918 because of the naval blockade imposed on Germany by the Allies, which prevented all exports. This led to a revival in the demand for natural indigo, but this demand proved to be short-lived. After the war, many dye companies in Switzerland, the United States of America, Japan and Great Britain began to produce their own Indigo, marking the end of natural indigo. Under such competitive pressure, BASF had to rationalise its own Indigo processes. In 1925, BASF developed an improved Indigo synthesis of *N*-phenylglycine from aniline, formaldehyde and hydrogen cyanide via phenylglycinonitrile fused with sodium amide under anhydrous alkali solutions (Fig. 21.8). Meanwhile, a new Degussa process led to a significant reduction in the price of sodium. In 1926, BASF took advantage of this process and secured a contract for low-cost supply. This made it possible to discontinue the practice of fusing *N*-phenylglycine and finely ground calcium oxide with its attendant dust problems. In 1926, the holding company I.G. Farbenindustrie AG was founded, and both companies produced synthetic Indigo using this improved Heumann-Pfleger process, which is still in use by BASF today [7, 26].

In the inter-war period, BASF's Indigo production grew steadily, peaking at 3,724 tons in 1928. The production fell sharply to an average of only 1,000 tons a

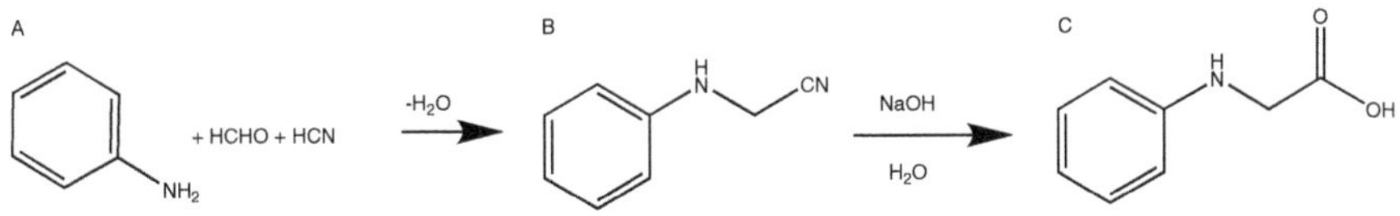


**Fig. 21.8** Improved synthesis of *N*-phenylglycine. (**A**) Aniline, (**B**) phenylglycinonitrile, and (**C**) *N*-phenylglycine

year in the 1930s. BASF responded by restructuring, disposing of unnecessary buildings and plants and merging its Indigo and other dye departments. A series of technical advances brought the necessary reduction in costs in the Farbwerke ML&B and BASF Indigo plants: continuous solvent extraction replaced steam distillation for removing the excess of aniline in the phenylglycinonitrile step; a high-speed Hoesch stirrer in the air-oxidation step resulted in shorter oxidation times, higher Indigo yields and greater purity; continuous rotary presses replaced the labour-intensive filter presses for filtering the Indigo [7].

## 21.13 World War II and a New Start at BASF

In 1939, Farbwerke ML&B stopped producing synthetic Indigo. The main reason was a shortage of monochloroacetic acid. BASF continued to produce a few hundred tons a year until 1944. When air raids caused considerable damage to buildings and equipment, the production had to be abandoned. In 1945, the BASF works were occupied by American troops and undamaged equipment was dismantled for war reparation purposes; these are compensation payments the Germans had to make after the Second World War. Nevertheless, the Indigo production resumed in October 1946, on an ad hoc and irregular basis. For the next few years, raw materials were often scarce and it was not until 1951 that pre-war production levels were restored. In that year, BASF produced more than 1,000 tons, much of which was sold to China. At Frankfurt-Hoechst, the Indigo production was not resumed and even BASF wanted to stop the production. In order to continue producing synthetic Indigo in the long term, the BASF plant had to be completely rebuilt, since repair and maintenance costs were too high. Between 1960 and 1966, different concepts were discussed, ranging from a completely new Indigo plant to ceasing production altogether after 1968. In late 1967, a model for long-term redevelopment finally won.

In 1997, BASF produced 7,000 tons of the global total of 20,000 tons of synthetic Indigo, Buffalo in the United States of America 3,500 tons, Bann of Brazil 2,000 and the rest was made in Japan and China. In 2002, 17,000 tons and in 2011 even 50,000 tons of synthetic Indigo were produced worldwide, almost all of it for the dyeing of blue jeans. Today, Indigo owes its popularity to the fashion for blue jeans and denim, a market which absorbs 99% of all the Indigo produced. For a pair of jeans, 3–12 g of synthetic Indigo, depending on the quality, is required [7].

## 21.14 Applications

Reduction of the insoluble synthetic Indigo to the soluble *leuco*-form gives the molecule an affinity for animal and plant fibres, which enables it to be used as a textile dye for wool, silk, cotton and linen. Apart from a few processes in which small amounts of synthetic Indigo are used to dye linen, wool and silk fibres (such as wax batik for fashionable hand-crafted objects), Indigo is used mainly for the dyeing of cotton yarns, especially today blue jeans and denim. In dye houses specialising in Indigo dyeing, continuously operated slasher or warp dyeing machines are used in which the cotton yarn is immersed successively in three to six dye baths of vatted Indigo. After each immersion, the yarn is squeezed and oxidised by air [45].

## 21.15 Properties

Indigo is a dark-blue, crystalline powder which dyes cotton and wool a splendid blue colour. It is insoluble in water, alcohol, or ether, but soluble in DMSO (dimethyl sulfoxide), DMF (dimethyl formamide), chloroform, nitrobenzene and concentrated sulfuric acid. Indigo is resistant to light and elevated temperature, even in the presence of air. It should be noted that the fastness of Indigo on wool is much better than on cotton. Lehne describes the fading process in his 1906 book[3]: 'After 14 days of exposure to the weather (August) dark wool and cotton dyeings are not changed' [46]. In the Colour Index, a light fastness for Indigo on wool of ISO 6 up to 8 is given, depending on the intensity of the colour (the darker the colour, the faster).

## 21.16 Recipes

Indigo is a vat dye which requires a special process for textile dyeing. As a matter of fact, the term vat dye was used for indigoids as a reduction reaction takes place in the dye vat. The main component of Indigo is indigotin. Indigotin is not soluble in water; to get it into solution it needs to be reduced into *leuco*-indigotin, also known as Indigo white. This can be done through fermentation, using micro-organisms [4, 47]. This is a rather laborious process; the fermentation takes at least a week and sometimes even longer. *Leuco*-indigotin is water soluble, resulting in a dye bath with a yellowish-green colour. The textile is immersed and the *leuco*-indigotin is absorbed by the textile fibres. Next, the textile is taken out of the dye bath and exposed to air. The *leuco*-indigotin molecules are then oxidised to

[3]Indigo: Echtheit: 'Nach 14 tägiger Einwirkung der Witterung (August) dunkle Woll- und Baumwollfärbung unverandert'.

indigotin molecules that adhere to the fibres. The intensity of the blue colour depends on the concentration of *leuco*-indigotin and the number of immersions.

This procedure changed in 1869 when the French chemist Paul Schützenberger (1829–1897) discovered that hydrosulfurous acid, an aqueous solution of sulfur dioxide ($SO_2$), in contact with zinc, reduced indigotin. This was already reported as early as 1854 by Christian Friedrich Schönbein (1799–1868), who believed that the combination of oxygen, zinc and $SO_2$ created ozone which subsequently destroyed indigo (and litmus) since the colour disappeared. However, Schützenberger noted that the blue colour returned when the solution was exposed to air, hence he concluded that a reduction took place rather than a destruction of indigotin. However, the hydrosulfurous acid solution was too unstable for the production of vats. Next, he discovered that sodium hydrosulfite ($NaHSO_3$) was a stronger and more stable reducing agent [48]. Later, sodium dithionite ($Na_2S_2O_4$) was used as a reducing agent, in a sodium hydroxide (NaOH) solution. The *leuco*-indigotin is formed and converted into its water-soluble sodium salt. The *leuco*-indigotin solution can then be used for dyeing and, when exposed to air, the textile will turn blue as the *leuco*-indigotin is oxidised to indigotin.

Lehne's 1893 book mentions hydrosulfite for reducing indigo extracted from plant material; pure synthetic Indigo was not yet commercially available [49]. He gives the following recipe for fabric printing[4]: 'In textile printing, probably only used in shirt printing instead of the reduced indigo previously printed for this purpose. Solid blue: seems to have been replaced in this field too by the indigo salt [K. = from Kalle & Co. in Biebrich am Rhein], i.e. the bisulphite compound of ortho-nitrophenyl lactic acid methyl ketone (cf. G. Ulrich. Färber-Zeitung 1893 94 p. 1.)—Printing ink: 1125 g propionic acid 25%, 110 g borax. 150 g sodium acetate, 250 cc calcium acetate of 22%, 500 g cream of tartar, 30 g turpentine oil. 1½ L of starch paste, ½–¾ L of mucilage of tragacanth (80: 1000), 500 g xanthogenic sodium bicarbonate and 50–125 cc acetic acid of 34%. The colour develops more quickly with less acetic acid, but is correspondingly less durable. It should be possible to store the ink for several days. After printing, [the textile] should be hung for 12 h at 35–40 °C in a well-ventilated room.'

In the book of 1899, Lehne also provided recipes for printing cotton and silk, using Indigo salt T from Kalle & Co, Biebrich am Rhein [50]. He described the use of sodium hydrogen sulfite as a reducing agent and a hot sodium hydroxide solution

---

[4] Im Zeugdruck wohl nur noch im Hemdendruck angewendet statt des früher zu diesem Zwecke in Form reducirten Indigos aufgedruckten. Solidblaus: scheint auch auf diesem Gebiete durch das Indigosalz [K.], d. h. die Bisulfitverbindung des Orthonitrophenylmilchsäuremethylketons ersetzt zu werden (vgl. G. Ulrich. Färber-Zeitung 1893 94 S. 1.)—Druckfarbe: 1125 g Propiolsäure 25%, 110 g Borax. 150 g essigsaures Natron, 250 cc. essigsaurer Kalk von 18° Bé, 500 g Brechweinstein, 30 g Terpentinöl. 1½ L Stärkekleister, ½–¾ L Traganthschleim (80: 1000), 500 g xanthogensaures Natron und 50–125 cc. Essigsäure von 7 °Bé. Mit wenig Essigsäure entwickelt sich die Farbe rascher, sie ist aber dem entsprechend weniger haltbar. Die Druckfarbe soll sich mehrere Tage aufbewahren lassen. Nach dem Druck wird 12 Stunden bei 35–40 °C in einem gut ventilirten Raum verhängt.

to rinse and wash the fabric after printing. He writes[5]: 'Cotton print. Printing ink: 600 g thickener, 25 g Indigo salt T, 77 cc sodium bisulfite 36%, 300 cc water. After printing, dry, pass through [a solution of] hot sodium hydroxide 11%, acidify and wash again.'

Recipes for dyeing with synthetic Indigo are not given until the 1906 volume [46], when it was available, but interestingly he mentions that for large quantities of textiles woad is still used (although indigo from the tropics was probably used instead). He provides the following recipe for dyeing wool[6]: 'Due to its high fastness, it [indigo] has always been widely used for dyeing wool, cotton and linen. The dye is converted into a "vat"—Indigo white dissolved in alkali—by using a reducing agent. The textile fabric is placed in the vat—which is heated to around 50 °C for wool—moved through [the vat], squeezed and then exposed to air, whereby the yellow colour changes to indigo blue through oxidation. Depending on the depth of the desired shade, different dippings must be made in the vat, i.e. the whole treatment must be repeated as required. For wool, especially for medium and dark colours, the fermentation vat is still primarily used today, e.g. the following woad vat for 10,000 L: 60 kg woad, 20 kg madder, 10 kg Solvay soda, 10 kg syrup, 25 kg wheat bran, 3 kg lime, 25 kg Indigo M.L.B. 20% or 4–5 kg Indigo M.L.B. powder. The hydrosulphite vat is particularly useful for dyeing light colours and for smaller companies where Indigo dyeing is not regularly performed.'

Note that by Indigo M.L.B., synthetic Indigo produced by Farbwerke ML&B was meant. For silk Lehne added a line stating[7]: 'Silk can be dyed in warm or cold vats that are not strongly alkaline.'

A recipe for linen and cotton is also provided[8]: 'Plant fibres are mainly dyed in an iron vitriol vat (iron vitriol + slaked lime) and in a zinc-lime vat (zinc dust +

[5] Baumwolldruck. Druckfarbe: 600 g Verdickung, 25 g Indigosalz T, 77 cc. Natriumbisulfite 36 °Bé, 300 cc. Wasser. Nach dem Druck trocknen, durch heisse Natronlauge 16 °Bé ziehen, säuren uns nogmals waschen.

[6] Wegen seiner hohen Echtheit von jeher viel verwendet zum Färben von Wolle, Baumwolle und Leinen. Der Färbstoff wird durch Reduktionsmittel in eine "Küpe"—in Alkali gelöstes Indigoweiß—übergeführt. Der Textilstoff wird in die—für Wolle auf etwa 50 °C angewärmte—Küpe eingebracht, umgezogen, abgequetscht und dann der Luft ausgesetzt, wobei durch Oxydation die gelbe Farbe in Indigoblau übergeht. Je nach der Tiefe des gewünschten Tons müssen verschiedene Züge auf der Küpe gegeben, d. h. die ganze Behandlung muß nach Bedarf wiederholt werden. Für Wolle, besonders für mittlere und dunkle Farben, dient auch heute noch vornehmlich die Gärungsküpe z. B. folgende Waidküpe für 10,000 L Inhalt: 60 kg Waid, 20 kg Krapp, 10 kg Solvay soda, 10 kg Sirup, 25 kg Weizenkleie, 3 kg Kalk, 25 kg Indigo MLB 20% oder 4–5 kg Indigo MLB-Pulver. Gute Dienste leistet die Hydrosulfitküpe besonders zum Färben in hellen Tönen und für kleineren Betrieb, wo nicht regelmäßig Indigo gefärbt wird.

[7] Seide kann in warmen oder kalten Küpen, welche nicht stark alkalisch sind, gefärbt werden.

[8] Zum Färben der Pflanzenfasern bei etwa 25 °C dienen vornehmlich die Eisenvitriolküpe (Eisenvitriol + gelöschter Kalk) und die Zink-Kalk-Küpe (Zinkstaub + gelöschter Kalk). Für lose Baumwolle, Cops u. dergl., welche auf Apparaten gefärbt werden, wird die Hydrosulfitküpe benutzt.

slaked lime) at around 25 °C. For loose cotton, cops etc., which are dyed with machines, the hydrosulfite vat is used'.

## 21.17 Identification Using UHPLC-PDA

The analysis of a reference of synthetic Indigotin (see Fig. 21.9) shows a clear signal. Note that based on chromatographic analysis, it is not possible to distinguish between natural and synthetic Indigo as the main compound, indigotin, is the same. Minor products can easily be missed, i.e. are often below the detection limit, and are not yet studied in detail to differentiate between the synthetic and the natural variant. As an example, the isomer indirubin is often found in natural indigo; however, our research showed that it can also be present in synthetic Indigo. When indigotin is present in high concentration and in a very pure form, one might conclude that synthetic Indigo is present; however, the dating of the object is often more helpful.

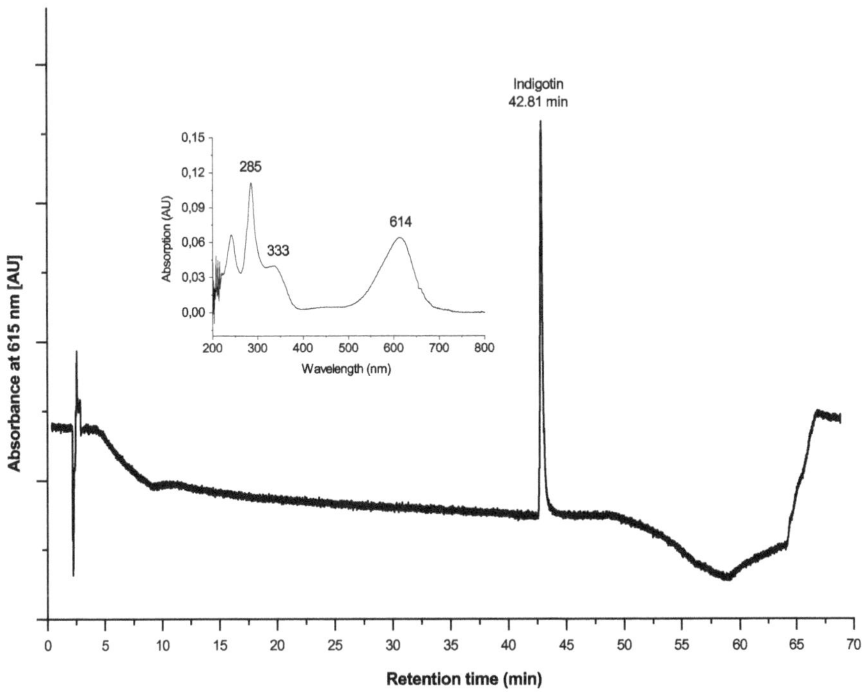


Fig. 21.9 Chromatogram and spectrum of synthetic indigotin

## 21.18 Denim Blue

In 1850, at the time of the gold rush, a twenty-year-old merchant named Levi Strauss (1829–1902) emigrated from Buttenheim (in Upper Franconia) to San Francisco. He made hard-wearing work clothes from indigo-dyed drill cloth. Blue jeans derive from 'Bleu de Gênes', for which two origins are given in the literature. On the one hand, Genoese sailors are said to have brought indigo-dyed drillich (drill) to America before Levi Strauss; on the other hand, the indigo needed for dyeing is said to have been exported to the United States of America by way of Genoa. The fabric was either dyed in America itself or imported already dyed from France, especially from the textile centre of Nimes. 'Bleu de Nîmes' thus became the Americanised 'blue denim', the name still commonly used today for this type of textile. In 1872, the well-known copper rivets for reinforcement were introduced and patented. Blue jeans, originally workwear, became clothing for the protesting youth in the 1960s. This gave rise to the blue jeans fashion worn in leisure and everyday life around the world today [51].

## References


1. Cardon, D.: Natural Dyes Sources, Tradition, Technology and Science. Archetype Publications Ltd., London (2007)
2. Reifarth, N., Völling, E.: Spuren aus dem Reich der Farben. Antike Welt. 44, 33–39 (2013)
3. Hofmann-de Keijzer, R., van Bommel, M.R., Joosten, I., Bogensperger, I.: Late Antique Textiles from the Papyrus Collection of the Austrian National Library—Scientific investigation of fibres, dyes and dyeing techniques. MPER XXXIV, Band 2. de Gruyter, Boston/Berlin (2024)
4. Hofenk de Graaf, J.H.: The Colourful Past, Origins, Chemistry and Identification of Natural Dyestuffs. Abegg Stiftung/Archetype Publications Ltd., Riggisberg / London (2004)
5. Balfour-Paul, J.: Indigo. Archetype Publications Ltd., London (2006)
6. Bäumler, E.: Ein Jahrhundert Chemie. Econ Verlag GmbH, Düsseldorf (1963)
7. Schmidt, H.: Indigo - 100 Jahre industrielle Synthese. In: Chemie in Unserer Zeit. pp. 121–128. Wiley-VCH Verlag (1997)
8. Cain, J.C., Thorpe, J.F.: The Synthetic Dyestuffs and the Intermediate Products from which they are Derived, second edition, revised. Charles Griffin and Company, Limited, London (1913)
9. Baeyer, A.: Ueber die Verbindungen der Indigogruppe. Vierte Abhandlung, Berichte der deutschen chemischen Gesellschaft zu Berlin. Ferd. Dümmler´s Verlagsbuchhandlung, Berlin. 16, 2188–2204 (1883)
10. Golding, B.T., Pierpoint, C.: Indigo blue. In: Education in Chemistry. pp. 71–73. RSC (1986)
11. Baeyer, A., Emmerling, A.: Synthese des Indols. Berichte der deutschen chemischen Gesellschaft zu Berlin. Ferd. Dümmler´s Verlagsbuchhandlung, Berlin. 2, 679–682 (1869)
12. Kekulé, A.: Ueber die Constitution des Isatins, der Isatinsäure und des Indols. Berichte der deutschen chemischen Gesellschaft zu Berlin. Ferd. Dümmler's Verlagsbuchhandlung, Berlin. 2, 748–749 (1869)
13. Baeyer, A., Emmerling, A.: Reduktion des Isatins zu Indigoblau. Berichte der deutschen chemischen Gesellschaft zu Berlin. Ferd. Dümmler´s Verlagsbuchhandlung, Berlin. 3, 514–517 (1870)

14. Baeyer, A., Emmerling, A.: Sur la réduction de l'isatine en indigo bleu. Bulletin de la Société Chimique de France, Paris. 14, 416–417 (1870)
15. Laurent, A.: Nouvelles recherches sur l'indigo. Comptes rendus l´Académie des sciences. 12, 537–539 (1841)
16. Laurent, A.: Recherches sur l'indigo. Annales de Chimie et de Physique. 3, 371–383 (1841)
17. Baeyer, A.: Synthèses de l´isatine et du bleu d´indigo. Bulletin de la Société Chimique de France, Paris. 31, 475–476 (1879)
18. Baeyer, A.: Synthese des Indigblaus. Berichte der deutschen chemischen Gesellschaft zu Berlin. Ferd. Dümmler´s Verlagsbuchhandlung, Berlin. 11, 1296–1297 (1878)
19. Baeyer, A.: Synthesis of isatin and indigo-blue. Journal of the Chemical Society, Abstracts. 34, 884–885 (1878)
20. Baeyer, A., Emmerling, A.: Synthese des Oxindols. Berichte der deutschen chemischen Gesellschaft zu Berlin. Ferd. Dümmler´s Verlagsbuchhandlung, Berlin. 11, 582–584 (1878)
21. Claisen, L., Shadwell, J.: Synthese des Isatins. Berichte der deutschen chemischen Gesellschaft zu Berlin. Ferd. Dümmler´s Verlagsbuchhandlung, Berlin. 12, 350–354 (1879)
22. Claisen, L., Shadwell, J.: Synthèse de l'isatine. Bulletin de la Société Chimique de France, Paris. 33, 322–323 (1880)
23. Adolf Baeyer, German patent 11.857, 19 March 1880, Darstellung von Derivaten der Orthonitrozimtsäure den Homologen und Substitutionsprodukten dieser Derivate und Umwandlung derselben in Indigblau und verwandte Farbstoffe
24. Friedlaender, P.: Synthesen von Indigblau, Isatin- und Indolderivaten, Fortschritte der Theerfarbenfabrikation und verwandter Industriezweige 1877-1887. Verlag von Julius Springer, Berlin (1888)
25. Dokumente aus Hoechster Archiven, Beiträge zur Geschichte der Chemie, Herbert von Meister, Der künstliche Indigo und seine Vorprodukte (1902), Heft 22, Farbwerke Hoechst Aktiengesellschaft vormals Meister Lucius & Brüning, Frankfurt (M)-Hoechst, 1967, 24
26. Von Nagel, A.: Fuchsin, Alizarin, Indigo, der Beginn eines Weltunternehmens. Schriftenreihe des Firmenarchivs der Badischen Anilin- & Soda-Fabrik. 4, 33–49 (1970)
27. Farbwerke vorm. Meister Lucius & Brüning 1863-1913. Hergestellt in der Hausdruckerei der Farbwerke vorm. Meister Lucius & Brüning in Höchst am Main, 31
28. Baeyer, A., Drewsen, V.: Darstellung von Indigblau aus Orthonitrobenzaldehyd. Berichte der Deutschen Chemische Gesellschaft zu Berlin. Ferd. Dümmler´s Verlagsbuchhandlung, Berlin. 14, 2856–2864 (1882)
29. Baeyer, A., Drewsen, V.: Darstellung von Indigblau aus Orthonitrobenzaldehyd. Berichte der deutschen chemischen Gesellschaft zu Berlin. Ferd. Dümmler's Verlagsbuchhandlung, Berlin. 15, 2856–2864 (1882)
30. Baeyer, A., Drewsen, V.: Einwirkung von Orthonitrobenzaldehyd auf Aldehyd. Berichte der deutschen chemischen Gesellschaft zu Berlin. Ferd. Dümmler's Verlagsbuchhandlung, Berlin. 16, 2205–2208 (1883)
31. Friedlaender, P.: Indigo und Akridinfarbstoffe. In: Fortschritte der Theerfarbenfabrikation und verwandter Industriezweige, Vierter Theil, 1894-1897. p. 1029. Verlag von Julius Springer, Berlin (1899)
32. Friedlaender, P.: Indigoblau (Isatin und Indolderivate). In: Fortschritte der Theerfarbenfabrikation und verwandter Industriezweige, Dritter Theil, 1890-1894. pp. 274–275. Verlag von Julius Springer, Berlin (1896)
33. Heumann, K.: Neue Synthesen des Indigos mit verwandter Farbstoffe. Berichte der deutschen chemischen Gesellschaft zu Berlin. Ferd. Dümmler´s Verlagsbuchhandlung, Berlin. 23, 3043, 3289 (1890)
34. Heumann, K.: Neue Synthesen des Indigos mit verwandter Farbstoffe, 2. Darstellung von Indigblau mittels Anthranilsäure. Berichte der deutschen chemischen Gesellschaft zu Berlin. Ferd. Dümmler´s Verlagsbuchhandlung, Berlin. 23, 3431–3435 (1890)
35. 1863 ML&B 1938 I.G. Farbenindustrie Aktiengesellschaft, Werke. In: Farbwerke vorm. Meister Lucius & Brüning. pp. 87–91. F. Bruckmann KG., München (1938)

36. Farbwerke vorm. Meister Lucius & Brüning 1863-1913. Hergestellt in der Hausdruckerei der Farbwerke vorm. Meister Lucius & Brüning in Höchst am Main, p. 19.
37. Dokumente aus Hoechster Archiven, Beiträge zur Geschichte der Chemie, Die entscheidenen Jahre der Indigo-Synthese, Heft 28, Farbwerke Hoechst Aktiengesellschaft vormals Meister Lucius & Brüning, Frankfurt (M)-Hoechst, 1967, p. 9
38. Friedlander, P.: Indigo. In: Fortschritte der Theerfarbenfabrikation und verwandter Industriezweige, Fünfter Theil 1897-1900. p. 387. Verlag von Julius Springer, Berlin (1901)
39. Dokumente aus Hoechster Archiven, Beiträge zur Geschichte der Chemie, Die entscheidenen Jahre der Indigo-Synthese, Heft 28, Farbwerke Hoechst Aktiengesellschaft vormals Meister Lucius & Brüning, Frankfurt (M)-Hoechst, 1967, 10-11
40. Dokumente aus Hoechster Archiven, Beiträge zur Geschichte der Chemie, Hoechst übernimmt das Pfleger´sche Indigo-Verfahren, Heft 30, Farbwerke Hoechst Aktiengesellschaft vormals Meister Lucius & Brüning, 1967, 9-11
41. Dokumente aus Hoechster Archiven, Beiträge zur Geschichte der Chemie, Herbert von Meister, Der künstliche Indigo und seine Vorprodukte (1902), Heft 22, Farbwerke Hoechst Aktiengesellschaft vormals Meister Lucius & Brüning, Frankfurt (M)-Hoechst, 1967, 9
42. Abelshauser, W.: Die BASF: eine Unternehmungsgeschichte, Zweite Auflage. Verlag C.H. Beck, München (2003)
43. Seefelder, M.: INDIGO. In: BASF Aktiengesellschaft, Ludwigshafen am Rhein. pp. 57–61. Kölnische Verlagsdruckerei GmbH, Köln (1982)
44. Paine, C.: The Development of the Dyestuff Industry. In: Perkin centenary London : 100 years of synthetic dyestuffs. pp. 47–48. Pergamon Press, London (1958)
45. Process Data Sheet: Continuous Dyeing of Cotton Warps with Indigo, BASF Aktiengesellschaft, Ludwigshafen am Rhein, 1976
46. Lehne, A.: Tabellarische Übersicht über die künstlichen organischen Farbstoffe und ihre Anwendung in Färberei und Zeugdruck, Zweiter Ergänzungsband. Verlag von Julius Springer, Berlin (1906)
47. Hartl, A., Gaibor, A.N.P., van Bommel, M.R., Hofmann-de Keijzer, R.: Searching for blue: Experiments with woad fermentation vats and an explanation of the colours through dye analysis. J Archaeol Sci Rep. 2, 9–39 (2015)
48. Wisniak, J.: Paul Schützenberger. Educ. quím. 26, 57–65 (2015)
49. Lehne, A.: Tabellarische Uebersicht über die künstlichen organischen Farbstoffe und ihre Anwendung in Färberei und Zeugdruck. Verlag von Julius Springer, Berlin (1893)
50. Lehne, A.: Tabellarische Uebersicht über die künstlichen organischen Farbstoffe und ihre Anwendung in Färberei und Zeugdruck. Verlag von Julius Springer, Berlin (1899)
51. Seefelder, M.: INDIGO. In: BASF Aktiengesellschaft, Ludwigshafen am Rhein. pp. 83–90. Kölnische Verlagsdruckerei GmbH, Köln (1982)

# Appendix: International Dye Producing Factories

**Table A.1** International dye producing factories [1–3]

| Abbreviation | Company | Location | Founded |
|---|---|---|---|
| **Dye companies in Germany** | | | |
| AGFA | Gesellschaft für Anilinfabrikation and Farbenfabrik Dr. Jordan | Rummelsburg near Berlin<br>Treptow-Berlin | 1867 |
| | Actien-Gesellschaft für Anilin-Fabrikation | Rummelsburg near Berlin | 1873 |
| | Part of the Union of Three (Dreibund) | Rummelsburg near Berlin | 1904 |
| | Union of Three and Association of Three merged to form Little IG. Farben | Rummelsburg near Berlin | 1916 |
| | Part of I.G. Farbenindustrie AG | Rummelsburg near Berlin | 1925 |
| BASF | Badische Anilin- & Soda-Fabrik | Mannheim | 1865 |
| | Badische Anilin- & Soda-Fabrik | Ludwigshafen am Rhein | 1867 |
| | Badische Anilin- & Soda-Fabrik (agreement with the companies Rudolf Knosp and Heinrich Siegle in Stuttgart) | Ludwigshafen am Rhein | 1873 |
| | Part of the Union of Three (Dreibund) | Ludwigshafen am Rhein | 1904 |
| | Union of Three and Association of Three merged to form Little IG. Farben | Ludwigshafen am Rhein | 1916 |
| | Part of I.G. Farbenindustrie AG | Ludwigshafen am Rhcin | 1925 |
| Bauer | C. vom Bauer | Elberfeld | 1869 |



M. R. van Bommel, M. de Keijzer, *Bright Colours from the Past*, Cultural Heritage Science, https://doi.org/10.1007/978-3-031-88664-5

| Abbreviation | Company | Location | Founded |
|---|---|---|---|
| Bayer | Friedr. Bayer und Co. | Barmen-Rittershausen near Wuppertal | 1863 |
| | Aktiengesellschaft Farbenfabriken vorm. Friedr. Bayer & Co. | Elberfeld near Wuppertal | 1881 |
| | Bayer bought the alizarin factory from Dr. Leverkus & Söhne | Elberfeld near Wuppertal | 1891 |
| | Part of the Union of Three (Dreibund) | Elberfeld near Wuppertal | 1904 |
| | Union of Three and Association of Three merged to form Little IG. Farben | Elberfeld near Wuppertal | 1916 |
| | Part of I.G. Farbenindustrie AG | Elberfeld near Wuppertal | 1925 |
| Beyer & Kegel | Leipziger Anilinfabrik Beyer & Kegel G.m.b.H. | Lindenau-Leipzig and Fürstenberg an der Oder | 1882 |
| Cassella | Farbwarenhandlung Cassel & Reiss | Frankfurt am Main | 1798 |
| | Leopold Cassella & Co. | Frankfurt am Main | 1828 |
| | Leopold Cassella & Co. (dye production) | Frankfurt am Main | 1870 |
| | Part of the Zweibund | Frankfurt am Main | 1904 |
| | Part of the Association of Three (Dreiverband) | Frankfurt am Main | 1907 |
| | Union of Three and Association of Three merged to form Little IG. Farben | Frankfurt am Main | 1916 |
| | Part of I.G. Farbenindustrie AG | Frankfurt am Main | 1925 |
| Dahl | Farbenfabrik A. Dahl & Co. | Barmen | 1842 |
| Wülfing | Anilinölfabrik A. Wülfing | Elberfeld | 1880 |
| Wülfing-Dahl | Farben- und Bleistift Fabriken Wülfing, Dahl & Co. Aktiengesellschaft | Barmen | 1906 |
| Degussa | Deutsche Gold- und Silber-Scheideanstalt vormals Roessler AG. | Frankfurt am Main | 1873 |
| Ewer & Pick | Firma Ewer & Pick | Berlin-Grünau | 1879 |
| Farbwerke ML&B | Theerfarbenfabrik Meister, Lucius & Co. | Hoechst am Main | 1862 |
| | Meister, Lucius & Brüning | Hoechst am Main | 1867 |
| | Farbwerke vorm. Meister, Lucius & Brüning AG. | Hoechst am Main | 1880 |
| | Part of the Zweibund | Hoechst am Main | 1904 |
| | Part of the Association of Three (Dreiverband) | Hoechst am Main | 1907 |
| | Union of Three and Association of Three merged to form Little IG. Farben | Hoechst am Main | 1916 |
| | Part of I.G. Farbenindustrie AG | Hoechst am Main | 1925 |
| Gans-Leonhardt | Frankfurter Anilinfarbenfabrik von Gans und Leonhardt | Fechenheim am Main | 1870 |
| Gessert | Farben- und Lackfabrik der Gebrüder Gessert | Elberfeld | |

| Abbreviation | Company | Location | Founded |
|---|---|---|---|
| Chem. Fabr. Griesheim | Actiengesellschaft für landwirtschaftlich chemische Fabrikate | Griesheim am Main | 1856 |
| | Chemische Fabrik Griesheim am Main (dye production) | Griesheim am Main | 1863 |
| Griesheim-Elektron | Chemische Fabrik Elektron | Griesheim am Main | 1893 |
| | Part of I.G. Farbenindustrie AG | Griesheim am Main | 1925 |
| de Haen | Chemische Fabrik Dr. Eugen de Haën & Cie. | Linden near Hanover | 1861 |
| | Chemische Fabrik Dr. Eugen de Haën & Cie. | List near Hanover | 1862 |
| | E. de Haën Chemische Fabrik List GmbH | List near Hanover | 1912 |
| | Aktiengesellschaft E. de Haën Chemische Fabrik Listt | | 1922 |
| | J.D. Riedel–E. de Haën Aktiengesellschaft | | 1928 |
| Hauff | J. Hauff | Feuerbach near Stuttgart | 1870 |
| von Heyden | Salicylsäure-Fabrik Dr. F. von Heyden | Radebeul near Dresden | 1874 |
| | Salicylsäure-Fabrik Dr. F. von Heyden GmbH | Radebeul near Dresden | 1896 |
| | Salicylsäure-Fabrik Dr. F. von Heyden AG. | Radebeul near Dresden | 1899 |
| Jäger | Carl Jäger, Anilinfarbenfabrik | Barmen and Düsseldorf-Derendorf | 1823 |
| Kalle | Chemische Fabrik Kalle & Co. | Biebrich am Rhein | 1863 |
| | Kalle & Co. Aktiengesellschaft | Biebrich am Rhein | 1904 |
| | Part of the Association of Three (Dreiverband) | Biebrich am Rhein | 1907 |
| | Union of Three and Association of Three merged to form Little IG. Farben | Biebrich am Rhein | 1916 |
| | Part of I.G. Farbenindustrie AG | Biebrich am Rhein | 1925 |
| Küchler-Buff | Küchler & Buff | Crefeld | 1855 |
| | Küchler & Buff G.m.b.H (Küchler, Buff and Eberhardt) | Crefeld | 1895 |
| Leonhardt | Farbwerk Mühlheim vorm. A. Leonhardt & Co., Chemische Fabrik Mühlheim am Main | Mühlheim am Main | 1879 |
| Leverkus | Ultramarinfabriken Carl Leverkus & Söhne | Wermelskirchen | 1834 |
| | Vereinigten Ultramarinfabriken vormals Leverkus, Zeltner und Consorten (production was later expanded to include an alizarin factory) | Leverkusen | 1890 |
| Bayer | Farbenfabriken vorm. Friedr. Bayer & Co. took over Leverkus | Leverkusen | 1918 |
| Neuhaus | Carl Neuhaus | Elberfeld | 1868 |
| Rathjen | A. Rathjen | Hamburg | |
| Sell | Sellsche Asphaltfabrik | Offenbach am Main | 1842 |

| Abbreviation | Company | Location | Founded |
|---|---|---|---|
| Oehler | K. Oehler Anilin- und Anilinfarbenfabrik took over Sellsche Asphaltfabrik | Offenbach am Main | 1850 |
| Griesheim-Elektron | Chemische Fabrik Griesheim-Elektron took over K. Oehler Anilin- und Anilinfarbenfabrik | Frankfurt am Main | 1905 |
| | Part of Little IG. Farben | Frankfurt am Main | 1917 |
| | Part of I.G. Farbenindustrie AG | Frankfurt am Main | 1925 |
| Tillmanns ter Meer | Farbwerk Krefeld (Heinrich Tillmanns) | Crefeld | 1862 |
| | Dr. E. ter Meer & Co. (E. and H. ter Meer) | Crefeld | 1877 |
| | Tillmanns, E ter Meer & Co. (Tillmanns, E. and H. ter Meer) | Uerdingen am Rhein | 1882 |
| | Farbwerk E. ter Meer & Co. (E. ter Meer, J. Weiler u.a.) | Uerdingen am Rhein | 1892 |
| Weiler-ter Meer | J.W. Weiler & Co. (J.W. Weiler and O. Jäger) | Uerdingen am Rhein | 1861 |
| | J.W. Weiler & Co. Aktiengesellschaft | Ehrenfeld near Cologne | 1889 |
| | Chemische Fabriken vorm. Weiler-ter Meer | Uerdingen am Rhein | 1896 |
| | Union of Three and Association of Three merged to form Little IG. Farben | Uerdingen am Rhein | 1916 |
| | Part of I.G. Farbenindustrie AG | Uerdingen am Rhein | 1925 |
| **Dye companies in Switzerland** | | | |
| Clavel | Alexandre Clavel | Basel | 1864 |
| Bindschedler & Busch | Firma Bindschedler & Busch took over Alexandre Clavel | Basel | 1873 |
| CIBA | Gesellschaft für Chemische Industrie took over Firma Bindschedler & Busch | Basel | 1885 |
| Durand | Louis Durand, Fabrique de Matieres | Basel | 1871 |
| Durand-Huguenin | Farbwerke vorm. L. Durand, Huguenin & Co. | Basel | 1872 |
| | Aktiengesellschaft Farbwerke vorm. L. Durand, Huguenin & Co. | Basel | 1900 |
| Geigy | Joh. Rud. Geigy und Co. | Basel | 1764 |
| | Anilinfarben- und Extract-Fabriken vorm. Joh. Rud. Geigy Co. | Basel | 1859 |
| | Anilinfarben- und Extract-Fabriken vorm. Joh. Rud. Geigy Co. with J.J. Müller & Co. | Basel | 1863 |
| Gerber | Anilinfabrik A. Gerber & Co. | Basel | 1864 |
| Kern & Sandoz | Chemische Fabrik Kern & Sandoz | Basel | 1886 |
| Sandoz | Chemische Fabrik vorm. Sandoz & Co., Aktiengesellschaft | Basel | 1895 |
| P. Monnet | Usine P. Monnet et Cie. | Dardagny-La Plaine near Geneva | 1860 |

| Abbreviation | Company | Location | Founded |
|---|---|---|---|
| **Dye companies in Britain** | | | |
| B&B | Burt & Bolton | London | 1848 |
| BB&H | Burt, Bolton & Haywood | London | 1874 |
| British Aliz. Co. | The British Alizarine Company, Limited took over Burt, Bolton & Haywood | Silvertown (London) | 1883 |
| BB&H | Burt, Bolton & Haywood | London | 1874 |
| British Aliz. Co. | The British Alizarine Company Limited took over Burt, Bolton & Haywood | Silvertown (London) | 1883 |
| British Aliz.Co. | The British Alizarine Company, Limited | Silvertown (London) | 1882 |
| BS&S | Brooke, Simpson & Spiller Limited | Hackney Wick in London | 1868 |
| BB&H | Burt, Bolton & Haywood took over Brooke, Simpson & Spiller | Greenford Green near Harrow | 1876 |
| British Aliz. Co. | The British Alizarine Company, Limited | Silvertown (London) | 1883 |
| Claus & Rée | Claus & Rée, Aniline Colour Manufacturers | Clayton near Manchester | 1891 |
| Claus & Rée Ltd. | Claus & Rée Limited | Clayton near Manchester | 1906 |
| Claus and Co. | Claus and Co. | Clayton near Manchester | 1908 |
| Levinstein | Levinstein Limited Blackley took over Claus & Co. | Clayton near Manchester | 1916 |
| Clayton | The Clayton Aniline Company, Ltd. | Clayton near Manchester | 1876 |
| Colne Vale | Colne Vale Chemical Company | Milnsbridge near Huddersfield | |
| Dawson | Dan Dawson Brothers | Milnsbridge near Huddersfield | 1858 |
| Graesser | Robert Graesser Chemical Works | Ruabon, North-Wales | 1870 |
| Levinstein | Levinstein Limited Blackley | Blackley near Manchester | 1864 |
| Lowe | Charles Lowe & Co. | Stockport near Manchester | 1860 |
| Perkons & Sons | Perkin & Sons | Greenford Green near Harrow | 1857 |
| BB&S | Brooke, Simpson & Spiller Limited took over Perkin & Sons | Greenford Green near Harrow | 1874 |
| Pullar | J. Pullar and Sons Limited | Perth, Scotland | 1824 |
| Read Holliday | Read Holliday & Sons Limited | Huddersfield | 1830 |
| Roberts-Dale | Roberts, Dale & Co. | Manchester and Warrington | 1852 |
| Rumney | Ardwick Chemical Works | Gorton | 1845 |
| Simpson-Maule | Simpson & Maule | Walworth in London | 1853 |
| Simpson, Maule & Nicholson | Simpson, Maule & Nicholson | Walworth in London | 1853 |

| Abbreviation | Company | Location | Founded |
|---|---|---|---|
| BS&S | Brooke, Simpson & Spiller Limited took over Simpson, Maule & Nicholson | Walworth in London | 1868 |
| Williams | Williams, Thomas & Dower resp. Thomas & Dower<br>Williams Brothers & Co. | Hounslow, Middlesex | 1877 |
| Wood & Bedford | Wood & Bedford | Leeds | 1850s |
| **Dye companies in France** | | | |
| Casthelaz | Laurent & Casthelaz | Belbeuf near Rouan | 1853 |
| | John Casthelaz, Bruère et Co. | Belbeuf near Rouan | 1885 |
| Fréderic Bayer | Produits Fréderic Bayer et Cie. | Flers near Roubaix | 1883 |
| Guinon | Guinon | Saint-Fons near Lyon | 1849 |
| | Guinon et Picard | Saint-Fons near Lyon | |
| GM&B | Guinon, Picard et Jay | Saint-Fons near Lyon | |
| | Lucien Picard et Co. | Saint-Fons near Lyon | 1892 |
| Guinon et Chabaud | Guinon et Chabaud | Lyon | 1831 |
| | Guinon | Lyon | 1842 |
| Guinon | Guinon, Marnas & Bonnet | Lyon | 1856 |
| Manufacture Lyonnaise | Manufacture Lyonnaise des Matières colorantes Société Anonyme | Lyon | 1885 |
| KSD&Cie | Koechlin Schmaltzer Dollfus & Cie. | Mulhouse | 1746 |
| DMC | Dolfuss-Mieg et Compagnie | Mulhouse | 1800 |
| Nicholas Dolfus | Nicholas Dolfus & Cie. | Mulhouse | 1790 |
| Hartmann-Risler | Hartmann, Risler & Cie. (name change) | Mulhouse | 1795 |
| | Hartmann, Risler & Cie. | Rixheim | 1797 |
| Jean Zuber | Zuber & Cie. acquired Hartmann, Risler & Cie. | Rixheim | 1802 |
| P. Monnet | Usine P. Monnet et Cie. | Saint-Fons near Lyon | 1882 |
| Gilliard-Monnet-Cartier | Gilliard, P. Monnet & Cartier took over Usine P. Monnet et Cie. | Saint-Fons near Lyon | 1886 |
| Soc. Chim. Us. Rhône | Société Chimique des Usines du Rhône took over Gilliard, P. Monnet & Cartier | Saint-Fons near Lyon | 1895 |
| Patry | Patry & Cie. | Puteaux near Paris | |
| George | A. George | Campdeville near Beauvais | 1891 |
| Picard | Lucien Picard & Co. | Saint-Fons near Lyon | 1892 |
| Poirrier & Chappat | Société Poirrier & Chappat fils | Saint-Denis | 1857 |
| Soc. Anon. St. Denis | Société Anonyme des Matières colorantes et Produits chimiques de St. Denis (establishments A. Poirrier and G. Dalsace) | Saint-Denis | 1881 |
| Raffard | Louis Raffard et Cie. | Givray near Péage de Roussilon | 1855 |
| Renard Frères | Renard Frères, Franc et Cie. | Pierre-Bénite near Lyon | 1859 |

| Abbreviation | Company | Location | Founded |
|---|---|---|---|
| Renard-Villet | Renard et Villet | Villeurbanne (Lyon) | 1866 |
| Renard-Villet-Bunand | Renard, Villet et Bunand | Villeurbanne (Lyon) | 1874 |
| Ruch | J. Ruch & Fils | Pantin | 1830 |
| Sévoz-Boasson | A. Sévoz & Boasson | Lyon-Vaise | 1885 |
| Société La Fuchsine | Société La Fuchsine | Lyon | 1863 |
| Steiner | Victor Steiner Fabrique de Produits chimiques | St. Marcel near Vernon | 1881 |
| Thann & de Mulhouse | Fabriques de Produits Chimiques de Thann et de Mulhouse | Mulhouse | 1808 |
| | Fabriques de Produits chimiques de Thann & de Mulhouse together with Courtois & Co. and d'Andiran and Wegelin | Mulhouse | 1883 |
| Thomas | Thomas Frères | Avignon | 1872 |
| Succursale BASF | Succursale de la Badische Anilin-& Soda-Fabrik | Neuville-sur-Saône near Lyon | 1882 |
| Us. Chim. Tremblay | L´usine chimique du Tremblay à Creil | Creil near Paris | 1859 |
| Coupier | Theodore Coupier & Cie. | Creil near Paris | 1873 |
| Farbwerke ML&B | Compagnie Parisienne de Couleurs d'Aniline | Creil near Paris | 1883 |
| **Dye companies in the United States of America** | | | |
| Albany Comp. | The Albany Aniline & Chemical Works | Albany, New York | 1868 |
| Am. Col. Chem. Co. | American Color & Chemical Co. | Rensselaer, New York | 1898 |
| American Aniline Co. | American Aniline Company | Albany, New Jersey | 1868 |
| Atlantic Dye. Comp. | Atlantic Dyestuff Company | Boston, Massachusetts | 1916 |
| Bulls Ferry | Bulls Ferry Chemical Company | Shady Side, New York | |
| Heller & Merz | The Heller & Merz Co. American Ultramarine and Globe Aniline Works | Newark, New York | 1869 |
| | Heller & Merz & Co., as a tar coal factory | Newark, New York | 1882 |
| Hudson River | Hudson River Aniline & Color Works | Albany, New Jersey | 1882 |
| Schoellkopf | Schoellkopf Aniline & Chemical Company | Buffalo, New York | 1879 |
| | Schoellkopf, Hartford & Hanna Co. | Buffalo, New York | 1900 |
| | Part of the National Aniline and Chemical Company | Buffalo, New York | 1917 |
| **Dye company in Belgium** | | | |
| Wiescher | L. Destrée, A. Wiescher & Co. | Haeren near Brussels | 1836 |
| | Wiescher & Co., Suc$^{rs}$ | Haeren near Brussels | |
| **Dye company in the Netherlands** | | | |
| Pick, Lange & Co. | Kleurstoffenfabriek Pick, Lange & Co. | Amersfoort | 1888 |
| Farbwerk Amersfoort | Farbwerk Amersfoort after liquidation of Kleurstoffenfabriek Pick, Lange & Co. | Amersfoort | 1904 |

| Abbreviation | Company | Location | Founded |
|---|---|---|---|
| NKI | Nederlandsche Kleurstoffen Industrie acquired Farbwerk AmersfoortA. | Amersfoort | 1918 |
| **Dye company in Italy** | | | |
| Lepetit | Lepetit, Dolfuß & Ganser | Susa | 1871 |
| **Dye company in Austria** | | | |
| Aussiger Verein | Österreichischer Verrein für chemische und metallurgische Produktion zu Aussig an der Elbe | Aussig, nowadays Ústí nad Labem, Czech Republic | 1856 |

# References


1. Schultz, G.: Die Chemie des Steinkohlenteers mit besonderer Berücksichtigung der künstlichen organischen Farbstoffe. Zweiter Band, Die Farbstoffe. Druck und Verlag von Friedrich Vieweg und Sohn, Braunschweig. Dritte Auflage, (1901)
2. Schultz, G.: Farbstofftabellen. Weidmannische Buchhandlung, Berlin. Fünfte Auflage (1914)
3. Schultz, G.: Farbstofftabellen, künstliche organische Farbstoffe bekannter Konstitution oder Herstellungsweise matürliche Farbstoffe und Farblacke aus Naturfarbstoffen anorganische Farbstoffe (Erd- und Mineralfarben). Akademische Verlagsgesellschaft M.B.H., Leipzig. Erster Band, Siebente Auflage (1931)

# Index – Keywords




M. R. van Bommel, M. de Keijzer, *Bright Colours from the Past*, Cultural Heritage Science, https://doi.org/10.1007/978-3-031-88664-5

**D**

E



F

## G

## Q



## R

# Index – Dyes




M. R. van Bommel, M. de Keijzer, *Bright Colours from the Past*, Cultural Heritage Science, https://doi.org/10.1007/978-3-031-88664-5

# Index – Colour Index Names and Numbers




M. R. van Bommel, M. de Keijzer, *Bright Colours from the Past*, Cultural Heritage Science, https://doi.org/10.1007/978-3-031-88664-5

# Index – Scientists, Dyers and Merchants





M. R. van Bommel, M. de Keijzer, *Bright Colours from the Past*, Cultural Heritage Science, https://doi.org/10.1007/978-3-031-88664-5

# Index – Manufacturers and Suppliers





M. R. van Bommel, M. de Keijzer, *Bright Colours from the Past*, Cultural Heritage Science, https://doi.org/10.1007/978-3-031-88664-5